Environmental and Low-Temperature Geochemistry

Environmental and Low-Temperature Geochemistry

Second Edition

Peter Ryan

Middlebury College, Middlebury, VT, USA, 05753

WILEY Blackwell

Registered Office(s)
John Wiley & Sons, Inc., 111 River Street, Hoboken, NJ 07030, USA
John Wiley & Sons Ltd, The Atrium, Southern Gate, Chichester, West Sussex, PO19 8SQ, UK

Editorial Office
9600 Garsington Road, Oxford, OX4 2DQ, UK

For details of our global editorial offices, customer services, and more information about Wiley products visit us at www.wiley.com.

Wiley also publishes its books in a variety of electronic formats and by print-on-demand. Some content that appears in standard print versions of this book may not be available in other formats.

Library of Congress Cataloging-in-Publication Data

Names: Ryan, Peter Crowley, 1966- author.
Title: Environmental and low-temperature geochemistry / Peter Ryan
    (Middlebury College, Middlebury, VT, USA).
Description: 2nd edition. | Hoboken, NJ : Wiley-Blackwell, 2019. | Includes
    index. |
Identifiers: LCCN 2019016029 (print) | LCCN 2019018347 (ebook) | ISBN
    9781119568629 (Adobe PDF) | ISBN 9781119568612 (ePub) | ISBN 9781119568582
    (pbk.)
Subjects: LCSH: Environmental geochemistry–Textbooks. |
    Geochemistry–Textbooks.
Classification: LCC QE516.4 (ebook) | LCC QE516.4 .R93 2019 (print) | DDC
    551.9–dc23
LC record available at https://lccn.loc.gov/2019016029

Cover Design: Wiley
Cover Image: © Peter Ryan

Set in 10/13pt PhotinaMTStd by SPi Global, Chennai, India
Printed and bound by CPI Group (UK) Ltd, Croydon, CR0 4YY

C9781119568582_150424

Contents

Preface

The title *Environmental and Low-Temperature Geochemistry* seeks to encompass the scope of this text, from topics commonly covered in traditional geology-based geochemistry texts as well as those in environmental chemistry and aqueous geochemistry texts. The "low-temperature" part indicates a focus on Earth surface systems ($< 50\,°C$ for most topics covered) rather than metamorphic or igneous environments, although information from these higher-temperature systems is also considered in some sections. The target audience is upper-level undergraduates and graduate students, as well as professionals in geology and environmental science. The goal is to explain basic concepts from biogeochemistry and geology as they apply to understanding and predicting behavior of natural and anthropogenic constituents in systems at and near the surface of the Earth. The importance of the geological record of environmental change informs our understanding of the natural environment and improves our ability to predict system behavior, so sediment (and glacier) records of paleoclimate and environmental change are presented and examined. Increasing knowledge of the role of organisms – from microbes to macroscale plants – in inorganic and organic geochemistry is reflected in examples of biomineralization, microbial mediation of reactions, influence of root exudates, and more.

The scope ranges from the atmosphere and oceans to streams, lakes, groundwater, soils, sediments, and shallow crust. It also spans global and regional-scale systems to pore spaces and nanoparticles. Understanding complex environmental systems requires factual, conceptual, and quantitative skills, so this book attempts to provide the background needed to objectively assess problems and questions through the lens of geochemistry. Inspiration for different sections comes from many excellent geochemistry and environmental chemistry texts, particularly those of Berner and Berner, Drever, Eby, Faure, Langmuir, Moore and Reynolds, Spiro and Stigliani, and Walther. The well-worn copies on my bookshelf attest to their influence on my understanding of geochemistry and how to write a university-level textbook that also can serve as a reference for earth and environmental scientists.

Chapter 1 presents basic principles of inorganic chemistry and geochemistry, including introductory kinetics and thermodynamics concepts meant to serve as a foundation for subsequent chapters. For most readers, parts of this chapter will serve as a review, but at the same time are useful for reinforcing concepts that are most applicable to environmental geochemistry. Minerals are the most abundant constituents of soils, sediments, and rocks, so Chapter 2 gives the reader foundational concepts in mineralogy and examples of minerals and mineral-influenced processes. Harmful amounts of organic compounds like pesticides, solvents, fuels, and pharmaceuticals are a common source of contamination in soils, water, and air, so Chapter 3 contains an introductory presentation of relevant principles from organic chemistry then focuses on structure, composition, environmental behavior, consequences, and remediation of organic contaminants.

Chapter 4 focuses on aqueous systems and the biogeochemical factors that influence the composition of natural and impacted waters, presenting many topics that form the focus of aqueous geochemistry texts. So too do Chapters 5 (carbonates and the C cycle) and 6 (N, P, and S cycles), with a systems approach that conceptually and quantitatively examines reservoirs, fluxes, and processes, from inorganic to microbially affected.

Chapter 7 examines the atmosphere at a global scale, including geological and ice core records of air composition (and climate) over time, structure and composition of the current atmosphere, infrared and ultraviolet radiation, and human-caused climate change, with up-to-date data on greenhouse gas sources and sinks. Urban and regional air pollution are the focus of Chapter 8, including oxygen chemistry and the role of free radicals, reactions involved in the formation and destruction of pollutants, the photochemical smog cycle, and origin and deposition of acid rain.

Chapter 9 focuses on the composition of soils, first considering processes and rates by which primary minerals transform to secondary minerals like clays, hydroxides, and carbonates. From there, Chapter 9 explores various topics, including: quantifying element mobility and how to apply it to paleoclimate analysis; thermodynamics and construction of mineral stability diagrams (and the utility of them); the geochemistry of the soil forming factors, soil profiles, and soil orders; how soils influence ecosystem response to acid rain; plant nutrients; toxic metals and metalloids; saline soils; and organic contaminants in soils.

Chapters 10 and 11 consider ways in which isotopes facilitate understanding of environmental systems as tracers, and in the case of radioactive isotopes, as environmental contaminants. Both chapters begin with fundamental concepts of isotope geochemistry. Stable isotopes are the focus of Chapter 10, with much attention paid to classic stable isotope pairs (e.g. ^{1}H–^{2}H, ^{12}C–^{13}C, ^{14}N–^{15}N, ^{16}O–^{18}O, ^{32}S–^{34}S) and how they serve as useful tracers of processes – natural and anthropogenically influenced – in the atmosphere, biosphere, hydrosphere, and lithosphere. Recent advances and new applications in the areas of metal stable isotopes (e.g. Cu, Fe, Hg) and clumped isotopes (e.g. ^{13}C–^{18}O) are also highlighted. Radioactive and radiogenic isotopes are useful in dating reservoirs (e.g. layers of sediment) and tracing environmental

processes, so these topics – along with radionuclides as contaminants – are the focal areas of Chapter 11.

Throughout the second edition of *Environmental and Low-Temperature Geochemistry* are *focus boxes* that highlight certain topics in the form of brief case studies, quantitative approaches, definitions, and examples. This is a response to student input indicating that varied modes of text and text-within-text would be easier to digest than the format of the first edition. Another feature of the second edition is a new detailed case study (Appendix II) of groundwater contamination by the industrial chemical PFOA (perfluorooctanoic acid), an emerging contaminant of which relatively little is known about, at least in terms of how it will behave in soils, groundwater, and surface water. The second edition contains some new end-of-chapter problems and updated knowledge on systems undergoing change, most prominently the atmosphere and oceans relative to carbon emissions (a worsening problem), and air quality relative to urban and regional air pollutants (with some notable improvements). Reader input contributed significantly to this improved second edition, and comments on this edition are very welcome.

Acknowledgements

This text would not be what it is if not for encouragement and insights from Margaret Crowley Ryan, especially logistical, motivational, and contractual. I wrote this second edition in Granada, Spain, with support, collegiality, and friendship of F. Javier Huertas of the CSIC Instituto Andaluz de Ciencias de la Tierra. So many other colleagues enrich my understanding and appreciation of geological and environmental systems; two in particular are Jon Kim of the Vermont Geological Survey and Chris Klyza of Middlebury College. I have a great cohort of colleagues at Middlebury College in the Geology Department and in the Environmental Studies Program. Many generations of students in GEOL 323 (Environmental Geochemistry) have provided inspiration, corrections, and recommendations – too many to name them all – but a few that stick out include Bayu Imaduddin Zulkifli Ahmad, Katharine Fortin, Andrew Hollyday, Josh Johnson, Emmet Norris, and Kate Porterfield. Seongnam An, a PhD candidate at Korea University and Korea Institute of Science and Technology (KIST) in Seoul, provided much excellent feedback, for which I am grateful. Editorial handling by Antony Sami, Vimali Joseph, Andrew Harrison and others at Wiley is also much appreciated.

Background and Basic Chemical Principles: Elements, Ions, Bonding, Reactions

1.1 AN OVERVIEW OF ENVIRONMENTAL GEOCHEMISTRY – HISTORY, SCOPE, QUESTIONS, APPROACHES, CHALLENGES FOR THE FUTURE

The best way to have a good idea is to have a lot of ideas. (Linus Pauling)

All my life through, the new sights of nature made me rejoice like a child. (Marie Curie)

I must speak of the soil hiding the rocks, of the lasting river destroyed by time. (Pablo Neruda)

Environmental and low-temperature geochemistry encompasses research at the intersection of geology, environmental studies, chemistry, and biology, and at its most basic level is designed to answer questions about the behavior of natural and anthropogenic substances at or near the surface of earth. The scope includes topics as diverse as trace metal pollution, soil formation, acid rain, sequestration of atmospheric carbon, and pathways of human–plant–wildlife exposure to contaminants. Most problems in environmental geochemistry require understanding of the relationships among aqueous solutions, geological processes, minerals, organic compounds, gases, thermodynamics and kinetics, and microbial influences, to name a few.

A good example to lay out some of the basic considerations in environmental and low-temperature geochemistry is the fate and transport of lead (Pb) in the environment. In many regions of the Earth, Pb in the earth surface environment is (was) derived from combustion of leaded gasoline, because even in places where it has been banned (most of the world), Pb tends to persist in soil. The original distribution of Pb was controlled at least in part by atmospheric processes ranging from advection (i.e. wind) to condensation and precipitation.

Once deposited on Earth's surface, the fate and transport of Pb is controlled by interactions and relationships among lead atoms, solids compounds (e.g. inorganic minerals or organic matter), plants or other organisms, and the composition of water in soils, lakes, or streams (including dissolved ions like Cl^- and $CO_3{}^{2-}$ and gases like O_2 and CO_2). In cases where Pb falls on soils bearing the carbonate anion ($CO_3{}^{2-}$), the formation of lead carbonate ($PbCO_3$) can result in sequestration of lead in a solid state where it is largely unavailable for uptake by organisms. If the $PbCO_3$ is thermodynamically stable, the lead can remain sequestered (i.e. in a solid state and relatively inert), but changes in chemical regime can destabilize carbonates. For example, acidic precipitation that lowers the pH of soil can cause dissolution of $PbCO_3$, but how much of the carbonate will dissolve? How rapidly? Much like the melting of snow on a spring day, geochemical processes are kinetically controlled (some more than others), so even in cases where phases exist out of equilibrium with their surroundings, we must know something about rate laws in order to predict how fast reactions (e.g. dissolution of $PbCO_3$) will occur.

If Pb is dissolved into an aqueous form (e.g. Pb^{2+}), additional questions of fate and transport must be addressed – will it remain in solution, thus facilitating

Environmental and Low-Temperature Geochemistry, Second Edition. Peter Ryan.
© 2020 John Wiley & Sons Ltd. Published 2020 by John Wiley & Sons Ltd.

Focus Box 1.1

Defining "Low-Temperature"

In the world of geochemistry and mineralogy, "low-temperature" environments generally refer to those at or near the Earth's surface, whereas "high-temperature" environments include those deeper in the crust, i.e. igneous and metamorphic systems. Most geochemists consider tropical soils with mean annual temperature of 25 °C, alpine or polar systems with mean annual temperature of <5 °C, and sediments buried 1 km below the surface with a temperature of 50 °C, all as low-temperature systems.

Focus Box 1.2

Lead (Pb) and Environmental Justice

Humans have known for thousands of years that lead is toxic, dating back at least to the Roman Empire, but it remains problematic because it relatively abundant and easy to work (e.g. to make pipes for water systems, or as an additive to paint or fuel). A recent case that highlights the connection of geochemistry and human health is that of Flint, Michigan (USA), where lead-bearing pipes were installed as part of the public water supply system in the early twentieth century. Beginning in 2014, a change in public water source from Lake Huron to the Flint River – and in part, the higher amounts of chloride anion (Cl^-) in Flint River water – began to leach lead from the old lead-bearing water mains (pipes) and fittings (Hanna-Attisha et al. 2016). Evidence for the corrosion was obvious to residents in the form of brownish particulates that clouded their water and made it taste bad, but slow response from government resulted in exposure via ingestion. Lead is one of those elements that is harmful at any level, especially to the cognitive development of children. The crisis in Flint serves as one of many examples of environmental injustice (see "environmental (in)justice" in the index for more examples).

---

uptake by plants? Or will it be carried in solution into a nearby surface water body, where it could be consumed by a fish or amphibian? Or will other soil solids play a role in its fate? Will it become adsorbed to the surface of a silicate clay or organic matter, transported downstream until it ultimately desorbs in lake sediments? We also need to consider the possibility that the $PbCO_3$ does not dissolve, but rather is physically eroded as a particulate (grain) into a stream or lake, where it might dissolve or remain a solid, possibly becoming consumed by a bottom feeder, from which point it could biomagnify up the food chain.

Environmental geochemistry has its origins in groundbreaking advances in chemistry and geology ushered in by the scientific breakthroughs of the late nineteenth and early twentieth centuries, particularly advances in instrumental analysis. The Norwegian Victor M. Goldschmidt is considered by many to be the founder of geochemistry, a reputation earned by his pioneering studies of mineral structures and compositions by X-ray diffraction and optical spectrograph studies. These studies led Goldschmidt to recognize the importance and prevalence of isomorphous substitution in crystals, a process where ions of similar radii and charges can substitute for each other in crystal lattices.

Goldschmidt's peer, the Russian Vladimir I. Vernadsky, had come to realize that minerals form as the result of chemical reactions, and furthermore that reactions at the Earth's surface are strongly mediated by biological processes. The orientation of geochemistry applied to environmental analysis mainly arose in the 1960s and 1970s with growing concern about contamination of water, air, and soil. The early 1960s saw publication of Rachel Carson's *Silent Spring*, and early research on acid rain at Hubbard Brook in New England (F.H. Bormann, G.E. Likens, N.M. Johnson, and R.S. Pierce) emphasized the interdisciplinary thinking required for problems that spanned atmospheric, hydrologic, soil, biotic, and geologic realms.

Current research in environmental geochemistry encompasses problems ranging from nanometer and micron scale (e.g. interactions between minerals and bacteria or X-ray absorption analysis of trace metal speciation), local scale (e.g. acid mine drainage, leaking fuel tanks, groundwater composition, behavior of minerals in nuclear waste repositories) to regional (acid rain, mercury deposition, dating of glacier retreat and advance) and global (climate change, ocean chemistry, ozone depletion) scale. Modern environmental geochemistry employs analytical approaches ranging from field mapping and spatial analysis to spectrometry and diffraction, geochemistry of radioactive and stable isotopes, and analysis of organic compounds and toxic trace metals. While the explosion of activity in this field makes it impossible to present all developments and to acknowledge the research of all investigators, numerous published articles are cited and highlighted throughout the text, and two case studies that integrate many concepts are presented in Appendices I and II.

1.2 THE NATURALLY OCCURRING ELEMENTS – ORIGINS AND ABUNDANCES

The chemical elements on Earth have been around for billions of years, thanks to nucleosynthesis, the Big Bang, approximately 12–15 billion years ago (especially the lighter elements), and processes in the interiors of stars later in the evolution of the universe (heavier elements). The early universe was extremely hot (billions of degrees) and for the first few seconds was comprised only of matter in its most basic form, quarks.

1.2.1 Origin of the light elements H and He (and Li)

Approximately 15 seconds after the Big Bang, the atomic building blocks known as neutrons, protons, electrons, positrons, photons, and neutrinos began to form from quarks, and within minutes after the Big Bang, the first actual atoms formed. Protons combined with neutrons and electrons to form hydrogen (^1_1H) and its isotope deuterium (^2_1H, or D), which rapidly began to form helium (He) through fusion, a process in which the nuclei of smaller atoms are joined to create larger, heavier atoms: $2H \rightarrow He + energy$, or to be more precise:

$$^1_1H + ^2_1H = ^3_2He + \gamma + E \qquad (1.1)$$

where γ is the symbol for gamma radiation emitted during nuclear fusion. (Note: basic principles of atomic theory are presented in Section 1.3.)

Small amounts of lithium were probably also produced in the first few minutes or hours after the Big Bang, also by fusion (in a simple sense represented as H + He = Li). From a graph of the abundances of elements in the universe (Figure 1.1), it is clear that H and He are the most abundant, and that, for the most part, element abundance decreases exponentially with increasing atomic number

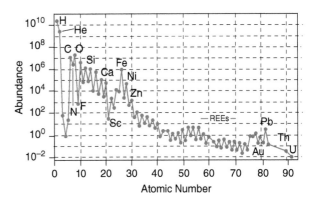

Fig. 1.1 Abundance of elements in the solar system normalized to Si = 10^6 on molar basis with a logarithmic y-axis – this is a standard means of plotting values for this type of data set. Selected elements plus the rare earth elements (REEs, La through Yb) are shown for reference.

up to atomic number 50; beyond that, elements are less abundant. The elemental composition of the solar system is similar to that of the universe, but with the caveat that inner planets (Earth included) are enriched in heavier elements relative to outer planets – note the differences in Figures 1.1 and 1.2.

It is also clear that some elements appear to be anomalously uncommon as compared to their neighbors (e.g. Li, Be, B, Sc), whereas others seem to be present in anomalously high concentrations (e.g. Fe, Ni, Pb). It is also interesting to note the sawtooth pattern produced by alternation of relatively abundant even-numbered elements as compared to neighboring odd-numbered elements (a phenomenon described by the Oddo-Harkins rule). How do these observed trends relate to the processes that formed the elements? The answer lies in basic principles of nuclear fusion.

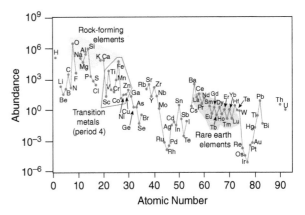

Fig. 1.2 Abundance of the elements in Earth's upper continental crust (molar basis) as a function of atomic number, and normalized to Si = 10^6 on a logarithmic y-axis.

1.2.2 Formation of heavier elements

Elements larger than He were formed by fusion in stars in the first few million to hundreds of millions of years after the Big Bang by a process generally referred to as *stellar nucleosynthesis*. Gravitational forces had produced a contracting, spinning disc-like mass of primordial H and He known as the solar nebula, which contained the energy necessary to form heavier elements by fusion as follows:

$$\ce{^4_2He} + \ce{^4_2He} = \ce{^8_4Be} + \gamma \qquad (1.2)$$

$$\ce{^4_2He} + \ce{^8_4Be} = \ce{^{12}_6C} + \gamma \qquad (1.3)$$

$$\ce{^{12}_6C} + \ce{^{12}_6C} = \ce{^{20}_{10}Ne} + \ce{^4_2He} \qquad (1.4)$$

Focus Box 1.3

Composition of the Universe vs. Earth

The vast majority of the mass in the universe (97–98%) consists of H and He that remain from the first few minutes following the Big Bang. These two elements formed the base fuel for subsequent stellar nucleosynthesis in the cores of stars that resulted in heavier elements. The composition of Earth is skewed toward relatively high abundance of these heavier elements (Table 1.1; Figure 1.2), as much of the lighter H and He were lost to outer space (H was retained to a far greater extent than He because H occurs in water, which is held by gravitational force in a way that gaseous He was not).

Table 1.1 Chemical differentiation of earth with major elements and selected trace elements.

	Granite	Basalt	Ultramafic	Sandstone	Shale	Carbonate
SiO_2	69.4	49.3	42.1	71.3	64.8	8.34
TiO_2	0.48	1.86	0.05	0.70	0.80	0.12
Al_2O_3	15.4	15.3	2.25	10.6	17.0	1.52
Fe_2O_3*	2.66	12.0	13.7	5.03	5.70	1.07
MnO	0.03	0.22	0.20	0.10	0.25	0.07
MgO	0.75	7.39	38.2	2.52	2.83	21.2
CaO	1.96	9.9	2.22	3.03	3.51	66.7
Na_2O	4.08	2.47	0.66	1.56	1.13	0.21
K_2O	4.48	0.98	0.02	4.61	3.97	0.55
P_2O_5	0.14	0.25	0.04	0.16	0.15	0.07
SUM	*99.4*	*99.7*	*99.4*	*99.7*	*100.1*	*99.8*
As	0.25	2.2	0.8[a]	1.0	28	1.6
Ba	1880	315	0.7	595	636	99
Cr	10*	185	1800	20	125	10
Co	4.6	47	175	2.3	20*	3.6
Cu	11	94	15	2.5	45*	1.3
Ni	10*	150	2000	10.1	58	3.3
Pb	30	7	0.5	8.3	20*	2.8
Sr	275	465	1	20	300	610
Th	13*	3.5	0.005	4.4	12.3	0.54
U	3*	0.75	0.002	1.3	2.7	1.6
Zn	86	120	40	50	95*	13

Major elements (SiO_2 through P_2O_5) are presented in units of wt % oxides and the trace elements are presented in concentrations of parts per million (ppm, or mg/kg). Data sources are as follows: granite is the United States Geological Survey granite standard "G-2"; basalt and ultramafic data are averages from Turekian and Wedepohl (1961) and Vinogradov (1962); sandstone and carbonate rock data are unpublished analyses of early Paleozoic sedimentary rocks from northwestern Vermont performed by the author; and shale is the North American Shale Composite (Gromet et al., 1984). Sr and the other trace element values with asterisks are averages from Turekian and Wedepohl (1961) and Vinogradov (1962).

[a]Indicates highly variable, from <1 ppm to >500 ppm, depending on metamorphic fluids. Additional resources include the text *The Continental Crust: its Composition and Evolution* by Taylor and McLennan (1985) and the chapter by Rudnick and Gao (2003) in *Treatise on Geochemistry*.

This process skipped over odd-numbered Li and B (atomic numbers 3 and 5) and the $_4^8$Be that formed was very unstable and was either rapidly transformed to ^{12}C before it decayed, (Eq. (1.3)) or was destroyed by radioactive decay.

Fusion was able to form elements up to iron ($_{26}^{56}$Fe), but beyond that no heat is released during fusion – that is, the process is no longer exothermic when fusing nuclei heavier than Fe. In fact, the iron nucleus is so stable that fusion reactions involving iron actually consume energy (i.e. an endothermic process), so without the heat needed to fuel fusion reactions, another process had to take over to form the heavier elements. This process is known as neutron capture and can be represented like this:

$$_{26}^{56}\text{Fe} +_0^1 \text{n} =_{26}^{57}\text{Fe} \qquad (1.5)$$

$$_{26}^{57}\text{Fe} +_0^1 \text{n} =_{26}^{58}\text{Fe} \qquad (1.6)$$

$$_{26}^{58}\text{Fe} +_0^1 \text{n} =_{26}^{59}\text{Fe} \qquad (1.7)$$

The Fe^{59} atom is unstable and undergoes spontaneous radioactive decay to cobalt by beta emission as follows:

$$_{26}^{59}\text{Fe} =_{27}^{59}\text{Co} + \beta^- \qquad (1.8)$$

In this case, the negatively charged beta particle (sometimes written as $_{-1}^0 e$ or e^-) can be thought of as the product of the transformation of a neutron ($_0^1$n) to a proton ($_1^1$p), which requires $_{-1}^0 e$ on the product side to balance the reaction. Neutron capture combined with radioactive decay then formed progressively heavier elements up to the heaviest naturally occurring element, uranium. Some elements such as ^{56}Ni and ^{56}Co are unstable and undergo radioactive decay to form stable ^{56}Fe, helping to explain the relative abundance of Fe as compared to elements with similar atomic number.

Returning to fusion, it is clear that progressive fusion reactions involving atoms with even numbers of protons will lead to the sawtooth pattern in Figure 1.1, but there is also another contributing factor to this pattern, and that is the *Oddo-Harkins rule*: atoms with an even number of protons in their nuclei are more stable than their odd-numbered counterparts. This is because, during nucleosynthesis, nuclei with an unpaired proton were more likely to capture an additional proton, producing a more stable proton arrangement, and this favors nuclei with even numbers of protons. For additional information on nucleosynthesis and the origin of the elements, the reader is referred to the accessible and more detailed presentation in Gunter Faure's text *Principles and Applications of Geochemistry*.

1.2.3 Formation of planets and compositional differentiation

As the universe continued to cool, galaxies and solar systems began to form. The solar nebula that was to form our solar system cooled and began to solidify into small masses known as chondrites and eventually larger masses known as planetesimals (on the order of 10s of km in diameter) approximately 5 billion years ago (the Earth is ~4.6 billion years old). Those closest to the early sun became enriched in heavier elements (especially Si, Al, Mg, Fe, Ca, Na, K), in part because centrifugal forces effectively flung lighter elements (especially H and He) preferentially to the farther reaches of the solar system (other important influences include temperature, pressure, redox conditions, and nebular density, but these factors are not covered here). The result is that the inner planets are terrestrial and rocky (Mercury through Mars) and enriched in heavier elements, whereas the outer planets are gaseous (Jupiter and beyond) and enriched in lighter elements (think of the possibility of methane oceans or methane ice on Saturn's moon Titan).

Planets ultimately formed when gravitational forces caused accretion of planetesimals. The accretion of

---

Focus Box 1.4

How Neutron Capture Works

Neutron capture (Eqs. (1.5)–(1.7)) takes place via two main mechanisms. The *r-process* ("r" is for rapid) takes place in core-collapse supernovae, where high neutron flux and extremely high temperatures (e.g. >10^9 K) facilitate nucleosynthesis involving a rapid series of neutron capture reactions starting (typically) with ^{56}Fe – the r-process explains the origin of ~50% of atoms heavier than Fe. The other main means by which heavy elements can be produced is known as the *s-process* ("s" is for secondary), in which nucleosynthesis occurs by means of slow neutron capture. The difference is that s-process neutron-capture nucleosynthesis occurs in asymptotic giant branch (AGB) stars, which have lower temperatures (e.g. 10^3–10^4 K) than supernovae, and thus the s-process requires preexisting (hence the "secondary" nature) heavy isotopes that can function as seed nuclei.

Compositional differentiation of the earth – and specifically with reference to different rock types – often exerts a strong control on the composition of soils. A good example is the weathering of ultramafic rocks (e.g. serpentine, peridotite) derived from the mantle – these ultramafic soils tend to be depleted in plant nutrients such as Ca and K, and enriched in the trace metals Ni and Cr (and in some places, arsenic [As]). Examine Table 1.1 to get an idea of the impact that different rock types can impose on composition of associated soils and waters.

---

what was to become Earth produced heat that left the proto-planet in a molten or semi-molten state and allowed relatively dense Fe and nickel (Ni) to sink to the *core of the Earth*, whereas relatively light silicon (Si), aluminum (Al), magnesium (Mg), calcium (Ca), sodium (Na), and potassium (K) floated to the top to form *Earth's crust*, leaving the Fe–Mg–Ni–Cr–Si *mantle* in between core and crust. While this is a broad generalization, the result is a differentiated Earth (Table 1.1), one where average continental crust (~ between 25 and 60 km thick) has a felsic composition much like that of granite, whereas oceanic crust (~ between 5 and 10 km thick) is, on average, compositionally mafic (or basic) and comprised mainly of basalt, rock that is less silica-rich (less SiO_2-rich) and relatively enriched in Fe, Mg, and Ca relative to continental crust. The mantle is comprised of rock types like peridotite and dunite and has an ultramafic (or ultrabasic) composition.

The primordial atmosphere of earth was comprised of CO_2, CH_4, H_2O, and N gases derived from volcanic eruptions. Oxygen in its form as O_2 gas is a relatively recent addition to earth's atmosphere, having begun to accumulate slowly and in a stepwise manner in the atmosphere after the appearance of photosynthetic algae 3 billion years ago. More on this topic is presented in Chapter 7.

1.3 ATOMS, ISOTOPES, AND VALENCE ELECTRONS

1.3.1 Atoms: protons, neutrons, electrons, isotopes

A pair of schematic sketches of the carbon atom, consisting of a central positively charged nucleus surrounded by a negatively charged "cloud" of electrons, is presented in Figure 1.3. All atoms consist of a nucleus that contains positively charged *protons* and neutral *neutrons*, both subatomic particles with a mass of 1 atomic mass unit (1 amu, or 1 Dalton [Da]).

The actual mass of a proton is 1.6726×10^{-24} g, so it is more convenient to say that the mass of a proton is 1 amu or 1 Da. The +1 charge on a proton $= 1.602 \times 10^{-19}$ coulomb. The number of protons in the nucleus is what distinguishes atoms of one element from another – hydrogen has 1 proton, helium has 2, carbon has 6, uranium has 92, and so on. If there is a nucleus with some other number of protons than 92, it is not uranium. The number of protons is commonly referred to as the *atomic number* (Z).

The mass of a neutron is effectively the same as the mass of a proton (1 amu; 1.6749×10^{-24} g), and sum of protons

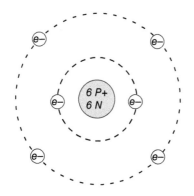
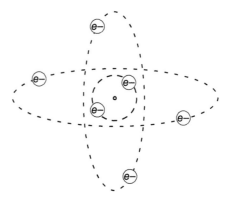

Fig. 1.3   Schematic sketches of Bohr models of a carbon atom (^{12}C) showing, on the left, 6 protons and 6 neutrons in the nucleus with 6 electrons in 2 separate orbitals, and on the right, an atom that attempts to show the actual size of the nucleus compared to the electron cloud, yet even here the nucleus is ~100 times larger than an actual nucleus. The example on the right also introduces the idea of separate orbitals in the outer electron "shell."

and neutrons is the *atomic mass* or *mass number* or *atomic weight* – all are essentially synonymous terms. Virtually all of the mass of an atom is contained in its nucleus. The carbon atom in Figure 1.3 contains a nucleus with 6 protons and 6 neutrons, a configuration represented in this text and in many other places as $^{12}_{6}C$. So, while the number of neutrons is not explicitly given in this notation, it is implied, and in the case of the most common form of uranium, $^{238}_{92}U$, there must be 146 neutrons in the nucleus along with 92 protons to produce the atomic mass of 238.

The carbon atom in Figure 1.3 contains 6 protons and likely also contains 6 neutrons, but it could contain 7 or 8 neutrons. Differences in the number of neutrons (6, 7, or 8) give rise to the three *isotopes* of carbon. Carbon-12, or ^{12}C, is the most abundant isotope of carbon and it contains 6 neutrons; ^{13}C contains 7 neutrons and it is much rarer than ^{12}C (Chapter 10). ^{12}C and ^{13}C are both stable isotopes of carbon and the ratios of these two isotopes in plants, rocks, waters, and sediments have proven very useful in environmental analysis. Carbon also has a *radioactive isotope*, ^{14}C, which is produced in the atmosphere in the presence of cosmic rays. Atoms of ^{14}C are unstable and undergo radioactive decay to nitrogen. This topic and many others in the field of isotope geochemistry are covered in Chapters 10 and 11.

1.3.2 Electrons and bonding

Electrons balance the positive charge of the protons. Electrons are located in specified positions, or energy levels, outside of the nucleus, and the charge on an electron is exactly opposite that of a proton, i.e. -1, or -1.602×10^{-19} coulomb, and neutrally charged atoms contain equal numbers of protons and electrons. Furthermore, while the "electron cloud" surrounding the nucleus is of a virtually negligible mass, it occupies a volume that is orders of magnitude larger than the volume of the nucleus (atoms are mostly open space). (This is shown schematically on the right side of Figure 1.3.)

For the purpose of this text and virtually all environmental geochemistry, the main concern to investigators is the outermost shell of electrons (the *valence shell*), because that is the part of the atom most intimately involved in bonding and ionization. One convenient way to depict valence electrons is with *Lewis electron dot diagrams* (Figure 1.4). Sodium, with one valence electron, and oxygen, with six, can be depicted as shown in Figure 1.4.

Sodium satisfies the octet rule by losing an electron, becoming Na$^+$. The neutral oxygen atom with six valence electrons can most easily satisfy the octet rule by gaining two electrons, producing O^{2-}, the most common form of oxygen in nature.

The *Aufbau principle* describes how electron orbitals are populated in a systematic manner. In brief, electrons occupy orbitals of fixed energy levels and tend to occupy the lowest energy levels possible to create a stable atom. Orbitals are filled in a relatively predictable, sequential manner, starting with the lowest quantum number ($n = 1$), for which there is only one orbital (the *s*-orbital) which can be occupied by two electrons (of opposite spin).

For a neutral hydrogen atom with only one electron, the symbol is $1s^1$. For neutral He with 2 electrons, the notation is $1s^2$. Quantum number 2 ($n = 2$) contains

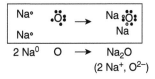

Fig. 1.4 Lewis electron dot diagrams showing valence electrons of sodium and oxygen in their ground states (left) and in the compound Na$_2$O (right), where each sodium has lost an electron to oxygen, resulting in two Na$^+$ and one O^{2-}. The oxygen atom on the left is an idealized representation because in reality it would likely be an O$_2$ molecule.

Focus Box 1.6

Octet Rule Considerations

Neutral atoms with atomic number (Z) ≤ 20 contain somewhere between 1 and 7 electrons in their valence (outermost) shell; the noble gases Ne and Ar contain 8 valence shell electrons. The most stable valence shell electron configuration for elements with Z ≤ 20 is one that consists of 8 electrons, so all atoms with Z ≤ 20 seek to produce ions or form bonds that result in eight valence electrons. They do this either by losing, gaining, or sharing electrons. For elements with Z > 20, the presence of d and f orbitals makes the octet rule inapplicable; nonetheless, heavier elements lose electrons in predictable manners to form more stable configurations, and the forms of these ions are commonly known (e.g. Fe^{2+} and Fe^{3+}; Ni^{2+}; U^{4+}, and U^{6+}) which enables prediction of their behaviors in environmental systems.

Focus Box 1.7

Aufbau Examples

Two useful examples of larger atoms in the Aufbau scheme are calcium and iron. The notation for calcium is: $1s^2$ $2s^2$ $2p^6$ $3s^2$ $3p^6$ $4s^2$. Note that the second and third quantum levels contain eight electrons, but that the fourth contains only two. What must calcium do to satisfy the octet rule when forming an ion? It loses two electrons, taking on an electron configuration like that of argon ($1s^2$ $2s^2$ $2p^6$ $3s^2$

$3p^6$) and becoming Ca^{2+}. (Note: the notation for Ca is often abbreviated as [Ar]$4s^2$.) The notation for iron is: $1s^2$ $2s^2$ $2p^6$ $3s^2$ $3p^6$ $3d^6$ $4s^2$ (26 electrons balance the 26 protons to produce a neutral Fe atom). At this point the octet rule does not exactly predict common oxidation states – iron tends to occur as Fe^{2+} or Fe^{3+} depending on redox conditions (more to come on redox later in this chapter).

---

both *s*- and *p*- orbitals, where the *s*-orbital again can be occupied by one electron pair and the *p*-orbital can host up to 3 electron pairs. For neutral lithium with 3 electrons, the notation is $1s^2$ $2s^1$. For neutral oxygen, the notation is $1s^2$ $2s^2$ $2p^4$ (to satisfy the octet rule, oxygen gains two electrons, producing the O^{2-} anion with an electron configuration like that of neon: $1s^2$ $2s^2$ $2p^6$). The third quantum level ($n = 3$) contains *s*-, *p*-, and *d*-orbitals, where the *s*- and *p*-orbitals can contain 1 and 3 electron pairs, respectively, and the *d*-orbital can host 5 electron pairs (10 e^-s). However, the $3d$ orbital exists at a higher energy level than the $4s$ orbital, so the $4s$ orbital is filled before the $3d$. The sequence in which these orbitals are filled is as follows:

$$1s \rightarrow 2s \rightarrow 2p \rightarrow 3s \rightarrow 3p \rightarrow 4s \rightarrow 3d \rightarrow 4p \rightarrow 5s$$

$$\rightarrow 4d \rightarrow 5p \rightarrow 6s \rightarrow 4f \rightarrow 5d \rightarrow 6p \rightarrow 7s \rightarrow 6d$$

1.4 MEASURING CONCENTRATIONS

Concentrations of elements or compounds are measured in a few common ways. Only in rare cases do scientists directly measure individual atoms/compounds, and this is because even microscopic crystals of minerals at the micrometer scale such as hydroxides and silicate clays typically contain millions or billions of atoms.

1.4.1 Mass-based concentrations

One common means of measuring concentrations is by units such as milligrams per kilogram (mg kg^{-1}) for solids and milligrams per liter (mg l^{-1}) for liquids, or sometimes in units of micrograms (µg kg^{-1} or µg l^{-1}) or nanograms (ng kg^{-1} or ng l^{-1}) for trace elements. Note that mg l^{-1} = µg ml^{-1}. Weight percent is a common mass-based approach for expressing element concentration in cases where elements are in high concentrations (e.g. Table 1.1), a good example being Si, which comprises approximately 28% by weight (or mass) of the continental crust, but as wt % oxide, approximately 59% of the crust – this conversion is presented here.

Weight percent oxide is a common means of expressing major (abundant) elements in soils and rocks. Note in Table 1.1 that Si, Al, Fe, and the other major elements are presented in units of SiO_2, Al_2O_3, Fe_2O_3, and so on. This is done partly by convention (or habit) and also because the major elements tend to occur in silicate minerals bonded to oxygen. The conversion factors for wt % element to wt % oxide for the common oxides are presented here in Table 1.2 (where C.F. = conversion factor).

In the case of converting Ca to CaO, the conversion is determined as the molar mass of CaO divided by the molar mass of Ca, i.e. $56.08 \div 40.08 = 1.399$.

Focus Box 1.8

A Few Useful Facts About Mass-Based Units of Concentration

For solids:
mg kg^{-1} is also known as parts per million (ppm), where mg is a milligram (10^{-3} g)
µg kg^{-1} is also known as parts per billion (ppb), where µg is a microgram (10^{-6} g)
ng kg^{-1} is also known as parts per trillion (ppt), where ng is a nanogram (10^{-9} g)

For liquids:
mg l^{-1} is also known as parts per million (ppm)
µg l^{-1} is also known as parts per billion (ppb)
ng l^{-1} is also known as parts per trillion (ppt)

Focus Box 1.9

Molar vs. Mass-Based Units

Mass-based measurements like µg l^{-1} or wt % are not always the best way to express concentrations. Consider for example groundwater with 98.7 µg l^{-1} each of nickel (^{59}Ni) and uranium (^{238}U); i.e. both elements are present in equal concentrations of 98.7 ppb. We could visualize this by evaporating the liter of water, which would leave 98.7 µg each of Ni and U. Consider this though: given that U atoms (238 g mol^{-1}) are ~4 times heavier than Ni (59 g mol^{-1}), there must be more Ni atoms; in fact, there are approximately four times as many Ni atoms as there are U atoms.

Table 1.2 Conversion factors for wt % element to wt % oxide.

Element	C.F.	Oxide
Al	1.889	Al_2O_3
Ca	1.399	CaO
Fe	1.286	FeO
Fe	1.430	Fe_2O_3
K	1.205	K_2O
Mg	1.658	MgO
Mn	1.291	MnO
Na	1.348	Na_2O
P	2.291	P_2O_5
Si	2.139	SiO_2
Ti	1.668	TiO_2

In the case of Al_2O_3, the conversion factor is determined as the molar mass of Al_2O_3 divided by the mass of the equivalent amount of Al in the oxide, i.e. $Al_2O_3 \div$ (2 × Al) = 101.957 ÷ (2 × 26.98) = 1.889. For Na to Na_2O, the conversion is the molar mass of Na_2O divided by the molar mass of 2×Na = 61.98 ÷ (2 × 22.99) = 1.348.

For elements that can occur in more than one oxidation state (e.g. Fe^{2+}, Fe^{3+}), values may be presented as either one of the oxidation states (i.e. either as FeO or Fe_2O_3), or as a combination of the two if the relative abundances of Fe^{2+} and Fe^{3+} are known. In many cases where iron oxidation state is not known, all iron is presented in terms of Fe_2O_3.

1.4.2 Molar concentrations

Quantifying concentrations on a molar basis has its roots in the work of Italian scientist Amadeo Avogadro, who in 1811 realized that equal volumes of gases at identical pressures and temperatures contain equal numbers of atoms (or molecules in the case of gases like N_2 and O_2), even though their atomic masses differ. The term *mole* (abbreviated mol) describes the number of atoms of a given element required to form a mass equal to the atomic mass of the substance, in grams. For C, this mass is 12.011 g. For U, this mass is 238.03 g, and so on. For all elements, the number of atoms required to form the atomic mass in grams is 6.0221×10^{23} atoms, a value known as *Avogadro's number*. What this means is that 238.03 g of uranium (1 mol of U) contains 6.0221×10^{23} atoms; similarly, 4.002 g of helium (1 mol of He) contains 6.0221×10^{23} atoms.

Focus Box 1.10

Converting Mass-Based to Molar Concentrations

The mole is a very useful concept in chemistry, especially considering that chemical equations are expressed in terms of moles of reactants and products. In order to express Ni and U concentrations in terms of moles per liter (mol l^{-1}), or for trace elements like these, µmol l^{-1}, the mass concentration must be multiplied by the inverse of the molar mass (and a conversion for g to µg) as follows:

For Ni: 98.7 µg l^{-1} × 1 mol/58.693 g × 1 g/10^6 µg = 1.68 × 10^{-6} mol l^{-1}

For U: 98.7 µg l^{-1} × 1 mol/238.03 g × 1 g/10^6 µg = 0.415 × 10^{-6} mol l^{-1}

It is often helpful to express units in easy to communicate terms, so in this case mol l^{-1} would probably be converted to micromoles per liter (µmol l^{-1}), by multiplying mol l^{-1} by 10^6 µmol mol^{-1}:

1.68 × 10^{-6} mol l^{-1} × 10^6 µmol mol^{-1} = 1.68 µmol l^{-1} of Ni in the groundwater

0.415 × 10^{-6} mol l^{-1} × 10^6 µmol mol^{-1} = 0.415 µmol l^{-1} of U in the groundwater

In a solid (e.g. sediment or rock) with $1 \, mol \, kg^{-1}$ each of Fe ($55.85 \, g \, mol^{-1}$) and Al ($26.98 \, g \, mol^{-1}$): (i) there is an equal number of atoms of Fe and Al in the soil; and (ii) Fe comprises a greater mass of the soil than Al (whether expressed as wt %, $g \, kg^{-1}$, or $mg \, kg^{-1}$). Given that mass units are a common way of expressing concentration, it may be necessary to convert from molar units to mass units. A few algebraic calculations allow conversion from $mol \, kg^{-1}$ to 3 common units, $g \, kg^{-1}$, $mg \, kg^{-1}$, and wt% ($0.1 \times g \, kg^{-1}$):

Given that 1 mol of Fe = 55.85 g:

$$55.85 \, g \, mol^{-1} \times 1 \, mol \, kg^{-1} = 55.85 \, g \, kg^{-1} \, Fe$$

... *or*

$$55.85 \, g \, kg^{-1} \times 1000 \, mg \, g^{-1} = 55\,850 \, mg \, kg^{-1} \, Fe$$

... *or*

$$55.85 \, g \, kg^{-1} \times 0.1 = 5.855\% \, Fe \, (by \, weight)$$

$$1 \, mol \, of \, Al = 26.98 \, g; \, 26.98 \, g \, mol^{-1} \times 1 \, mol \, kg^{-1}$$
$$= 26.98 \, g \, kg^{-1} \, Al$$

... *or*

$$26.98 \, g \, kg^{-1} \times 1000 \, mg \, g^{-1} = 26\,980 \, mg \, kg^{-1} \, Al$$

... *or*

$$26.98 \, g \, kg^{-1} \times 0.1 = 2.698\% \, Al \, (by \, weight)$$

1.4.3 Concentrations of gases

Atmospheric gas concentrations are typically expressed as the proportion of the total volume accounted for by a given gas. For example, the current atmospheric concentration of CO_2 is $\sim 410 \, ppmv$, indicating that 410 out of every one million molecules of gas in earth's atmosphere is CO_2. At the onset of the industrial revolution atmospheric CO_2 was 280 ppmv. Less-abundant gases are often expressed in terms of pptv or ppbv (parts per billion or trillion, volumetrically), and the major components of the atmosphere like the fixed gases N_2, O_2, and Ar, are expressed in terms of percent (by vol): $N_2 = 78.1\%$, $O_2 = 20.9\%$, and $Ar = 0.9\%$ (the amounts vary depending on the amount of H_2O vapor in the air, which can range from 0% to 4% by volume). Expressed in this way, CO_2 comprises approximately 0.041% percent of the atmosphere (vol %), but clearly units of ppmv are more useful for a gas like CO_2. Gases also dissolve in liquids, and units of concentration in these cases are commonly $mg \, l^{-1}$ or $mmol \, l^{-1}$.

1.4.4 Notes on precision and accuracy, significant figures, and scientific notation

A few important considerations related to data analysis and presentation of results are embodied in the concepts of precision and accuracy. Simply stated, *accuracy* describes how closely a measured value agrees with the actual value. The accuracy of chemical analyses can be tested by analyzing standards of known concentration. Consider a certified standard solution that contains $250 \, mg \, l^{-1}$

Focus Box 1.11

Expressing Uncertainty or Error

One way to express scatter or uncertainty is to determine the standard deviation of the data (many scientists use spreadsheets for this, for which there are many tutorials on the internet). The formula for standard deviation and details on its appropriate use can be found in a statistics textbook, but using the Microsoft Excel ® standard deviation formula produces σ values as follows:
For the values 237, 271, 244, 262, and $240 \, mg \, l^{-1}$,
   σ = 14.9
For the values 277, 281, 274, 278, and $275 \, mg \, l^{-1}$, σ = 2.7
For the first set, we could report results as: 251 ± 14.9 $mg \, l^{-1}$ (1 sd)
For the second set, we could report the results as: $277 \pm 2.7 \, mg \, l^{-1}$ (1 sd)

If these data are for a 250.0 ppm standard, which do you consider better, analyses with greater accuracy but less precision ($251 \pm 14.9 \, mg \, l^{-1}$), or greater precision but less accuracy ($277 \pm 2.74 \, mg \, l^{-1}$)? Precise values are easier to correct because there is less uncertainty than if you have to deal with accurate mean values plagued by low precision. This emphasizes one of the reasons it is important to run standards when making analytical measurements. Of course, the ideal situation is to use a well-calibrated instrument for which accuracy and precision are both high, but regardless, it is imperative that researchers seek to quantify both parameters when making measurements.

Focus Box 1.12

Thoughts on Significant Figures

When considering significant figures, it is important to understand the significance of zeros. Consider the following values: 7200, 700 043, 0.0436, 0.043600. How many significant figures are reported for each, and why?

1. 7200 is assumed to have 2 significant figures because any zero at the end of a number and before a decimal point is assumed to not be significant. How many significant figures does 847 000 possess? Three. 7200.0 contains 5 significant figures.
2. 700 043 has 6 sig figs. Any zeros within a number are significant.
3. 0.0436 has 3 sig figs because any zeros after a decimal point and before the first non-zero digit are not significant.

4. 0.043600 has 5 significant figures because zeros that follow a non-zero digit after a decimal point are considered significant. (as in 7200.0 example above).

Some of these examples serve to illustrate why scientists tend to express values in terms of **scientific notation**. Using scientific notation, the values above become:

7.2×10^3 (2 sig figs)
7.2000×10^3 (if we needed to report 7200 to 5 sig figs, this is how it would look)
7.00043×10^5 (6 sig figs)
4.36×10^{-2} (3 sig figs)
4.3600×10^{-2} (5 sig figs)

---

of aluminum (Al), and five analyses of this standard on your spectrometer produces results of 237, 271, 244, 262, and 240 mg/l. The mean of those five values is $251\ \mathrm{mg\,l^{-1}}$ – the average value is very close to the certified value of $250\ \mathrm{mg\,l^{-1}}$. One way to express the accuracy of this test is as a percent difference from the certified value:

$$[(251\ \mathrm{mg\,l^{-1}} - 250\ \mathrm{mg\,l^{-1}}) \div 250\ \mathrm{mg\,l^{-1}}]$$
$$\times 100 = 0.4\%$$

However, the five results are somewhat lacking in *precision*, which is basically a measure of the reproducibility of results – how closely do measured results agree with each other?

It is conceivable to produce results with a high degree of precision that are lacking in accuracy. For example, after recalibrating the spectrometer and re-analyzing the Al standard, values now are 277, 281, 274, 278, and $275\ \mathrm{mg\,l^{-1}}$. The mean value of $277\ \mathrm{mg\,l^{-1}}$ is farther from the certified value of $250\ \mathrm{mg\,l^{-1}}$ (the difference from the certified value is 10.8%), but the results are definitely more precise.

In addition to quantifying uncertainty, it is very important to present numerical results in a manner that relates to the sensitivity of the measurement, or the degree of confidence associated with that measurement, while also trying to avoid propagation of error. Every measurement has some limited number of *significant digits* (or *significant figures*). Measurements of pH made using litmus paper can only be reported to 1 significant figure (e.g. pH = 6) whereas measurements made with well-calibrated probes may be reported to 3 significant figures (e.g. pH = 5.87).

As a rule, calculations should be carried out using all figures with each of their representative significant figures, and then the final result should be rounded at the end of the calculation. Using an example where the average concentration of uranium in groundwater in an aquifer is $14.3\ \mathrm{\mu g\,l^{-1}}$, the aquifer volume is $1.241 \times 10^8\ \mathrm{m^3}$, and average porosity (and thus % water in the saturated zone) is 1.8%, resulting mass of U in the aquifer is:

$$14.3\ \mathrm{\mu g\,l^{-1}} \times 1.241 \times 10^8\ \mathrm{m^3} \times 1000\,\mathrm{l\,m^{-3}} \times 0.018$$
$$= 3.2 \times 10^{11}\ \mathrm{\mu g\ U}$$

The final result is limited by the two significant figures in 1.8% (note that conversions involving liters to cubic meters, cm^3 to ml, etc. do not limit sig figs).

As a closing thought on this topic, the precision of the road sign in southwestern Ecuador shown in Figure 1.5 is notable. Being literal about significant figures, the distance to Engunga is precisely indicated as being between 14.995 and 15.005 km, a level of precision down to $\leq 10\ \mathrm{m}$ (note: it is a very small town, so the precision may be justified).

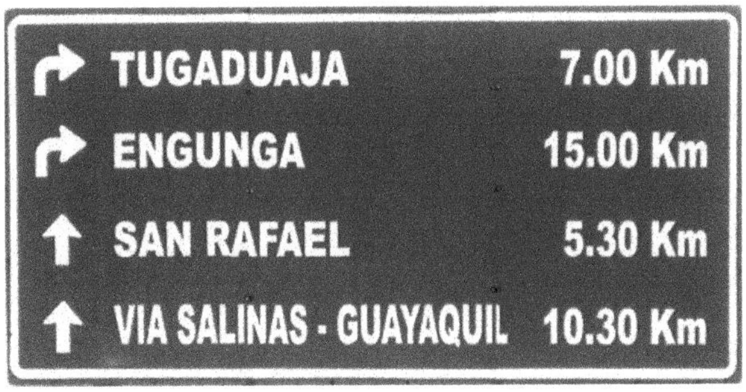

↱ TUGADUAJA	7.00 Km
↱ ENGUNGA	15.00 Km
↑ SAN RAFAEL	5.30 Km
↑ VIA SALINAS - GUAYAQUIL	10.30 Km

Fig. 1.5 Road sign in coastal Ecuador showing high level of precision.

1.5 PERIODIC TABLE

The Periodic Table of the Elements (*inside front book cover*) is a very useful tool for researchers and students of chemistry, geochemistry, and biochemistry. First developed by the Russian chemist Dmitri Mendeleev in 1869 and refined ever since, it lists elements in order of atomic number (from left to right in each row, and also from top to bottom in each column) and by similarities in chemical properties. For example, row 3 of the periodic table begins with sodium (atomic number $Z = 11$) and progresses to the right with increasing Z all the way up to argon ($Z = 18$).

1.5.1 Predicting behavior of elements using the periodic table

Columns (or groups) generally contain elements with similar valence electron configurations, that is, the outermost shell of electrons tends to be arranged in a similar manner for elements of a given group. Good examples are group 1, the alkali metals, all of which lose one electron when they form chemical bonds, resulting in a series of 1+ charged ions including Na^+ and K^+; group 2, the alkaline earth metals that form divalent cations (e.g. Mg^{2+}, Ca^{2+}, Sr^{2+}); group 3 ions include B^{3+} and Al^{3+}. In group 4, silicon occurs as Si^{4+} but carbon is much more complex, ranging from C^{4+} to C^{4-} and various oxidation states in between (Chapter 5, Section 5.2.1). On the right side of the periodic table, the group 8 noble gases include elements like argon and neon with complete valence shells (as a result, they do not form chemical bonds in nature). The group 7 halogens tend to gain one electron when forming chemical bonds, resulting in halide anions like F^- and Cl^-. Oxygen in group 6 is O^{2-} in nearly all compounds, and in reducing environments sulfur occurs as S^{2-}.

It may not come as a surprise that alkali metals tend to form bonds with halides, such as:

$$Na^+ + Cl^- \rightarrow NaCl \qquad (1.9)$$

The point to understand here is that the periodic table presents information in a systematic way that can help to predict the behavior of elements in environmental systems. Like Na^+, K^+ can also form bonds with chloride (Cl^-) to form a different salt, KCl (a substitute for NaCl in reduced-sodium diets). In fact, K^+ and Na^+ substitute for one another in many minerals. The periodic table also implies that arsenic (As) might substitute for phosphorus (P) in molecules, which it does.

1.5.2 The earth scientist's periodic table

In 2003, the environmental geochemist Bruce Railsback from the University of Georgia developed an innovative new periodic table for geologists known as *An Earth Scientist's Periodic Table of the Elements and Their Ions*,

---

Focus Box 1.13

Atomic Mass of Sulfur

The atomic mass for each element represents a weighted average of the mass of the isotopes of that element – the following is a mass-balance calculation that takes into account the weighted average of ^{32}S and the three naturally occurring heavier isotopes of S (data from Faure 1986):

Atomic masses for elements are expressed in terms gram molecular weights (i.e. the gram molecular weight, or mass of 1 mol of sulfur, is 32.06 g). Note that here the atomic mass is represented with 4 significant figures. The reason that each of the isotopes is not an integer value (e.g. ^{32}S is stated as 31.97 rather than 32.0 g mol^{-1}) is that all atomic masses are normalized to the mass of ^{12}C.

$^{32}S = 95.02\%$ $^{33}S = 0.75\%$ $^{34}S = 4.21\%$ $^{36}S = 0.02\%$

^{32}S: 0.9502×31.97 g mol^{-1} = 30.377894 g mol^{-1}
^{33}S: 0.0075×32.97 g mol^{-1} = 0.247275 g mol^{-1}
^{34}S: 0.0421×33.97 g mol^{-1} = 1.430137 g mol^{-1}
^{36}S: 0.0002×35.97 g mol^{-1} = 0.007194 g mol^{-1}
$\qquad\qquad\qquad\qquad$ Sum = 32.062500 g mol^{-1}

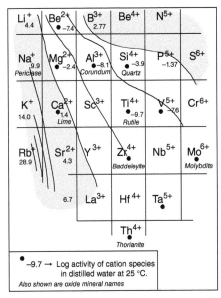

Fig. 1.6 Solubilities of ions as a function of ionic charge, from Inset 4 of *An Earth Scientist's Periodic Table of the Elements and Their Ions* (Railsback, 2003). Oxides of the elements are used as reference (e.g. lime, periclase, etc.). Note that low-charge base cations/alkali metals and cations of the alkaline earth metals are relatively soluble, as are high-charge cations such as S^{6+} and N^{5+}, which form polyatomic anions (SO_4^{2-}, NO_3^-). Cations with +3 and +4 charges (e.g. Al^{3+}, Ti^{4+}, Zr^{4+}) are insoluble in most surficial environments, with activities in H_2O at 25 °C of $\sim 10^{-8}$ to 10^{-10}.

one where elements are organized according to their occurrences in geological environments (http://www .gly.uga.edu/railsback/PT.html). This new formulation is designed to predict how elements and ions behave in the environment. Unlike the conventional periodic table originally envisioned by Mendeleev, the earth scientist's periodic table organizes elements by charge, so it shows many elements multiple times because many elements have numerous oxidation states. It also contains plentiful information on abundance of different elements in soils, seawater, mantle vs. crust, ionic radii in crystals, and much more. For example, Figure 1.6 shows the relative solubilities of oxides of various ions, and from this inset it is apparent that Al and Ti form insoluble oxides (corundum, Al_2O_3; rutile, TiO_2) but that Na, K, N, and S do not (they are more likely to occur as soluble ions such as Na^+, K^+, NO_3^-, and SO_4^{2-}).

The earth scientist's periodic table also shows the numerous species of nitrogen that exist in natural systems (e.g. oxidation states and common molecules of N), predicted behaviors, attributes and affinities of the abundant and trace cations in natural systems, and more (Railsback 2003).

1.6 IONS, MOLECULES, VALENCE, BONDING, CHEMICAL REACTIONS

It is relatively rare to find elements in their native, neutral state in nature. Some elements do occur in neutral states, including the diatomic gases hydrogen (H_2), oxygen (O_2), and nitrogen (N_2), solids like gold (Au), graphite, and diamond (both forms of C), and the noble gases (He, Ne, Ar ...). Most elements, however, occur as ions, molecules, or compounds in liquids (petroleum, alcohols, H_2O, and the dissolved species it contains), solids (minerals, proteins, humus), and gases (CO_2, CH_4, NO_2). *Ions* are charged atoms, atoms that have either gained or lost one or more electrons from their neutral state. Examples of ions include the cation Na^+ and the anion S^{2-}. Most of the elements form cations, because metals generally lose electrons to form cations, and most of the elements are metals. A few common metal cations are Mg^{2+}, Al^{3+}, $Cr^{3+,6+}$, $Fe^{2+,3+}$ – note that some elements have more than one oxidation state. Generally only those elements in the upper right of the periodic table form anions (e.g. O^{2-}, N^{3-}, F^-, etc). *Oxidation state* is a term for the charge on an ion: "the oxidation state of iron in swamps is generally 2+ (divalent), whereas in streams, iron tends to occur in the 3+ (trivalent) oxidation state."

Molecules form when two or more atoms are joined by a chemical bond, and *compounds* are a specific type of molecule formed when two or more atoms of *different* elements are joined by chemical bonds. The gases H_2, O_2, and N_2 occur as molecules but they are not compounds (they are diatomic gases). Examples of compounds include NaCl, H_2O, and C_8H_{18} (octane). Most substances in nature, other than some gases and a few metals, are compounds.

There are two main types of *chemical bonds* responsible for forming molecules and compounds, ionic bonds and covalent bonds. These two bond types represent polar ends of the bonding spectrum, but it is useful to consider each type separately, and then examine intermediate cases. We will also examine metallic bonds and van de Waals bonds, but first will begin with ionic bonds.

1.6.1 Ionic bonding

Ionic bonds occur between atoms of elements with very different valence electron configurations. Coulomb forces describe interactions between charged atoms or molecules (i.e. ions), and when the charges are opposite (i.e. involving cations and anions), the result is attraction (conversely,

---

Focus Box 1.14

Oxidation State According to IUPAC

IUPAC is the International Union of Pure and Applied Chemistry and is the authority on questions of nomenclature in chemistry. IUPAC nomenclature (Karen et al. 2016) indicates that "the oxidation state of an atom is the charge of this atom after ionic approximation of its heteronuclear bonds," and that "oxidation state" and "oxidation number" are effectively synonymous terms. In terms applicable to geochemistry, oxidation state is the charge on an atom in a solution (e.g. Ca^{2+} in water) or in a compound (e.g. Ca^{2+} in calcite). It is easier to determine for ionically bonded compounds (Na^+ and Cl^- in halite), but even in compounds dominated by covalent bonds, oxidation state

can be assigned to atoms, e.g. in methane (CH_4), C and H can be indicated as C^{4-} and H^+. Often "valence state" and oxidation state are used synonymously, but it is best to not use the term valence state to indicate oxidation state. You might say that chlorine has seven valence electrons in the neutral state, but in terms of charge you would say that Cl can adopt many oxidation states, e.g. Cl^- (in NaCl), Cl^0 (Cl_2 gas), Cl^{7+} (in the perchlorate anion ClO_4^-), and more. Also according to IUPAC, oxidation states are best represented as e.g. Al^{3+} and O^{2-} (rather than e.g. Al^{+3} and O^{-2}), although it is common to see both used.

Focus Box 1.15

Coulombic Interaction Example

The classic example of Coulombic attraction is the bond between an alkali metal cation (e.g. Na^+) and a halogen anion (e.g. Cl^-). The cation and anion are electrostatically attracted to each other and the result is formation of an ionic bond producing a solid, in this case cubic vitreous crystals of halite (NaCl).

Loss of an electron by sodium and its incorporation into the valence shell of chlorine can be viewed in the sense of a Lewis electron dot diagram (Figure 1.7).

One way this type of bond forms in the natural world is when evaporation of salty water (e.g. marine water or brines) drives Na^+ and Cl^- concentrations up to levels

Na⁺ :Cl:

Fig. 1.7 Schematic representation of formation of an ionic bond by transfer of the Na valence electron to Cl. Resulting Na^+ and Cl^- ions form a strong electrostatic attraction. The empty circle next to the Na atom represents the electron lost to the Cl atom.

sufficiently high to allow formation of solid crystals. (More details about the controls on aqueous processes are covered in detail in Chapters 4, 5, and 9.)

---

the Coulombic interaction between like-charged particles, e.g. 2 cations, causes repulsion).

1.6.2 Determining ionic bond strength

The strength of an ionic bond is largely controlled by the charges on the ions and by the radii of the ions involved in the bonding. This is a concept that should be intuitively apparent because ions with higher charges will be more attracted (provided that they have opposite charges), and the closer the spacing of the ions, generally the stronger the bond.

The attraction of two ions can be quantified by Coulomb's Law:

$$F_c = k \times (q_1 \times q_2)/(\varepsilon \times r) \tag{1.10}$$

where k is a constant (described below), q_1 and q_2 are the values of ionic charges on the ions, ε is the dielectric constant, and r is the distance between the nuclei of the two ions joined by the bond.

When F_c is negative the ions are attracted; the negative sign indicates that the system (the two ions) have shifted to a lower (more stable) energy state than is the case when the ions are separated. Given that all other terms (k, ε, r) are positive, negative F_c results from two ions with opposite charge, i.e. a cation and an anion.

The constant k is expressed as:

$$k = 1/(4\pi\varepsilon_0) \tag{1.11}$$

where ε_0 is the permittivity constant (sometimes known as P_0 or D) – it equals 8.854×10^{-12} C^2J^{-1} m^{-1} and results in a value for the constant $k = 8.998 \times 10^9$ J·m C^{-2}, where J = joules and C = 1.602×10^{-19} C, a measure of charge.

The dielectric constant ε expresses the effect of the ambient environment on the strength of the bond, the best example being the difference in ionic bond strength in dry air as compared to water. At 20 °C and 1 atm, the *dry air* value for $\varepsilon = 1.0$ whereas the value for ε in *water* at 20 °C = 88. Therefore, all other factors being equal, an ionic bond is 88 times weaker in water than in dry air.

For Na^+ and Cl^- ions, q_1 and q_2 are +1 and −1. The ionic radius of Na^+ in a crystal of NaCl is 1.16Å and the ionic radius for Cl^- is 1.67 Å (where an angstrom is 10^{-10} m). The reason that the Cl^- anion is larger than the Na^+ cation is that Cl^- contains an additional shell of electrons. (In general, anions have larger radii than cations because anions gain electrons when they form ions.)

The bond distance between nuclei will be:

$$1.16 \times 10^{-10} \text{ m} + 1.67 \text{ Å} \times 10^{-10} \text{ m} = 2.83 \times 10^{-10} \text{ m}$$

so the attractive energy of the NaCl bond in air ($\varepsilon = 1.0$) is:

$$F_c = 8.998 \times 10^9 \text{ J·m C}^{-2} \times (-1.602 \times 10^{-19})$$
$$\times (+1.602 \times 10^{-19}) \div 2.83 \times 10^{-10} \text{ m}$$
$$= -8.15 \times 10^{-19} \text{ J}$$

Note that units of charge in the numerator (C^2, Coulombs squared) cancel with C^2 units from the constant, and also that m (meters, from ionic radius r) in the denominator cancel with m in the numerator leaving us with units of J (joules), which express energy lost (if J < 0) or gained (if J > 0), where negative values indicate transition to a more stable state.

This calculation applies to a single Na–Cl atom pair. If we wished to compute this for a mole we would multiply the result by Avogadro's number (1 mol of NaCl would contain 6.022×10^{23} atoms each of Na and Cl):

$$8.15 \times 10^{-19} \text{ J/atom} \times 6.022 \times 10^{23} \text{ atoms mol}^{-1}$$
$$= 4.91 \times 10^5 \text{ J mol}^{-1} = -491 \text{ kJ mol}^{-1}$$

If we compare this to the bond between K^+ (ionic radius = 1.52 Å) and Cl^-, a bond that results in formation of KCl (sylvite or potassium chloride), we find that:

$$F_c = 8.998 \times 10^9 \text{ J·m C}^{-2} \times (-1.602 \times 10^{-19})$$
$$\times (+1.602 \times 10^{-19}) \div 3.19 \times 10^{-10} \text{ m}$$
$$= -7.23 \times 10^{-19} \text{ J}$$

and

$$-7.23 \times 10^{-19} \text{ J/atom} \times 6.022 \times 10^{23} \text{ atoms mol}^{-1}$$
$$= 4.35 \times 10^5 \text{ J mol}^{-1} = -435 \text{ kJ mol}^{-1}$$

Considering the attraction of cation and anion, NaCl forms a stronger bond than KCl because the smaller ionic radius of Na allows the Na and Cl nuclei to be held closer.

In the case of magnesium chloride ($MgCl_2$), the ionic charge on Mg is 2+ and its ionic radius in $MgCl_2$ is 0.86 Å. So in this case,

$$F_c = 8.998 \times 10^9 \text{ J·m/C}^2 \times (-1.602 \times 10^{-19})$$
$$\times (2 + \times 1.602 \times 10^{-19}) \div 2.53 \times 10^{-10} \text{ m}$$
$$= -1.82 \times 10^{-18} \text{ J}$$

and

$$-1.82 \times 10^{-18} \text{ J/atom} \times 6.022 \times 10^{23} \text{ atoms mol}^{-1}$$
$$= 1.10 \times 10^6 \text{ J mol}^{-1} = -1100 \text{ kJ mol}^{-1}$$

Based on Coulombic attraction, the higher charge and smaller radius of Mg^{2+} compared to K^+ and Na^+ make the $MgCl_2$ lattice energy approximately twice that of KCl and NaCl.

For ionic solids in water ($\varepsilon = 88$ in the denominator), the lattice energies of NaCl, KCl, and $MgCl_2$ are 5.58, 4.95, and 12.5 kJ mol^{-1}, respectively, which is a quantitative way of saying that ionic bonds are approximately two orders of magnitude weaker in water than in dry air.

The treatment presented above only considers the attraction between two ions and does not consider other factors associated with the strength of ionic bonds, two of which include (i) the effect of other ions and the geometry of the lattice structure (this can be assessed using the Madelung constant, which is not covered here), and (ii) temperature; for example, melting points for NaCl, KCl, and $MgCl_2$, respectively, are 801, 770, and 714 °C, differences which are not explained by Coulomb's Law.

1.6.3 Covalent bonding

Covalent bonds involve overlap of electron orbitals, i.e. they are bonds that form by sharing electrons between atoms. Unlike ionic bonding, where atoms have very different attractions to their respective valence electrons, atoms that form covalent bonds have similar attractions to their valence electrons; therefore, because their valence electrons are attracted to their nuclei with similar strength, it

---

Focus Box 1.16

Covalent Bonding: Diatomic Gas Example

A classic example of covalent bonding involves the attraction of atoms in the diatomic gases H_2, N_2, and O_2. When two oxygen atoms combine to form O_2, there is clearly no difference in the valence electron configuration or in the attraction of each oxygen atom for its electrons. The atoms are effectively equal in structure, so the solution to the octet rule for O_2 involves reorganization and overlapping of electrons to produce valence shells that each contain eight electrons (Figure 1.8).

Both O atoms have satisfied the octet rule the same way, by overlapping electron orbitals to produce a stable configuration. So, while each O atom has six valence electrons, two of them are shared with the adjacent O atom by orbital

Fig. 1.8 Lewis electron dot diagram of an oxygen molecule (O_2) shown using two different notations. Note double bond between oxygen atoms consisting of two electron pairs depicted by the double dashed lines in the lower example.

overlap to produce a strong chemical bond. In this case the bond between the oxygen atoms is a *double bond* because it involves two pairs of electrons, as contrasted to a typical *single bond* that involves only one pair of electrons (refer to the water molecules shown in Figure 1.11).

---

is impossible for one atom to lose its valence electron(s) to another atom to form a bond.

Covalent bonds occur between elements with identical or similar electronegativities (Figure 1.9), including the diatomic gases, between H and O in H_2O, between C and O in CO_2, and Si and O in SiO_2. Chemical bonds that are a mix of ionic and covalent (e.g. SiO_2) are termed polar covalent bonds.

1.6.4 Electronegativity and predicting bond type

Electronegativity is a measure of the attraction of a nucleus for its valence electrons. Na and K have low electronegativities because they readily lose their valence electron to satisfy the octet rule, resulting in +1-charged cations. Conversely, fluorine and chlorine have strong attractions to the electrons in their valence shell and furthermore have the ability to pull valence electrons away from nearby atoms to form −1-charged anions and satisfy the octet rule – these elements have high electronegativity values. Electronegativity is important because it helps to predict the type of chemical bond that will form between elements – elements with large differences in electronegativity form dominantly ionic bonds, whereas elements with low differences in electronegativity form covalent or polar covalent bonds.

Figure 1.9 presents electronegativities of the elements and Figure 1.10 shows the % ionic character of

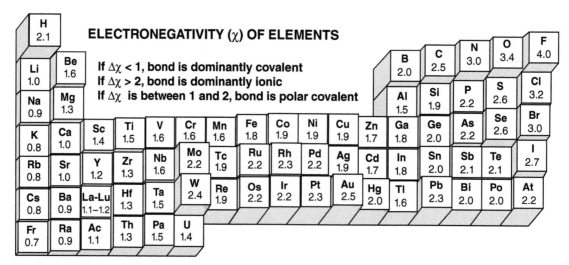

Fig. 1.9 Electronegativity of elements based on Linus Pauling's early twentieth-century research.

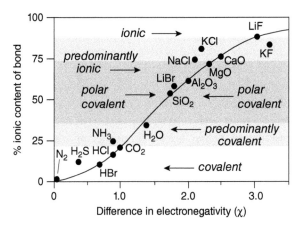

Fig. 1.10 Plot of relationship between difference in electronegativity of two atoms in a bond and the percent ionic character of the bond. Note that no bond shown here is 100% ionic; however, diatomic gases (e.g. N_2, Cl_2, O_2) and the C-C bonds in diamond are 100% covalent. Based on the early and middle twentieth-century work of Linus Pauling and more recent research by Lu et al. (2006).

some selected chemical bonds expressed by difference in electronegativity. Nearly all bonds involving atoms of two elements (i.e. not including diatomic gases like H_2 or O_2 native elements such as Au or Ag) involve some aspect of electron transfer (ionic character) and orbital overlap (covalent character) – the greater the difference in electronegativity, the greater the ionic character (and usually, the greater the tendency to dissociate and dissolve in water).

The electronegativity difference ($\Delta\chi$) for NaCl is 3.0 (Cl) – 0.9 (Na) = 2.1, resulting in a predominantly ionically bonded compound. For a purely covalent molecule like N_2, O_2, Cl_2, or diamond (pure C), $\Delta\chi$ is zero. Methane (CH_4) and hydrogen sulfide (H_2S) are strongly covalent ($\Delta\chi = 0.4$ and 0.5, respectively). SiO_2 is a mix of covalent and ionic (Figure 1.10) and the term *polar covalent bond* (or *polar bond*) is used to describe this type.

$CaCO_3$ (calcium carbonate, e.g. the mineral calcite) contains bonds between Ca–O and C–O. The Ca–O bonds are predominantly (~70%) ionic ($\Delta\chi = 3.4–1.0 = 2.4$), whereas the C–O bond is predominantly (~80%) covalent ($\Delta\chi = 3.4–2.5 = 0.9$). Understanding this concept is important when predicting solubilities of minerals; for example, in water, calcite dissolves to produce Ca^{2+} ions and the polyatomic CO_3^{2-} anion in solution (see Chapter 5 for details of carbonate geochemistry). The covalent character preserves the C–O bond and the carbonate anion (CO_3^{2-}) is a common constituent of natural waters.

1.6.5 Metallic bonds, hydrogen bonds, and van der Waals forces

Metallic bonds occur among metals such as Cr, Cu, Fe, Ni, and Zn in solid molecules, but unlike ionic or covalent bonds, once valence electrons are released by a metal atom, they are not fixed with a specific atom, but rather migrate through the crystal structure. This type of bond occurs in sulfide minerals such as pyrite, and also in native elements such as Cu, gold (Au), and silver (Ag). These bonds are weaker than covalent and ionic bonds and are part of the reason why metal-bearing sulfide minerals are relatively unstable at the Earth surface.

Dipole bonds, also known as *van der Waals* bonds (and sometimes called van der Waals *forces*), exist between electrically neutral molecules or compounds with some unequal distribution of charge. A great example of a dipole bond exists between water molecules. Water is a dipolar compound, with a positively charged pole and a negatively charged pole (remember that the molecule as a whole is neutral). Figure 1.11 presents a schematic sketch of three adjacent water molecules attracted by dipolar bonds – in this case, the bond occurs between hydrogen atoms at the positively charged pole (δ^+) of water molecules and the negatively charged pole (δ^-) produced by valence

---

Focus Box 1.17

Ionic vs. Covalent Compounds

NaCl and SiO_2 serve as good examples of the differences between solids dominated by ionic bonds and those dominated by polar covalent bonds. Ionically bonded halite is highly soluble in water, but quartz in very insoluble, helping to explain the common occurrence of quartz sand beaches – if the covalent bonds in quartz were readily destroyed by water we would have no sandy beaches. In fact, we probably wouldn't have mountains or canyons either, for the strong bonds in rock-forming minerals are primarily polar covalent. In the absence of water, ionic bonds and covalent bonds are generally both strong, and appreciably stronger than the other types of chemical bonds in nature such as metallic bonds and dipole bonds (e.g. hydrogen bonds, van der Waals bonds).

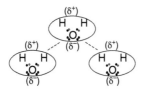

Fig. 1.11 Diagram of three water molecules showing polarity of the water molecule (δ^+, δ^-) as well as hydrogen bonds (dashed lines) between H and O in adjacent molecules. Ovals are used here to highlight the three individual water molecules. The small empty circles between H and O atoms represent electrons from the H atoms; the solid circles represent electrons from oxygen atoms.

electrons from the highly electronegative oxygen atom in the adjacent water molecule.

This specific type of dipolar bond is known as a *hydrogen bond*, and although shown between adjacent H_2O molecules in this case, hydrogen bonds can also occur between H and O (or N) within a molecule. Note that the dots surrounding the O atoms represent electrons – oxygen satisfies the octet rule by sharing an electron with an electron from each of the H atoms, and the H–O bonds within the H_2O molecules (within the ovals) are single bonds (dominantly covalent) involving one pair of electrons each.

Water has a permanent dipole, but *van der Waals* bonds also exist between electrically neutral molecules or compounds where the electrostatic attractions are temporary, commonly existing as transitory states involving alternating positive and negative charge distributions that minimize repulsion. The example in Figure 1.12 represents two different transitory states of adjacent atoms where the dots are a schematic representation of an electron cloud. The delta symbol (δ^-) represents the side of

the atom or molecule with the greater concentration of negative charge (Figure 1.12).

While only schematic, this diagram indicates how alternating electron distributions can lead to attraction between adjacent atoms or nonpolar compounds. Both A and B are transitory states, and in fact, both are unstable states – as soon as the electrons take on the configuration represented in A, they are driven in the opposite direction by repulsive forces, causing configuration B. This then drives the electron cloud back toward configuration A, and the alternating transitory states facilitate atomic or molecular attraction. Although these bonds are far weaker than covalent, ionic, or metallic bonds, they are important to understand because they control melting and boiling points of many compounds, particularly nonpolar organic compounds including pesticides, fuels, and solvents.

1.7 ACID–BASE EQUILIBRIA, PH, K VALUES

Chemical reactions in natural systems commonly occur in the presence of water, and understanding acid–base chemistry is crucial to understanding a large proportion of issues in geochemistry, including the solubility of minerals and trace metals, chemical weathering, the decomposition of organic matter, speciation of chemical elements, and reactions in the atmosphere.

1.7.1 Definitions of acids and bases

Following the convention established by the Swedish chemist Svante Arrhenius, an *acid* is a compound that releases hydrogen ions. The dissociated hydrogen ion is often written as H^+, but be aware that H^+ is effectively a proton, and also be aware that the tendency of H^+ to bond to water molecules is why we often see H_3O^+ (the "hydronium ion") used to represent the dissociated proton in solution (i.e. when dissolved in water). A few examples of classic acids are the inorganic acids hydrochloric acid (HCl), sulfuric acid (H_2SO_4), and nitric acid (HNO_3), and organic acids such as formic acid (HCOOH, also known as methanoic acid, which is what an ants stings with) and acetic acid (CH_3COOH also known as ethanoic acid, or more commonly as vinegar).

The Arrhenius definition of a *base* is a compound that, when dissolved in water, releases hydroxyl anions

State 1   **State 2**

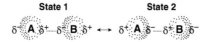

Fig. 1.12 Van der Waals bonds between adjacent atoms. A and B represent nuclei of adjacent atoms and dots are electrons (shown schematically). Note that in the pair of atoms in "State 1," electron clouds are shifted to the left, causing attraction between the opposite poles ($\delta +$ and δ-). At a subsequent moment in time ("State 2"), electron clouds are shifted to the right, also enhancing attraction of the atoms. If this type of alternating motion is synchronized, the atoms will be weakly attracted by the van der Waals bonds indicated by the dashed line.

Focus Box 1.18

Acids in Solution

Nitric acid is a simple example of how acids behave in solution

$$HNO_3 \rightarrow H^+_{(aq)} + NO_3^-_{(aq)} \qquad (1.12)$$

In water, HNO_3 dissociates to release hydrogen ions (and an equal amount of nitrate anions) into solution. The subscript (aq) is used here to indicate that the ions are dissolved in water (i.e. are aqueous). The extent to which an acid like HNO_3 dissolves determines its strength, and Table 1.3 contains data that allow you to predict this. Weak acids tend to remain undissociated in solution, the result being low amounts of H^+ in solution; the opposite is true for strong acids.

Table 1.3 Selected acids and their Ka values, pKa values, and conjugate bases.

Acid				Conjugate base	
Name	Formula	Ka	pKa	Formula	Name
Hydroiodic acid	HI	3.2×10^9	−9.5	I⁻	Iodide
Hydrobromic acid	HBr	1.0×10^9	−9.0	Br⁻	Bromide
Hydrochloric acid	HCl	1.3×10^6	−5.1	Cl⁻	Chloride
Sulfuric acid	H_2SO_4	1.0×10^3	−3.0	HSO_4^-	Hydrogen sulfate anion
Hydrogen sulfate ion	HSO_4^-	1.0×10^{-2}	+2.0	SO_4^{2-}	Sulfate
Sulfurous acid	H_2SO_3	1.3×10^{-2}	+1.9	HSO_3^-	Hydrogen sulfite anion
Nitric acid	HNO_3	2.4×10^1	−1.4	NO_3^-	Nitrate
Phosphoric acid	H_3PO_4	7.1×10^{-3}	+2.2	$H_2PO_4^-$	Dihydrogen phosphate anion
Dihydrogen phosphate ion	$H_2PO_4^-$	6.3×10^{-8}	+7.2	HPO_4^{2-}	Hydrogen phosphate anion
Hydrogen phosphate ion	HPO_4^{-2}	4.2×10^{-13}	+12.4	PO_4^{3-}	Phosphate anion
Nitrous acid	HNO_2	7.2×10^{-4}	+3.1	NO_3^-	Nitrite anion
Hydrofluoric acid	HF	6.8×10^{-4}	+3.2	F⁻	Fluoride
Methanoic (formic) acid	HCOOH	1.8×10^{-4}	+3.7	$HCOO^-$	Methanoate (formate) anion
Benzoic acid	C_6H_5COOH	6.3×10^{-5}	+4.2	$C_6H_5COO^-$	Benzoate anion
Ethanoic acid	CH_3COOH	1.8×10^{-5}	+4.7	CH_3COO-	Ethanoate (acetate) anion
Carbonic acid	H_2CO_3	4.4×10^{-7}	+6.4	HCO_3^-	Bicarbonate
Hydrogen carbonate ion	HCO_3^-	4.7×10^{-11}	+10.3	CO_3^{2-}	Carbonate anion
Hydrogen sulfide	H_2S	1.1×10^{-7}	+7.0	HS^-	Hydrogen sulfide anion
Ammonium ion	NH_4^+	5.8×10^{-10}	+9.2	NH_3	Ammonia

(OH^-) in solution. A few classic examples of bases are sodium hydroxide (NaOH), calcium hydroxide ($Ca[OH]_2$), and ammonium hydroxide (NH_4OH). In solutions, bases behave as follows:

$$NaOH = Na^+_{(aq)} + OH^-_{(aq)} \qquad (1.13)$$

Strong bases – like strong acids – dissociate completely, or nearly so. A strong base like NaOH may almost completely dissolve in solution, producing a high concentration of hydroxyl anions, thus producing a very basic solution. A weak base like NH_4OH (ammonium hydroxide) is much less soluble, so in solution it produces a much lower concentration of OH^-. Similarly, the mineral acids listed above are strong acids, while the organic acids (formic and acetic acid) are relatively weak and produce much lower concentrations of H^+ in solution.

Two additional definitions of acids and bases include Brønsted and Lewis classifications.

A *Brønsted acid* is a substance that can donate a proton (i.e. H^+) to another substance, and a *Brønsted base* is a substance that can accept a proton from another substance. The following chemical reaction illustrates this

---

Focus Box 1.19

Defining pH

The pH scale is the conventional means of expressing the acidity or alkalinity of a solution. The concentration[a] of H^+ ions in solution defines pH according to the equation

$$pH = -\log[H^+] \quad (1.14)$$

where $[H^+]$ is the concentration of hydrogen ions in solution, in mol l^{-1}. In a highly acidic solution with 10^{-2} mol l^{-1} of H^+, the pH = 2. In an alkaline solution with 10^{-11} mol l^{-1} of H^+, the pH = 11. In aqueous solutions, the product of the concentrations of H^+ and OH^- is 10^{-14} ($[H^+] \times [OH^-] = 10^{-14}$). What this translates to is that if the concentration of $H^+ = 10^{-4}$, the concentration of $OH^- = 10^{-10}$. In

this case, pH = 4 and qualitatively, it makes sense that the solution is acidic because there is far more H^+ than OH^-. In an alkaline solution with $[OH^-] = 10^{-2}$, the concentration of $[H^+] = 10^{-12}$ and the pH = 12.

[a]Note: this section discusses aqueous species (e.g. H^+, NO_3^-) in terms of concentration (e.g. $[NO_3^-]$), yet often species in solution are represented by activity $a(NO_3^-)$, a term which takes into account effects of other ions in solution on the effective concentration (often activities are less than actual concentrations). In this introductory section and in many other resources, concentrations are used; Chapter 4 (Section 4.6) discusses the concept of activity relative to concentration.

---

relationship:

$$HCl + NH_3 = Cl^- + NH_4^+ \quad (1.15)$$

where HCl (the Brønsted acid) donates a proton that is accepted by the Brønsted base NH_3 (ammonia). The result is formation of a chloride anion and ammonium, where Cl^- can be termed a Brønsted base (it can accept a proton) and NH_4^+ is considered a Brønsted acid. Most *minerals* can be viewed as Brønsted bases because they consume H^+ when they undergo chemical weathering in soils. A good example is the weathering of gibbsite, $Al(OH)_3$, in acidic soils:

$$Al(OH)_3 + 3 H^+ \rightarrow Al^{3+} + 3 H_2O \quad (1.16)$$

Consider also that NaOH, described as an Arrhenius base above because it yields dissolved OH^- in solution, is also considered a base by the Brønsted definition:

$$NaOH + H^+ \rightarrow Na^+ + H_2O \quad (1.17)$$

Lewis acids are substances that can accept electron pairs when forming bonds (H^+ is a good example), whereas *Lewis bases* are electron pair donors (OH^- is a good example of this type of substance).

1.7.2 The law of mass action and quantifying acid dissociation

Consider now the dissociation of two acids, one a strong acid (nitric acid, HNO_3) and one a relatively weak acid (carbonic acid, H_2CO_3).

$$(1)\, HNO_3 \rightarrow H^+ + NO_3^- \quad (1.18)$$

$$(2)\, H_2CO_3 \rightarrow H^+ + HCO_3^- \quad (1.19)$$

The *Law of Mass Action* quantifies the dissociation of nitric acid as follows:

$$Ka_{HNO3} = [H^+] \times [NO_3^-] \div [HNO_3] = 10^{-1.3} \quad (1.20)$$

The concentrations of H^+ and NO_3^- are equal when HNO_3 dissolves and if we assume a HNO_3 concentration of 1 mol l^{-1}, $[H^+] = [NO_3^-] = \sqrt{10^{-1.3}} = 0.22$ or 2.2×10^{-1} mol l^{-1}.

The same treatment for carbonic acid produces:

$$Ka_{H2CO3} = [H^+] \times [HCO_3^-] \div [H_2CO_3] = 10^{-6.37} \quad (1.21)$$

If, like was done with HNO_3, we assume an H_2CO_3 concentration of 1 mol l^{-1}, $[H^+] = [HCO_3^-] = \sqrt{10^{-6.37}} = 6.5 \times 10^{-4}$ mol l^{-1}. In other words, the concentration of H^+ produced by nitric acid dissociated in water is approximately one thousand (10^3) times greater than H^+ produced by an equivalent amount of carbonic acid in water.

K_a values for some relatively common acids are presented in Table 1.3.

1.8 FUNDAMENTALS OF REDOX CHEMISTRY

Reduction–oxidation (redox) chemistry refers to processes that take place when atoms gain or lose electrons, and often involves reactions where O_2 is a reactant or product. Electron transfer facilitates exchange of energy that is crucial to processes across the chemical spectrum, from aquifer and soil dynamics to photosynthesis and degradation of toxic organic compounds. In nature, redox reactions often involve changes to the oxidation state of elements like carbon, nitrogen, oxygen, sulfur, manganese, and iron that can exist in different oxidation states (e.g. carbon exists in many oxidation states, including C^{4-}, C^0, C^{2+}, C^{4+}), where the change from one oxidation state to another involves gain or loss of electrons. Electron transfer associated with redox reactions is energy for microbial organisms, so when we think about redox reactions, we must consider the potential influence of microbial activity.

1.8.1 Defining oxidation and reduction

Oxidation refers to the loss of electrons by an atom. Two common examples are the oxidation of iron from its ferrous state (Fe^{2+}) to its ferric state (Fe^{3+}) by loss of one electron, and of nitrogen from N^{3+} to N^{5+} by loss of two electrons. These two reactions are represented as follows:

$$Fe^{2+} = Fe^{3+} + e^- \qquad (1.22)$$

$$N^{3+} = N^{5+} + 2e^- \qquad (1.23)$$

Where do those liberated electrons go? They probably were pulled away from the oxidized atom because a neighboring atom had a greater attraction for those valence electrons. The oxidation of one atom cannot occur without a corresponding change to another atom. This change is *reduction* and it takes place when an atom gains electrons.

Two common examples are the reduction of oxygen gas (where the oxidation state of oxygen is O^0) to oxygen anions (O^{2-}), and of C^{4+} (e.g. the C in CO_2 and $CaCO_3$) to molecular carbon, C^0 (e.g. the oxidation state of C in some forms of organic matter).

$$O_2 + 4e^- = 2O^{2-} \qquad (1.24)$$

$$C^{4+} + 4e^- = C^0 \qquad (1.25)$$

1.8.2 Redox reactions

While it is useful to examine individual examples of reduction or oxidation, loss of electrons from one atom results in gain of electrons for another atom. A simple example is the oxidation of iron metal (Fe^0) to iron oxide (Fe_2O_3) where Fe occurs in its trivalent or ferric state (Fe^{3+}).

$$4Fe^0 + 3O_2 = 2Fe_2O_3 \qquad (1.26)$$

Examining individual reduction and oxidation pairs helps to see where and how the exchange of electrons

takes place:

$$Fe^0 = Fe^{3+} + 3e^- \qquad (1.27)$$

$$\tfrac{1}{2}O_2 + 2e^- = O^{2-} \qquad (1.28)$$

All elements in their pure state, like the Fe atom and O atom (represented as $\tfrac{1}{2}O_2$) shown above, have an oxidation state of zero.

The oxidation of iron metal (Fe^0) by oxygen gas (O_2) involves the loss of three electrons from a neutral iron atom and gain of two electrons by a neutral oxygen atom. Clearly, one oxygen atom cannot cause the oxidation of one iron atom from Fe^0 to Fe^{3+}, and this brings up the need to balance redox reactions, as follows:

$$2 \times (Fe \rightarrow Fe^{3+} + 3e^-)$$

$$3 \times (\tfrac{1}{2}O_2 + 2e^- \rightarrow O^{2-})$$

This results in a balanced pair of reactions (with respect to electrons) where two iron atoms lose a sum of six electrons and three oxygen atoms gain a sum of 6 electrons:

$$2Fe \rightarrow 2Fe^{3+} + 6e^-$$

$$1.5O_2 + 6e^- \rightarrow 3O^{2-}$$

... and the paired redox reaction can be expressed as:

$$2Fe + 1.5O_2 \rightarrow 2Fe^{3+} + 3O^{2-} \qquad (1.29)$$

The electrons are excluded from this redox reaction because there are $6e^-$ on both sides (products and reactants) and thus cancel each other out, yet *electron flux is implied by the change in oxidation states of Fe and O atoms*. Terminologically, iron is the reducing agent that donates electrons to oxygen, causing oxygen to become reduced (and iron to be oxidized); oxygen is the oxidizing agent that pulls electrons from iron, which results in reduction of oxygen and oxidation of iron. In reality, the oxidation of iron ends up producing iron oxide, shown here as the mineral hematite:

$$2Fe + 1.5O_2 \rightarrow Fe_2O_3 \qquad (1.30)$$

Hematite consists of Fe in its most oxidized state (Fe^{3+}, ferric iron) and oxygen in the form that it takes in virtually all compounds, O^{2-}.

Redox chemistry comprises some of the most important reactions in the realm of geochemistry and biochemistry. Microbial activity often plays an important role in redox chemistry because electron transfer is an energy source – a classic example is the microbially mediated decomposition

---

Focus Box 1.20

"Oxidizing" or "Oxidized"? "Reducing" or "Reduced"?

This is a question of terminology. In effect, the terms "oxidizing" and "oxidized" are synonymous. An oxidized soil will likely contain abundant available O_2 as well as minerals that are stable in an oxidized (or oxidizing) environment, e.g. iron oxides. This soil will be an oxidizing environment because if a chemically reduced substance like an organic compound or sulfide were to be transported into the soil, an oxidation reaction would likely lead to decomposition of

the reduced substance. Similarly, "reduced" and "reducing" environments are effectively synonymous. The terms can also be viewed in this way: "oxidizing" and "reducing" are often used to describe the environment whereas "oxidized" and "reduced" might be more likely to be used for species or elements, as in "Is the iron in an oxidized or reduced state in that part of the aquifer?"

---

of fuels and solvents in soils, where the oxidation of organic carbon provides energy to the microbe and results in the transformation of leaked fuel into H_2O and CO_2. Refer to Chapter 3 for organic compounds in the environment, and to Chapter 4 for more on redox reactions and aqueous geochemistry (Sections 4.5 and 4.9).

1.9 CHEMICAL REACTIONS

Reactions among elements and compounds have been presented in a few ways in this chapter thus far, including reference to chemical bonding, the formation of elements, and redox chemistry. Chemical equations are algebraic expressions that represent the masses and charges of constituents involved in chemical reactions, and in some ways are the language of geochemistry, or at least one of the languages. Accordingly, the following section will present a few fundamental concepts about chemical reactions, what they represent, how to balance them, and how to interpret them.

First, a few general rules. Chemical reactions commonly take place in the presence of water, but if water is not produced or is not consumed by the reaction, it is not listed in

---

Focus Box 1.21

Determining and Balancing Chemical Equations

How to determine a balanced chemical reaction for the combustion of octane: First, knowing that combustion is the reaction of a substance with oxygen, you can identify O_2 as a reactant. Looking up the composition of octane (C_8H_{18}, see Chapter 3) gives you the other reactant. Products of hydrocarbon combustion are water and carbon dioxide[a], so knowing reactant and products, the unbalanced equation is:

$$C_8H_{18} + O_2 = H_2O + CO_2$$

Rule #1: Perhaps this goes without saying, but in a balanced chemical equation there must be equal moles (or atoms) of each element on the reactant and product side; in this example, on the reactant side there are 8 mol of C and 18 mol of H (comprising 1 mol of octane), and 2 mol of O (in 1 mol of oxygen gas, O_2).

Rule #2: Always save the single element (in this case, O_2) or the least complex compound for last – it is easiest to adjust at the end.

Step #1: Add coefficients to adjust upward elements that are lacking. In this case, coefficients need to be placed in

front of H_2O and CO_2 to raise molar values of H and C on the reactants side – to balance H, we need 9 mol of H_2O (i.e. 18 mol of H), and to balance C, we need 8 mol of CO_2 (equal to 8 mol of C).

Step #2: This gives us an intermediate-stage reaction, i.e.

$$C_8H_{18} + O_2 = 9\,H_2O + 8\,CO_2$$

A quick glance makes it obvious that oxygen is not balanced. The products side now has 9 mol of O from H_2O and 16 mol of O from CO_2.

Step #3: To balance O_2, we need 25 mol of O on the reactants side, which equals 25/2 or 12.5 mol of O_2.

The balanced reaction is: $C_8H_{18} + 12.5\,O_2 = 9\,H_2O + 8\,CO_2$

Equal signs (=) are often used rather than arrows in chemical equations in cases where the reaction is reversible. This is a simple example, but there are others in the end-of-chapter problem set.

([a] Note that in many cases, fuel-rich combustion can also produce carbon monoxide [CO].)

the reaction. Given that the First Law of Thermodynamics states that matter can neither be created nor destroyed, but rather can only change forms, chemical reactions should not give the illusion that the first law is being violated – what this means is that all chemical reactions must be balanced. If there are four oxygen *atoms* (or moles of oxygen) expressed on the reactants side of the equation, then there also must be four oxygen *atoms* (or moles of oxygen) on the products side. Similarly, if the reaction is of the redox variety, the charges (sum of + and −) should be equal on both sides. If the reaction involves nuclear fusion or fission, then energy and mass should be equal on both sides of the equation.

1.10 EQUILIBRIUM, THERMODYNAMICS, AND DRIVING FORCES FOR REACTIONS: SYSTEMS, GIBBS ENERGIES, ENTHALPY AND HEAT CAPACITY, ENTROPY, VOLUME

1.10.1 Pyrite oxidation as an introductory example

This treatment of thermodynamics begins with the example of pyrite (FeS_2 or iron disulfide), a mineral that forms in O_2-poor systems (reducing environments) that include deep crustal levels and anoxic surface environments like swampy muds (technically iron sulfide that forms at earth surface temperatures is a poorly ordered form of $\sim FeS$, e.g. mackinawite). Iron, in the Fe^{2+} state, and sulfur, in a combination of S^{2-} and S^0 states (average $= S^-$), both occur in chemically reduced forms in the mineral pyrite – these oxidation states are stable in reducing/anoxic/anaerobic/O_2-poor environments.

Given the conditions under which pyrite forms, it is possible to predict its fate in oxidizing (O_2-rich) environments, common examples being soils exposed to the oxygen-rich, water vapor-bearing atmosphere, or an O_2-rich bubbling stream, where the stable forms of iron and sulfur are Fe^{3+} and S^{6+}. Under oxidizing conditions, pyrite undergoes (bio)chemical oxidation, producing iron hydroxide with its characteristic rusty orange stains typical of many rock outcrops and soils. The reaction of pyrite, water, and oxygen to produce iron hydroxide, sulfuric acid, and free electrons (oxidation!) can be expressed as follows:

$$2FeS_2 + 7H_2O + {}^{15}/_2O_2 \rightarrow 2Fe(OH)_3 + 4H_2SO_4 \tag{1.31}$$

(In reality this reaction occurs in two or more steps, often in the presence of sulfur-oxidizing bacteria). Viewing this reaction in terms of the extent of iron oxidation would

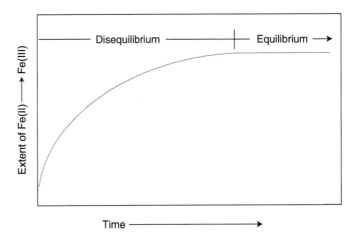

Fig. 1.13 Typical progression of a geochemical system toward equilibrium. A high degree of disequilibrium in the early stages causes rapid rates of change, but as the system approaches equilibrium, rates decrease logarithmically.

likely result in a characteristic pattern observed for many reactions in nature (Figure 1.13).

In Figure 1.13, the initial system (pyrite in contact with atmosphere) is out of equilibrium – both the Fe and S are unstable in reduced forms when exposed to O_2-rich air (e.g. by a landslide that exposes fresh unoxidized rock) and the pyrite begins to react. This reaction is expressed as Fe(II) → Fe(III) in the graph, and note that initially the reaction occurs rapidly but as the reaction progresses the rate steadily decreases and ultimately ceases altogether. There are two probable explanations:

1 the system has reached *equilibrium*, a condition where net concentrations of products and reactants do not change. *Dynamic equilibrium* sometimes occurs in natural systems, when the rate of formation of products equals the rate of formation of reactants. If the $Fe(OH)_3$ and H_2SO_4 produced by pyrite oxidation are not leached out of the system, a dynamic equilibrium may be established where the rate of the forward reaction shown above is equal to the rate of the reverse reaction – i.e. FeS_2, O_2, and H_2O are produced at the same rate as are $Fe(OH)_3$ and H_2SO_4. At dynamic equilibrium, reactions are taking place, namely "products" are dissolving to produce "reactants" at the same rate that "reactants" produce "products," but the net concentrations of products and reactants do not change with time. However, any change to a variable involved in a dynamic equilibrium (e.g. concentration of reactant or product, volume, pressure, temperature) will cause a shift to counter the change. For example, loss of a reactant (e.g. burial under a subsequent landslide and water-logging limits exposure to O_2) will shift the reaction toward the direction of reactants – the rate will slow, or reactants will

be produced at the expense of products. Or, if products are lost from the system (e.g. soluble H_2SO_4 is leached away), the reaction will continue to form products, which leads to the other possibility:

2 The reaction has run to completion. In some natural systems, where products are lost due to leaching (e.g. of H_2SO_4 in the case above) or degasification (e.g. of CO_2 with dissolution of carbonates), a dynamic equilibrium cannot be established. If the pyrite reaction above stopped because all Fe(II) had been consumed to produce Fe(III) (i.e. if the reaction were to run to completion because products are being lost from the system), that system will not reach dynamic equilibrium. The natural environment differs from the laboratory in that reactants are often lost from soils, rocks, and ground waters, and dynamic equilibrium may not apply. (In this case, we might use the term *steady state* to describe the static condition.) In other cases, reactions in nature may not reach equilibrium because reaction rates are very slow – the reaction never proceeds past the early convex part of the disequilibrium curve. This often occurs in soils, where igneous minerals such as amphiboles and pyroxenes that are stable in high-temperature, low-O_2 environments persist in a state of disequilibrium because the rates at which they decompose in weathering environments are relatively slow.

1.10.2 Systems, species, phases, and components

In spite of certain limitations, examining environmental systems through the lens of *equilibrium thermodynamics* can be very useful. It can help determine the direction in which chemical changes will take place (e.g. pyrite is unstable and will oxidize in contact with the atmosphere) and also to infer rates because the farther a system is from equilibrium, the faster it will react to reach equilibrium. A common term used in thermodynamics is *system*, which is a somewhat arbitrary definition of the components we

wish to consider. Depending on the question, a system might be an aquifer, or a pore within an aquifer, or the entire hydrologic cycle; it could be an entire granitic pluton, or it might be a micron-sized fluid inclusion in a quartz crystal. It really depends on the scale of study. If the question is climate change, the entire troposphere might be considered as the system, or a smaller system comprised only of a landfill might be defined as the system if the main concern is a single source of carbon (e.g. CH_4).

In geochemistry, *species* are microscopic entities, commonly ions or gases such as Ca, CO_3^{2-}, SO_4^{2-}, CO_2, or H_2S, whereas *phases* are physically separable parts of a system, typically minerals, liquids (e.g. H_2O), and distinct gases (dissolved CO_2 and O_2 in stream water, for example). Phases are comprised of species – for example, the *phase* calcite is comprised of the *species* Ca^{2+} and CO_3^{2-}, or of the *species* CaO and CO_2 (calcite can be defined in either way). While species are generally substances than can or do exist in nature, *components* do not necessarily exist in nature. Sometimes they are similar to species, but in other cases they may be mathematical expressions useful in thermodynamic calculations, one example being KNa_{-1}, a mathematical operator used to indicate gain of K and loss of Na (e.g. by substitution in a phase such as smectite – refer to Chapter 2, Section 2.4.1). So, a soil with the mineral dolomite ($CaMg[CO_3]_2$), water, dissolved Ca^{2+}, dissolved Mg^{2+} and CO_3^{2-}, CO_2 gas, and quartz could be defined as having four phases (dolomite, water, quartz, CO_2 gas) and 6 species or components (Ca^{2+}, Mg^{2+}, CO_3^{2-}, CO_2, H_2O, and SiO_2).

Lastly, the *phase rule* (sometimes referred to as *Gibbs'* phase rule) relates components (C), phases (P), and degrees of freedom (F) according to this simple equation:

$$F = C - P + 2 \qquad (1.32)$$

Degrees of freedom represent tangible changes to a system, typically temperature and pressure. In a system with 2 degrees of freedom, temperature and pressure can both

---

Focus Box 1.22

What is a "System" in Geochemistry?

Systems can be *open* (e.g. a stream, the atmosphere, a leaking landfill), where material is added or lost, or *closed*, where flow of material is restricted (e.g. tiny pores within impermeable fine-grained sediments, and where chemical reactions can be modeled with no gain or loss of elements).

In some cases, systems are closed with respect to solids but open with respect to gases or heat, and in other cases systems can be defined as closed to physical and thermal flux, in which case they are *isolated*. Systems are comprised of components, phases, and species.

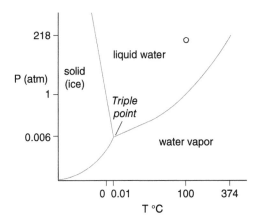

Fig. 1.14 Phase diagram for H_2O depicting one component (H_2O) and three phases (solid, liquid, and gaseous water). The circle in the upper right is plotted at $100\,°C$ and $200\,atm$ pressure, an example of a point where there exist two degrees of freedom.

change without producing a change in the state of the system. The classic example involves a simple system involving only one component, H_2O, which can exist in three phases (solid, liquid, and vapor forms of H_2O), as displayed in the H_2O phase diagram (Figure 1.14).

There is one point in the H_2O phase diagram where all three phases coexist – the triple point. At the triple point, there is one component (H_2O; $C = 1$) and three phases (solid, liquid and vapor; $P = 3$). The phase rule indicates that for this system,

$$F = 1 - 3 + 2 = 0$$

In other words, there are no degrees of freedom. Any change in either T or P will produce a change in the state of the system; for example, increasing T at constant P will cause ice to melt and liquid to vaporize. Increasing P at constant T will cause vapor to condense and ice to melt, producing a system with only one phase. Consider for a moment a system with only one phase, liquid water at $100\,°C$ and $200\,atm$ of pressure (indicated by the circle in Figure 1.14). In this case, the phase rule indicates that (for a system where $C = 1$ and $P = 1$) there will be two degrees of freedom, and this is borne out in the diagram – if either T or P changes, or even if both T and P simultaneously change, there will be no change to the state of the system. It will remain liquid water until the system either cools or decreases in pressure to the point where P-T reaches the phase boundary (line) between liquid and solid or between liquid and vapor.

1.10.3 First law of thermodynamics

The term thermodynamics implies an approach based on changes in *heat*, and the field did originate in studies of the transformation of heat energy to mechanical energy during the 1800s, much of it related to the industrial revolution. Energy can change forms, e.g. *potential energy* to heat (or vice versa), as in the combustion of organic matter, a process where *stored chemical potential energy* is converted to heat energy. Thermodynamics deals with transfers of energy, and one of the fundamental principles is the

---

Focus Box 1.23

Example of Heat, Work, and Systems

Consider an experiment you could perform while sitting in your overstuffed chair while reading this book. This experiment will illustrate the example of heat (Q) produced by the work (W) of rubbing hands together. The system can be defined as your body and its immediate surroundings. The energy you need to move your shoulders and arms to produce friction comes from the stored energy in the chemical bonds in the food you recently ate, and in a simplified approximation, this chemical energy is transformed to mechanical energy. The friction between your hands converts mechanical energy to heat energy, causing your hands to warm up (and raising the temperature of the system). This heat will soon be lost to the surroundings (an issue to be addressed with the Second Law), but if we consider the net energy of the system (body and immediate surroundings), it has not increased or decreased, but rather merely has changed forms. Of course, the heat will eventually escape from your surroundings and ultimately you will need to eat more food to provide the chemical energy needed to reinitiate the cycle (conversion of chemical energy in food to heat energy is an irreversible reaction in our bodies). In terms of the equation of the First Law, the work (W) done will have a positive sign, as will the heat term Q (heat is gained by the system, so the term Q will have a positive sign, because in geochemistry, the flow of heat is considered positive for any system that gains heat). So, in a semi-quantitative sense, W and Q will both be positive and the values will be equal, and $\Delta E = Q - W = 0$, reflecting the fact that there is no change in the net energy of our body and its surroundings (at least in the short term).

First Law of Thermodynamics, often known as the *Law of Conservation of Matter and Energy*, which can be stated as follows:

$$\Delta E = Q - W \qquad (1.33)$$

Or, for small increments of change, this equation can be stated as:

$$dE = dq - dW \qquad (1.34)$$

E is the *internal energy* of a system, Q or dQ represent *heat* flux, and W represents *work* done on the system. The first law essentially states that energy is neither created nor destroyed – matter and energy may change forms, new phases may be produced, gases may convert to solids, or solids may dissolve, but the net energy of a closed system never changes (or, for an open system, the net energy of the open system and its surroundings never changes). The kinetic energy of friction along a geological fault may produce heat, or the chemical potential energy stored in the covalent bonds in hydrocarbons may release heat during combustion, but the net change in energy is zero.

Work can be defined as follows:

$$dw = P \times dV \qquad (1.35)$$

where P is external *pressure* on a system and dV is the change in *volume* of the system. A typical way to consider work relative to volume is to quantify the work done by expansion during the change of state from liquid water to water vapor. At constant P, a positive dV term (expansion increases V) will produce a positive dw, indicating that work has been done on the surroundings by the system. If you substitute the PdV term for W into Eq. (1.34) of the first law, you will arrive at this equation for the first law, expressed in terms of change in *internal energy*:

$$dE = dq - PdV \qquad (1.36)$$

1.10.4 Second law of thermodynamics

The *Second Law of Thermodynamics* deals with *entropy*, a measure of the degree of *disorder* within a system. Any system tends toward a state of increasing randomness unless energy is added to the system to increase order. Increasing entropy is a *spontaneous* process and energy must be added to produce order. In geochemistry, an example of entropy as a spontaneous process is chemical weathering of a basalt, for example, where minerals with ordered crystal lattice structures (a low entropy state) are decomposed into soluble aqueous species such as Na^+, Ca^{2+}, and $Si(OH)_4$ that are then scattered across the globe. Only with addition of energy (e.g. the internal heat of the earth) can some order be restored (i.e. entropy decreased). An everyday example of entropy is this: the natural state of a kitchen or living room will progress toward a state of disorder (greater entropy) characterized by dirty dishes, potato chip bags, music scores, and old newspapers scattered about unless we expend energy to restore order.

Entropy (S) can be represented by the following equation, where q is heat and T is temperature and the process is reversible:

$$dS = dq/dT \qquad (1.37)$$

Entropy can be quantified in terms of change in heat content per change in temperature, and units of S are joules/degree (non-Si units are cal/deg.; 1 cal = 4.18 J). A useful example is to consider the entropy of liquid water compared to water vapor at 25 °C and a pressure of 1 atm (1.01325×10^5 Pascals, or 1.01325 bar). S (entropy) for liquid water (H_2O_l) is 292 cal/mol/deg, and for H_2O_g it is 789 cal/mol/deg (Dean 1979). Intuitively, the value of S is greater for the gaseous, more disordered state of H_2O than for the liquid state.

Focus Box 1.24

Notes on Standard Temperature and Pressure

For thermodynamic data in low-temperature geochemistry (Appendix IV), standard conditions are 25 °C (298.15 K, 77 °F) and 1 atm pressure (1.01325×10^5 Pascals [Pa] or 1.01325 bar) – IUPAC refers to 25 °C and 1 atm as standard ambient temperature and pressure (SATP), consistent with standard values used by the US EPA. The US National Institute of Standards and Technology (NIST) uses 20 °C and 1 atm (often referred to as normal temperature and pressure, NTP) and IUPAC defines standard temperature and pressure (STP) as 273.15 K (0 °C) and 10^5 Pa (100 kPa, 1 bar, 0.99 atm).

1.10.5 Enthalpy

Another important consideration in thermodynamics is *enthalpy (H)*, the *heat content* of a system. In some cases enthalpy is expressed in units of calories (cal) or kilocalories (kcal), which makes sense because calories are a common measure of heat in everyday life; however, the SI units are joules (J) (or kilojoules, kJ).

Enthalpy is typically expressed as follows:

$$H = E + PV \qquad (1.38)$$

or

$$dH = dE + PdV \qquad (1.39)$$

But often the most important consideration related to enthalpy is the change in H ($\Delta H = H_2 - H_1$) during a reversible reaction, where the two values of H represent enthalpies associated with different states of matter, like liquid water and water vapor, or with elements in different bonding arrangements, e.g. Fe and O_2 vs. Fe_2O_3. ΔH is an important parameter in geochemistry because it expresses heat absorbed or released during changes of state (e.g. evaporation) or during chemical reactions that produce minerals, ions, or molecules. If heat is gained during a reaction (i.e. ΔH is positive), the reaction is endothermic; conversely, if heat is lost (i.e. ΔH is negative), the reaction is exothermic.

Given that reactions either consume or produce heat, we can calculate the difference in enthalpy between the reactants and products and determine the amount of heat produced (released) or consumed (absorbed) by the reaction. This is important because it is one of the ways that thermodynamics can help to predict the behavior of environmental systems such as soils and waters; typically, spontaneous processes produce (i.e. release) heat, i.e. spontaneous processes usually are exothermic (caveat:

while this is generally true it is not always the case – for example, when some salts dissolve in water the solution gets colder. Dissolution of some salts absorbs heat from the water yet is a spontaneous process because the increase in entropy is more important than the positive ΔH).

Enthalpies of reactions (ΔH^o_R) are determined by summing the *enthalpies of formation* (ΔH^o_f) of all reactants and subtracting this term from the sum of ΔH^o_f values of all products (standard state conditions):

$$\Delta H^o_R = \Sigma\, n_x \times H^o f_x\ (\text{products}) - \Sigma\, n_x \times H^o f_x\ (\text{reactants}) \qquad (1.40)$$

Enthalpies of formation are available from various sources and selected examples are presented in Appendix IV. By convention, $H^o f = 0$ for elements in their pure state (e.g. Fe, Si, Na) and for gases such as H_2, N_2, and O_2. In the ΔH^o_R reaction above, the $H^o f$ for each reactant or product (represented by the variable x) is multiplied by the number of moles (n) expressed in the reaction.

We can examine the reaction of Fe and O_2 to form hematite (Fe_2O_3) by the chemical reaction $2\,Fe + 1.5\,O_2 = Fe_2O_3$. Values of $H^o f$ (in kJ mol^{-1}):

$$H^o f\ (\text{Fe}) = 0$$

$$H^o f\ (O_2) = 0$$

$$H^o f\ (Fe_2O_3) = -824.2$$

It is crucial to remember to multiply $H^o f$ values by molar abundances presented in the chemical reaction.

$$\Delta H^o_R = (1 \times -824.3) - (2 \times 0 + 1.5 \times 0)$$

$$= -824.3\ \text{kJ}\ mol^{-1}.$$

The negative ΔH^o_R for this reaction implies that it is spontaneous at standard ambient temperature and pressure, i.e. that iron will oxidize to form iron oxide.

Focus Box 1.25

Examples of Exothermic, Endothermic, Q, and H

Combustion of organic matter is clearly exothermic – we burn firewood and hydrocarbons to produce heat. Boiling of water, the transformation of andalusite to sillimanite during prograde metamorphism, and the maturation of petroleum in sedimentary basins are all endothermic processes – they absorb heat. The stored heat can then later be released; for example, stored heat in petroleum is released during the exothermic reaction known as combustion. It is worth pointing out the difference between two terms that represent heat, Q and H. Q represents flow of heat or heat transfer, for example from hot Hawaiian lava into cool ocean water; H represents heat stored within a system, such as the stored heat in petroleum, water vapor, or sillimanite.

Focus Box 1.26

Example of Enthalpy of Reaction

Under certain metamorphic conditions, calcite ($CaCO_3$) and quartz (SiO_2) react to form wollastonite ($CaSiO_3$) plus CO_2 according to this reaction:

$$CaCO_3 + SiO_2 = CaSiO_3 + CO_2 \qquad (1.41)$$

Is this a spontaneous process at the earth's surface (i.e., 25 °C and 1 atm)? Values of $H^{\circ}f$ (in $kJ\,mol^{-1}$) are (Appendix IV):

$$H^{\circ}f\,(CaCO_3) = -1207.4$$

$$H^{\circ}f\,(SiO_2) = -910.7$$

$$H^{\circ}f\,(CaSiO_3) = -1630$$

$$H^{\circ}f\,(CO_2) = -393.5$$

The enthalpy of the reaction can then be determined as follows:

$$\Delta H^{\circ}_R = [(1\ mol \times -1630\ kJ\,mol^{-1})$$

$$+ (1\ mol \times -393.5\ kJ\,mol^{-1})]$$

$$- [(1\ mol \times -1207.4\ kJ\,mol^{-1})$$

$$+ (1\ mol \times -910.7\ kJ\,mol^{-1})] = +94.6\ kJ$$

This reaction requires addition of 94.6 kJ of heat (per mol of each reactant given the stoichiometry of the reaction), suggesting that it is not spontaneous (without considering entropy we cannot be sure – more on this in Section 1.10.6). Evidence from geology is consistent with this conclusion because calcite and quartz do not react to form wollastonite and carbon dioxide until systems reach medium-grade metamorphic temperatures and pressures (~500 °C, >1 kb).

1.10.6 Heat capacity

Heat capacity is the amount of heat required to raise the temperature of a given amount (1 mol or 1 g) of a substance by 1 °C. It is defined as the ratio of heat added relative to the extent of temperature increase, where the greater amount of heat required to cause an increase in temperature corresponds to higher heat capacity.

$$C = dq/dT \qquad (1.42)$$

In this equation, C is heat capacity, and dq and dT are changes in heat and temperature. Rearranging emphasizes the point that, for a given amount of heat added (dq), the magnitude of temperature increase will be lower if the heat capacity (C) is high.

$$dT = dq/C \qquad (1.43)$$

At constant volume, heat capacity can be expressed as:

$$C_v = (\delta q/\delta T)_v \qquad (1.44)$$

Given Eq. (1.36), and realizing that PdV = 0 at constant volume, $\delta E = \delta q$ (at constant P), so δE can be substituted for δq:

$$C_v = (\delta E/\delta T)_v \qquad (1.45)$$

Heat capacity at constant pressure (C_p) is defined as:

$$C_p = (\delta q/\delta T)_p \qquad (1.46)$$

and Eq. (1.36) can be rearranged slightly to give:

$$dq = dE + PdV \qquad (1.47)$$

Now, substituting the right side of Eq. (1.47) into the numerator of the right side of Eq. (1.46) gives:

$$C_p = (\delta E/\delta T)_p + P(\delta V/\delta T)_p \qquad (1.48)$$

And given Eq. (1.39) (effectively, $\delta H = \delta E + P\delta V$),

$$C_p = (\delta H/\delta T)_p \qquad (1.49)$$

Focus Box 1.27

Denoting "Change in" Using the Symbols d vs. Δ and δ in Equations

The symbols Δ, δ, and d are commonly used to signify change in quantity, e.g. dx/dy. From Eqs. (1.43) to (1.44), the notation shifted from d to δ. As used here, the symbol δ indicates that the derivation is performed with a restriction, in this case constant volume. The subscript v indicates that volume of the system is constant.

Examples of Heat Capacities of Substances

Comparing the heat capacity (at 25 °C) of liquid water (a relatively high value, C = 4.19 J/g·K) to dry rock (relatively low values in the range of 0.7 to 0.9 J/g·K) provides some insight into how heat capacity affects environment. The paucity of water in deserts means that rock is the dominant control on temperature change and this results in drastic temperature swings, both diurnally and seasonally.

In moister regions such as tropical forests or temperate coastal regions, temperature extremes are minimized by the high heat capacity of water present in places such as lakes (or seas), air (as clouds or water vapor), vegetation, and soils. The high heat capacity of water allows it to function as a temperature buffer.

Heat capacity is an *extensive property*, i.e. it is dependent on the amount of the substance in question – the greater the amount of substance, the greater the amount of heat needed to be added to achieve the same change in temperature. That said, values of heat capacity are typically normalized to a mass of 1 g, meaning that effectively heat capacity is an *intensive property*, i.e. when normalized (divided by mass), it reflects a characteristic of a given substance independent of amount (i.e. mass).

1.10.7 Gibbs free energy and predicting stability

The *Gibbs free energy* (G) of a system accounts for changes in both enthalpy and entropy during reactions and is a useful tool in predicting stability – it considers change in enthalpy as well as change in entropy.

$$\Delta G^o_R = \Delta H^o_R - T\Delta S^o_R \qquad (1.50)$$

Any reaction that produces a decrease in the Gibbs free energy is spontaneous – that is to say, any reaction for which ΔG^o_R is negative is spontaneous. Systems tend toward lower energy states in the absence of new addition of energy, so a decrease in G will produce a more *stable* system. In a schematic way (Figure 1.15), stability of a physical system can be used to illustrate this point. The ball at point A is unstable with respect to location, and to decrease this instability (high energy state), it will roll down to point B, and if it can overcome the slight energy barrier at point C, it will eventually achieve its most stable configuration by rolling down to point D.

Point B might be referred to as a *metastable* condition, one that is not the most stable configuration (that would be D), but one that may play a strong role in system behavior.

Determining Gibbs free energies for geochemical systems allows prediction of their behavior much like the

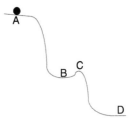

Fig. 1.15 Schematic diagram illustrating relative stability conditions, from unstable at points A and C, to metastable at point B and stable at point D. Note that the metastable point B can be reached from points A or C, and that to shift from metastable (B) to stable (D), some energy must be introduced. If not, the metastable condition may persist for a long time.

simple analysis of the ball between points A and D. For example, if magnetite (Fe_3O_4), water, and oxygen gas exist in a soil at standard conditions, we might ask, is this assemblage stable or of out of equilibrium? Is magnetite stable in this soil, or would we predict that it will eventually transform into an iron hydroxide such as goethite? The possibility that the system will react to form goethite (FeOOH) to achieve a lower energy state could be assessed according to this reaction:

$$2\,Fe_3O_4 + 0.5\,O_2 + 3\,H_2O = 6\,FeOOH \qquad (1.51)$$

ΔG^o_R for the reaction is computed according to this equation:

$$\Delta G^o_R = \Sigma\,n_x \times G^o f_x\,(\text{products}) - \Sigma\,n_x \times G^o f_x\,(\text{reactants}) \qquad (1.52)$$

Values for $G^o f$ (in kJ mol^{-1}) are (Appendix IV):

$$G^o f\,(Fe_3O_4) = -1015.5$$

$$G^o f\,(O_2) = 0$$

$$G^o f\,(H_2O_l) = -237.2$$

$$G^o f\,(FeOOH) = -488.6$$

Focus Box 1.29

Examples of "Metastability" in the Geochemical Realm

The mineral halloysite is a disordered form of kaolinite that forms in tropical soils when unstable igneous minerals such as olivine and pyroxene rapidly dissolve (over hundreds to thousands of years). Rapidly dissolution may lead to formation of metastable halloysite rather than kaolinite due to kinetic factors; once formed, halloysite may require hundreds of thousands of years to finally transform into thermodynamically stable kaolinite. Another

example is aragonite ($CaCO_3$), which forms when marine organisms obtain dissolved calcium and bicarbonate from seawater to form shells. Aragonite is unstable relative to calcite but may persist for thousands of years or longer before increased temperature (e.g. burial in a sedimentary basin) or changing solution chemistry (e.g. in the pores of sediments) trigger the aragonite → calcite transformation.

These values produce the following equation:

$$\Delta G^o_R = [6 \, mol \times -488.6 \, kJ \, mol^{-1}]$$
$$- [(2 \, mol \times -1015.5 \, kJ \, mol^{-1})$$
$$+ (0.5 \, mol \times 0 \, kJ \, mol^{-1})$$
$$+ (3 \, mol \times -237.2 \, kJ \, mol^{-1})] = -189.0 \, kJ$$

What this negative Gibbs free energy value demonstrates is that magnetite is unstable in the presence of water and oxygen at the earth's surface and will react to produce goethite, a common soil mineral. What this reaction does not tell us is how fast this reaction will take place. Is there an energy barrier? Does the system need to overcome an activation energy like that at point C in the ball example of Figure 1.15? Does diffusion of O_2 or H_2O to magnetite surfaces limit reaction rate? In reality, magnetite can persist in soils in a metastable state for at least thousands of years, and that fact is not evident from this ΔG^o_R calculation. Section 1.11 on kinetics addresses some questions related to rates of processes.

1.10.8 The Van't Hoff equation: relating gibbs free energy to the equilibrium constant (K_{eq})

The Gibbs free energy of a reaction can also be expressed in terms of the equilibrium constant (K_{eq}), which is much like the K_a discussed earlier in this chapter when considering acids. Consider a general reaction at equilibrium where:

$$aA + bB = cC + dD \qquad (1.53)$$

A and B, and C and D are reactants and products, respectively, and the lowercase a, b, c, and d represent molar fractions of reactant or product. In this general case:

$$K_{eq} = [C]^c \times [D]^d / [A]^a \times [B]^b \qquad (1.54)$$

$[C]^c$ represents the concentration of reactant C raised to the c power, and so on. Products are in the numerator and reactants in the denominator. Or, for a more tangible example, we can write the K_{eq} for oxidation of ferrous iron (Fe^{2+}). The chemical reaction is:

$$4Fe^{2+} + O_2 + 10H_2O = 4Fe(OH)_3 + 8H^+ \qquad (1.55)$$

and:

$$K_{eq} = [Fe(OH)_3]^4 \times [H^+]^8 / ([Fe^{2+}]^4 \times [O_2] \times [H_2O]^{10}) \qquad (1.56)$$

However, by convention, the concentration of water is given the value of 1, and pure solid phases like ferrihydrite ($Fe(OH)_3$) are similarly assigned values of 1, so in this case the K_{eq} simplifies to:

$$K_{eq} = [H^+]^8 / ([Fe^{2+}]^4 \times [O_2]) \qquad (1.57)$$

which means that the main controls on the reaction are the concentrations of H^+ and Fe^{2+} and availability and reactivity (fugacity) of O_2. A high value for the K_{eq} indicates that the reaction will proceed in the direction of products. In other words, for reactions that favor products, the numerator will be greater than the denominator, producing a high K_{eq}. In cases where reactants are more favored (i.e. reactants are predicted to occur in higher concentrations than products), K_{eq} values are small.

Focus Box 1.30

Using the K_{eq} to Predict Reactivity

The dissolution of gibbsite in the presence of acidic soil water is represented by the following reaction,

$$Al(OH)_3 + 3H^+ = Al^{3+} + 3H_2O \qquad (1.58)$$

and the $K_{eq} = 10^{8.11}$ at 25 °C and 1 bar, which is often expressed in log form as log $K_{eq} = 8.11$ (Langmuir 1997). Expressed in K_{eq} form:

$$K_{eq} = [Al^{3+}]/[H^+]^3 = 10^{8.11} \qquad (1.59)$$

The high value of the K_{eq} indicates that the system strongly favors dissolution of $Al(OH)_3$ and formation of dissolved Al^{3+}. (Of course, pH of the solution will strongly control the probability of this reaction.) Conversely, the dissolution of fluorite ($CaF_2 = Ca^{2+} + 2F^-$) has a log $K_{eq} = -10.6$ (i.e. a K_{eq} of $10^{-10.6}$), and this very low value indicates that fluorite is insoluble in water. The fact that the product of the concentrations of $[Ca^{2+}]$ and $[F^-]^2$ is extremely low ($K_{eq} = 10^{-10.6}$) indicates that the reactant side of the equation is favored, indicating that solid CaF_2 is the favored form of these components.

$\Delta G^o{}_R$ for a reaction is related to the K_{eq} and can be expressed as:

$$\Delta G^o{}_R = -RTlnK_{eq} \qquad (1.60)$$

This is the van 't Hoff equation, where R (the gas constant) is equal to $8.314\,J\,mol^{-1}\,K^{-1}$ ($8.314 \times 10^{-3}\,kJ\,mol^{-1}\,K^{-1}$) and T is temperature (in K), which produces units of $J\,mol^{-1}$ (or $kJ\,mol^{-1}$) for $\Delta G^o{}_R$.

In most cases, Gibbs free energy data are determined for systems at 25 °C (298.15 K), so for systems at 25 °C (reasonably representative of surface environments), $\Delta G^o{}_R$ can be further simplified and converted to log_{10} as:

$$\Delta G^o{}_R = -5.708 \times log(K_{eq}) \qquad (1.61)$$

Typically enthalpy and entropy changes are not strongly temperature dependent, so given Eq. (1.50), the Gibbs free energy, ΔG, should vary in a linear manner with temperature. Rearranging Eq. (1.50) slightly and presenting the variables in a generic sense gives us:

$$\Delta G/T = \Delta H/T - \Delta S \qquad (1.62)$$

By then substituting the right side of Eq. (1.60) for ΔG:

$$RTlnK/T = -\Delta H/T + \Delta S \qquad (1.63)$$

Then dividing and rearranging:

$$lnK = -\Delta H/RT + \Delta S/R \qquad (1.64)$$

Expressed in this way, Eq. (1.64) is an equation of a line: $\ln K = y$, $-\Delta H/R$ is m (slope), $1/T$ is x and $\Delta S/R$ is b

(constant). Two characteristic types of plots (Figure 1.16) result when different values for T are plugged in to Eq. (1.64).

Taking the derivative of Eq. (1.64) with respect to temperature gives a different formulation of the van 't Hoff equation:

$$d(lnK)/dT = \Delta H/RT^2 \qquad (1.65)$$

Note that the term $\Delta S/R$ disappears because both terms are constants (assuming S does not change with temperature).

Integrating Eq. (1.65) between temperatures T_1 and T_2 produces Eq. (1.66), another expression of the van 't Hoff equation, one which makes it possible, given a K value (e.g. K_1) at standard temperature (e.g. T_1), to determine an unknown value of K (e.g. K_2) at a different temperature (i.e. T_2).

$$ln(K_2/K_1) = (-\Delta H/R) \times (1/T_2 - 1/T_1) \qquad (1.66)$$

For calculations using this equation, R is the gas constant ($8.314\,J\,mol^{-1}\,K^{-1}$) and temperature is in Kelvins. It is straightforward and can be useful.

1.11 KINETICS AND REACTION RATES

The thermodynamic approaches just presented are often insufficient to understand behaviors or states of ions and compounds in nature, especially in low-temperature systems like soils, waters, and the atmosphere. Determining rates of processes are addressed in the field of *kinetics*.

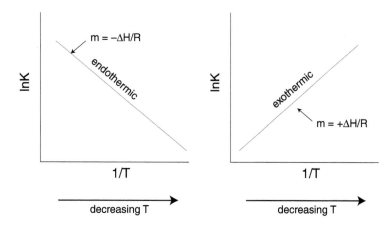

Fig. 1.16 Arrhenius plots showing relationship between equilibrium constant (K) and temperature for endothermic and exothermic reactions. For endothermic reactions, increasing temperature increases K, i.e. adding heat increases reaction rate; for exothermic reactions, the opposite is true.

Focus Box 1.31

Soils and Kinetics

Soils, for example, almost always contain minerals that are (i) thermodynamically unstable at the earth's surface, such as the olivine, augite, and plagioclase that form at $T > 800\,°C$ in cooling basalt lava, as well as (ii) minerals that are thermodynamically stable at low temperatures ($25\,°C$, for example), such as calcite, kaolinite, and hematite.

While the latter three can come to equilibrium with surface waters, olivine, augite, and plagioclase do not. Thermodynamics tells us that. What thermodynamics does not quantify is rate. Consider magnetite in tropical soils; thermodynamics (e.g. ΔG) indicates that it will transform to goethite, but it does not indicate how fast.

Reactions that occur slowly, are not reversible, or do not take place in a system at equilibrium for whatever reason (e.g. soils) are often best understood using kinetic approaches. The same can be said for *heterogeneous reactions*, those that involve various states of matter such as solid minerals and amorphous solids mixed with liquid and gas phases, conditions that tend to occur in soils, sediments, streams, aquifers, and the atmosphere.

1.11.1 Factors controlling reaction rate

Rates of reactions are controlled by numerous factors, including temperature, pressure, redox conditions, pH, mineral composition, abundance of organic matter, pore water composition, diffusion rates, bond types, biotic factors, system composition, and more. In many cases, reaction rates have been determined in laboratory studies that require extrapolation to natural environments and are often prone to large uncertainties (in some cases, an order of magnitude or more; refer to Section 9.1.2 and Table 9.1). The multitude of potential biotic factors can be harder to quantify than the main inorganic controls, yet it is important because biotic effects can be very influential. For example, plants can alter soil pH by exuding (releasing) H^+ from their roots into soil

pores to enhance dissolution of nutrient-bearing minerals; another way that biological factors influence rates and stabilities is the oxidation and reduction reactions mediated by microbes. Thus, the complex array of variables influencing reaction rates can complicate kinetic analysis.

Biogeochemical processes are often controlled by a *rate-limiting step*, a step in a process that is much slower than others. Consider the *dissolution* of a mineral grain, a process that could involve five steps: (i) diffusion of reactant(s) toward the mineral surface (where a common example of a reactant is H^+); (ii) sorption of reactants onto the mineral surface; (iii) formation of a bond between the reactant and the part of the mineral grain under attack (perhaps the O atom bonded to a K^+ ion in a feldspar); (iv) desorption of the newly formed complex between reactant and mineral component (e.g. OH^-), and finally (v) transport of the newly formed product away from the mineral surface by diffusion. If any of one of the steps 1 through 5 is slower than others, it will be the rate-limiting step. In clay-rich sediments, diffusion (steps 1 or 5) can be a rate-limiting step, whereas in highly-insoluble minerals, step 3 (and/or 4) can be the rate-limiting step. Kinetic limitations on *crystal growth* are similar, and can be envisioned by reversing steps 1 through five.

1.11.2 Reaction rate, reaction order

Reaction rates are partly controlled by the order of the reaction, where reaction order is defined as the dependence of reaction rate on concentrations or ratios of species involved in reactions. Reactions can be of zero, first, second, or third order (first and second are most common in environmental geochemistry). Rates of zero-order reactions occur at rates independent of the concentration of the reactant(s); rates of first order reactions are controlled by the concentration of one reactant or product; rates of second-order reactions are controlled by concentrations of two reactants or products, or a reactant (or product) squared.

1.11.2.1 Zero-order reactions

In a *zero-order reaction*, rate is independent of the concentration of reactants. This reaction (sometimes called "zeroth-order" reaction) is one where reactant A undergoes transformation to product P, and can be represented by the simple chemical equation A $\rightarrow$ P. The following equation describes change in concentration of A (represented as [A]) with time (dt):

$$d[A]/dt = -k \qquad (1.67)$$

The term k here is a constant, most likely determined by laboratory experiments or by studies of natural systems, and the negative sign indicates decreasing abundance of A with time. In integrated form, this equation is:

$$[A] = [A]_o - kt \qquad (1.68)$$

$[A]_o$ represents the amount of A at time zero (where examples of t_o could be when sediments were deposited in a floodplain or when a gaseous compound was emitted into the atmosphere), and the negative sign for kt correlates to decreasing reactant with time. So the abundance of reactant A at any time, i.e. [A], is merely determined

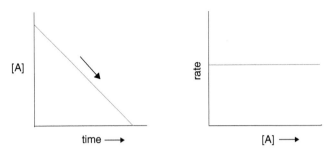

Fig. 1.17 Two graphical representations of zeroth (0th) order chemical reactions, where [A] represents concentration of a chemical species (e.g. ion, mineral, or compound). In the example on the left, decrease in concentration of A is linear, and on the right, rate of change is constant.

by the amount of A at t_o, i.e. $[A]_o$, as well as how much time has passed (t), and the rate constant (k) that governs reaction of A into P. The rate of reaction is not controlled by the amount of A. Plotted graphically (Figure 1.17), it becomes clear that, for 0th order processes, [A] decreases in a linear fashion with time, and rate does not vary with time.

1.11.2.2 First-order reactions

A *first-order reaction* is one where rate depends on the concentration of a reactant (or product) raised to the first power – that is, first-order reactions are typically proportional to the concentration of a reactant, a good example being radioactive decay. The equation for a first-order reaction representing decreasing A with time is:

$$d[A]/dt = -kA \qquad (1.69)$$

Literally, the rate of decrease of A depends on the rate constant (k) multiplied by the amount of A present at time t. In its integrated form, this equation is:

$$[A] = [A]_o \times e^{-kt} \qquad (1.70)$$

Focus Box 1.32

Examples of Zero-Order Reactions

In nature, zero-order reactions are uncommon, because for most reactions, rate increases as the probability of interactions among reactants increases; stated in another way, reaction rate usually increases with increasing concentration of reactants, at least in part because more reactant means more likelihood of interacting with other reactants to foster the reaction (this concept is covered in first- and second-order reaction kinetics). Nonetheless, dissolution of salts like halite (NaCl $\rightarrow$ Na$^+$ + Cl$^-$) and fluorite (CaF$_2$ $\rightarrow$ Ca^{2+} + 2F$^-$) in dilute solutions has been described as zero-order, or pseudo-zero-order (Posey-Dowty et al. 1986). The "pseudo-zero-order" qualifies the observation that rate is slightly nonlinear compared to the idealized plots in Figure 1.17.

Focus Box 1.33

Example of a First-Order Reaction

In the case of radioactive decay, the amount of daughter isotope (e.g. ^{222}Rn produced by decay of ^{226}Ra) produced per unit time decreases as the amount of parent isotope (e.g. ^{226}Ra) decreases; in other words, as reactant is used up during the reaction, the rate of formation of products decreases (which is not to say that the amount of product decreases – it continues to increase, but at a progressively slower rate). In Figure 1.18, [A] decreases exponentially (and the product increases parabolically, following the dashed line) while the rate of change is controlled by the amount of A (reactant) – less A equals lower rate. In Figure 1.18, the solid line would represent amount of parent isotope (and the dashed line amount of daughter isotope) in a radioactive decay reaction.

Graphically, first-order reactions depict nonlinear changes in concentration with time as well as rates that vary as a function of the concentration of reactants (Figure 1.18) – note that the rate (expressed as, e.g., mol yr^{-1}) varies but the rate constant (expressed as a percent or proportion, does not).

1.11.2.3 Second-order reactions

Rates of *second-order reactions* are proportional to the concentration of a reactant squared (dependent on A where $2A \rightarrow B$), or in some cases to the product of the molar concentrations of two reactants (where $A + B \rightarrow C$).

$$d[A]/dt = -kA^2 \ or \ d[A]/dt = -kAB \qquad (1.71)$$

and integrated, the first equation can be expressed as:

$$1/[A] = k \times t + C \qquad (1.72)$$

And given that $[A] = [A_o]$ when $t = 0$,

$$1/[A] - 1/[A_o] = k \times t \qquad (1.73)$$

When depicted graphically (Figure 1.19), second-order reactions exhibit rapid initial changes in concentrations

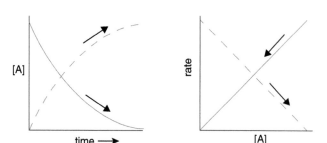

Fig. 1.18 Two graphical representations of first (1st) order chemical reactions, where [A] represents concentration of a chemical species. On the left, change in concentration of A is logarithmic (solid line) or exponential (dashed line); on the right, for both cases (increasing or decreasing concentration), rate of change decreases in a linear fashion with time. Arrows in diagrams indicate direction of forward reaction (emphasizing that, on the right, the solid line represents decreasing [A] with time).

and progressively slower rates with time, and rate increases exponentially with higher concentrations of reactant(s).

Kinetics and reaction order can be summarized with this equation:

$$2A + B = A_2B \qquad (1.75)$$

Focus Box 1.34

Examples of Second-Order Reactions

The reaction of nitrogen dioxide to nitrogen monoxide plus oxygen – one that plays an important role in atmospheric chemistry – is modeled as a second-order reaction:

$$2NO_2 = 2NO + O_2 \qquad (1.74)$$

In this case, rate is dependent on abundance of NO_2, and given that the concentration of the reactant NO_2 is squared (Eq. (1.71)), that rate will be high initially but will decay rapidly as shown in Figure 1.19. Another example of a second-order reaction is the desorption and leaching of soluble cations (e.g. K^+, Ca^{2+}) from tropical soils over time. Rate is initially very high but then decays rapidly as abundance of the reactants decreases (Fisher and Ryan, 2006). Ion exchange of heavy metals also has been modeled as second-order kinetics (Lee et al. 2007; Zewail and Yousef 2015).

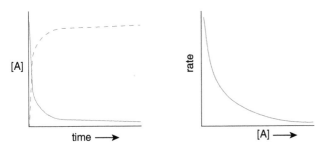

Fig. 1.19 Two graphical representations of second (2nd) order chemical reactions, where [A] represents concentration of a chemical species. In the example on the left, [A] decreases exponentially, and on the right, we see that rate of change decreases exponentially with time.

If the reaction rate depends on the concentration of none of the components in Eq. (1.75), then it is a zero-order reaction. If rate depends on the concentration of B in Eq. (1.75), then it is first-order, and if it depends on the concentration of A, it is second-order. If it depends on the concentration of A, B, and the product A_2B, it is third-order (and first order with respect to B and A_2B, and second order with respect to A), but this is a rare occurrence in geochemical systems.

1.11.3 Temperature and the Arrhenius equation

Temperature also plays an important role in reaction rate – higher temperatures usually foster higher reaction rates. In biogeochemical systems, reaction rates tend to double, triple, or quadruple for every increase of $10\,°C$. The greater energy imparted by higher temperature tends to enhance interactions between reactants, thus enhancing the probability of productive interactions (and here we must be referring to first- and second-order reactions). Higher temperature can also overcome *activation energies* that can sometimes serve as barriers like the one depicted schematically in Figure 1.15. The Arrhenius equation relates temperature to the rate constant:

$$k = A \times e^{-E_a/RT} \qquad (1.76)$$

where E_a is activation energy, R is the gas constant, T is temperature (K), and A is the temperature-independent term known as the pre-exponential factor or the A factor – it serves to convert the product term into values appropriate for k of different reaction orders. In a qualitative sense, and because e is raised to the *negative* E_a/RT, increasing temperature will increase k, speeding up the reaction. Plots of the effect of temperature on reaction

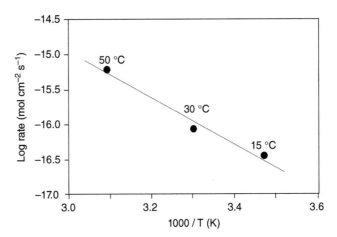

Fig. 1.20 Effect of temperature on chemical weathering rate of rhyolite as a function of temperature. Source: Adopted from Yokoyama and Banfield 2002.

rate are typically plotted as inverse of T (1/T, where T is in Kelvins) versus log of the rate constant k, a type of graph known as an Arrhenius plot. Figure 1.20 depicts dissolution rate of powdered rhyolite (compositionally similar to granite; Table 1.1) as a function of temperature.

The rates shown in Figure 1.20 increase more than 10-fold with a temperature increase of approximately $35\,°C$ (note log scale of y-axis).

This introduction to equilibrium thermodynamics and kinetic constraints is designed to present some of the concepts, approaches, and limitations contained within these two important fields. In the following chapters some of these approaches are applied to predicting and understanding behaviors of environmental systems. For more detailed treatments of this topic, the reader is referred to excellent and much more thorough treatments of these topics in geochemistry texts by James Drever (*The Geochemistry of Natural Waters*), Gunter Faure (*Principles of Geochemistry*), and Donald Langmuir (*Aqueous Environmental Geochemistry*). The *Kinetics of Geochemical Processes* issue of *Reviews in Mineralogy* edited by Lasaga and Kirkpatrick (1981) also presents various perspectives on this topic.

QUESTIONS

1.1 Magnetite, hematite, and goethite all occur in tropical soils. Rank them in order of thermodynamic stability in oxidized soil at $25\,°C$. Which is more likely to be released into soil water, a trace element (e.g. arsenic) substituted into the structure of hematite or goethite? Explain.

1.2 What is (are) the oxidation state(s) of manganese in each of the following compounds?

$$MnO_2 \quad Mn_3O_4 \quad MnO \quad MnCO_3$$

1.3 Recalculate the chemical compositions of the igneous rocks given below, one mafic (#1) and one felsic (#2). First determine weight percent concentrations as elements (i.e. convert the oxide values to elemental values) and rank them in terms of their abundance. Also calculate concentrations of each element in terms of $mg\ kg^{-1}$ and $mol\ kg^{-1}$. Using a spreadsheet would be a time-effective way to do this problem compared to using a calculator.

	#1	#2
SiO_2	45.0	68.0
Al_2O_3	3.5	15.0
FeO	8.0	3.2
Fe_2O_3	0.0	1.3
MgO	39.0	1.7
CaO	3.2	3.4
Na_2O	0.32	3.1
K_2O	0.04	3.6
Cr_2O_3	0.42	1.1×10^{-3}
NiO	0.25	9.2×10^{-4}

1.4 Why are elements with even atomic numbers more abundant than their neighbors with odd atomic numbers? Explain with appropriate nuclear reactions.

1.5 Why is Fe more abundant than neighbors with similar atomic #? Describe with appropriate nuclear reactions.

1.6 Why is Pb more abundant than neighbors with similar atomic #? Describe with appropriate nuclear reactions.

1.7 List the units of concentration that are commonly used in the following cases (consider mass-based and molar units):
A. Acids in water.
B. Salts in water.
C. Metals in soil.
D. Metals in water.
E. SiO_2 in rock.

1.8 Do the following calculations related to concentrations of solutions:

What is the concentration of Ca^{2+} (in ppm, i.e. $mg\ l^{-1}$) in a $4.7 \times 10^{-5}\ M$ solution of $CaCl_2$? What is the concentration of $CaCl_2$ (in ppm) in the same solution?

You need to prepare 100 mL of a $10.0\ mg\ l^{-1}$ solution of NO_3^- from KNO_3 powder. How much KNO_3 do you need to add? What would be the resulting concentration of K^+ (in $mol\ l^{-1}$ and $mg\ l^{-1}$)?

1.9 A. Write the chemical reaction for the dissolution of barite.
B. Calculate the change in enthalpy associated with the dissolution of $BaSO_4$ (barite).
C. Based on your result, predict how the solubility of barite varies with temperature.
D. Calculate the Gibbs free energy for the dissolution of barite into ions.
E. Calculate the solubility of barite at $25\,^{\circ}C$ and 1 atm.
F. Given that Ra coprecipitates with Ba, predict the solubility of Ra in groundwaters (a) ~lacking Ba (e.g. 1 ppb); and (b) rich in Ba (e.g. 100 ppb).

1.10 Which compound is more soluble in water, MgO, or CaO? Explain your reasoning.

1.11 Examine the kinetics of dissolution of a salt like NaCl by experimentation using an approach like that described by Velbel (2004).

1.12 Balance the following chemical reactions:

$$C_3H_8 + O_2 = H_2O + CO_2$$

$$CaCO_3 + H_2SO_4 = Ca^{2+} + SO_4{}^{2-} + H_2O + CO_2$$

$$Al(OH)_3 + H_4SiO_4 = Al_2Si_2O_5(OH)_4 + H_2O$$

$$CaAl_2Si_2O_8 = Ca_3Al_2Si_3O_{12} + Al_2SiO_5 + SiO_2$$

REFERENCES

Dean, J.A. (1979). *Lange's Handbook of Chemistry*, 12e. McGraw-Hill: New York.

Faure, G. (1986). *Principles of Isotope Geology*. Canada: Wiley.

Fisher, G.B. and Ryan, P.C. (2006). The smectite to disordered kaolinite transition in a tropical soil chronosequence, Pacific Coast, Costa Rica. *Clays and Clay Minerals* (5): 571–586.

Gromet, L.P., Dymek, R.F., Haskin, L.A., and Korotev, R.L. (1984). The "North American shale composite": its compilation, major and trace element characteristics. *Geochimica et Comochimica Acta* 48: 2469–2482.

Hanna-Attisha, M., LaChance, J., Sadler, R.C., and Champney Schnepp, A. (2016). Elevated blood lead levels in children associated with the flint drinking water crisis: a spatial analysis of risk and public health response. *American Journal of Public Health* 106: 283–290. https://doi.org/10.2105/AJPH.2015.303003.

Karen, P., McArdle, P., Takats, J. (2016). Comprehensive definition of oxidation state (IUPAC Recommendations 2016).: https://doi.org/10.1515/pac-2015-1204.

Langmuir, D.L. (1997). *Aqueous Environmental Geochemistry*. Upper Saddle River, New Jersey, USA: Prentice Hall.

Lasaga, A.C. and Kirkpatrick, R.J. (eds.) (1981). *Kinetics of geochemical processes: Reviews in Mineralogy*, vol. 8. Washington, DC, USA: Mineralogical Society of America.

Lee, I.-H., Kuan, Y.-C., and Chern, J.-M. (2007). Equilibrium and kinetics of heavy metal ion exchange. *Journal of Chinese Institute Chemical Engineers* 38: 71–84.

Lu, H., Dai, D., Yang, P., and Li, L. (2006). Atomic orbitals in molecules: general electronegativity and improvement of Mulliken population analysis. *Physical Chemistry Chemical Physics* 8: 340–346. https://doi.org/10.1039/B511516G.

Posey-Dowty, J., Crerar, D., Hellmann, R., and Chang, C.D. (1986). Kinetics of mineral-water reactions: theory, design and application of circulating hydrothermal equipment. *American Mineralogist* 71: 85–94.

Railsback, L.B. (2003). An earth scientist's periodic table of the elements and their ions. *Geology* 31: 737–740. (Versions in English and other languages are available at: http://www.gly.uga.edu/railsback/PT.html).

Rudnick, R.L. and Gao, S. (2003). The composition of the continental crust. In: *Treatise on Geochemistry*, vol. 3 (ed. H.D. Holland and K.K. Turekian), 1–64. Elsevier-Pergamon: Oxford, England.

Taylor, S.R. and McLennan, S.M. (1985). *The Continental Crust: Its Composition and Evolution*. Oxford, England: Blackwell Scientific Pub.

Turekian, K.K. and Wedepohl, K.H. (1961). Distribution of the elements in some major units of the earth's crust. *Geological Society of America Bulletin* 72: 175–192.

Velbel, M.A. (2004). Laboratory and homework exercises in the geochemical kinetics of mineral-water reaction: rate law, Arrhenius activation energy, and the rate-determining step in the dissolution of halite. *Journal of Geoscience Education* 52: 52–59.

Vinogradov, A.P. (1962). Average contents of chemical elements in the principal types of igneous rocks of the Earth's crust. *Geochemistry* 7: 641–664.

Yokoyama, T. and Banfield, J.F. (2002). Direct determinations of the rates of rhyolite dissolution and clay formation over 52,000 years and comparison with laboratory measurements. *Geochimica et Comochimica Acta* 66: 2665–2681.

Zewail, T.M. and Yousef, N.S. (2015). Kinetic study of heavy metal ions removal by ion exchange in batch conical air spouted bed. *Alexandria Engineering Journal* 54: 83–90.

2

Surficial and Environmental Mineralogy

By virtue of their abundance in terms of mass and volume – and their reactivity in soils, sediments, rocks, and aqueous systems – minerals play a vital role in determining partitioning and speciation of elements and compounds. Without considering relationships between water and minerals it is often difficult to understand or predict the composition of water in aquifers, soils, or streams. The importance of minerals ranges from their potential to sorb toxic substances and remove them from water (or desorb and release them to water), to the fact that some minerals have structures or compositions make them directly hazardous to human health.

In order to understand whether a given contaminant or nutrient will remain in solution, or whether it will adsorb to a mineral surface, you must know the quantities and types of minerals present. Some minerals are very effective at ion exchange and play important roles in contaminant retention and nutrient cycling, while others are virtually nonreactive. So, whether the geochemical question relates to aquifer geochemistry or paleoclimate analysis, nutrient cycling or buffering of acid rain, fate and transport of organic contaminants or sediment diagenesis, minerals and related solids are bound to play an important role. Thus, the following chapter first systematically presents information on minerals in a "textbook" manner, then highlights minerals that play important roles in surface and shallow crustal environments.

Various minerals have been referred to thus far in this text – quartz, hematite, magnetite, kaolinite, and calcite are a few examples – but now consider what defines a mineral (after Klein and Hurlbut 1999):

1 Minerals are *naturally occurring* solids formed in soils, volcanoes, sedimentary, igneous and metamorphic rocks, and other geological environments;

2 Minerals are *inorganic* – there are numerous naturally occurring organic solids (e.g. cellulose, organic salts such as sodium acetate), but they are not minerals. At the same time mineralogy can be strongly controlled by biotic factors;

3 Minerals possess ordered *crystalline* lattice structures (repeated arrangement of ordered atoms). One consequence of this is that obsidian (disordered volcanic glass) is not a mineral;

4 They also possess *definite (but not fixed) chemical compositions* – some minerals (e.g. quartz, SiO_2) have nearly fixed compositions, but many minerals vary within certain limits because ions with similar charge and radii (e.g. Na^+ for K^+ or As^{3+} for Fe^{3+}) substitute in crystalline lattice structures;

5 Minerals are *homogeneous solids* with characteristic physical properties – the ordered crystalline lattice expresses itself from atomic-scale to macroscopic-scale and results in characteristic physical properties such as density, hardness, color, and crystal shape.

2.1 INTRODUCTION TO MINERALS AND UNIT CELLS

Minerals generally form when chemical bonds develop between ions dissolved in aqueous solutions (although they can also form by other processes, such as direct

---

Focus Box 2.1

Reactivity of Minerals

Minerals and the solutions associated with them are dynamic:
1. certain conditions (including redox, pH, dissolved solids, microbial community) may cause a given mineral to dissolve, releasing into solution components that can range from major ions to trace elements;
2. under a different set of conditions, a given mineral will crystallize from solution, and in doing so may remove from solution major or trace components;
3. the surfaces of many minerals such as phyllosilicate clays and hydroxides – as well as non-mineral inorganic

and organic solids (e.g. humus) – are reactive and can participate in exchange reactions that strongly influence water chemistry.

Some minerals, such as asbestiform amphiboles, are hazardous by mere virtue of their crystal form, whereas others contain uranium and other radioactive elements that can affect the quality of soil, air, and drinking water. Minerals are the primary source of most plant nutrients and often strongly control behavior of toxic substances at the earth surface.

Focus Box 2.2

Calcite, the (Often) Biochemically Precipitated Mineral

For every rule, there is likely to be an exception, and for minerals, perhaps the best example is calcite that occurs in carbonate shells of marine organisms. Calcite often forms by inorganic processes (via evaporation of water in soils, or crystallization from oversaturated metamorphic fluids) and satisfies all of the criteria for "mineral" listed above, but clearly the biogenic form fails the "inorganically formed"

criterion. However, because the calcite in seashells is identical in composition and atomic structure to inorganic calcite, it is given a waiver of sorts and is considered a "mineral." This also applies to minerals whose growth is facilitated by conditions created on or near the surfaces of microbial cells (Section 2.8).

---

solid-state transformations from other minerals). Aqueous solutions range from low-temperature environments like soils and surface waters to mid- and high-temperature fluids in sedimentary, metamorphic, and igneous rocks and hydrothermal systems. In terms of thermodynamics, minerals crystallize to lower the internal energy of a system.

At its most basic, irreducible level, the 3D arrangement of atoms in a mineral structure produces an entity known as the *unit cell*, which is defined as the smallest unit that can be repeated in three dimensions to produce the lattice structure of the mineral. The unit cell consists of a specific group of atoms bonded to each another in a specific ratio and geometric arrangement. Repeated in three dimensions, the unit cell produces a crystalline lattice that manifests itself in the form of a *crystal*, and typically the macroscopic crystal that you can hold in your hand possesses the same *form* and *symmetry* as that exhibited by the unit cell. Examples of the cubic halite unit cell are shown in Figure 2.1, in two dimensions on the left and in three dimensions on the right.

In these diagrams, the Cl^- anion is larger than the Na^+ cation – Na^+ has one less valence electron shell than does

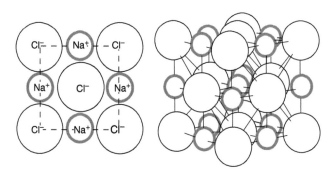

Fig. 2.1 Sketches of halite (NaCl), with a two-dimensional example on the left and a three-dimensional example on the right. Note the relative sizes of cations (Na^+ in this case) compared to anions (Cl^- in this case) as well as the ordered crystalline lattice structure.

Cl^-. For this reason, in almost all cases, the radius of the anion is much larger than that of the cation.

Minerals are electrically neutral – the sum of positive charges imparted by cations balances the negative charges of anions. This is clearly illustrated in Figure 2.1, where the dashed line represents the boundaries of a unit cell

Evaporites, Tangible Examples of Mineral Growth

Mineral crystallization can occur in soils and lakes of arid regions, where evaporation increases concentrations of salt ions (e.g. sodium and chloride, or calcium and sulfate) to the point that they become oversaturated and combine to form minerals like halite (NaCl) or gypsum ($CaSO_4 \cdot 2H_2O$). The chemical bonds that form among ions of Na^+ and Cl^- or Ca^{2+} and SO_4^{2-} result in ordered three-dimensional arrangements of atoms.

that contains the equivalent of two full chloride anions and two full sodium cations (one complete Cl^- and four quarters of Cl^- atoms, and four half Na^+ atoms). The composition of this unit cell is thus Na_2Cl_2, with a net charge of zero. These same concepts are illustrated in three dimensions on the right of Figure 2.1, where the halite unit cell possesses six one-half Cl atoms on each face of the cube, and eight one-eighth Cl atoms on each corner:

$$(6 \times {}^1\!/_2) + (8 \times {}^1\!/_8) = 4 \text{ complete Cl atoms.}$$

The center of this unit cell contains one complete Na atom and 12 Na atoms on the edges of the unit cell (one quarter of each of these 12 atoms rests inside the unit cell):

$$(12 \times {}^1\!/_4) + 1 = 4 \text{ complete Na atoms.}$$

This produces a *stoichiometry* of the 3D unit cell of Na_4Cl_4, but chemical compositions of minerals are typically reduced to simplest integer form, so the conventional form of the chemical composition of halite is NaCl. In

Fig. 2.2 Schematic two-dimensional sketch of a unit cell of halite. Small spheres (circles) are Na^+ and large ones are Cl^-.

terms of relationships between unit cells and macroscopic crystals, halite unit cells are cubic, and so are halite crystals (examine table salt under a microscope and you will find small cubes) – the lattice structure exhibited in the unit cell is carried upward in scale to cubic crystals visible to the naked eye.

Figure 2.2 depicts a schematic 2D representation of four adjacent halite unit cells.

This 2D representation contains the equivalent of eight Na atoms and eight Cl atoms. The charges balance and the stoichiometry is correct. Repeated billions of times in three dimensions, the halite unit cell will produce a macroscopic crystal of halite.

2.2 ION COORDINATION, PAULING'S RULES, AND IONIC SUBSTITUTION

Ions in crystals occur in various configurations depending on charges and radii of cations and anions – the most common anion by far is oxygen (O comprises ~47% of rocks and sediments of the earth's crust; Si is second at 28%), although occasionally other anions such as Cl^-, F^-, and S^{2-} occur in minerals.

2.2.1 Coordination and radius ratio

Figure 2.3 contains two examples of geometrical arrangements of cations coordinated with anions, one in trigonal coordination and the other in octahedral coordination.

On the left of Figure 2.3, the cation occurs in *trigonal coordination*, a planar configuration with three oxygen atoms and a cation in the same plane (effectively, the plane of the page). The *coordination number* (CN) for the cation is three. If you set three relatively large spheres (e.g. tennis balls) on a table as close as possible with a small sphere (e.g. a marble) in the center you will have a marble (representing a cation) in trigonal coordination with three tennis balls representing anions.

In the diagram on the right of Figure 2.3, the cation occurs in *octahedral coordination*, an arrangement that

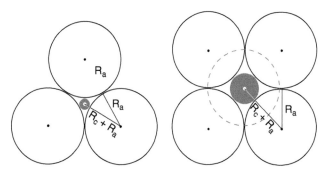

Fig. 2.3 Sketch of cations in threefold (trigonal, on left) and sixfold (octahedral, on right) coordination. The cation in each case is represented by the small dark blue circle with a white center representing the nucleus, and the anions (oxygen in most minerals) are the large white circles (not to scale). The dashed circle in the sketch on the right represents one anion above the plane of the four anions, as well as one anion below the plane of the four anions.

consists of three planes of atoms, one with four O atoms that form a square around a central cation (in the plane of the page), and two planes that each contain one O atom, one above and one below the page – these O atoms fit in the dimple created by the four O atoms in the square. These basal and apical O atoms are represented in the diagram by the dashed circle. The CN for octahedral coordination is six. You would need six tennis balls and a larger marble to visualize octahedral coordination.

Geochemists and mineralogists commonly express relationships between anions and cations by the *radius ratio*, an approach devised by Linus Pauling in the 1920s and defined as the ratio of the radius of the cation to the radius of the anion. Table 2.1 presents crystal radii and ionic radii for common cations and anions. Pauling used X-ray diffraction of minerals (Appendix III) to determine the radii of ions in crystals, where the radius is defined as the distance from nucleus to the valence electron shell, and the atom conceptualized as a ~ hard sphere. (Note in Table 2.1 that the radius of ions varies a little depending on the CN).

In both diagrams in Figure 2.3, R_a represents the *ionic* radius of the anion and R_c is the cation *ionic* radius, but because the cation is so small compared to the anion, conventional diagrams show $R_c + R_a$. Regardless, in these diagrams R_c is identified by the white line segment that crosses the cation (the segment becomes black inside the nucleus for visibility).

For a cation in trigonal threefold coordination, the range of ionic radius ratios (R_c/R_a) is 0.155–0.225. If the ionic radius ratio is smaller than 0.155, the cation is too small, bond lengths between cation and anions are too great, and the bond is unstable; in a sense, the cation would rattle around inside the interstice created by the anions, and anions would be too close causing repulsion

and destabilization. Conversely, if the radius ratio is greater than 0.225, the cation is too large for trigonal coordination – it would distort the triangle to such an extent that the bonds would be unstable (the coordinating anions would be separated too far from each other). In this case, the cation would likely be better accommodated in fourfold tetrahedral coordination.

The radius ratio of NaCl (1.10 Å/1.72 Å) is 0.64, indicating that Na atoms should occur in sixfold coordination with Cl. In Figure 2.1 (3D diagram on right), one Cl atom is directly above the central Na atom, one is directly below, and four Cl atoms are in the same horizontal plane as the Na atom, so visually and by radius ratio we see that the CN for Na in halite is six.

Figure 2.4 presents some common examples of cation sites and geometries in minerals. Not all possibilities are shown, and note that there are three common cubic geometries. The one presented is body-centered cubic, but other varieties include face-centered cubic (e.g. halite) and edge-centered (also known as simple) cubic. Furthermore, large cations can exist in sites with a CN of 12, a good example being the hexagonal close-packed potassium ions (K^+) that occur in interlayers of micas.

2.2.2 Bond strength considerations

Close-packing of ions in crystals is important because bond strength, expressed as the energy of electrostatic attraction (F), is determined by the product of charges (absolute values) of the ions involved in the bond (e_1, e_2) divided by the radius (r) squared, where the radius is the distance between nuclei of adjacent atoms in a chemical bond (Faure 1998):

$$F = (e_1 \times e_2)/r^2 \qquad (2.1)$$

From this equation it is clear that minimizing distance between nuclei of cations and anions enhances the strength of the bond, thus ions strive to coordinate themselves in close-packing arrangements. Chemical bonds also tend to be stronger when they involve ions of greater charge magnitude, so a bond between Na^+ and Cl^- would be expected to be weaker than a $Si^{4+} - O^{2-}$ bond based on charge considerations alone. If we also consider radius of the Na–Cl bond (2.8 Å) compared to the radius of the Si–O bond (1.6 Å), we can arrive at a quantitative comparison of bond strengths:

$$F_{Si-O} = (4 \times 2)/(1.6^2) = 3.1$$

$$F_{Na-Cl} = (1 \times 1)/(2.8^2) = 0.13$$

Table 2.1 Radii of selected major and trace element ions in minerals.

Ion	CN	Geometry	Crystal radius (Å)	Ionic radius (Å)
O^{2-}	4	Tet	1.24	1.38
O^{2-}	6	Oct	1.26	1.40
K^+	8	Cubic, dodec	1.65	1.51
K^+	12	CP	1.78	1.64
Na^+	6	Oct	1.16	1.02
Na^+	8	Cubic, dodec	1.32	1.18
Mg^{2+}	6	Oct	0.86	0.72
Ca^{2+}	6	Oct	1.14	1.00
Ca^{2+}	8	Cubic	1.26	1.12
Fe^{2+}	6	Oct	0.92	0.78
Fe^{3+}	6	Oct	0.79	0.65
Al^{3+}	4	Tet	0.53	0.39
Al^{3+}	6	Oct	0.68	0.54
Si^{4+}	4	Tet	0.40	0.26
Mn^{2+}	6	Oct	0.97	0.83
Mn^{4+}	6	Oct	0.67	0.53
Co^{2+}	6	Oct	0.89	0.75
Ni^{2+}	6	Oct	0.83	0.69
Cu^{2+}	6	Oct	0.87	0.73
Zn^{2+}	6	Oct	0.88	0.74
As^{3+}	4	Tet	0.54	0.40
As^{3+}	6	Oct	0.72	0.58
As^{5+}	4	Tet	0.48	0.34
As^{5+}	6	Oct	0.60	0.46
Cr^{3+}	6	Oct	0.76	0.62
Cr^{6+}	4	Oct	0.40	0.26
Ti^{4+}	6	Oct	0.75	0.61
S^{2-}	6	Oct	1.7	1.84
Cl^-	6	Oct	1.67	1.81
Rb^+	8 to 12	Cubic, dodec, CP	1.75–1.86	1.61–1.72
Sr^{2+}	8 to 12	Cubic, dodec, CP	1.40–1.58	1.26–1.44
Ba^{2+}	8 to 12	Cubic, dodec, CP	1.56–1.75	1.42–1.61
Pb^{2+}	8 to 12	Cubic, dodec, CP	1.43–1.63	1.29–1.49
Ra^{2+}	8 to 12	Cubic, dodec, CP	1.62–1.84	1.48–1.70
Zr^{4+}	6	Oct	0.86	0.72
Th^{4+}	8	Cubic, dodec	1.19	1.05
U^{4+}	6 to 8	Oct, cubic	1.03–1.14	0.89–1.0
U^{6+}	6	Oct	0.87	0.73

Oct, octahedral; dodec, dodecahedral; Tet, tetrahedral; CP, closed-packed; CN, coordination number.

For some elements with multiple oxidation states (e.g. Cr, Mn), this table presents selected oxidation states that commonly occur in minerals. Ions are arranged from lowest-charge (O^{2-}) to highest-charge, first for common rock-forming minerals (O through Si), followed by heavy metals plus As, then other useful ions (S through U). Note that coordination number (CN) corresponds to geometry, i.e. CN of 4 produces tetrahedral geometry, CN of 6 produces octahedral geometry, CN of 8 produces cubic or dodecahedral geometry, and CN of 12 results in closed-packed (CP) geometry as explained in the text.

Transition metals can occur in high-spin or low-spin orbital configurations. At the Earth's surface and shallow crust, high-spin states occur and thus high-spin values are given where applicable (e.g. $Fe^{2+,3+}$, Co^{2+}, Mn^{2+}).

Source: Shannon (1976).

CN	Ion arrangement	Radius ratio		Example
2	Linear	<0.155		Carbon dioxide
3	Trigonal	0.155–0.225		C in carbonate
4	Tetrahedral	0.225–0.414		Si, Al in silicates (As, P)
6	Octahedral	0.414–0.732		Mg, Fe, Al, Ni, other heavy metals, As
8	Cubic e.g. body-centered	0.732–1.00		Large ions, e.g. Ca, K, Na (U, Th, Pb)

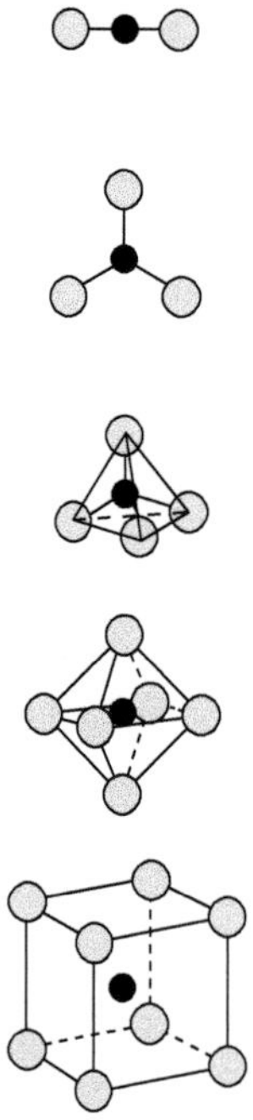

Fig. 2.4 Examples of geometric arrangement of cations and anions in minerals (or gases, e.g. CO$_2$), showing range of radius ratio (ratio of ionic radius of cation: anion) for each configuration.

Focus Box 2.4

Ionic Radius vs. Crystal Radius and Å vs. nm

Note that Table 2.1 presents data for both crystal radii and ionic radii. Crystal radii reflect the space occupied by an ion in a crystal lattice whereas ionic radii correspond to the inherent radius of the ion unaffected by distortions associated with bonding in a crystal. Both are used in geochemistry, and unfortunately it is often not clearly defined which is used. Pauling used ionic radii when formulating his radius ratio ideas (although in the case of tetrahedral Si it appears that he used a value [0.37] closer to the crystal radius than the ionic radius given in Table 2.1). Also worth noting are the units used for measuring ionic or crystal radii, as well as dimensions of unit cells in minerals. The SI unit is the nanometer (nm = 10^{-9} m), but for historical reasons, the angstrom (Å) is still commonly used in mineralogy (1 Å = 10^{-10} m). For example, mineralogists are more likely to indicate that trioctahedral clay minerals have b-axis dimensions of 9.2 Å (rather than saying that it is 0.92 nm). It is just a convention in the field.

Focus Box 2.5

The Silica Tetrahedron, Radius Ratio, and the Si–O Bond

The radius ratio of Si^{4+} (ionic radius = 0.26 Å) in fourfold coordination relative to O^{2-} (ionic radius = 1.38 Å) is 0.19, a value less than the ideal range of radius ratio for tetrahedral coordination (0.225–0.414; Figure 2.4). If we use crystal radii, the radius ratio is 0.40 Å/1.24 Å = 0.32, which fits perfectly in the 0.225–0.414 range for tetrahedral coordination. Examining a primary source (e.g. Pauling 1960), it appears that radii were normalized to octahedral coordination, making the ionic radii larger than those shown in Table 2.1 and affecting Si more than most given its smaller size. Some sources indicate that Si^{4+} in tetrahedral coordination has a radius of 0.34 (Whittaker and Muntus 1970), which is intermediate to the ionic and crystal radii reported in Table 2.1 and gives a radius ratio of 0.25 for tetrahedral Si. Another factor to consider is the [Ne]3s^{2}3p^2 electronic structure of Si^{4+} (Section 1.3.2) that favors a tetrahedral sp^3 hybrid orbital configuration of four valence electrons (akin to sp^3 carbon, Section 3.1.2). In other words, Si^{4+} is most stable with valence electrons arranged in a tetrahedral configuration, and when bonded with four O^{2-} anions stabilizes the tetrahedral SiO$_4^{4-}$ anion that is the building block for all silicate minerals.

Refer also to the example of ionic bond strength presented in Eq. (1.10) in Chapter 1.

2.2.3 Pauling and goldschmidt rules of ionic solids

The brilliant twentieth-century mineralogist, chemist, and two-time Nobel Prize winner Linus Pauling devised a systematic approach to predict crystal chemistry and ionic arrangements in minerals (Pauling 1960). Sections 2.2.3.1–2.2.3.5 present what are commonly referred to as Pauling's Rules for Ionic Crystals.

2.2.3.1 Rule 1: the coordination principle

A polyhedron of coordinated anions (typically oxygen) forms around each cation, and the cation–anion distance and type of coordination polyhedron that results is determined by the *radius ratio*.

2.2.3.2 Rule 2: the principle of local charge balance

The most stable ionic structures are those for whom the sum of the strengths of the electrostatic bonds in a polyhedron equals the charge on the anion (typically, O^{2-}). In short, charges must balance. The first equation that governs this rule determines the value for S, the strength of an electrostatic bond,

$$S = e_{cation}/CN \tag{2.2a}$$

where e_{cation} is the charge on the cation, and CN is coordination number, and;

$$e_{anion} = \Sigma S_i \tag{2.2b}$$

where e_{anion} is the charge on the anion and S_i is the sum of the strengths of the electrostatic bonds. If $e_{anion} = \Sigma S_i$, then the crystal (polyhedron) is stable in terms of charge balance. As ΣS_i values become progressively distant from 1, the crystal becomes progressively less stable, or it requires additional components to counter the charge imbalance (see calcite, below). If we return to the example of NaCl (Figure 2.5), each Cl^- ion is surrounded by six Na ions, so $S = 1/6$, and if we sum a bond strength (S) of 1/6 over six Na–Cl bonds, we see that $\Sigma S_i = 1 = e_{anion}$.

The coordination of 6 Na about a Cl anion produces a stable configuration. Now if we consider C^{4+} in trigonal coordination with O^{2-} (Figure 2.6), we observe that $S = 4/3$, and $S_i = 4/3 \times 3 = 4$; $e_{anion} = -2$.

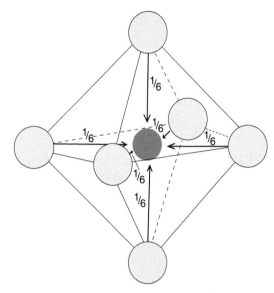

Fig. 2.5 Halite showing coordination of six Cl^- ions around a central Na^+ ion.

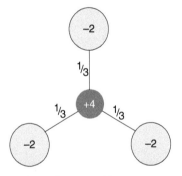

Fig. 2.6 Trigonal coordination of three oxygen atoms (O^{2-}) around a central carbon (C^{4+}) atom.

Clearly, trigonal coordination of CO_3 does not produce a charge-balanced ionic crystal. Three O^{2-} atoms combined with one C^{4+} atom produces $+4 - [3 \times (-2)] = -2$; the -2 charge produces the CO_3^{2-} (carbonate) anion. In minerals, the charge imbalance is satisfied by inclusion of a divalent cation such as Ca^{2+}, Mg^{2+}, or Fe^{2+} in the crystal structure to produce carbonate mineral such as calcite ($CaCO_3$), magnesite ($MgCO_3$), or siderite ($FeCO_3$). Other examples of complex anions that combine with cations to form minerals include silica (SiO_4^{4-}), phosphate (PO_4^{3-}), and sulfate (SO_4^{2-}).

2.2.3.3 Rule 3: sharing of polyhedral edges and faces

Sharing of faces of tetrahedral or octahedral polyhedrons reduces the stability of ionic crystals. This is so because sharing of faces and, to an even greater extent, edges,

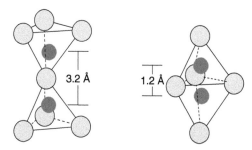

Fig. 2.7 Two potential arrangements of adjacent tetrahedra with cations (e.g. Si^{4+}) as dark blue spheres and anions (e.g. O^{2-}) as light blue spheres. Note that the arrangement on the left produces a greater distance between cations (3.2 Å) compared to the arrangement on the right (1.2 Å), so Pauling's Rule #3 indicates that the structure on the left is more stable.

Table 2.2 Composition of the earth's crust presented as both wt % element (left) and wt % oxide (right).

Element (Wt %)		Oxide (Wt %)	
Oxygen	46.6	NA	
Silicon	27.7	SiO_2	59.3
Aluminum	8.1	Al_2O_3	15.3
Iron	5.0	Fe_2O_3	7.1
Calcium	3.6	CaO	5.0
Sodium	2.8	Na_2O	3.8
Potassium	2.6	K_2O	3.1
Magnesium	2.1	MgO	3.5
Titanium	0.60	TiO_2	1.0
Phosphorus	0.13	P_2O_5	0.30
Manganese	0.1	MnO	0.15
SUM	*99.3*	*SUM*	*98.6*

The difference between the sums listed and 100% reflects the abundance of all other elements.

brings cations closer together. Cations strongly repel other cations, and the electrostatic repulsion between neighboring cations causes destabilization. In reality, sharing of polyhedra is common, but the most stable of these sharing arrangements are those where only one anion per polyhedron is shared with an adjacent polyhedron. The two examples in Figure 2.7 illustrate this principle with tetrahedra.

The shorter cation–cation distance in the potential crystal on the right of Figure 2.7 emphasizes the point that shared faces produce unstable arrangements due to cation repulsion.

2.2.3.4 Rule 4: valency and sharing of polyhedral components

Cations with high charge and small radius tend not to share polyhedra because of ensuing repulsion and because the high valency of the cation will limit the amount of charge available to the anion to distribute among other polyhedral (e.g. C^{4+} in CO_3^{2-}, P^{5+} in PO_4^{3-}, and S^{6+} in SO_4^{2-}). This rule is basically an extension of Rule 3.

2.2.3.5 Rule 5: the principle of parsimony

Although numerous different types of cation–anion arrangements, like those presented in Figure 2.4 and Table 2.1, occur in crystalline solids, the number of essentially different kinds of constituents in a crystal tends to be small. For example, quartz and halite have only one structural arrangement each, and even complex minerals like the phyllosilicate clays possess only two to three cation–anion arrangements (commonly, tetrahedral, octahedral, and 12-fold close-packed interlayer sites).

An additional implication of Pauling's Rules is that *ionic substitutions in crystals tend to occur when ions are of*

similar radii, ionic charge, and electronegativity. Na^+ and K^+ commonly substitute for each other in minerals; Mg^{2+} and Fe^{2+} commonly can occupy the same sites in minerals, as can the pairs Na^+ and Ca^{2+}, Fe^{3+} and Al^{3+}, Ni^{2+} and Mg^{2+}, Cr^{3+} and Fe^{3+}, and Al^{3+} and Si^{4+}. Note that in some cases, cations of slightly different charge can substitute for each other, provided that the charge imbalance is rectified elsewhere in the crystal. For example, in certain silicates, Ca^{2+} can substitute for Na^+, resulting in a net gain of +1, provided that a paired substitution occurs, e.g. Al^{3+} for Si^{4+} or Mg^{2+} for Al^{3+} (to balance the gain of +1).

2.3 SILICATES

Silicates are by far the most abundant class of minerals in the crust and at the surface of the earth. The two most abundant elements in the crust, including rocks, soils, and sediments, are oxygen (46.6% by weight) and silicon (27.7% by weight), with Al, Fe, Ca, Mg, Na, K, and Ti comprising nearly all of the remainder. Stated in terms of oxide weight percents, SiO_2 comprises approximately ~59% of the crust (Table 2.2) and it is for this reason, but not this reason alone, that silicate minerals are so abundant (refer to Focus Box 2.5). The polar covalent bond between Si^{4+} and O^{2-} forms the *silica tetrahedron*, which serves as the basis for all silicate minerals, whether they be high-temperature igneous minerals or low-temperature clay minerals that form in soils, sediments, and sedimentary rocks.

The silica tetrahedron is comprised of a central Si atom surrounded by four O atoms; stoichiometrically, it is SiO_4^{4-}. The tetrahedral configuration maximizes

Focus Box 2.6

Rules of Ionic Substitution

The factors influencing ionic substitution (isomorphous substitution) were devised by Victor M. Goldschmidt and are as follows:
- ionic radii should differ by ≤15%;
- ionic charge can differ by ±1 if balanced by a paired substitution (Ca^{2+} for Na^+, balanced by Al^{3+} for Si^{4+}), and rarely by ±2 if balanced by paired substitution;
- ions with higher ionic potential (ionic charge: ionic radius) tend to occupy sites preferentially over those with lower ionic potential;
- the substituting ions must be able to form similar chemical bonds (e.g. they should have similar values of electronegativity).

These rules can predict trace elements that might occur in common minerals. Consider iron oxides and hydroxides, minerals where ferric iron (Fe^{3+}) occurs in octahedral coordination. Examining Table 2.1 and Figure 1.9, we could predict that Ni^{2+}, Cu^{2+}, Co^{2+}, Zn^{2+}, As^{3+}, and Cr^{3+} may occur in the structure of Fe oxide or hydroxide. Those that would not be predicted to substitute for Fe^{3+} include Cr^{6+} (charge too high and radius too low) and Pb^{2+} (radius too large). A similar treatment could be applied to the potential for trace metals to substitute for Fe^{2+} in pyrite or for Ca^{2+} in calcite, and many more.

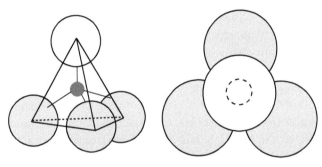

Fig. 2.8 Representations of the silica tetrahedron (SiO_4^{4-}). The three basal oxygen atoms are shaded light blue, the apical oxygen is clear, and the silicon atom is dark blue. The view of the example on the right is from above, looking down at the apical oxygen with three basal oxygens below (hidden Si atom is represented by dashed circle).

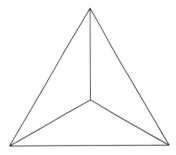

Fig. 2.9 Schematic representation of the silica tetrahedron.

distance between negatively charged O^{2-} atoms while also minimizing distance between O^{2-} and Si^{4+} atoms (Figure 2.8).

Atoms are not to scale in the left side of Figure 2.8 to emphasize the tetrahedral arrangement of the atoms; in the diagram on the right, atoms are approximately to scale and emphasize (i) close-packing of atoms, and (ii) the fact that the small Si atom rests snugly in the interstice produced by the four O atoms. Four equal-sized spheres can help to visualize the arrangement of the O atoms as well as the space occupied by the Si atom.

To simplify, the silica tetrahedron is often represented by the sketch in Figure 2.9, where the four edges of the pyramid represent O atoms and the Si atom is implicitly in the center.

The following presentation of silicate minerals will use this symbol. This section on silicates will be presented

systematically, beginning with the relatively simple nesosilicates, skipping a pair of groups (sorosilicates, cyclosilicates), and progressing to groups with progressively greater sharing of silica tetrahedra (inosilicates and phyllosilicates), eventually to the highly polymerized tectosilicates (where *polymerization* consists of extensive sharing of oxygen atoms among silica tetrahedra).

2.3.1 Nesosilicates

The simplest silicates in terms of structure and composition are the *nesosilicates* (also referred to as *orthosilicates*), a group that consists of isolated silica tetrahedra whose negative charge is balanced by cations located between tetrahedra. Unlike the other silicate groups, there is no polymerization of silica tetrahedra in nesosilicates – no oxygen atoms are shared between tetrahedra.

For the purpose of illustration, *olivine* is presented as an example of this group (Figure 2.10). The olivine group is comprised of two end-members, forsterite (Mg_2SiO_4) and fayalite (Fe_2SiO_4). Note that the tetrahedra appear to

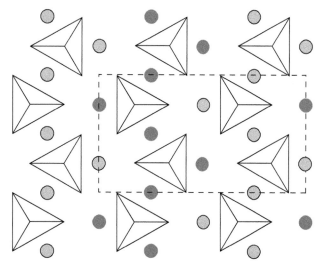

Fig. 2.10 Sketch of an olivine crystal showing isolated silica tetrahedra (triangles) as well as atoms of Mg or Fe represented by dark and light blue circles. Dotted line indicates parameters of an olivine unit cell.

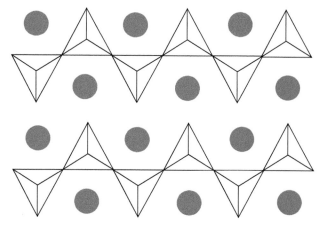

Fig. 2.11 Sketch of pyroxene group structure. Note linked silica tetrahedra that form chain structures, and occurrence of cations (e.g. Ca^{2+}, Mg^{2+}, Fe^{2+}, Na^+) in interstices.

exist as islands amongst divalent Mg (or Fe) cations. The cations are indicated by dark and light blue circles, where the difference indicates an aspect of the three-dimensional character of the olivine structure – in order to minimize cation–cation repulsion, the light blue cations are located above the plane of the page while the dark blue ones are below the plane of the page. One way to visualize an olivine unit cell is indicated by the dashed rectangle, within which are four complete silica tetrahedra, four complete Mg atoms, and eight one-half Mg atoms, for a sum composition of $Mg_8(SiO_4)_4$, which is simplified to a formula unit of Mg_2SiO_4.

Olivines are high-temperature minerals that commonly form in basaltic magmas and tend to decompose rapidly in weathering environments (e.g. soils). Examples of other nesosilicates include *zircon* ($ZrSiO_4$), *garnet* group minerals (which have the general formula $X_3Y_2(SiO_4)_3$, where X and Y are divalent and trivalent cations, respectively), and the Al_2SiO_5 polymorphs *andalusite*, *kyanite*, and *silliminite* (*polymorphs* have identical composition but different structures).

2.3.2 Inosilicates

Linear chains of linked silica tetrahedra form the foundation of the *inosilicate* class of minerals, including: the *pyroxenes*, which consist of single chains of silica tetrahedra; and *amphiboles*, which are composed of double chains of silica tetrahedra. In the single chains (Figure 2.11), each silica tetrahedron shares two basal oxygen atoms, one with each adjacent tetrahedron, producing the fundamental

unit $SiO_3{}^{2-}$ – if two oxygen atoms are equally shared with adjacent tetrahedra, each tetrahedron possesses two full O atoms and two one-half O atoms. The -2 charge is balanced by divalent cations such as Ca, Mg, Fe and in some cases, pairs of (Na^+ and Al^{3+}) or (Na^+ and Fe^{3+}).

Common pyroxenes include the clinopyroxenes augite ([Na,Ca,Mg,Fe]SiO_3), diopside ($CaMgSi_2O_6$), and jadeite ($NaAlSi_2O_6$), and the orthopyroxenes enstatite ($MgSiO_3$) and ferrosilite ($FeSiO_3$). Pyroxenes are common rock-forming minerals, comprising significant parts of numerous igneous and metamorphic rock types. A schematic sketch of a generalized pyroxene is presented in Figure 2.11, where blue spheres are cations and two single chains of tetrahedra are shown.

Note how cation–cation repulsion is minimized by positioning of cations relative to each other and to silica tetrahedra.

In *amphibole double chains*, one half of the tetrahedra share two oxygen atoms (like those in pyroxenes), while the other half share three oxygen atoms (all but the apical oxygen). The result is linear double chains forming hexagonal rings that produce the $Si_4O_{11}{}^{6-}$ fundamental unit (often expressed as Si_8O_{22} in many structural formulae). Shown in Figure 2.12 is a schematic sketch of an amphibole depicting a double chain with relatively large cations within the hexagonal rings and smaller cations in sites outside of the double chain. What are not shown in this picture are the numerous different cation sites, positions of hydroxyl (OH^-) groups, and any real information about three-dimensional structure.

Amphiboles are among the most complex minerals in terms of stoichiometry. Two relatively simple amphiboles, in terms of composition, are *tremolite*, $Ca_2Mg_5Si_8O_{22}$

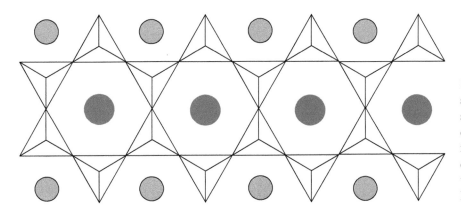

Fig. 2.12 Sketch of amphibole group structure. Note cross-linked chains of silica tetrahedra and occurrence of cations (e.g. Ca^{2+}, Mg^{2+}, Fe^{2+}, Na^+) in interstices. Larger sites in between the chains (four are shown) hold Na^+ or Ca^{2+} whereas smaller sites (eight are shown) hold Al^{3+} or Mg^{2+}.

(OH)$_2$, and *anthophyllite*, $(Mg,Fe)_7Si_8O_{22}(OH)_2$. The most common amphibole, *hornblende*, has the more complex formula $(Na,Ca)_{2-3}(Mg,Fe,Al)_5(Al,Si)_8O_{22}(OH)_2$, where the range from two to three cations in the Na–Ca site depends on the ratio of Na : Ca.

2.3.3 Phyllosilicates

As the name implies (the Greek word for leaf is φύλλο [phyllo]), *phyllosilicates* are sheet silicates. They consist of extensive two-dimensional linkages of silica tetrahedral sheets that extend virtually infinitely. All three basal oxygens are shared, leaving only the apical oxygen not involved in silica polymerization, thus producing a fundamental stoichiometric unit of $SiO_{2.5}^-$, which is expressed in mineral formulas as $(Si_2O_5^{2-})$ or $(Si_4O_{10}^{4-})$. Al^{3+} can substitute for Si^{4+} in phyllosilicates in up to 50% of tetrahedral sites, although typically Al for Si substitution occurs in ≤25% of total tetrahedral sites. As the radius ratio approach would indicate, Al^{3+} also occurs in octahedral sites (Section 2.2.1); in fact, it is the dominant cation in dioctahedral clay minerals.

Figure 2.14 is a view looking down the crystallographic c-axis onto a sheet of silica tetrahedra. Three O atoms per tetrahedron are in the same plane as the page, and the fourth (the apical oxygen) sits above the plane of the page. By virtue of the fact that this tetrahedral sheet is not charge-balanced, it must form chemical bonds to satisfy the residual negative charge. Consequently, all phyllosilicates contain two main structural elements, tetrahedral sheets and octahedral sheets.

Tetrahedral sheets are comprised of silica tetrahedra (often with some Al substituted for Si) like the one depicted in Figure 2.14, whereas *octahedral sheets* are comprised of cations (commonly, Al, Fe, and Mg, but many trace metals also fit – see Table 2.1) in sixfold coordination with a combination of oxygen atoms and OH^- (hydroxyl) anions; three anions form a triangle above the cation and three form a triangle below the cation, producing sixfold octahedral geometry.

Octahedral configurations are shown in Figure 2.15 emphasizing (i) the existence of two planes of trigonally coordinated anions (e.g. O^{2-}) sandwiching a cation (e.g. Al^{3+}, Mg^{2+}), and (ii) the way in which lines can be drawn between anions to emphasize the eight sides from which the term octahedron is derived (for reference, diamonds are octahedral).

The minerals brucite $[Mg_3(OH)_6]$ and gibbsite $[Al_2(OH)_6]$ have structures that are very similar to phyllosilicate octahedral sheets, except for the fact that in these minerals all anions are OH^-, whereas in phyllosilicate octahedral sheets 1/3 to 2/3 of octahedral sheet anions

Focus Box 2.8

Toxic Amphiboles

Some amphiboles, particularly *crocidolite* ($Na_2Fe^{2+}_3Fe^{3+}_2$ $Si_8O_{22}(OH)_2$) and also to a significant extent tremolite, amosite, anthophyllite, and actinolite, are a human health hazard when inhaled as dust. The toxicity results from: (i) the shape of the fibers (Figure 2.13) which makes them difficult to cough out; (ii) the fact that amphiboles are relatively stable in lungs and thus persist for long periods; and (iii) factors 1 and 2 can result in tissue damage that leads to development of asbestosis and cancer, e.g. mesothelioma of the lungs or other organs. Two examples of toxicity and mortality include (i) the South African crocidolite mining district, and (ii) a former vermiculite mine in Libby, Montana

(USA), where the vermiculite ore contained asbestos amphibole.

Asbestiform amphibole was not a target of mining operations at the former WR Grace Mine in Libby, Montana – vermiculite was (for heating insulation and fertilizer products) – but a contact metamorphic event in the Cretaceous Period that altered biotite into vermiculite also altered pyroxene to amphibole (sodic–calcic compositions with characteristics similar to tremolite and actinolite). The intermixing of asbestiform amphiboles and vermiculite led to exposure of miners and residents in the area to asbestos-laden dust that has resulted in negative health outcomes.

Fig. 2.13 Morphology of asbestiform amphibole. Image C on the right is of intermixed vermiculite (platy) and asbestiform amphibole (fibrous) from the WR Grace mine in Libby, Montana (USA). Images A and B are from the US Geological Survey (USGS) Microbeam Lab (http://usgsprobe.cr.usgs.gov/picts2.html) and image C is from USGS Bulletin 2192 (Van Gosen et al. 2002).

are apical oxygen atoms from the tetrahedral sheet and the rest are OH^-.

The diagram in Figure 2.16 presents a view looking down the b-axis (edge-on), showing the relationship of tetrahedral and octahedral sheets in mica. In Figure 2.14, the tetrahedral sheet shown is akin to looking down upon the upper tetrahedral sheet in Figure 2.16. The hexagonal ring formed by linked tetrahedra is visible in both.

On the upper left of Figure 2.16, Si (or Al) atoms in tetrahedral sheets are small blue circles, while octahedral cations are slightly larger blue circles. O atoms are small white circles and OH^- anions are light blue circles. Figure 2.16 depicts two *2:1 layers* on the left and two 2:1 layers on the right. The 2:1 layers are comprised of a pair of tetrahedral sheets sandwiched around an octahedral sheet. The upper 2:1 layer on the left is sketched in a more transparent manner than the lower one in an attempt to show cation positions and cation–anion geometry. The 2:1 layers, in

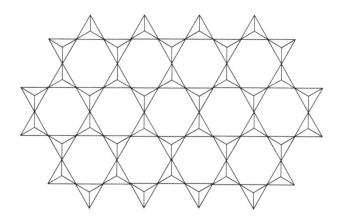

Fig. 2.14 Arrangement of silica tetrahedral in phyllosilicates, showing laterally extensive development of sheet-like structures. This diagram shows a top-down view onto a tetrahedral sheet of a phyllosilicate.

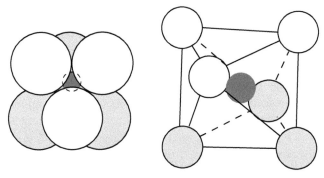

Fig. 2.15 Sketches of octahedral coordination of anions (typically O^{2-} or OH^-) around a cation (dark blue). Anions are light blue in the lower plane and white in the upper plane of both sketches to allow the viewer to correlate the planes above and below the cation.

turn, are sandwiched around an *interlayer cation*, in this case K^+. Clearly the atoms are not drawn to scale – for purposes of illustration, the K ion is drawn larger to show how it fits into the hexagonal holes of bounding tetrahedral sheets. The simplified sketches on the right were developed by Pauling to represent phyllosilicate structures.

2.3.3.1 The 2 : 1 phyllosilicates

Pyrophyllite forms in Al-rich, low-grade metamorphic rocks and has the formula $Al_2Si_4O_{10}(OH)_2$. Its trioctahedral analog *talc* is $Mg_3Si_4O_{10}(OH)_2$. These two minerals will serve as a starting point for presenting different types of phyllosilicates. Pyrophyllite and talc can be represented schematically (Figure 2.17), with cations per unit cell shown for tetrahedral and octahedral sheets.

Positive charges ($2\times Al^{3+} + 4\times Si^{4+} = 22$; *or* $3\times Mg^{2+} + 4\times Si^{4+} = 22$) and the negative charges ($10\times O^{2-} + 2\times OH^- = -22$) balance. Note that the interlayer is vacant, and the main force bonding the 2 : 1 layers is van der Waals bonds. Figure 2.18 depicts pyrophyllite compared to a mica (with interlayer K^+) – note difference in $d(001)$, the spacing along the c-axis.

The *micas* can be developed conceptually by starting with a pyrophyllite unit cell and substituting one Al atom for one Si atom. The stoichiometry change is:

$$Al_2(Si_4)O_{10}(OH)_2 + Al^{3+} = [Al_2(Si_3Al)O_{10}(OH)_2]^- + Si^{4+} \tag{2.3}$$

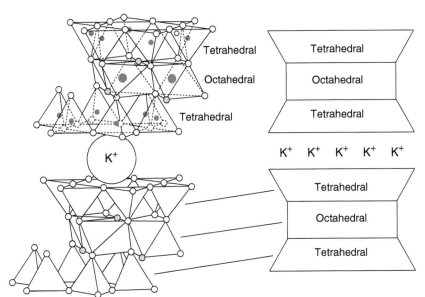

Fig. 2.16 Two representations of a mica, both showing tetrahedral and octahedral sheets as well as interlayer potassium. The schematic sketches on the right are often known as crown-and-girdle diagrams, following early work of Pauling. Source: Adapted from Moore and Reynolds (1997).

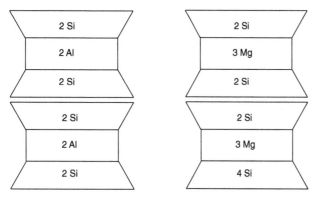

Fig. 2.17 Schematic sketch of pyrophyllite (left) and talc (right). Note (a) difference of 3 Mg (talc) for 2 Al (pyrophyllite) in octahedral sheet, and (b) absence of interlayer cations in both.

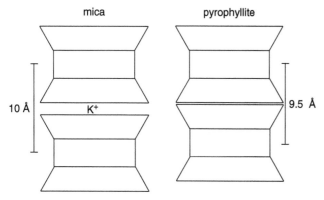

Fig. 2.18 Schematic sketches of mica and pyrophyllite. Note effect of presence/absence of interlayer potassium on the thickness of the unit cells in the c × direction.

This produces a net charge of -1 of the 2 : 1 layer. No mineral can exist if it is not charge-balanced, so for phyllosilicates with a -1 charge on the 2 : 1 layer, cations must be introduced into the *interlayer* for charge balance. Phyllosilicates with a charge of -1 on the 2 : 1 layer and with interlayers containing alkali cations (mainly K^+) are *micas*. The aluminous dioctahedral mica $KAl_2(Si_3Al)O_{10}(OH)_2$ is *muscovite*, a common mineral in granites and various metamorphic rock types. The trioctahedral mica *biotite* has a unit cell composition $K(Mg,Fe)_3(Si_3Al)O_{10}(OH)_2$, with 3 Mg occurring instead of 2 Al (note that real-world biotite

has some substitution of Fe^{2+} for Mg^{2+} in the octahedral layer). By convention, interlayer, octahedral, and tetrahedral cations are presented separately in mineral formulas.

The *interlayer* K^+ ion in micas fits very snugly into the hexagonal rings formed by linked silica tetrahedra – when extended across the interlayer space, opposed hexagonal rings produce a 12-fold coordination site (i.e. from the hexagonal rings of 6 O atoms above and 6 O atoms below), and the electrostatic bond between -1-charged 2 : 1 layers and $+1$-charged interlayer cations holds 2 : 1 layers together. However, the interlayer bond is still the weakest in the mineral, so micas and other phyllosilicates tend to *cleave* along this plane, producing sheets. In cases where the charge on the 2 : 1 layer is < 1, the result is a series of clay minerals including *smectite* (layer charge (LC) ranging from -0.3 to -0.5), *vermiculite* (LC ~ -0.5 to -0.7), and *illite* (LC -0.7 to -0.9) – these and other environmentally significant clay minerals are covered separately in Section 2.4.

Chlorites comprise a group of phyllosilicates with a 2 : 1 layer, but instead of an interlayer cation like micas, they contain an octahedral sheet in the interlayer. (For this reason, sometimes chlorites are known as 2 : 1 : 1 minerals). The magnitude of the negative charge on the 2 : 1 layer is equal to the positive charge on the octahedral interlayer, e.g. by substitution of Al^{3+} for Mg^{2+} in a $(Mg,Fe^{2+})_3(OH)_6$ octahedral sheet, which would produce a representative unit formula of:

$$[(Mg, Fe^{2+})_2Al] (Mg, Fe^{2+})_3(Si_3Al)O_{10}(OH)_8$$

The first set of brackets $[(Mg,Fe^{2+})_2Al]$ is the cations in the octahedral interlayer, the second set $(Mg,Fe^{2+})_3$ is the cations in the octahedral sheet within the 2 : 1 layers, and the third set (Si_3Al) is the cations in the two tetrahedral sheets. Chlorites can be visualized as shown in Figure 2.19.

Chlorites form in Mg–Fe-rich sedimentary environments and soils in arid environments. In some cases (e.g. highly alkaline or highly acidic soils), aluminous dioctahedral chlorites form. Fe–Mg rich chlorites also form during low-grade metamorphism of basalts and other mafic (Mg–Fe-rich, K-poor) rocks.

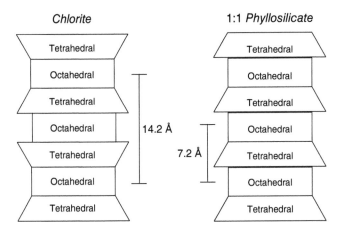

Chlorite   **1:1 Phyllosilicate**

| Tetrahedral |
| Octahedral |
| Tetrahedral |
| Octahedral | 14.2 Å
| Tetrahedral | 7.2 Å
| Octahedral |
| Tetrahedral |

Fig. 2.19 Chlorite (left) and a 1:1 phyllosilicate (e.g. serpentine or kaolinite). Note that the interlayer in chlorite is an octahedral sheet, and that both chlorite and 1:1 phyllosilicates consist of alternating octahedral and tetrahedral sheets, the difference being the orientation of every other tetrahedral sheet.

The presence of hydroxyl groups makes phyllosilicates *hydrous*; that is, they possess water in their structures in the form of hydroxyl (OH^-) groups. If heated to $\geq 500\,°C$, the hydroxyl anions in hydrous minerals combine as follows to produce H_2O:

$$2OH^- \rightarrow H_2O + O^{2-} \qquad (2.4)$$

Prior to chemical analysis, samples are often heated to $1050\,°C$ to drive off water (including hydroxyl water) as well as other gas-forming elements (carbon and sulfur mainly). The term loss-on-ignition (LOI) includes all of these "ignitable" components, and often is listed as a constituent in chemical analysis reports.

2.3.3.2 The 1:1 phyllosilicates

Phyllosilicates that contain repetitions of one tetrahedral sheet and one octahedral sheet (T-O) are known as 1:1 phyllosilicates (Figure 2.19). The dioctahedral forms have the nominal composition $Al_2Si_2O_5(OH)_4$ and are known as *kaolin group* minerals; trioctahedral 1:1 minerals form the *serpentine group*, with the general composition $Mg_3Si_2O_5(OH)_4$.

The most common kaolin group mineral is *kaolinite*, a common constituent of soils, especially in well-leached ones like those of warm temperate and tropical climates. The other main varieties, *halloysite*, *dickite*, and *nacrite*, differ from kaolinite in the way that adjacent $Al_2Si_2O_5(OH)_4$ 1:1 layers are stacked (or rotated) with respect to each other. Halloysite is metastable relative to kaolinite and often forms during rapid crystallization in soils and hydrothermal environments, whereas dickite and nacrite are higher-temperature 1:1 dioctahedral minerals (e.g. $> 150\,°C$) that form in hydrothermal or diagenetic environments.

Serpentines may crystallize in soils and sediments – one example being berthierine, which has the general formula $(Mg,Fe,Al)_{2.5-3}[Si,Al]_{2-2.5}O_5[OH]_4$ – and they also occur in Mg-rich low-grade metamorphic rocks. Serpentine has three main polytypes, *antigorite*, *lizardite*, and *chrysotile*, each with the approximate composition $Mg_3Si_2O_5(OH)_4$, and each of which is differentiated from the others by stacking arrangements.

Chrysotile occurs in an *asbestiform* morphology. The cause is tetrahedral-octahedral mismatch – in all serpentines, the octahedral sheet is slightly larger than the tetrahedral sheet, and this mismatch is addressed by slight curvature of the two sheets in order to form shorter and stronger hydrogen bonds. If the apical oxygen atoms in adjacent tetrahedra rotate away from each other a few degrees they become better-suited to bond with the octahedra that have been stretched by the fact that they need to accommodate 3 Mg atoms per unit cell. Figure 2.20 shows how chrysotile deals with the mismatch by forming rolled 1:1 layers, whereas antigorite contains periodic inversions of 1:1 layers, known as a modulated structure, that results in a wavy pattern.

The 1:1 layers in chrysotile are like rolled-up carpet, each layer of carpet being equivalent to a 1:1 layer, and this results in the fibrous or asbestiform character of chrysotile. The gently rolling waves of antigorite 1:1 layers produce a platy morphology that is typical of most phyllosilicates.

The two examples in Figure 2.21 exhibit the characteristic morphology of chrysotile – on the left is a transmission electron microscope image, and on the right is a cm-scale hand specimen of chrysotile from the Belvidere Mine in northern Vermont, USA. The images show:

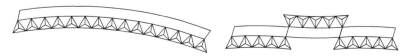

Fig. 2.20   The two ways that serpentines deal with the "mismatch problem." On the left, the slightly larger octahedral sheet in chrysotile is draped over the curved tetrahedral sheet, ultimately forming rolled fibers. In antigorite (on the right), the mismatch is solved by short-period alternations (often of ~17 tetrahedra in length) of tetrahedral and octahedral sheets.

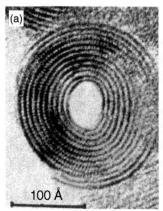

Fig. 2.21 Rolled tetrahedral and octahedral sheets viewed down long axis of chrysotile from a transmission electron microscope image (A, Yada 1971), and fibrous white chrysotile on a serpentine rock surface (B, image is grayscale).

1 individual 1 : 1 layers in the transmission electron microscopy (TEM) image, in which each 1 : 1 layer consists of a white and black band (there are ten or eleven 1 : 1 layers counting from the center out toward the left), and;

2 distinctive white chrysotile fibers in the image on the right (note the 1 cm scale bar). The Belvidere Mine specimen also contains green antigorite (on the bottom right of the image) that exhibits platy morphology (parallel to the rock face) produced by the periodic alternations sketched above.

The fibrous shape of chrysotile classifies it as a type of *asbestos*, and as with asbestiform amphiboles this raises concerns about lung disease. Due to its higher solubility in pulmonary fluids, chrysotile may be less toxic than asbestiform amphibole but often amphibole asbestos (e.g. tremolite) is intermixed with chrysotile in ore deposits so the effects are hard to isolate.

2.3.4 Tectosilicates

Tectosilicates are the most highly polymerized silicates – all oxygen atoms are shared among adjacent tetrahedra. Tectosilicates, also known as framework silicates, include some of the most common minerals in the earth's crust,

soils, and sediments, including *quartz* and the *feldspar group*, which together comprise about 75% of the earth's crust. The tectosilicates also include less common groups such as the *feldspathoids* (~ akin to Si-poor feldspars) and *zeolites*. With cavities (~ 2–10 Å) in their structures, zeolites have strong sorption capacity and are used extensively in industrial and municipal operations for removal of toxins.

The sharing of all four oxygen atoms in SiO_4^{4-} tetrahedra results in the charge-balanced fundamental compositional unit of SiO_2 and unlike the other silicate groups, results in a mineral that requires no other cations – this mineral is quartz. Altering quartz slightly helps to understand feldspars: starting with a hypothetical Si_4O_8 component and substituting an Al^{3+} ion for an Si^{4+} ion results in $AlSi_3O_8^-$. The -1 charge imbalance can be satisfied by inclusion of Na^+ in the structure, producing the common igneous mineral *albite* ($NaAlSi_3O_8$), or K^+ to produce *potassium feldspars* (*microcline, orthoclase,* and *sanidine*). In Figure 2.22, some tetrahedra have edges (apical oxygen atoms) facing toward the reader; the remaining ones (blue) have a face toward the reader. Note that there are two planes of K atoms, and although Si and Al tetrahedra are not differentiated on the diagram, $1/4$ of the tetrahedral sites contain Al and the remainder contain Si.

Arsenic in Phyllosilicates

Serpentine can incorporate arsenic to concentrations of hundreds of $mg\,kg^{-1}$ in serpentinites (Hattori et al. 2005; Ryan et al. 2011). The crystal radii (Table 2.1) and charges of two arsenic cations, As^{3+} (CR = 0.54 Å) and As^{5+} (CR = 0.48 Å), are sufficiently similar to Si^{4+} (CR = 0.40 Å) that Pauling and Goldschmidt rules would predict substitution. This is even more evident when noting the 0.53 Å crystal radius of Al^{3+}, a common tetrahedral ion. One way to accommodate incorporation of As^{3+} into

a Si^{4+} tetrahedral site and create charge-balance is paired substitution of Al^{3+} into an Mg^{2+} octahedral site.

$$As^{3+}_{(tet)} + Al^{3+}_{(oct)} = Si^{4+}_{(tet)} + Mg^{2+}_{(oct)} \qquad (2.5)$$

Substitutions of divalent and trivalent trace metals (particularly Cu^{2+}, Cr^{3+}, Ni^{2+}, and Zn^{2+}) occur in many phyllosilicates, particularly in the octahedral sheet of many ferromagnesian clay minerals (Section 2.4).

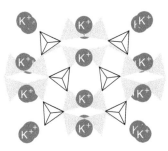

Fig. 2.22 Sketch of potassium feldspar showing potassium ions and silica tetrahedra. Blue triangles represent silica tetrahedral whose apices point away from the viewer. Recall also that ¼ of tetrahedral sites are occupied by Al.

Feldspars form in metamorphic, igneous, and hydrothermal systems but they are common in soils and sediments where they often decompose into soil clays (Chapter 9).

2.4 CLAY MINERALS (1 : 1 AND 2 : 1 MINERALS, INTERSTRATIFIED CLAYS)

The term "clay" can mean two things. One is a specific mineral group known as the *clay minerals* or *silicate clays*, a class of phyllosilicates that tend to occur as particles finer than 2 μm, and it is this mineral group that comprises the focus of this section. The other designation of "clay" is the *clay-size fraction*, a size class commonly defined as < 2 μm, but also sometimes defined as < 4 or < 5 μm (where a μm is 10^{-6} m). Particles in this size range tend to be highly reactive due to their fine grain size and high ratio of surface area : volume, but in theory the *clay-size fraction* could include any mineral, whether quartz, calcite, feldspar, smectite, amphibole, iron oxides, or olivine; however, given their tendency to occur as crystals < 2 μm, clay minerals often dominate the clay fraction.

Some clay minerals are finer-grained, more disordered versions of coarser, higher-temperature phyllosilicates

(e.g. micas and chlorite), but most clay minerals (e.g. smectite, kaolinite, halloysite, and palygroskite) have no real coarsely crystalline, metamorphic, or igneous analogues. They tend to form at low temperatures (e.g. <100 °C) and tend to occur in soils, sediments, sedimentary rocks, and other surficial or near-surface environments. Clay minerals are structurally distinct layer silicates with (generally) platy morphology, they possess extensive compositional diversity, and they are the most abundant reactive phases in low-temperature systems like soils and sediments.

This section on clay minerals will first present the 2 : 1 layer clays, focusing on smectites, a diverse group of expandable clays, as well as other 2 : 1 clays (illite, glauconite, vermiculite, and chlorite), followed by discussion of the 1 : 1 clays (i.e. kaolin group and serpentines) and the interstratified clays (e.g. illite-smectite, kaolinite-smectite). The information presented on phyllosilicates in Section 2.3.3 serves as the starting point for this treatment of the clay minerals.

2.4.1 Smectite

Smectites are a compositionally diverse group of clay minerals that are abundant in a wide array of natural systems, from soils in temperate and arid climates to sediments, shales, marine and terrestrial sediments, and weathered volcanic ash. They are particularly important environmentally because of their high cation exchange capacity (CEC) and their ability to expand in the presence of water (Figure 2.23).

The structure and layer charge (LC) of smectites can be derived conceptually from 2 : 1 phyllosilicates with zero LC, pyrophyllite or talc. Phyllosilicates derive LC from ionic substitutions in the tetrahedral or octahedral sheet, common examples being Al^{3+} for Si^{4+} in the tetrahedral

---

Focus Box 2.13

Environmental Significance of Clay Minerals Enumerated

Silicate clays are significant factors in environmental and low-temperature systems:

1. They are much more reactive than most minerals (largely due to their fine grain size), and for this reason they play a strong role in adsorption (or release) of trace metals, organic compounds, and other potential contaminants; they also are the main minerals responsible for uptake and release of plant nutrients and they have the potential to buffer changes in pH, e.g. in response to acid precipitation;
2. Chemical weathering tends to produce clay minerals and thus clays are major components of soils; in addition to their role described in point 1, their composition often reflects the environment in which the soil formed and therefore can be used in paleoclimate analysis;
3. Their fine-grained character means that they are physically mobile, so when considering point 1, physical transport of clays can result in transport of adsorbed contaminants or plant nutrients;
4. Clays like illite and smectite are common cements in sedimentary rocks and thus control rheology and fluid flow in sedimentary aquifers and deeper realms within sedimentary basins where hydrocarbons occur.

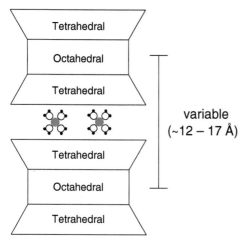

Fig. 2.23 Two smectite 2 : 1 layers with hydrated cations (blue spheres) occupying the interlayer site. Layer charge on each of the 2 : 1 layers (T-O-T) is usually −0.3 to −0.5 and this negative charge attracts cations to the interlayer. Variability in c-axis spacing is due to ability to intercalate different amounts of interlayer water (where the water is bonded to cations).

sheet and Mg^{2+} for Al^{3+} in the octahedral sheet; both result in residual negative charge.

Smectites have LCs of approximately −0.3 to −0.5, in contrast to the −1 LC of micas. Also in contrast to the micas, the interlayer cations that balance the negative 2 : 1 LC in smectites are not fixed, and instead consist of hydrated, low-charge cations such as Na^+, Ca^{2+}, and Mg^{2+}, and less commonly, K^+ or various transition metals (e.g. Ni^{2+}, Zn^{2+}, Cu^{2+}) or hydroxy-aluminum complexes (e.g. $Al(OH)^{2+}$). Organic compounds, some of them contaminants (e.g. pesticides) can also occupy smectite interlayers. The composition of the smectite interlayer typically reflects the composition of ambient pore water – if the water in the pores of smectite-rich sediment is rich in Na^+, so too will be the smectite interlayer. In a soil contaminated with soluble nickel, much of the nickel may be taken up in the smectite interlayer as exchangeable Ni^{2+}.

2.4.1.1 Smectites with tetrahedrally derived layer charge

Substituting 0.33 Al atoms into the *tetrahedral* sheet of pyrophyllite results in a 2 : 1 *beidellite* layer with −0.33 charge: $Al_2(Si_{3.67}Al_{0.33})O_{10}(OH)_2$. The negative LC is shown balanced here by hydrated interlayer Ca^{2+}:

$$Ca_{0.165}(Al, Fe^{3+})_2(Si_{3.67}Al_{0.33})O_{10}(OH)_2$$

Note that this example contains some ferric iron in the interlayer. Some beidellites are very aluminous, with nearly no Fe or Mg substitution in the octahedral layer, yet others have extensive Fe for Al substitution – these ferruginous beidellites are common soil minerals, including this example from a Holocene tropical soil (Ryan and Huertas 2009):

$$(Ca_{0.18})(Al_{1.27}Fe_{0.51}Mg_{0.41})(Si_{3.44}Al_{0.56})O_{10}(OH)_2$$

Two idealized compositions of other varieties of smectite with charge derived from the tetrahedral layer include *saponite*

$$M^+_{0.33}Mg_3(Si_{3.67}Al_{0.33})O_{10}(OH)_2$$

and *nontronite*

$$M^+_{0.33}Fe^{3+}_2(Si_{3.67}Al_{0.33})O_{10}(OH)_2$$

In both of these formulas, the interlayer cation is represented by the variable M^+.

2.4.1.2 Smectites with octahedrally derived layer charge

Substitution of 0.33 Mg atoms into the *octahedral* layer of pyrophyllite results in $(Al_{1.67}Mg_{0.33})Si_4O_{10}(OH)_2$, a 2 : 1 layer with a negative 0.33 charge balanced by hydrated cations, in this case Na^+ (Table 2.3). Two common ways to represent the formula unit are:

$$[(Na)_{0.33} \cdot H_2O](Al_{1.67}Mg_{0.33})Si_4O_{10}(OH)_2, \, or$$

$$Na_{0.33}(Al_{1.67}Mg_{0.33})Si_4O_{10}(OH)_2$$

Aluminous smectite with octahedrally derived charge is known as *montmorillonite*, a common smectite that forms by alteration of volcanic ash in marine environments – montmorillonite is very abundant in marine sediments and low-grade shales and soils eroded from these sources (i.e. many terrestrial soils and sediments).

Stevensite is a trioctahedral clay that derives its LC from vacancies in the octahedral sheet and, in some cases, from minor substitution of Al for Si in the tetrahedral sheet, a concept illustrated by these data adopted from Hay et al. (1995):

$$(Na_{0.15}Ca_{0.05}K_{0.05})(Mg_{2.8})(Si_{3.9}Al_{0.1})O_{10}(OH)_2$$

Extensive substitution results in smectites that can vary appreciably from the idealized compositions, with

1 numerous intermediaries between smectite end-members; and

2 smectites with LC derived from substitutions into both tetrahedral and octahedral layers.

Furthermore, the ionic radii of many trace metals differ by $\leq 15\%$ from those of the common trioctahedral cations ferrous iron and magnesium ($Fe^{2+} = 0.78\,Å$,

Table 2.3 Classification of smectites based on octahedral occupancy and origin of 2 : 1 layer charge (i.e. octahedral vs. tetrahedral sheets).

	Dominant origin of 2 : 1 layer charge	
	Octahedral sheet	Tetrahedral sheet
Dioctahedral Smectites		
Al-rich	Montmorillonite	Beidellite
Fe-rich	N/A	Nontronite
Trioctahedral Smectites		
Mg-rich	Hectorite (Mg + Li)	Saponite

$Mg^{2+} = 0.72\,Å$), so metals such as Co^{2+} (0.79 Å), Cr^{3+} (0.76 Å), Ni^{2+} (0.69 Å), and Zn^{2+} (0.74 Å) are known to occur in clay octahedral sheets (Table 2.1). In some cases they are trace components, but in other cases they are major components of octahedral sheets, a good example being Ni-rich stevensite formed by the weathering of Ni-rich ultramafic rock in Greece (Christidis and Mitsis 2006) with a composition of

$$Mg_{0.12}(Mg_{1.44}Ni_{1.41}Co_{0.03})Si_{4.0}O_{10}(OH)_2$$

Given that trace metals can occur in both octahedral sheets of trioctahedral clays and interlayers of smectites, clays play an important role in fate and transport of these elements.

2.4.2 Vermiculite

Vermiculite is structurally similar to smectite, but with higher LC (−0.6 to −0.9) and often much coarser particles. In fact, there is little distinction between high-charge smectites (those with −0.5 to −0.6 LC) and some vermiculites, at least in terms of LC and CEC. Vermiculite often forms in soil by weathering of biotite, so we will begin by examining biotite and work toward vermiculite. An illustrative biotite composition (structural formula) is:

$$K(Mg_{2.7}Fe^{2+}_{0.3})(Si_3Al)O_{10}(OH)_2$$

During chemical weathering, ferrous iron typically oxidizes to ferric iron and this can have a destabilizing effect on many minerals. Assuming that all ferrous iron oxidizes, the result is:

$$K(Mg_{2.7}Fe^{3+}_{0.3})(Si_3Al)O_{10}(OH)_2$$

This produces a −0.7 charge on the 2 : 1 layer (it was −1.0 in the unoxidized biotite), which is too low (i.e. insufficiently negative) to hold the K^+ ion in the interlayer. The result is that K^+ diffuses out of the interlayer (as a potential plant nutrient), and hydrated ions (commonly Mg but sometimes Al or other cations) migrate into the interlayer, producing this representative trioctahedral vermiculite:

$$(Mg_{0.35} \cdot H_2O)(Mg_{2.7}Fe^{3+}_{0.3})(Si_3Al)O_{10}(OH)_2$$

The +0.7 charge on the interlayer ($Mg_{0.35} \cdot H_2O$) balances the 2 : 1 layer charge of −0.7.

Dioctahedral vermiculite forms by weathering of micas in acidic soils with high concentrations of dissolved Al. The following formula is adopted from data presented in

Focus Box 2.14

Shrink–Swell Capacity of Smectites

Depending on available moisture, *smectite interlayers* can contain one layer of hydrated cations (a 2.8 Å one-layer complex), or two (a 5.6 Å two-layer complex), or three water layers (a 8.4 Å three-layer complex). This swelling capacity is a distinctive characteristic of smectites, also sometimes referred to as *expandable* clays or *shrink–swell* clays. When the interlayer cation is Na^+, the amount of interlayer water can increase by orders of magnitude due to the very high hydration potential of Na^+. This can have negative consequences for agriculture, engineered structures, and land stability. In climates with alternating rainy and dry seasons (e.g. monsoonal climates), smectite-rich soils expand and shrink seasonally, wreaking havoc on buildings and roads and causing problems for farmers due

to alternating periods of impermeability and waterlogging (when saturated) and dessication cracking (dry periods) of farm fields. It is equally important to emphasize that the negative aspects of smectites are balanced by the fact that they are widely used in landfills and hazardous waste repositories where water-borne contaminants must be contained. Hydrated smectites are virtually impenetrable to flow of water and their high potential to chemically adsorb contaminants means that they form a physical and chemical barrier to contaminant flow. This is true is both engineered systems (e.g. landfills) and in natural systems (e.g. clay-rich layers known as confining layers between shallow potentially contaminated aquifers and deeper, confined aquifers).

April et al. (1986) and, for purposes of illustration, does not include trace amounts (<0.07 atoms per formula unit) of octahedral Mg and Ti.

$$(Al_{0.2}K_{0.1}·H_2O)(Al_{1.7}Fe^{3+}_{0.3})_3(Si_{3.3}Al_{0.7})O_{10}(OH)_2$$

Note that (i) 2 : 1 LC is −0.7, in this case entirely derived from the tetrahedral sheet, and (ii) not only is Al a major component of the octahedral sheet, but also the vermiculite interlayer, reflecting high Al concentrations in soil water.

Vermiculites possess very high CECs and can play important roles in nutrient cycling. This was one of the uses of the vermiculite mined in Libby, Montana – see Focus Box 2.8 in Section 2.3.3.

2.4.3 Illite

Illite can be considered as a low-temperature analogue muscovite. Muscovite contains a LC of −1 and is characterized by relatively large crystals (typically visible without a microscope), whereas illite possesses a LC of −0.7 to −0.9 and typically occurs in crystals only visible under a microscope. A typical illite composition is:

$$K_{0.8}(Al_{1.8}Fe^{3+}_{0.1}Mg_{0.1})(Si_{3.3}Al_{0.7})O_{10}(OH)_2$$

Note that the 2 : 1 LC is −0.8, with −0.7 derived from the tetrahedral sheet (Al^{3+} for Si^{4+}) and −0.1 derived from Mg^{2+} for Al^{3+} substitution in the octahedral sheet. The interlayer contains fixed ions of K^+ (they are not exchangeable as are the interlayer cations in smectites).

Illite commonly forms in sedimentary basins from precursor smectite (commonly montmorillonite), a reaction that is largely driven by temperature but also time (i.e. thermodynamics and kinetics both play a role in the smectite → illite reaction). Shallowly buried marine shales (depths <2 km, < 75 °C) are dominated by smectite whereas deeper shales are dominated by illite (~150–200 °C); intermediate shales contain a mixture of the two (more on this later in Section 2.4.6). If shales are further buried or overthrusted so that they reach low-grade metamorphic temperatures of ≥250 °C, illite transforms to muscovite, a point at which shales often transform to slates or phyllites. Because of the predictable transformation from smectite to illite to muscovite, this series can be used as a *geothermometer* to interpret burial and metamorphic histories of sedimentary rocks (factoring in compositional and kinetic controls on the smectite to illite transformation).

Illite is also a common soil mineral, most commonly derived from *detrital* or physical weathering processes but in some cases from *pedogenic* processes. It is stable in many weathering environments and can be inherited from parent material like granite and schist or via eolian (wind) or fluvial (river-borne) processes. In other cases, where soils are potassium-rich, illite can form in the soil environment by, for example, weathering of potassium feldspar.

2.4.4 Chlorite and berthierine

These two mineral types are grouped together here because of similarities in structure, composition, and occurrence in soils, sediments, and sedimentary rocks. Chlorites are generally trioctahedral, as is berthierine, and they form in environments rich in Mg and Fe. The composition of berthierine is much like that of chlorite:

$$(Al_{0.8}Fe^{2+}_{1.5}Mg_{0.6})(Si_{1.4}Al_{0.6})O_5(OH)_4$$

By doubling the composition of berthierine, you arrive at a reasonable chlorite composition (Section 2.3).

Now compare the structures of the two minerals, as shown in Figure 2.19. The only appreciable difference is the orientation of every other tetrahedral sheet. In the case of berthierine, all tetrahedra are oriented in the same direction, producing a repeat distance along the c-axis of 7.2 Å; in the case of chlorite, orientations of tetrahedral sheets alternate, producing a repeat distance almost exactly double that of berthierine.

2.4.5 Kaolin group (kaolinite and halloysite)

Kaolinite and halloysite are the two polytypes of the 1 : 1 kaolin group that commonly occur in low-temperature environments like soils, sediments, and sedimentary rocks. Both are nominally $Al_2 Si_2 O_5 (OH)_4$ although Fe^{3+} (and to a lesser extent Mg^{2+} or other divalent or trivalent cations) can substitute to a limited extent into octahedral sites.

Kaolinite and halloysite form in environments with high dissolved Al content and lower concentrations of Si and base cations (Na, K, Mg, Ca) relative to environments that foster formation of smectite, vermiculite, chlorite,

and illite. Tropical soils are often leached of base cations, resulting in Al–Si-rich solutions that foster formation of kaolin or halloysite.

One very important difference between the kaolin group minerals and most other clays is that *kaolin group clays have no LC* (or very low LC e.g. in the case of some halloysites). The consequence is that kaolinites and its polytypes have very low CECs compared to other clay minerals, i.e. they have very limited attraction for many nutrients as well as trace metals and other cationic forms (refer to Section 2.5). This factor contributes to the nutrient-poor status of many tropical soils (Chapter 9).

2.4.6 Interstratified clay minerals

Also known as mixed-layer clay minerals, interstratified clays are common in soils and sedimentary rocks, and the types, extent, and changes in interstratification are controlled by important environmental factors such as composition, climate, temperature, and time. The most common and widely studied interstratified clay series is illite/smectite (I/S). Low-temperature shales (e.g. 50 °C, 1 km burial) are dominated by smectite, and high-temperature shales (e.g. 200 °C, 6 km burial) are dominated by illite. Between these two end-members exists a progressive series of interstratified clays ranging from smectite-rich I/S to intermediate I/S and illite-rich I/S (Figure 2.24). In tropical soils, leaching depletes

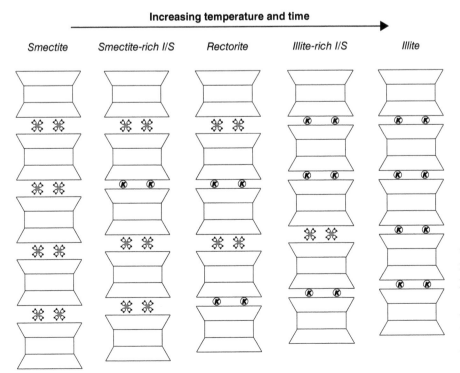

Fig. 2.24 The transition from smectite to illite via interstratified illite/smectite, including the ordered I/S phase rectorite, a mineral in which illite and smectite layers alternate in an ordered manner. In sedimentary basins, this transition typically takes place progressively with increasing burial over the temperature range of 50–200 °C.

Focus Box 2.15

Glauconite, Berthierine, and Paleodepositional Analysis

Two minerals that occur in sandstones are useful tools in paleoenvironmental analysis. The Fe-rich mica *glauconite* crystallizes on the sea floor at depths of ~250 m (near the edge of the continental shelf) within about 60° of the equator, and because it is syndepositional (i.e. crystallized when sediments were being deposited), dating of glauconite by K–Ar and Rb–Sr radioactive decay methods can provide depositional ages. An average chemical composition obtained from chemical analyses of Thompson and Hower (1975) is:

$$K_{0.7}(Al_{0.6}Fe^{3+}_{0.9}Fe^{2+}_{0.2}Mg_{0.4})(Si_{3.6}Al_{0.4})O_{10}(OH)_2$$

Note that glauconite derives its −0.7 LC from both the tetrahedral and octahedral sheets.

Berthierine and its disordered analogue *odinite* form at or near the sediment–water interface in shallow marine lagoons, estuaries, and river mouths located within 30° of the equator. In the humid tropics, rivers transport iron and silicate materials derived from intense tropical weathering into shallow marine environments. Chemical reactions in the marginal oxidizing-reducing conditions of these environments foster formation of Mg–Fe-rich clays (odinite or berthierine); with sedimentary burial and increasing temperature, odinite or berthierine transforms into chlorite, providing a geothermometer in Fe-rich sedimentary rocks (Hillier 1994; Ryan and Reynolds 1996).

---

the soil solution of dissolved Si, and the Si-depleted soil water destabilizes tetrahedral sheets, causing stripping of tetrahedral sheets and resulting in transformation of 2 : 1 layers to 1 : 1 layers. This layer-by-layer transformation of smectite layers to interstratified kaolinite/smectite and then to kaolinite occurs over thousands to tens of thousands of years.

2.4.7 Trace metals and metalloids in clay minerals

Trace metals such as Cr^{3+}, Ni^{2+}, and Zn^{2+} (and less frequently Co^{2+} and Cu^{2+}) are known to occur in octahedral sheets of clay minerals, particularly smectites, serpentines, and chlorites (Newman and Brown 1987). The literature contains numerous studies which document elevated concentrations of trace metals in clay structures; for example, Brindley and Maksimovic (1974) and Gaudin et al. (2004) describe Ni-rich layer silicates in soils and other low-temperature environments, and Oze et al. (2004) report Cr-rich chlorites in serpentine soils. Manceau et al. (2004) determined that, in a soil with $128\,mg\,kg^{-1}$ Zn, >80% of the Zn is contained in interlayers and octahedral sheets of magnesian smectite, and Dähn et al. (2002) showed that Ni-rich smectite layers can crystallize on surfaces of pre-existing montmorillonite crystals in Ni-rich solutions – in these Ni-smectites, the nickel occurs as Ni^{2+} in octahedral sites.

2.5 CRYSTAL CHEMISTRY OF ADSORPTION AND CATION EXCHANGE

"Adsorption is probably the most important chemical process affecting the movement of contaminants in groundwater" (Drever 1997). The composition of soil water and ground water that feeds streams and lakes is similarly strongly controlled by adsorption–desorption reactions. Drever's statement could also be extended to nutrient availability in natural and agricultural systems, as well as compositions of surface water. The adsorption and desorption of ions and other substances from surfaces of minerals and natural organic matter exerts a very strong control on cycling of plant nutrients by regulating nutrient uptake and leaching. An important consequence of adsorption may be that while an ion or molecule is very soluble in groundwater or other aqueous systems, if it is prone to adsorption onto surfaces of solid phases, its concentration in solution will be less than predicted when only considering dissolution-precipitation reactions (e.g. mineral stability).

Adsorption is a surface effect or process. Adsorbed ions or molecules can be released into solution without mineral dissolution, and adsorption–desorption reactions may only require slight changes in the concentrations of non-adsorbed ions or molecules in solution. Conversely, ions incorporated (or adsorbed) *into* mineral structures are generally only released into solution when the mineral dissolves. Adsorption–desorption reactions are *reversible* and

---

Focus Box 2.16

The Nomenclature of Interstratified Clay Minerals

Using the proper terminology is important; in the case of mixed-layer clays, the component with the lower d-spacing is listed first, followed by either a slash (used here) or dash, and then the name of the component with higher d-spacing. This gives us names like illite/smectite and kaolinite/smectite, or abbreviated terms like I/S and K/S. Where proportions are known (e.g. 75% kaolinite layers and 25% smectite layers in kaolinite/smectite), you will see terms like kaolinite(0.75)/smectite, abbreviated as

K(0.75)/S. In some cases where ordered interstratifications occur with 50% of each layer type, a new mineral name results, e.g. corrensite for 50–50 chlorite/smectite, and rectorite for 50–50 illite/smectite.

In soils, interstratified kaolinite/smectite (progressing from smectite → kaolinite/smectite → kaolinite with leaching over time) and biotite/vermiculite (progressing from B → B/V → V with leaching over time) are common interstratified series.

---

take place at rates that are typically orders of magnitude more rapid than mineral dissolution–precipitation reactions. In fact, ion exchange reactions can occur so rapidly that their rates are quite difficult to measure by conventional methods (McBride 1994). To reiterate an important point, *adsorption* describes the attraction of substances (adsorbates) to the surface of a solid particle (adsorbent); *absorption*, on the other hand, is a process by which a substance enters into the internal structure of a particle or other absorbent. The term *sorption* includes both adsorption and absorption.

2.5.1 Cation exchange

Cation exchange is a particular type of adsorption characterized by attraction of cations to particle surfaces.

2.5.1.1 Mechanisms by which cations are attracted to surfaces

1 *chemical attraction*, which occurs when a dissolved phase forms a chemical bond with an atom (or atoms) exposed at the surface of the solid – an example might be a cation attracted to the localized negative charge of an exposed oxygen atom at crystal edge;
2 *electrostatic attraction*, where a dissolved phase is attracted to the surface of a charged particle – a good example is hydrated interlayer cations being attracted to the negatively charged 2 : 1 layer in smectite;
3 *physical attraction*, the weakest form of adsorption that is due primarily to van der Waals forces.

These factors are regulated by a combination of attributes related to (i) the mineral itself, (ii) the substance that may be sorbed, and (iii) properties of the solution (e.g. soil water) in which the solid resides. The remainder of this section will consider attributes of minerals as they relate to surface exchange. Details of solution chemistry and its relationship to ion exchange and water chemistry are presented in Section 4.12.

2.5.1.2 Particle attributes that influence ion exchange

Ions in solution are attracted to clays and other fine-grained particles by four main attributes:
1 *Charge on the particle.* Charged particles will electrostatically attract ions of the opposite charge – this is particularly true of smectite, vermiculite, and other clay minerals with negative LCs, each of which are strong cation exchangers.
2 *Particle size.* Exposed atoms on mineral surfaces are potential adsorption sites, and the greater the amount of exposed surface area per volume of mineral, the greater the ability to adsorb. Fine-grained particles have *high ratios of exposed surface area to volume*, i.e. enormously high surface area relative to volume as compared to coarse minerals, and thus contain much higher proportions of potential exchangeable sites relative to the amount of mineral.
3 *Interlayers or other exchange sites in minerals.* Interlayers in smectites and channels in zeolites are examples of additional adsorption sites. Comparing the available exchangeable sites in smectite and kaolinite (Figure 2.25) emphasizes that each smectite unit cell contains an interlayer that can sorb cations, whereas kaolinite does not (nor do minerals like illite, chlorite, and oxyhydroxides).

This concept becomes even more apparent when considering that the a–b plane extends virtually infinitely in clays like smectite and vermiculite; their interlayers are very extensive due their platy textures, meaning that smectites and vermiculites have extensive interlayer exchange sites.

4 *Edge sites, dangling bonds, surface effects.* These effects are derived from surface complexation, a process whereby hydrated ions attach to ions exposed at mineral surfaces. Edges of crystals commonly contain unsatisfied chemical bonds, a good example being a metal cation in octahedral coordination at the edge of a crystal that is bonded to less than six oxygen atoms or hydroxyls (OH^-). In acidic solutions, mineral surfaces tend to be dominated by exposed H^+ ions, making the H^+-charged surface attractive to anions, whereas in alkaline solutions, surfaces are negatively charged and attract cations. These effects are most pronounced in oxides, hydroxides, and silicate clays like kaolinite and serpentine with little or no layer charge.

2.5.1.3 Point of zero charge and isoelectric point

The pH at which the surface charge of a particle is zero in natural solutions with a range of ions is the *point of zero charge* (PZC); in an idealized solution in which only H^+ and OH^- sorb to the mineral surface, this concept is known as the *isoelectric point* (IEP). At values less than the PZC or IEP, the surface charge is positive (dominated by adsorbed H^+) and so the particle attracts anions. As pH increases and H^+ (protons) are drawn away from the surface and into solution, the surface becomes negatively charged and acts as a cation exchanger. The values of PZCs and IEPs presented in Table 2.4 indicate the following:

1 Smectite adsorbs cations in all situations except for very acidic (pH < 3) environments;

2 Kaolinite possesses weak anion exchange capacity at pH < 3 to 4; at pH > 4, kaolinite possesses weak CEC.

3 The IEP of iron oxides and hydroxides occurs at pH 7 ± 2, meaning that in acidic solutions they are anion exchangers, in alkaline solutions they are cation exchangers, and in neutral solutions they have little or

Table 2.4 Isoelectric points (IEP) and points of zero charge (PZC) of selected minerals (values = pH) and non-minerals[a].

Mineral	Composition	IEP	PZC
Smectite	$Ca_{0.17}(Al,Fe^{3+})_2$ $(Si_{3.66}Al_{0.34})O_{10}(OH)_2$	1–3	2–3
Kaolin	$Al_2(Si_2)O_5(OH)_4$	4	2–4
Humus[a]	~ $C_{20}H_{20}O_{15}N$	~ 2–3	~ 2–3
Allophane[a]	~ $Si_{1.5}Al_2O_4$ $(OH)_4 \cdot 4H_2O$	5–6	~5
Hematite	Fe_2O_3	6–8	4–7
Goethite	$FeOOH$	6–7	6–7
Gibbsite	$Al(OH)_3$	~9	~10

Note that IEP and PZC values are very similar, but unlike IEPs, which are only a function of protonation and deprotonation, PZCs are also controlled by solution chemistry (e.g. adsorption–desorption of metals and other ions). At pH values below the IEP or PZC, the mineral is an anion exchanger; above this value, the mineral sorbs cations.

[a]Note: all of the above substances are minerals with the exceptions of humus and allophane. Humus is organic matter and allophane lacks ordered internal structure.

no preference for cations or anions. These relationships can have consequences for cycling of nutrients such as the NO_3^- (nitrate) anion, for sorption of anionic contaminants (e.g. certain pesticides and some forms of arsenic), and for sorption of trace metals and other aqueous species. Figure 2.26 depicts a *goethite* crystal in two different solutions, one with a pH = 5 (i.e. below the IEP) and the other at a pH = 9 (above the IEP). Note the adsorption of protons at pH = 5 that causes *goethite* to adsorb anions.

It is important to note here that the most abundant minerals in soils that contribute to cation exchange (clay minerals such as smectite) have IEPs at pH < 3, meaning that at nearly all pH values in natural systems, the potential for cation exchange is far greater than the potential for anion exchange. Anions such as nitrate (NO_3^-) and sulfate (SO_4^{2-}) tend to be poorly retained by soils except in highly evaporitic environments (where evaporation prevents leaching).

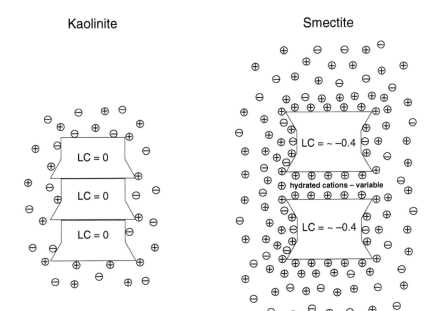

Kaolinite   Smectite

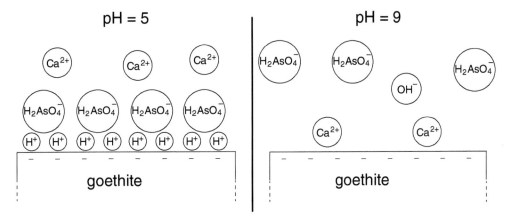

Fig. 2.25 Schematic representation of the differences in cation exchange capacity (CEC) between kaolinite and smectite. "LC" indicates layer charge, which is zero in kaolinite but is usually -0.3 to -0.5 per unit cell in smectite. Expandable interlayer spaces and smaller crystal size further enhance smectite CEC.

Fig. 2.26 Sketch of a goethite surface in solutions at pH = 5 (left) and pH = 9 (right), demonstrating the effect of IEP or PZC on ion exchange capabilities of soil solids. Note occurrence of anionic arsenate ($H_2AsO_4^-$) as inner sphere complex at pH = 5 (facilitated by H^+) compared to occurrence of Ca^{2+} as inner sphere complex at pH = 9. Negative signs near goethite surface indicate net negative charge at mineral surface.

2.5.1.4 Double-layer complexes

Exchangeable ions occur on particle surfaces in a *double layer* consisting of *inner-sphere* and *outer-sphere* complexes. *Inner-sphere complexes*, also known as the fixed-layer or Stern layer, are typically held by electrostatic forces directly on the surface of the solid; for this reason, the inner-sphere complex often consists entirely of cations (assuming pH > PZC). *Outer-sphere complexes* (also known as diffuse or Gouy layers) are farther from the crystal surface. At the boundary between inner-sphere and outer-sphere complexes, cations > anions, but the concentration of anions increases exponentially away from this boundary until anions and cations are balanced

(chargewise) – this is the outer limit of the exchange complex, and these concepts are illustrated in Figure 2.25.

2.5.1.5 Units of concentration and measurement of CEC

CEC is expressed as molar units of exchangeable cation per mass of solid (where the solid is a given mineral or other type of solid such as organic matter). Units are centimoles of charge per kg ($cmol_c$ kg^{-1}), although sometimes CEC is expressed in units of millimoles of charge per kg ($mmol_c$ kg^{-1}). The outdated unit milli-equivalents per 100 g (meq 100 g^{-1}) produces values that are identical to $cmol_c$ kg^{-1}, but $cmol_c$ kg^{-1} is preferred because it is consistent with modern terminology.

To illustrate the concept of $cmol_c$ kg^{-1}, let us consider three different exchangeable cations, Na^+, Ca^{2+}, and Al^{3+} and the quantity of charge contained in one mole of each:

1 One mole of Na^+ (22.98 g) contains one mole of positive charge, and 1 centimole (cmol) of Na^{+1} (0.2298 g) contains 1 cmol of positive charge (1 $cmol_c$);

2 One mole of Ca^{2+} (40.08 g) contains two moles of positive charge, so 1 cmol of Ca^{2+} (0.4008 g) contains 2 $cmol_c$; and

3 One mole of Al^{3+} (26.98 g) contains three moles of positive charge, so 1 cmol Al^{3+} (0.2698 g) contains 3 $cmol_c$.

Now consider a hypothetical example of an extremely small crystal of smectite with cumulative layer charge of −6 (equivalent to 20 smectite unit cells, each with −0.3 charge). Envision that this smectite crystal occurs in sodium-rich soil water, and that the −6 charge is balanced by six Na^+ ions exchanged onto its exposed surfaces (mainly interlayers). If the composition of the soil water were to change to Ca-rich, an exchange reaction would take place, and Na^+ would be replaced by Ca^{2+}. How many calcium ions would replace the six Na^+ ions? In order to maintain charge balance, three Ca^{2+} ions would exchange with six Na^+ ions, a reaction that can be depicted as:

$$6\,Na^{+\cdot}[MINERAL] = 3\,Ca^{2+\cdot}[MINERAL] \qquad (2.6a)$$

Continuing, if the soil water were to become dominated by dissolved aluminum (this occurs in soils affected by acid rain), the following exchange reaction would likely occur:

$$3\,Ca^{2+\cdot}[MINERAL] = 2\,Al^{3+\cdot}[MINERAL] \qquad (2.6b)$$

In other words, 6 mol of Na^+ are released for each 3 mol of Ca^{2+} or 2 mol of Al^{3+} adsorbed.

Analytical measurement of CEC is often done by exchange reactions in the lab. The objective is to swap the cations occupying exchange sites for another cation that is strongly attracted to the mineral surface. One common approach is to use an ammonium salt (e.g. NH_4^+-acetate at pH = 7), beginning with a 1 M NH_4^+ solution to which 2.5 g of soil or mineral is added, shaken well, and allowed to equilibrate overnight. The concentration gradient of NH_4^+ ions in solution (high) to Na^+, Ca^{2+}, K^+ or other ions in exchange sites (relatively low) will cause desorption of Na^+, Ca^{2+}, K^+, etc. as NH_4^+ ideally occupies all exchange sites. Desorbed cations can be measured by inductively coupled plasma-optical emission spectrometry (ICP-OES) or some other method, and results converted to $cmol_c$ kg^{-1} (e.g. Pincus et al. 2017). However, as Dohrmann and Kaufhold (2009) report, NH_4^+-acetate

contributes to sulfate and carbonate dissolution, so this approach cannot be used on many soils (e.g. arid and semi-arid soils). Dohrmann and Kaufhold (2009) propose a similar approach but with different salts, e.g. cobalt(III) hexamine (CoHex) and copper(II) triethylenetetramine (Cu-trien) (Table 2.5).

2.6 LOW-TEMPERATURE NON-SILICATE MINERALS: CARBONATES, OXIDES AND HYDROXIDES, SULFIDES, SULFATES, SALTS

Silicates are the most abundant mineral type in most soils and sediments in terms of mass and volume, but non-silicate minerals often play important roles in contaminant fate and transport, plant nutrient cycling, the carbon cycle, and soil chemistry, and in many cases these minerals are controlled by climate, and hence are important to consider with respect to analysis of paleoclimate and modern climate change.

2.6.1 Carbonates

All carbonate minerals contain the carbonate anion (CO_3^{2-}), formed by trigonal planar arrangement of one C^{4+} and three O^{2-} atoms (Figure 2.4). The −2 charge is typically balanced by divalent cations, and by far the two most common carbonate minerals are calcite ($CaCO_3$) and dolomite ($CaMg[CO_3]_2$). In calcite the −2 charge on the carbonate anion is satisfied by Ca^{2+}; in dolomite, the −4 charge of two CO_3^{2-} anions is balanced by one Ca^{2+} and one Mg^{2+}. A small sampling of the dozens of additional carbonate minerals include cerussite ($PbCO_3$), siderite ($FeCO_3$), rhodochrosite ($MnCO_3$), smithsonite ($ZnCO_3$), nahcolite ($NaHCO_3$), and malachite ($Cu_2CO_3[OH]_2$). There are three $CaCO_3$ polymorphs – calcite, aragonite, and vaterite.

Carbonates are soluble in acid so tend not to occur in acidic soils like those of tropical, temperate, or moist alpine regions; rather, they tend to form in semi-arid to arid soils and lakes as well as in marine environments, sedimentary and metamorphic rocks, and limestone caves. The dissolution of carbonates has the potential to buffer acid deposition or acid mine drainage:

$$CaCO_3 + H_3O^+ = Ca^{2+} + HCO_3^- + H_2O \qquad (2.7)$$

Note that the hydronium ion is consumed by calcite dissolution and the result is a circum-neutral solution with dissolved calcium buffered by bicarbonate (HCO_3^-).

Focus Box 2.18

Determining CEC from a Unit Formula

The role of LC in cation exchange can be examined for this smectite (ferruginous beidellite) from a Holocene soil in Costa Rica, whose composition was determined by TEM-AEM analysis (see Appendix III for instrumental analysis):

$$(Ca_{0.2})(Al_{1.0}Fe_{0.8}Mg_{0.5})(Si_{3.2}Al_{0.8})O_{10}(OH)_2$$

The layer charge is −0.4 (balanced by 0.2 Ca^{2+}) and the molecular mass is 401.48 g. The CEC of this smectite produced solely by LC can be determined as follows:

$$CEC = |^-0.4| \times mol_c / 401.48\,g \times 100\,cmol_c/mol_c$$

$$\times\, 1000\,g/kg = 99.6\,cmol_c/kg$$

$$\rightarrow 100\,cmol_c/kg$$

Note that the result 99.6 $cmol_c$ kg^{-1} is reported to only one significant figure (i.e. 100 $cmol_c$ kg^{-1}), limited by imprecision in determination of − 0.4 LC. This value is a slight underestimate because it does not include contributions from complexation to unsatisfied bonds – the actual CEC of this mineral determined by ammonium acetate exchange analysis is ~105 $cmol_c$ kg^{-1}, but the similarity in calculated vs. measured in this example illustrates the overwhelming control of LC on CEC in clay minerals.

One kg of smectite with CEC = 100 $cmol_c$ kg^{-1} could adsorb 100 cmol of Na^+, 50 cmol of Ca^{2+}, or 33.3 cmol of Al^{3+}. Stated in terms of mass, 1 kg of this smectite could absorb 22.98 g of Na^+ (100 cmol×0.2298 g $cmol^{-1}$), 20.04 g of Ca^+, 8.98 g of Al^{3+}, or any charge-balanced mixture of these three as long as the sum of charges amounts to 100 $cmol_c$ kg^{-1} (e.g. 10 $cmol_c$ of Na^+, 60 $cmol_c$ of Ca^{2+}, and 30 $cmol_c$ of Al).

Table 2.5 Cation Exchange Capacities of Selected Minerals (note that oxyhydroxides are anion exchangers at pH values approximately < 7).

Mineral/phase	CEC ($cmol_c$ kg^{-1})	S_A (m^2 g^{-1})	Origin of CEC
Organic matter (humus)	100–500	300–1000	Sub-micron, localized charge
Vermiculite	100–200		High (−) layer charge, interlayer, often sub-micron
Smectites	80–150	500–800	(−) layer charge, sub-micron, interlayer
Illite	10–40	50–100	(−) layer charge, often sub-micron
Halloysite	10–40	20–100	Sub-micron, minor layer charge, interlayer
Kaolinite	1–10	10–40	Sub-micron, possibly minor layer charge in disordered kaolinite
Fe and Al oxyhydroxides	1–10	50–300	Dangling bonds, sub-micron CEC at pH < 7, AEC at pH > 7
Quartz, feldspars	1–2	< 1	Weak van der Waals forces

These values were determined at pH = 7; values vary as a function of pH and aqueous composition.

Source: From Sparks 1995; Drever 1997, Langmuir 1997.

Ca^{2+} and CO_3^{2-} (often in the form HCO_3^-) ions are fixed from seawater by many marine organisms in order to form shells, and when these organisms die their shells fall to the sea floor to someday comprise parts of limestones. Limestones are also comprised of carbonate mud that directly precipitates out of saturated seawater, especially in tropical and subtropical shallow seas.

Carbonate formation effectively removes CO_2 from the atmosphere and sequesters it in sediments, soils, or rocks, contributing to long-term regulation of atmospheric CO_2. In the atmosphere, CO_2 dissolves in H_2O and falls out of the sky as carbonic acid (H_2CO_3), which then dissolves in water to produce the bicarbonate anion:

$$H_2O + CO_2 = H_2CO_3 \qquad (2.8)$$

$$H_2CO_3 = H^+ + HCO_3^- \qquad (2.9)$$

Dissolved calcium ions then combine with dissolved bicarbonate to form calcite and dissolved H^+:

$$Ca^{2+} + HCO_3^- = CaCO_3 + H^+ \qquad (2.10)$$

The CO_2 molecule in Eq. (2.8) is transferred to carbonic acid, then to dissolved bicarbonate, and finally into a solid carbonate mineral. This reaction occurs in any solution that is saturated with respect to calcium carbonate, whether in a marine environment, soil, or lake. A greater level of detail of carbonate geochemistry is presented in Chapter 5.

Most carbonates are rhombohedral, although some, including aragonite, are in the orthorhombic crystal class. Figure 2.27 illustrates the rhombohedral crystal structure of carbonates, in this case, calcite.

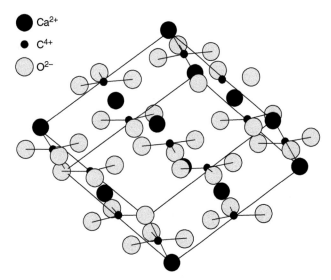

Fig. 2.27 Molecule of calcite. Note rhombohedral structure that is also expressed in macroscopic crystals. Lines between oxygen atoms are used to indicate individual CO_3^{-2} molecules within the mineral structure.

Cerussite and smithsonite are ores of lead and zinc, respectively, and while effectively insoluble in neutral solutions, they are soluble in weak acids. Cerussite was used as a white pigment in paint until the early 1900s. By the 1930s Pb in paint had been banned in many countries (e.g. Australia, France, Greece, Spain, Sweden, Tunisia, United Kingdom). Sale of lead in residential paint was finally banned in the United States in 1978. Cerussite in the form of *lead paint* is still present in many residential environments, where chipped and peeling paint in old buildings can cause lead poisoning (especially in children), and cerussite also occurs in lead-contaminated soils. When ingested, it dissolves in stomach acids ($PbCO_3 + H^+$ $\rightarrow Pb^{2+} + HCO_3^-$). From there it is incorporated into the bloodstream where it ultimately interferes with the development of the central nervous system – it is a potent neurotoxin, responsible for diminished intelligence and other neurological problems (particularly in young children), and it also impairs functioning of the kidney and blood-forming organs.

2.6.2 Oxides and hydroxides

A landscape with reddish or orange soils, sediments, or rocks is visual evidence of iron oxides or iron hydroxides in a natural environment. These hues are produced by various configurations of iron with O and/or (OH). The most common *oxides* of iron are *hematite (Fe₂O₃)* and *magnetite (Fe₃O₄)*. In hematite, both Fe atoms in the unit formula are ferric (Fe^{3+}); in magnetite, two Fe atoms are ferric and one is ferrous (Fe^{2+}), indicating that magnetite

occurs in less-oxidizing environments than hematite. Iron *hydroxides* include *ferrihydrite* ($Fe_5HO_8 \bullet 4H_2O$ or $5Fe_2O_3 \bullet 9H_2O$ or $Fe_2O_3 \bullet 2FeOOH \bullet 2.6H_2O$), *goethite* ($FeOOH$ or $Fe_2O_3 \bullet 3H_2O$), and *lepidocrocite* ($Fe(OH)_3$), all of which can be considered metastable precursors to hematite. In each of the iron hydroxides shown here, iron is ferric (Fe^{3+}). When the formulas of hydroxides are written as $Fe_2O_3 \bullet xH_2O$, it becomes evident that increasing temperature and expulsion of H_2O from hydroxides will lead to the formation of hematite. Whereas hematite imparts a distinctly reddish-orange color to soils and rocks, the hydroxides impart more muted brownish yellow to brownish orange colors (Plate 1).

Trace metals substitute for iron in oxide and hydroxide structures, so iron oxides and hydroxides often exert strong controls on trace metal partitioning and mobility. Some of these trace metals also form their own oxide minerals; for example, uranium (as U^{4+}) forms uraninite (UO_2) and copper (as Cu^+) forms cuprite, Cu_2O.

Al and Ti are other major elements that form oxides; for example, Al_2O_3 is the metamorphic mineral corundum and Al also forms $Al(OH)_3$ (gibbsite), a common mineral in well-leached soils of the tropics and in some acidic hydrothermal systems. There are three polymorphs of TiO_2, rutile, anatase, and brookite, and $FeTiO_3$ is the igneous-metamorphic mineral ilmenite.

Oxides and hydroxides form in oxidized (or oxidizing) environments, which generally include well-aerated soils and sediments, including well-drained soils (as opposed to saturated wetland soils) and sandy sediments at or near the earth's surface, or rocks with high amounts of free O_2. Oxides tend to be quite stable in these environments, but in O_2-poor reducing environments, oxides tend to dissolve, driven by reduction of iron ($Fe^{3+} \rightarrow Fe^{2+}$), a reaction often catalyzed by microbes.

2.6.3 Sulfides and sulfates

Sulfides and sulfates are sulfur-based minerals that are sensitive to redox environment and are generally highly reactive when outside of their environments of formation. Sulfides contain the *sulfide* (S^{2-}) anion, a common species in chemically reduced systems such as wetlands, anoxic sediments (in deep lakes, estuaries, and marine environments), and O_2-poor deeper crustal environments. Sulfates, on the other hand, contain the *sulfate* anion (SO_4^{2-}) and form in oxidizing environments, especially where evaporation rates are high, as well as in certain sedimentary and low-grade metamorphic environments.

1 The most common *sulfides* are the iron sulfides like pyrite (FeS$_2$; Plate 2) and its polymorph marcasite, as well as pyrrhotite (~FeS) and mackinawite (~FeS), a disordered low-temperature sulfide that forms in anoxic surficial environments and imparts a black hue to chemically reduced sediments (Plate 6). In pyrite, the FeS$_2$ stoichiometry and the +2 charge on ferrous iron indicates that each sulfur atom has an average −1 charge. The sulfur atoms more accurately occur as pairs of S^{2-} and S^0 atoms that are represented as a disulfide anion S$_2{}^{2-}$.

Given the similarity in ionic radii and charge among ferrous iron (Fe^{2+}) and many transition metals, the sulfide anion forms chemical bonds with a wide array of trace metal cations, resulting in:

1 extensive substitution of metals such As, Cu, Ni, Pb, and Zn for iron in pyrite;

2 the occurrence of dozens of trace metal-sulfide minerals, including arsenopyrite (FeAsS), chalcopyrite (CuFeS$_2$), cinnabar (HgS), galena (PbS), sphalerite (ZnFeS$_2$), and many others.

Sulfides are common ores of heavy metals, and for millennia have been mined by civilizations ranging from Roman to Aztec (Triple Alliance) to those of the modern world. The dissolution of sulfides is a common source of trace metal contamination and acidic drainage, whether naturally derived or exacerbated by human activity. When removed from their reducing environments of formation and exposed to oxidizing environments (such as most soils, river sediments, freshly exposed rock surfaces), sulfides are unstable and oxidize rapidly. For example, when pyrite is exposed by erosion, uplift, or mining to air with O$_2$, it reacts in the following manner to release iron to solution and produce 2 moles of sulfuric acid (2 H$_2$SO$_4$, shown here as 4H$^+$ + 2SO$_4{}^{2-}$):

$$FeS_2 + 3O_2 + 2H_2O = Fe^{2+} + 4H^+ + 2SO_4{}^{2-} + 2e^- \tag{2.11}$$

Note that oxidation here is driven by oxidation of the S$_2{}^{2-}$ *disulfide* anion to *sulfate* (SO$_4{}^{2-}$), a process that liberates eight electrons per mole of disulfide, and which is worth examining individually here:

$$S_2{}^{2-} + 2O_2 = SO_4{}^{2-} + S^0 \tag{2.12}$$

Or considering only the oxidation of the more-reduced S^{2-} atom:

$$S^{2-} + 2O_2 = SO_4{}^{2-} \tag{2.13}$$

The eight electrons are consumed by the four oxygen atoms (or, stated differently, by the two O$_2$ molecules) because each O atom consumes two e^- and transforms

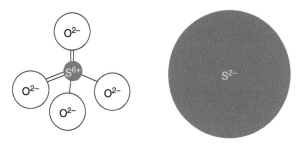

Fig. 2.28 Schematic diagram of sulfate (left) and sulfide (right) anions, with S ions drawn to scale (the radius of S^{2-} is approximately six times greater than S^{6+} in tetrahedral coordination).

from zero-charged O$_2$ (i.e. O$^{\cdot}$) to O^{2-} oxygen anions. The electron flow associated with this and other oxidation reactions is a significant source of energy to microbes, and although not depicted in this reaction, sulfide oxidation is typically very strongly driven by microbial activity, a topic that is covered in Section 2.8.

The S atom in the sulfate anion is S^{6+}. Examples of the sulfide and sulfate anions are shown in Figure 2.28. The sulfate anion is an example of a polyatomic anion, because it is comprised of more than one atom.

Sulfide and sulfate minerals are structurally very different. In sulfides, the S atom occurs as the large sulfide anion (S^{2-}) with an crystal radius of 1.7 Å covalently bonded to octahedrally coordinated metal cations, whereas in sulphate minerals, sulphur is a small S^{6+} cation tetrahedrally coordinated with four O^{2-} ions. (The S^{6+} crystal radius = 0.26 Å). The SO$_4{}^{2-}$ anion is octahedrally coordinated about metal cations, but the bonds are ionic, a significant difference between and sulfates and sulfides (which are dominated by covalent metal-S bonds).

The metal cations in sulfates are commonly divalent (Ca^{2+}, Mg^{2+}, Fe^{2+}, Ba^{2+}) as well as monovalent (Na$^+$, K$^+$) and trivalent (Al^{3+}, Fe^{3+}), and some sulfates contain additional components like H$_2$O and OH$^-$. They form when solutions become saturated with respect to the metal cation and sulfate anion, environments that include arid region S-bearing soils and lakes, sulfidic mine tailings and hydrothermal systems. By far the most common sulfate is the hydrated calcium sulfate *gypsum*, CaSO$_4$•2H$_2$O, a component of some arid-region soils where evaporation drives up concentrations of Ca^{2+} and SO$_4{}^{2-}$ in solution to the point that gypsum precipitates directly from solution.

In acidic soils, rocks, and mine tailings, where oxidation of sulfides produces dissolved SO$_4{}^{2-}$ and low pH values and evaporation increases Al and Fe increase to the point of saturation, sulfates like *alunite, jarosite, melanterite, rozenite,* and *chalcanthite* precipitate out of solution (Plate 2).

Fig. 2.29 Salt precipitates on dessicated shoreline of Salt Creek, Death Valley, USA. White crusty coating is halite and other soluble salts precipitated at the surface when evaporation caused solute concentrations to rise above the saturation point. Dessication polygons are ~0.5 m in diameter. Source: U.S. Geological Survey Photographic Library, http://libraryphoto.cr.usgs.gov/index.html. Photographer: J.R. Stacy.

Focus Box 2.19

Sulfates and Metal Mobility

Because the bonds between cations and sulfate anions in sulfate minerals are ionic (and thus generally soluble in water), sulfates play an important yet often ephemeral role in trace metal transport and availability in certain environments. One such environment is sulfide-rich mine tailings and their surroundings. The chalcanthite depicted in Plate 3 occurs on the surface of a sulfide-rich rock in copper mine tailings where oxidation of chalcopyrite ($CuFeS_2$) – triggered by sulfide oxidation and promoted by bacterium of the Alicyclobacillus species (Lesley-Ann Giddings, pers. comm.) – releases Cu^{2+} and SO_4^{2-} to solution. Evaporation on hot summer days drives concentrations of these reactants to saturation and chalcanthite ($CuSO_4 \cdot 5H_2O$) precipitates on exposed surfaces. This serves to sequester copper into the solid state, but during rainy periods, the ionically bonded chalcanthite dissolves, releasing copper (Cu^{2+}) to solution, where it eventually flows downstream and impairs aquatic habitat.

2.6.3.1 Halide and nitrate salts

Salts formed by ionic bonds between ions of alkali metals (e.g. Na^+, K^+) or alkaline earth metals (e.g. Mg^{2+}, Ca^{2+}) and the halide anion chloride (Cl^-) are extremely soluble in water, so halide salts only tend to crystallize and persist in arid environments. The most common halide salt is *halite*, which forms when salty seas evaporate to produce sedimentary salt beds like those of the Gulf of Mexico and Saudi Arabia. The salty crusts that form when playa lakes evaporate in arid regions are halite (and/or carbonates or sulfates) in a natural surficial environment (Figure 2.29).

Nitrate salts contain a metal cation that is ionically bonded to the nitrate anion NO_3^- – they are very soluble at the earth surface and thus only exist in very arid environments like the Atacama Desert and Death Valley. Examples include KNO_3 (known variously as potassium nitrate, niter, or saltpeter) and $NaNO_3$ (known variously as sodium nitrate, sodium niter, or chile saltpeter), both of which are used in explosives and fertilizers.

2.7 MINERAL GROWTH AND DISSOLUTION

Crystallization and dissolution of minerals occur in the presence of solutions, most commonly aqueous solutions in pores among mineral grains in soils and sediments, or water in fractures of crystalline bedrock. In some cases minerals grow directly from aqueous solutions, a process referred to as mineral *precipitation* or *neoformation* or *authigenesis*. In other cases, minerals form via transformation from pre-existing minerals and incorporate parts of crystallographic structures inherited from the parent mineral, a process referred to as *transformation* (the interstratified clay mineral reactions profiled in Section 2.4.6 are considered transformation reactions). Mineral dissolution can be *congruous*, the term for the complete dissolution of a mineral, or *incongruous*, a partial dissolution process where only part of a mineral goes into solution and the remainder persists in the solid state, forming a new mineral via transformation.

Many external factors influence mineral formation or dissolution, including:

1 *Temperature.* Minerals such as smectite, goethite, and kaolinite form in low-temperature environments, i.e. soils, sediments, and sedimentary rocks. Minerals like olivine, pyroxene, and Ca-feldspar form at elevated temperatures associated with igneous or metamorphic environments. Some minerals (e.g. quartz, hematite, calcite) can form across a wide range of temperature, from surface conditions to higher-temperature crustal regimes;

2 *Redox conditions.* Ferrous iron (Fe^{2+})-bearing minerals (including many silicates) typically are stable in reducing environments of wetlands or deeper crustal realms but are unstable when exposed to the atmosphere. Sulfide minerals will crystallize in reducing systems but are unstable (and thus will dissolve) in oxidizing systems (e.g. shallow groundwater, most streams and soils); the opposite is true for iron oxides, which tend to be stable in soils but unstable in reducing environments of wetlands and O_2-poor ground water systems;

3 *Time.* Mineral growth and dissolution are always controlled to some extent by kinetic factors, some of which are inherent to specific mineral characteristics, some of which are related to the interplay of temperature, solution chemistry, redox conditions, physical conditions, and biological factors;

4 *Biological factors.* The scale of biological control on mineral growth ranges from macroscopic to microscopic. Plant roots exude chelating compounds that extract K^+ from illite, thus enhancing its transformation to smectite, and the microorganism *thiobacillus ferrooxidans* enhances pyrite dissolution in acidic systems by fostering sulfide oxidation;

5 *Physical factors.* The porosity and permeability of sediment, soil, or rock control mineral growth by limiting (or enhancing) flow of reactants and products. For example, mineral reactivity can be limited in clays and shales where transport of reactants to the mineral surface is limited by low permeability; and

6 *Solution chemistry* – Calcite will not form in the presence of solutions that lack dissolved calcium and carbonate (or bicarbonate), nor will calcite form in acidic systems.

Mineral growth or dissolution is ultimately controlled by conditions at the interface of mineral surface and aqueous solution.

Advection is the movement of matter by physical flow in response to a force. A good example of advective transport is flow of water in a river channel that transports silt particles carried in suspension – the water and silt are moving by advection, flowing down gradient in response to gravity. Dissolved ions in the same river are also moved by advection, as are water vapor and dissolved CO_2 in moving air masses, or immiscible organic liquid contaminants (or ions or particles) in migrating ground water.

In terms of mineral growth and dissolution, advection is important because it is a mechanism for migration of dissolved ions to and from the mineral surface, whether the aqueous solution is seeping through soil pores and carrying Ca^{2+} and HCO_3^- to a surface where calcite can precipitate, or whether it is water emanating from a ground water spring that delivers dissolved Fe^{2+} to the land surface, where it can oxidize to form iron hydroxide.

Examples of flow rates in various geological materials are presented in Table 2.6.

Examples of crystalline rocks include basalt, granite, gneiss, marble, crystalline (metamorphosed) limestone, slate, and schist. Overlap of K values of different grain sizes is the result of textural differences, mainly compaction. Well-compacted sand can be less permeable than loosely compacted silt, for example. Note that flow rates differ by many orders of magnitude as a function of grain size or rock type. Rates also vary by orders of magnitude even within a given sediment or rock type, again accounting for textural differences.

Darcy's Law, formulated by French engineer Henri Darcy in the mid-1850s, is useful in illustrating controls on flow rate (Q) in porous media:

$$Q = K \cdot A \cdot dh/_{dl} \qquad (2.14)$$

Table 2.6 Hydraulic conductivity of selected rocks and sediments (from Heath 1983).

Shale	10^{-8}–10^{-4}
Sandstone	10^{-4}–10^0
Porous Limestone	10^{-3}–10^0
Cavernous Limestone	10^1–10^4
Crystalline rocks (unfractured)	10^{-8}–10^{-6}
Crystalline rocks (fractured)	10^{-3}–10^{-1}
Clay	10^{-7}–10^{-3}
Silt	10^{-3}–10
Sand	1–10^2
Gravel	10^2–10^4

Values are in units of meters per day ($m\,d^{-1}$).

Focus Box 2.20

Formation (or Dissolution) of Minerals in Three Steps

Mineral crystallization can be viewed as a three-step process involving:
1. movement of reactants (ions) to surface of crystal growth by advection or diffusion;
2. reaction on the mineral surface, involving, sequentially, attachment/adsorption of ions to surface, diffusion of ions to bonding sites, and formation of chemical bonds with reaction surface; and
3. desorption and diffusion or advection of reaction products away from reaction surface.

Mineral dissolution can be conceptualized in the opposite direction, involving:

1. advection or diffusion of species such as H^+ or H_2O to mineral surface;
2. attachment of H^+ or H_2O to cations (and breaking of cation–mineral chemical bond), and
3. diffusion or advection of complexed cations away from mineral surface.

Chemical reactions that foster mineral growth and dissolution clearly require movement of aqueous species (e.g. cations and anions), and the main processes governing ion transport are advection and diffusion.

The variable K is hydraulic conductivity (commonly in $m\,s^{-1}$ or $m\,day^{-1}$), a value inherent to the soil, sediment, or rock in question (Table 2.6), the variable A is the cross-sectional area under consideration (commonly in m^2); and dh/dl is the gradient. When combined, the terms produce units of $m^3\,s^{-1}$ or $m^3\,day^{-1}$. This equation and its derivatives are commonly applied to predicting flow rates and, by extrapolation, advective transport in aquifers and other *saturated* media. It is important to note that Darcy's Law does not apply to the type of unsaturated flow that occurs in systems such as soils and sediments above the water table.

Often when analyzing advection pertaining to chemical reactions, rate of advective flow is a useful parameter. Rate of flow (V) in saturated soils and ground waters can be determined by the following equation, where n is porosity; n = void space/total volume expressed as a decimal (after Heath 1983):

$$V = (K/n) \cdot {}^{dh}/_{dl} \qquad (2.15)$$

In saturated *sand* with $K = 50\,m\,day^{-1}$, porosity $= 0.20$ (20%), and a gradient of 1.0 m per 1000 m ($^{dh}/_{dl} = 10^{-3}$), advective flow rate is:

$$V = (50\,m\,d^{-1}/0.20) \times 10^{-3} = 0.25\,m/d$$

In saturated *clay* with $K = 5 \times 10^{-5}\,m\,day^{-1}$, porosity $= 0.4$, and the same gradient used in the sand calculation ($^{dh}/_{dl} = 10^{-3}$), advective flow rate is:

$$V = (5 \times 10^{-5}\,m\,d^{-1}/0.40) \times 10^{-3} = 1.25 \times 10^{-7}\,m\,d^{-1}$$

When ions in solution experiencing transport by advection are being removed from solution by mineral growth

(or adsorption), this type of transport is termed *nonconservative*. The same is true for ions added to solution by mineral dissolution, or degassing of CO_2 or O_2 from aqueous systems – if concentrations are altered by addition or removal of aqueous species, their behavior is considered nonconservative. Conversely, when ions or other species in solution are not removed or added by reactions involving minerals or other processes, the behavior of the components is termed *conservative*. Soluble species that tend to remain in solution and not participate in exchange reactions or mineral dissolution–precipitation reactions are conservative species. Often Na^+ and Cl^- are conservative, whereas Al^{3+} and PO_4^{3-} are not.

Diffusion is the migration of ions or particles from areas of high concentration to areas of low concentration. It is a spontaneous process in response to the tendency of systems to progress from ordered to more disordered, i.e. in the direction of greater entropy. Diffusion differs from advection in two main ways. The first is that movement is in response to a chemical gradient, rather than to a physical force. The second is that it can occur in the absence of movement of the solution that contains the ion or particle undergoing diffusion. Ions can diffuse in completely stationary water bodies, or even in solids.

In some cases, diffusion and advection occur simultaneously – consider advection of $H_2AsO_4^-$-poor ground water through a sandy layer in contact with a clay layer containing $H_2AsO_4^-$-rich pore water. Because the clay is virtually impermeable, water effectively does not migrate across the sand–clay boundary, but $H_2AsO_4^-$ ions will diffuse into the sandy layer water across a diffusion gradient. The diffusion of ions raises an important point because charge balance must be maintained, either by exchange of ions of the same charge (i.e. one cation for another cation,

Early

Later

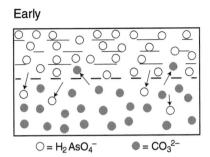
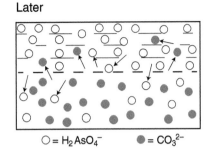

○ = $H_2AsO_4^-$ ● = CO_3^{2-} ○ = $H_2AsO_4^-$ ● = CO_3^{2-}

Fig. 2.30 Schematic representation of diffusion across the semi-permeable boundary between a clay-rich layer (upper) and sandy layer (lower). The pronounced difference in concentration of the two species across the boundary drives diffusion over time, from "Early" to "Later" (this system has not reached dynamic equilibrium). Loss of arsenate from the upper layer could drive further dissolution or desorption of the As-bearing minerals. According to Fick's Law, as time passes in this example, rate of diffusion will decrease as the chemical gradient decreases. Also, note charge balance, where two monovalent arsenate ions exchange for each divalent carbonate ion.

or one anion for another anion), or by paired migration of charge-balanced cation–anion pairs. One possible mechanism for maintaining charge balance in the sand–clay diffusion example above is that the water in the sand layer may be quite dilute compared to, e.g. Na^+–$H_2AsO_4^-$-rich pore water in the clay, in which case Na^+ and $H_2AsO_4^-$ would diffuse out of the clay into the sand, satisfying two constraints: (i) spontaneous diffusion in response to a concentration gradient, and (ii) maintenance of electrical neutrality of ionic charge in solution.

If the groundwater in the sand layer is CO_3^{2-}-rich compared to the clay pore water, CO_3^{2-} will likely diffuse into the clay layer and $H_2AsO_4^-$ into the sand layer to satisfy gradients and balance charge (Figure 2.30).

The 1850s apparently was a time of great thinking about flux, because in addition to the work of Henry Darcy on advection, the German physicist Adolf Fick formulated the following equation to represent the net movement of a diffusing particle proportional to the concentration gradient:

$$F = -D \frac{dc}{dx} \quad (2.16)$$

Fick's Law states that the rate of movement per unit cross-sectional area (i.e. the flux, F) toward lower an area of lower concentration is controlled by the difference in concentration ($\frac{dc}{dx}$) between two points – the greater the difference, the greater the rate of diffusion. The concentration gradient ($\frac{dc}{dx}$) is expressed in units of $mol\,cm^{-4}$, which is arrived at by dividing difference in concentration (dc, which is expressed in $mol\,cm^{-3}$) by distance (dx, in cm). Dimensional analysis tells us that units of F are $mol\,cm^{-2}\cdot sec^{-1}$ as follows:

$$-D\,(cm^2\cdot sec^{-1})\frac{dc}{dx}\,(mol\cdot cm^{-4}) = F\,(mol\cdot cm^{-2}\cdot sec^{-1}) \quad (2.17)$$

F is the flux of a particle or ion (molar concentration) through a cross-sectional area (expressed in cm^2) per second. The negative sign on the right side of Eq. (2.16) reflects flux in the opposite direction of the concentration gradient, that is, toward areas of lower concentration.

If there is no difference in concentration, there will be no net transfer. Motion will continue to occur, but migration of ions or molecules from point A to point B will equal migration from B to A, the classic case of a system in a state of dynamic equilibrium. The term D is the diffusion coefficient, values of which depend on the ion or molecule in question. In aqueous solutions, D is typically on the order of $10^{-5}\,cm^2\,sec^{-1}$; in solids, D is on the order of $10^{-10}\,cm^2\,sec^{-1}$. Ions with higher charge (absolute value) tend to diffuse more slowly than similar ions with higher charge (e.g. K^+ vs. Al^{3+}; $H_2PO_4^-$ vs. PO_4^{3-}).

The process whereby finer-grained crystals dissolve and the dissolution products recrystallize on the surfaces of larger, compositionally similar crystals is known as *Ostwald ripening*. The greater surface area : volume ratio of finer-grained crystals makes them less stable than coarser crystals. Ostwald ripening tends to occur as systems progress from relatively disordered to more ordered, a common example being the transformation over time from the disordered iron hydroxide ferrihydrite ($\sim 5Fe_2O_3\cdot 9H_2O$) to more-ordered goethite ($Fe_2O_3\cdot H_2O = 2FeOOH$) and eventually hematite (Fe_2O_3). The disordered, kinetically favored ferrihydrite tends to sorb trace elements (e.g. arsenate, phosphate, trace metals) to a much greater extent than goethite or hematite; thus, in this case the result of Ostwald ripening of iron oxyhydroxides from ferrihydrite → goethite → hematite may result in release of potential contaminants to soil water or groundwater.

2.8 BIOMINERALIZATION

In low-temperature systems like soils and groundwaters, mineral reactions, whether precipitation or dissolution, are often strongly controlled by organisms. Biomineralization is the term for mineral growth aided by organisms. It has been known for well over a century that marine organisms such as bivalves, corals, gastropods, microscopic coccoliths, and many other organisms remove dissolved calcite and carbonate from sea water to form shells or exoskeletons – this is the origin of much of the calcite in limestone. More recently research has focused on the role of microbes in biomineralization and this is the focus of this section.

The most-abundant biominerals are carbonates followed by phosphates, iron hydroxides and iron sulfides, sulfates (e.g. gypsum), and various other less-common groups. Amorphous silica ($SiO_2 \cdot nH_2O$) also forms in the tests of diatoms, but the amorphous structure means that it is technically not a mineral; in reality, biominerals are often disordered and lack the crystal morphology (Figure 2.31a) that tends to characterize minerals formed by more "typical" inorganic processes (Figure 2.31b).

Biomineralization is typically divided into two main types, *biologically induced mineralization (BIM)* and *biologically controlled mineralization (BCM)*. BIM describes mineralization that is mediated by an organism but occurs outside of the organism; BCM, on the other hand, occurs when mineral growth takes place within the organism (e.g. within vesicles inside cells, as in many marine organisms that foster calcite growth).

Given that BIM occurs outside of organisms, it is more similar to inorganic mineral growth, and the main mechanism by which organisms foster mineral growth is by locally causing supersaturation of elements. Effectively, the organisms produce elevated concentrations of mineral-forming elements in solution (e.g. immediately outside of a cell of a bacterium) to the extent that chemical bonds form and crystal lattices (or disordered solids) begin to take shape. Organisms have the potential to alter pH (e.g. by releasing H^+ or OH^-) and to alter redox conditions (e.g. by exuding O_2 into pore water as a product of photosynthesis). They also have the potential to release to solution constituents of minerals (e.g. cations or CO_2), and to provide a surface (e.g. cell wall or biofilms exuded by organism) upon which nucleation can occur. This latter attribute is important because often the lack of a condensation nucleus will limit crystallization, so cell surfaces or exudates can provide the starting point for crystal growth.

Two sources of much more information and case studies of biomineralization are the *Reviews in Mineralogy and Geochemistry* issue on this topic edited by Dove et al. (2003) and the issue of *Elements* magazine devoted to this topic and edited by Lu and Dong (2012). The latter of these two focuses more explicitly on environmentally related topics.

QUESTIONS

2.1 There is evidence that uranium can substitute into the structure of gypsum as well as apatite. In what oxidation state, and for which element, would U likely substitute? Again, justify your response using data and pertinent rules from Pauling and Goldschmidt.

2.2 Diagram schematic structures of olivine, a trioctahedral mica (2:1 phyllosilicate) and a pyroxene. For each, indicate points or planes where water can attack the crystal and cause dissolution. Why are these points prone to attack? To help you get started, compare the bond types between Si–O and (larger cations)-O (e.g. Fe–O, Mg–O).

2.3 Determine the CEC for two different soils:
   A. Soil that contains 25% quartz, 20% Fe(OH)$_3$, 50% kaolinite, and 5% humus (organic material).
   B. Soil that contains 35% quartz, 10% illite, and 55% smectite.

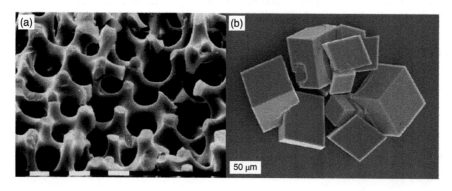

Fig. 2.31 Biogenic calcite (left, formed by an echinoderm) and inorganically precipitated calcite (right). Source: From Dove et al. (2003), used with permission of the Mineralogical Society of America.

Focus Box 2.21

An Example of Bacteria, Redox, and BIM

The bacteria *Galljanella* and *Leptothrix* are capable of altering redox conditions in a way that results in crystallization of ferrihydrite even in reduced/anaerobic sediments. In reducing/anaerobic sediments, iron oxidation is limited by lack of available O_2, and at first consideration, we might not expect an oxidized phase like iron hydroxide to form in a reducing environment. In order to form ferrihydrite, some process would need to foster oxidation of iron from ferrous to ferric ($Fe^{2+} \rightarrow Fe^{3+} + e^-$), and this is where these species of bacteria play a role. The electron released by iron oxidation is an important energy source at the microbial level, so the bacteria promote iron oxidation to obtain a source of energy. The ferric iron that is produced by microbially mediated oxidation results in ferrihydrite formation, which

can be represented as:

$$10\,Fe^{3+} + 24\,H_2O \rightarrow 5\,Fe_2O_3 \cdot 9H_2O + 30\,H^+ \quad (2.18)$$

According to Le Chatelier's principle, increasing a reactant such as Fe^{3+} (or O_2 or H_2O) will drive the reaction to the right, so effectively what these iron-oxidizing bacteria do is serve to increase concentrations of the reactant Fe^{3+}, forcing the system (at a small scale, i.e. adjacent to the cell) to supersaturation. The ferrihydrite forms around the perimeter of the cell and over time may evolve to more crystalline minerals such as goethite and hematite, perhaps by Ostwald ripening as described above.

C. Given that many plant nutrients exist as cations in solution, what implications does this have for the fertility of soil A vs. soil B?

2.4 List chemical elements that can substitute for Fe and Mg in the octahedral sheet of chlorite or smectite. Constrain yourself to rows 3 and 4 of the periodic table.

2.5 What is the overall charge on the 2 : 1 layer in a smectite with a 2 : 1 layer composition of $Al_{1.8}$ $Mg_{0.2}Si_{3.8}Al_{0.2}O_{10}(OH)_2$? What is the overall charge on the 1 : 1 layer of kaolinite with a composition of $Al_2Si_2O_5(OH)_4$? What implications does this have for the CEC of each ($\leq$3 sent.)?

2.6 What is the CN of a cation in octahedral coordination? Why is it termed octahedral coordination?

2.7 List three elements and their oxidation state(s) that could potentially substitute for Si in tetrahedral coordination. Justify your response using data and pertinent rules from Pauling and Goldschmidt.

2.8 Determine the structural (unit cell) formula for a smectite with the following composition (determined by TEM-AEM analysis, Ryan and Huertas 2009). [For a hint, a structural formula for muscovite, e.g. is $KAl_2(Si_3Al)O_{10}(OH)_2$]

SiO_2	**56.04**
Al_2O_3	25.50
Fe_2O_3	11.15
MgO	4.48
CaO	2.80
Na_2O	< 0.12
K_2O	< 0.12

2.9 Describe the main differences between BIM and BCM and explain three ways by which organisms can enhance mineral growth.

2.10 Express the composition of calcite in terms of wt % CaO and CO_2, and also in terms of $mg\,kg^{-1}$ and $mol\,kg^{-1}$ of C, O, and Ca.

2.11 Will a mineral dissolve more rapidly, or more slowly, if products of the dissolution reaction diffuse away from the mineral surface after dissolution? Provide an example of a mineral dissolution reaction as reference and use equations from both Le Châtelier's Principle and Fick's Law to justify your response.

REFERENCES

April, R., Hluchy, M., and Newton, R. (1986). The nature of vermiculite in Adirondack soils and till. *Clays and Clay Minerals* 34: 549–556.

Brindley, G.W. and Maksimovic, Z. (1974). The nature and nomenclature of hydrous nickel-containing silicates. *Clay Minerals* 10: 271–277.

Christidis, G.E. and Mitsis, I. (2006). A new Ni-rich stevensite from the ophiolite complex of Othrys, central Greece. *Clays and Clay Minerals* 54: 653–666.

Dähn, R., Scheidegger, A.M., Manceau, A. et al. (2002). Neoformation of Ni phyllosilicate upon Ni uptake on montmorillonite: a kinetics study by powder and polarized extended X-ray absorption fine structure spectroscopy. *Geochimica et Cosmochimica Acta* 66: 2335–2347.

Dohrmann, R. and Kaufhold, S. (2009). Three new, quick CEC methods for determining the amounts of exchangeable calcium cations in calcareous clays. *Clays and Clay Minerals* 57: 338–352. https://doi.org/10.1346/CCMN.2009.0570306.

Dove, P.M., De Yoreo, J.J., and Weiner, S. (eds.) (2003). *Biomineralization*, Reviews in Mineralogy and Geochemistry, vol. 54. Mineralogical Society of America.

Drever, J.I. (1997). *The Geochemistry of Natural Water Surface and Groundwater Environments*, thirde. Upper Saddle River, New Jersey, USA: Prentice-Hall.

Faure, G. (1998). *Principles and Applications of Geochemistry*, seconde. Upper Saddle River, New Jersey, USA: Prentice-Hall.

Gaudin, A., Petit, S., Rose, J. et al. (2004). Accurate crystal chemistry of ferric smectites from the lateritic nickel ore of Murrin Murrin (Western Australia); I, XRD and multi-scale chemical approaches. *Clay Minerals* 39: 301–315.

Hattori, K., Takahashi, Y., Guillot, S., and Johanson, B. (2005). Occurrence of arsenic (V) in forearc mantle serpentinites based on X-ray absorption spectroscopy study. *Geochimica et Cosmochimica Acta* 69: 5585–5596.

Hay, R.L., Hughes, R.E., Kyser, T.K. et al. (1995). Magnesium-rich clays of the Meerschaum mines in the Amboseli Basin, Tanzania and Kenya. *Clays and Clay Minerals* 43: 455–466.

Heath, R.C. (1983). *Basic Ground-Water Hydrology*. Reston, Virginia, USA: United States Geological Survey, Water Supply Paper 2220.

Hillier, S. (1994). Pore-lining chlorites in siliciclastic reservoir sandstones: Electron microscope, SEM, and XRD data, and implications for their origin. *Clay Minerals* 29: 665–679.

Klein, C. and Hurlbut, C.S. (1999). *Manual of Mineralogy*: (after James D. Dana), 681. New York: Wiley.

Langmuir, D. (1997). *Aqueous Environmental Geochemistry*. Upper Saddle River, New Jersey, USA: Prentice-Hall.

Lu, A. and Dong, H. (eds.) (2012). *Minerals, Microbes, and Remediation*, Elements: An International Magazine of Mineralogy, Geochemistry and Petrology, vol. 8.

Manceau, A., Marcus, M.A., Tamura, N. et al. (2004). Natural speciation of Zn at the micrometer scale in a clayey soil using X-ray fluorescence, absorption, and diffraction. *Geochimica et Cosmochimica Acta* 68: 2467–2483.

McBride, M.B. (1994). *Environmental Chemistry of Soils*. New York: Oxford University Press.

Moore, D. and Reynolds, R.C. Jr. (1997). *X-Ray Diffraction and the Identification and Analysis of Clay Minerals*, 2e. New York: Oxford University Press.

Newman, A.C.D. and Brown, G. (1987). The chemical constitution of clays. In: *The Chemistry of Clays and Clay Minerals* (ed. A.C.D. Newman), 1–128. London: Mineralogical Society.

Oze, C., Fendorf, S.E., Bird, D.K., and Coleman, R. (2004). Chromium geochemistry in serpentinized ultramafic rocks and serpentine soils in the Franciscan complex of California. *American Journal of Science* 304: 67–101.

Pauling, L. (1960). *The Nature of the Chemical Bond - An Introduction to Modern Structural Chemistry*, seconde. Ithaca, New York: Cornell University Press.

Pincus, L.N., Ryan, P.C., Huertas, F.J., and Alvarado, G.E. (2017). The influence of soil age and regional climate on clay mineralogy and cation exchange capacity of moist tropical soils: a case study from Late Quaternary chronosequences in Costa Rica. *Geoderma* 308: 130–148.

Ryan, P.C. and Huertas, F.J. (2009). The temporal evolution of pedogenic Fe-smectite to Fe-kaolin via interstratified kaolin-smectite in a moist tropical soil chronosequence. *Geoderma* 151: 1–15.

Ryan, P.C., Kim, J., Wall, A.J. et al. (2011). Ultramafic-derived arsenic in a fractured bedrock aquifer. *Applied Geochemistry* 26: 444–457.

Ryan, P.C. and Reynolds, R.C. Jr. (1996). The origin and diagenesis of grain-coating serpentine/chlorite in Tuscaloosa Formation sandstone, U.S. Gulf Coast. *American Mineralogist* 81: 213–225.

Shannon, R.D. (1976). Revised effective ionic radii and systematic studies of interatomic distances in halides and chalcogenides. *Acta Crystallographica* A32: 751–767.

Sparks, D.L. (1995). *Environmental Soil Chemistry*. San Diego, USA: Academic Press.

Thompson, G.R. and Hower, J. (1975). The mineralogy of glauconite. *Clays and Clay Minerals* 23: 289–300.

Van Gosen, B.S., Lowers, H.A., Bush, A.L., et al. (2002). Reconnaissance study of the geology of U.S. vermiculite deposits - are asbestos minerals common constituents? *U.S. Geological Survey Bulletin 2192*, accessed 24-October-2018 at https://pubs.usgs.gov/bul/b2192/b2192.pdf.

Whittaker, E.J.W. and Muntus, R. (1970). Ionic radii for use in geochemistry. *Geochimica et Cosmochimica Acta* 34: 945–956.

Yada, K. (1971). Study of microstructure of chrysotile asbestos by high-resolution electron microscopy. *Acta Crystallographica* A27: 659–664.

3

Organic Compounds in the Environment

Two main areas of concern provide the impetus for understanding the role of organic biogeochemistry in environmental science. The first is the fate, transport, and degradation of synthetic organic contaminants in soils, surface waters, aquifers, and the atmosphere. These synthetic compounds are typically petroleum-based fuels, solvents, pesticides, industrial chemicals, and pharmaceuticals. The second involves origin and behavior in the environment of a wide array of natural organic compounds including (but not limited to) humic and fulvic substances; methane and other nonindustrial hydrocarbons (from deep crustal and mantle origins to atmospheric reactions); dimethyl sulfide and other natural marine carbon-sulfur compounds; amino acids in sedimentary rocks; and charcoal produced during forest fires. It is an interdisciplinary field including contributions from geology, soil science, biogeochemistry, environmental geochemistry, hydrogeology, limnology, oceanography, and atmospheric chemistry, among others. At its most fundamental level, organic biogeochemistry is concerned with carbon-based compounds studied in organic chemistry classes, so the first part of this chapter focuses on selected concepts from the field of organic chemistry, including the structures and compositions of some compounds that are environmental contaminants.

3.1 INTRODUCTION TO ORGANIC CHEMISTRY: CHAINS AND RINGS, SINGLE, DOUBLE, AND TRIPLE BONDS, FUNCTIONAL GROUPS, CLASSES OF ORGANIC COMPOUNDS, ORGANIC NOMENCLATURE

3.1.1 Definition of organic compounds

Chains or rings of carbon atoms are the fundamental units of organic compounds. Hydrogen is the element that most commonly bonds with the C chains or rings, but it is not uncommon for organic compounds to also contain, nitrogen, oxygen, sulfur, chlorine, phosphorus, and less commonly silicon, iron, trace metals, or other elements. For reasons based on historical classification and their often-inorganic origins, there are a few carbon-based compounds that are considered "inorganic carbon," including the carbonate minerals (e.g. calcite, dolomite), bicarbonate and carbonate anions (HCO_3^- and CO_3^{2-}), carbon dioxide, and the carbon polymorphs diamond and graphite – they are both pure carbon (allowing for trace impurities) but they possess different atomic structures. Diamonds occur in kimberlite, a rare igneous rock that has its origins in the upper mantle, so its origins are inorganic.

Environmental and Low-Temperature Geochemistry, Second Edition. Peter Ryan.
© 2020 John Wiley & Sons Ltd. Published 2020 by John Wiley & Sons Ltd.

Graphite in metamorphic rocks is produced by burial, compaction, and heating of algae and other organisms that existed millions to billions of years ago. Although the origin of the carbon in graphite is often anaerobic decay of organisms (clearly organic), the mineral is considered inorganic because of its metamorphic origin.

3.1.2 Hybridization of carbon atoms in organic compounds

Carbon atoms have four valence electrons, which can be configured in three different geometric arrangements known as sp^3, sp^2, and sp hybridizations. Sp^3-hybridized carbon atoms can form four single bonds known as σ (sigma) bonds, and to minimize repulsion, the electrons spread themselves as far apart as possible, meaning they are tetrahedrally coordinated around the nucleus (Figure 3.1). Sp^2 carbon forms a double bond comprised of σ and π bonds (using two of carbon's four valence electrons) as well as two single (σ) bonds, each using one valence electron. The resulting structure of the sp^2-hybridized atom is trigonal planar (triangular). Carbon atoms can also occur in sp hybridizations that form linear arrangements and feature a σ bond on one side and a triple bond (comprised of one σ bond and two π bonds and making use of three valence electrons) on the other. Each of these three C atom configurations is illustrated in Figure 3.1 using Lewis electron dot diagrams and simplified Lewis structure diagrams featuring single lines for single bonds, double lines for double bonds, and triple lines for triple bonds.

Examples of simple organic compounds, each featuring two C hybridized atoms of sp^3, sp^2, or sp arrangements, are also diagrammed.

Ethane is comprised of two sp^3 atoms joined by a single bond. Note in the Lewis diagrams (Figure 3.1) that each of the two sp^3 atoms has three unpaired electrons that can be satisfied by bonding to the single valence electron in three H atoms, producing a chemical formula of C_2H_6. Ethene (commonly known as "ethylene") is comprised of two sp^2 atoms joined by a double bond; the Lewis diagrams depict two unpaired electrons in each C atom that are free to bond with two H atoms to make ethene, producing the chemical formula C_2H_4. Finally, ethyne (commonly known as "acetylene," the fuel for welding torches) consists of two sp-hybridized atoms joined by a triple bond, and in the Lewis diagrams, you will see that each C atom has only one unpaired electron available to bond to H, producing the ethyne chemical formula C_2H_2.

Hydrocarbons like alkanes with only sp^3 orbital arrangement and single bonds are sometimes referred to as *saturated* hydrocarbons. Alkenes, which contain at least one double bond, are considered *unsaturated*, as are alkynes. Here the term saturation refers to the capacity for bonding to H atoms – ethane is saturated with respect to H atoms but the electrons in ethene could be rearranged to accommodate two more H atoms (in effect transforming ethene to ethane). The concept of saturation becomes important when considering fate and transport because it affects natural degradation of organic pollutants in soils, groundwater, and other environmental systems. Saturated organic compounds tend to degrade more slowly, meaning that their toxic effects can persist for longer in the natural environment.

3.1.3 Alkanes

This section presents a systematic progression through some common organic compounds in the group of hydrocarbons (i.e. those compounds that contain only C and H) known as *alkanes*, a relatively simple group of hydrocarbons in terms of composition and structure. Alkanes, also sometimes known as paraffins, are comprised of a single chain of sp^3-hybridized carbon atoms bonded to each other and to H atoms by single (σ) bonds; there are no sp^2- or sp-hybridized C atoms.

Ethane, with its short single chain of two sp^3-hybridized C atoms bonded to six H atoms, is a simple example of an alkane. Its *structural formula*, which is a graphical representation of structure and composition, is shown in Figure 3.1, as is its *chemical formula* (or molecular

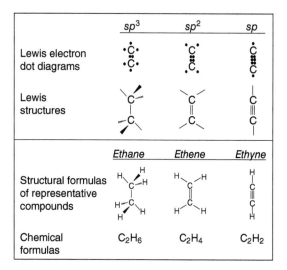

Fig. 3.1 Hybridization of carbon atoms, with ethane, ethene (common name: ethylene), and ethyne (common name: acetylene) shown as examples of compounds containing the different hybridizations.

H H H H H H H H
H-C-C-C-C-C-C-C-C-H
H H H H H H H H

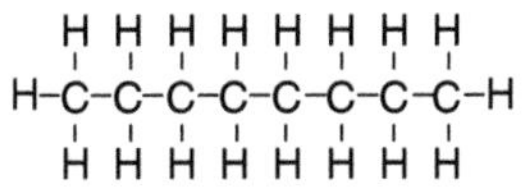

Fig. 3.2 Structural formula of octane.

formula), C_2H_6. *Octane*, a common word in the transportation lexicon, is another good example of an alkane. Its chemical formula is C_8H_{18} and it consists of eight linearly arranged sp^3-hybridized C atoms bonded to 18 carbon atoms, as illustrated in Figure 3.2. Note that the example in Figure 3.2 presents octane in simplified Lewis structure form, which does not depict the kinks and bends in the structure imparted by the tetrahedral nature of the sp^3-hybridized C atoms.

C_nH_{2n+2} is the general chemical formula for the alkane group, where n represents the number of C atoms in the chemical formula; the alkane with 15 C atoms, for example, will have 32 H atoms and have the chemical formula $C_{15}H_{32}$. Table 3.1 presents some common alkanes and makes clear that alkanes are primarily used as fuels. The prefixes presented in this table apply to all forms of hydrocarbons, particularly alkenes and cyclic hydrocarbons in addition to alkanes.

The physical properties of alkanes are strongly controlled by the length of the hydrocarbon chain. For example, melting and boiling points both increase as a function of the number of C atoms, with the exception of the slight irregularities in the melting point curve, which occurs because even-numbered alkanes pack more closely in the solid phase, and thus more energy is needed to break bonds in the solid → liquid transition.

Figure 3.3 allows us to make a few interesting observations, including (i) alkanes with C ≤ 3 occur in the gaseous

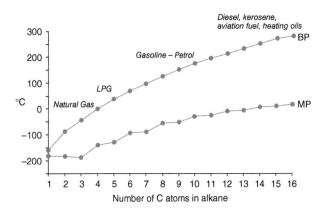

Fig. 3.3 Melting point (MP) and boiling point (BP) both increase with increasing length of alkane chain. LPG = liquid petroleum gas. Methane would plot as "1" on the x-axis, octane would plot as "8," etc.

Table 3.1 Prefixes used in IUPAC nomenclature applied to common alkanes.

No. of C atoms	Prefix	Name	Chem. formula	Common use
1	Meth-	Methane	CH_4	Natural gas
2	Eth-	Ethane	C_2H_6	Natural gas
3	Prop-	Propane	C_3H_8	Liquefied petroleum gas (LPG)
4	But-	Butane	C_4H_{10}	LPG – lighter fluid
5	Pent-	Pentane	C_5H_{12}	Solvent, fuel
6	Hex-	Hexane	C_6H_{14}	Solvent, gasoline
7	Hept-	Heptane	C_7H_{16}	Gasoline/petrol
8	Oct-	Octane	C_8H_{18}	Gasoline/petrol
9	Non-	Nonane	C_9H_{20}	Gasoline/petrol
10	Dec-	Decane	$C_{10}H_{22}$	Gasoline/diesel
16	Hexadeca-	Hexadecane	$C_{16}H_{34}$	Diesel, jet fuel, heating oil

---

Focus Box 3.2

Halogenated Alkanes: The CFCs

When Cl_2 and F_2 gas are reacted with methane or ethane (e.g. in a chemical manufacturing plant), the product is chlorofluorocarbons (CFCs). These compounds are relatively nontoxic liquids historically used in air conditioners, refrigerators, aerosol cans, and foam plastic (e.g. fast-food boxes). Compared to their predecessors, they are chemically stable (e.g. they don't explode) and are chemically and structurally simple. Shown in Figure 3.4 are two representative examples, CFC-12 (Freon-12) and CFC-113 (Freon-113).

The problem with CFCs is their tendency to leak into the atmosphere as gases, where their resistance to decomposition gives them long residence times. In the

presence of ultraviolet radiation, chlorine atoms break apart from the CFC molecule, where they become catalysts in feedbacks loops that destroy stratospheric ozone (O_3) (Section 8.10).

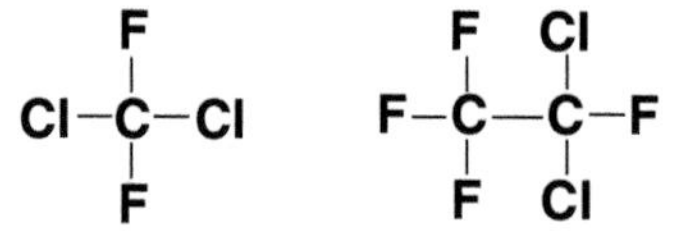

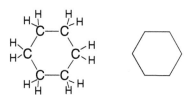

Fig. 3.4 Structural formulas for two common chlorofluorocarbons.

---

state at effectively all earth-surface conditions; (ii) butane and pentane have boiling points in the range of typical air temperatures, making them transitional between gas and liquid (hence the need to compress them to produce liquid petroleum gas [LPG]); and (iii) components of diesel and jet fuel (approximately from C = 12 through 16) will solidify at temperatures of around 0 °C. While this can create problems with these types of "heavier" fuels in cold climates, this information is useful when it comes to the refining of crude petroleum (crude oil), which comes out of reservoir rocks possessing a wide array of alkanes of various length as well as alkenes, aromatic hydrocarbons, and other compounds. The basic principle of petroleum refining involves first heating crude oil and then cooling it; as it cools, and with the use of a complex piping system, progressively lighter (smaller molecules) compounds condense into liquid (Figure 3.3) and are separated from the remaining crude. The first compounds to be removed are the tars and heaviest oils, progressing through lighter and lighter compounds, ultimately leaving only methane and ethane.

Alkanes that consist of a hydrocarbon ring rather than a single chain are *cycloalkanes*. Also sometimes known as napthalenes, cycloalkanes occur in crude petroleum and have the general formula C_nH_{2n}. Their nomenclature is similar to alkanes but with the prefix "cyclo-," so the six-C cycloalkane diagrammed is known as cyclohexane (C_6H_{12}) (Figure 3.5).

Note that the abbreviated symbol for cyclohexane is a hexagon – similarly, cyclopentane would be represented by a pentagon and cyclooctane by an octagon.

Fig. 3.5 Structural formula for cyclohexane (left) and the hexagon which is the commonly used symbol for cyclohexane.

3.1.4 Alkenes

When a hydrocarbon chain contains at least one double bond it is termed an alkene. The common alkene known as ethene is commonly referred to as *ethylene* in environmental geochemistry. Similarly, propene, diagrammed below with ethylene, is commonly known as *propylene*. Industrial reaction of ethylene with Cl_2 gas produces trichloroethylene (TCE), a solvent used in cleaning circuit boards and in dry cleaning, among other uses (IUPAC term = trichloroethene) It is a common groundwater contaminant and probable carcinogen.

Alkenes are derived from crude oil and by transformation from alkanes in petrochemical plants – they are used as reactants in the production of many petroleum-based products including solvents like TCE and perchloroethylene (PCE; C_2Cl_4, ethylene with four Cl atoms replacing the four H atoms in Figure 3.6) as well as polymers such as polystyrene.

Ethylene   *Trichloroethylene*   *Propylene*
C_2H_4 C_2HCl_3 C_3H_6

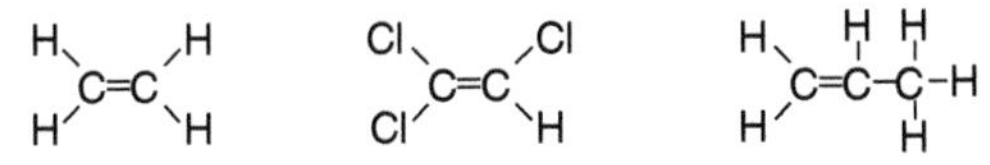

Fig. 3.6 Common ethenes (commonly referred to as ethylenes).

3.1.5 Functional groups

Small groups of atoms – or even single atoms in some cases – bonded to organic molecules are known as functional groups. They impart important characteristics to compounds, including solubility, toxicity, volatility, and reactivity. Examples of functional groups include hydroxyls, alkanes, alkenes, aromatic hydrocarbons, metals, halides, and many other components. Functional groups typically are comprised of atoms such as hydrogen, oxygen, nitrogen, chlorine, and sulfur, and carbon is often a constituent as well (e.g. the C atoms in a phenyl group).

Many functional groups (Table 3.2), particularly those containing oxygen, nitrogen, or halides (F, Cl, Br, I), can create polar molecules from compounds that are otherwise nonpolar (and hence insoluble in water) – this clearly

has industrial applications (e.g. increased effectiveness of solvents), but equally as importantly can have environmental consequences related to contaminant fate and transport. As a rule, organic compounds are insoluble in water (picture an oil slick on the ocean), but increased solubility produced by many functional groups means that organic compounds with O, N, and Cl (certain pesticides and solvents, for example) will be more soluble in groundwater and present greater challenges for remediation.

Functional groups appear again in the following sections, but for purposes of illustration, consider the effect of two functional groups on an ethane skeleton (Figure 3.7). If one H atom is replaced with a hydroxyl

Ethane   *Ethanol*   *Acetic Acid*
C_2H_6 C_2H_5OH CH_5COOH

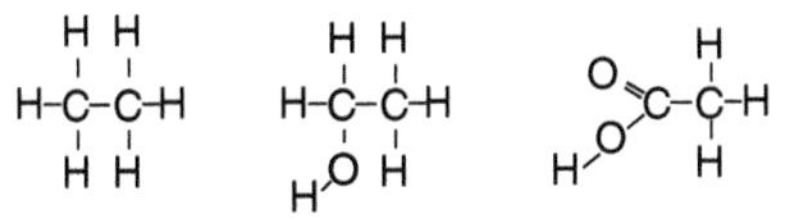

Fig. 3.7 Examples of two functional groups attached to an ethane chain. Ethanol contains a hydroxyl group (OH) in place of a hydrogen atom. Acetic acid contains a carboxylic acid functional group (COOH).

Table 3.2 Examples of some relatively common functional groups.

Group name	Formula	Structure	Prefix	Suffix	Example
Methyl	CH_3	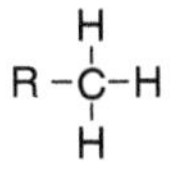	Methyl-	-methane	Methyl bromide
Ethyl	$C_2H_5^-$	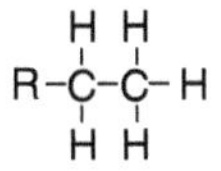	Ethyl-	-ethane	Ethylbenzene
Hydroxyl (alcohol)	OH^-	R–O–H	Hydroxy-	-ol	Ethanol
Carboxyl	COOH	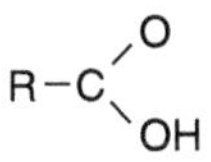	Carboxy-	-oic acid	Acetic acid
Phenyl	$C_6H_5^-$		Phen-	-benzene	Polychlorinated biphenyls (PCBs)
Chloro	Cl^-, Br^-, F^-, I		Phenyl- Chloro- Bromo- Fluoro- Iodo-	-chlorine -bromine -fluorine -iodine	Trichloroethylene (TCE)
Amine	NH_2	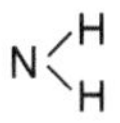	Amino-	-amine	N-nitrosamine
Sulfide	S^{2-}	–S–	—	-sulfide	Dimethylsulfide (DMS)

Focus Box 3.3

Naturally Occurring Organic Acids: Carboxylic and Amino Acids

Carboxylic acids such as formic acid (IUPAC = methanoic acid) and acetic acid (ethanoic acid) occur naturally (ants inject formic acid when they bite), and carboxylic groups are the source of the acidity derived from humic and fulvic acids in soils and surface waters (Table 3.3). Typical pKa values of carboxylic acids are 2–5, meaning that they are relatively weak acids similar in strength to carbonic acid. In natural waters where inorganic acids (e.g. nitric acid or sulfuric acid) are lacking or are low in abundance, carboxylic acids buffer aqueous systems at pH of approximately 4–5. Amino acids contain amine (NH_2) functional groups, and the simplest one is glycine, with a composition of NH_2CH_2COOH. In organisms, amino acids are cell-signaling molecules, regulators of gene expression, and precursors for synthesis of hormones and proteins; in soils, amino acids are a source of nitrogen for plants.

Table 3.3 Compositional ranges of some natural organic compounds (by wt%).

	Humin (%)	Humic acids (%)	Fulvic acids (%)
Carbon	>62	50–60	40–50
Hydrogen	<3	3–6	4–7
Oxygen	<30	30–40	40–50
Nitrogen	>5	2–5	1–4
Sulfur	<0.5	0.1–1	0.1–4

(OH^-) group, we end up with the compound known as *ethanol*, one that possesses very different properties than ethane. For example, unlike ethane, ethanol is highly soluble in water (the hydroxyl group makes ethanol a polar molecule); it is also less volatile and, in moderate doses, is a nontoxic component of beverages like wine, beer, and other alcohol-bearing drinks.

If a carbon atom in the ethane skeleton is part of a carboxyl group (COOH) rather than a methyl group (CH_3, as in ethane), the result is acetic acid (also known as ethanoic acid), a compound with markedly different properties than ethane. For example, the H atom bonded to the O atom can dissociate from the rest of the compound according to this reaction, which produces the acetate anion (often abbreviated OAc):

$$CH_3COOH \rightarrow CH_3COO^- + H^+ \qquad (3.1)$$

3.1.6 Aromatic hydrocarbons and related compounds

Cyclic hydrocarbons characterized by six C atoms bonded to each other by alternating σ and π bonds are aromatic hydrocarbons. The alternating bonds were once thought to be fixed in space (Figure 3.8), but it is now known that delocalized electrons that effectively produce alternation of single and double bonds form the bonds between carbon atoms. Effectively, the bonds between C atoms in aromatic hydrocarbons can be viewed as a hybrid of σ and π bonds.

A hexagon around a circle (Figure 3.8, example c) is the currently accepted symbol for the benzene ring. If no atoms are shown attached to the ring, it is assumed that the only atoms bonded to the carbon atoms are hydrogen atoms, as is shown in example d. All four examples in Figure 3.8 depict benzene (C_6H_6), with a, b, and d each depicting the compound at an instant in time.

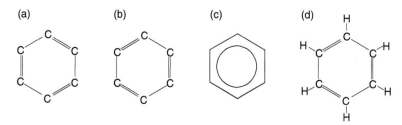

Fig. 3.8 Representations of benzene rings. Examples a and b indicate alternation of sigma and pi bonds. At one instant in time, the double bonds are located as is shown in example a; the next instant in time they switch to the configuration in example b. This rapid alternation characterizes the delocalized single and double bonds in aromatic hydrocarbons. The circle-and-hexagon symbol under c is the commonly used, pared-down structural formula for benzene, and a more detailed structural formula is presented in d.

Many types of functional groups can attach to the benzene ring to produce a variety of compounds, some of which are covered in the following section.

When polymerized (Figure 3.9), benzene rings produce the family of compounds known as polycyclic aromatic hydrocarbons, or polyaromatic hydrocarbons, or polynuclear aromatic hydrocarbons (abbreviated PAHs in any case). Naphthalene, a two-ring PAH, is a component of coal tar and occurs as a soil and aquifer contaminant near coal tar facilities. Benzo(a)pyrene, a five-ring PAH, is a product of incomplete combustion and a known indoor air contaminant, particularly in buildings with cigarette smoke or woodstove smoke.

The term "aromatic hydrocarbon" comes from the sweet odor that they emit, one that you have probably caught a whiff of at a gasoline filling station. However, not all aromatic hydrocarbons produce those distinctive aromas – the term originated in the nineteenth century – and technically what makes a compound "aromatic" is the cyclic ring structure with delocalized σ and π bonds.

The *BTEX* group of aromatic hydrocarbons – benzene, toluene, ethylbenzene, and xylene – are common environmental contaminants associated with petroleum spills or leaks. They have similar molecular mass to octane, so when crude petroleum is refined for market, BTEX ends up in gasoline. The BTEX compounds are much more soluble than octane (and other alkanes in gasoline) and can present significant groundwater contamination

problems in areas with leaking fuel storage tanks. *Toluene* (IUPAC = methylbenzene) contains a methyl group on a benzene ring, and *xylene* contains two methyl groups on a benzene ring. Xylene can have three different geometric arrangements, where o-xylene signifies the ortho geometry (IUPAC = 1,2-dimethylbenzene), m signifies meta geometry (1,3-dimethylbenzene) and p signifies para geometry (1,4-dimethylbenzene). *Ethylbenzene*, as the name strongly suggests, contains an ethyl group on a benzene ring (Figure 3.10).

Phenols (short "e," accent on second syllable) are a group of compounds derived from benzene rings by addition of at least once hydroxyl group in place of an H atom. If all five H atoms are replaced by Cl atoms, the result is a toxic liquid known as pentachlorophenol (and by now you should be starting to appreciate the systematic nomenclature for organic compounds). "Penta," as it is often known, is used as a wood preservative and is the target of many groundwater remediation efforts (Figure 3.11).

Two phenyl groups, each with one chlorine atom in place of a hydrogen atom, attached to a chlorinated ethane group, comprise the molecule known variably as dichlorodiphenyltrichloroethane, *DDT*, or (in IUPAC nomenclature) 4,4′-(2,2,2-trichloroethane-1,1-diyl)bis(chlorobenzene) (Figure 3.12).

The Swiss chemist P.H. Muller won the Nobel Prize in 1948 for his discovery of the usefulness of DDT as an insecticide, which resulted in great success at combating

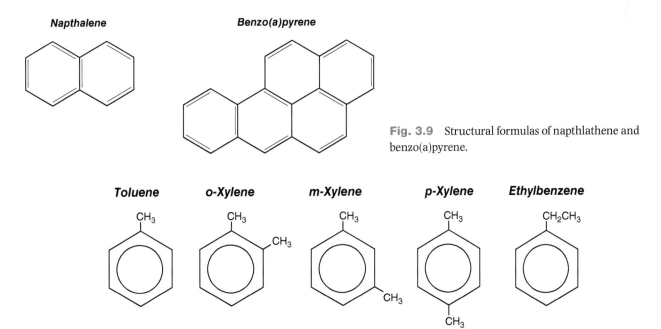

Fig. 3.9 Structural formulas of napthlathene and benzo(a)pyrene.

Fig. 3.10 Structural formulas for toluene, the three isomers of xylene, and ethylbenzene. Composition, abundance, and positioning of functional groups distinguish these compounds. The terms shown here are the common names used in the environmental field; IUPAC names are given in the text.

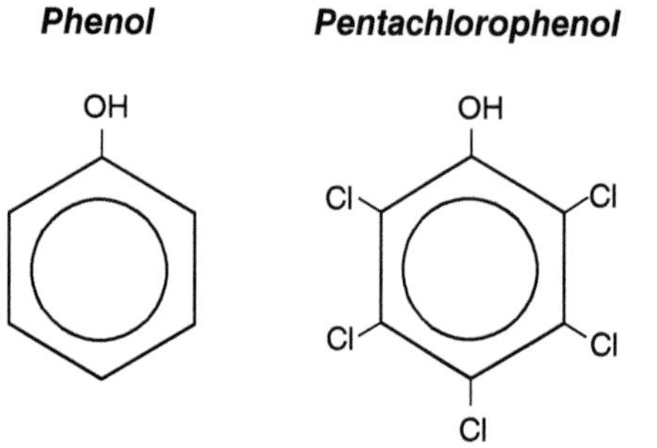

Phenol   **Pentachlorophenol**

Fig. 3.11 Structural formulas of phenol (C_6H_5OH) and pentachlorophenol (C_6Cl_5OH).

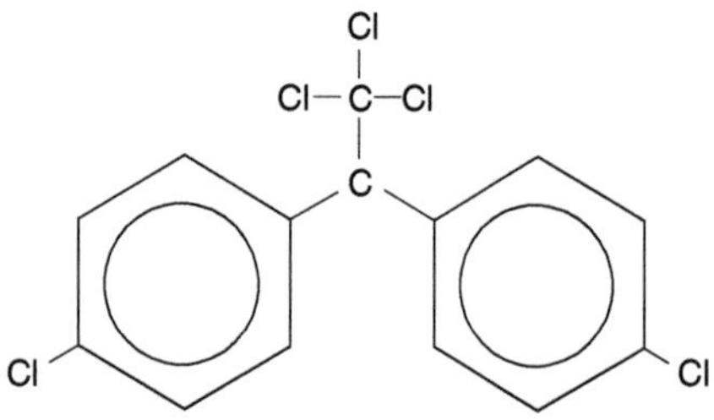

Fig. 3.12 Structural formula of the compound variably known as dichlorodiphenyltrichloroethane (*DDT*) and 4,4'-(2,2,2-trichloroethane-1,1-diyl)bis(chlorobenzene).

insect-borne diseases like malaria. However, DDT is a persistent organic pollutant that decomposes very slowly in the natural environment, and thus bioaccumulates and biomagnifies up the food chain. It was responsible for drastic declines in eagle populations in the 1960s and 1970s because of the way that it causes thinning

of egg shells, early hatching, and high mortality rates of chicks. Because of their slow decomposition rate, DDT and its metabolites dichlorodiphenyldichloroethylene (DDE) and dichlorodiphenyldichloroethane (DDD) remain pollutants in many aquatic ecosystems. In fact, DDE and DDD are more toxic and more resistant to decomposition than DDT.

Polychlorinated biphenyls (PCBs) were used extensively in the twentieth century because they are very stable liquids that are excellent electrical insulators. They are relatively simple chemically, containing two phenyl groups bonded across C atoms with variable amounts of Cl in place of H at any of the sites 2 through 6 and 2' through 6' on the benzene rings (phenyl groups) (Figure 3.13).

PCBs are carcinogenic, and although no longer used, their resistance to biodegradation means they are persistent organic pollutants that remain trapped in sediments in aquatic environments and also are transported atmospherically – they have been detected in regions far from manufacture and use, including the polar regions. The Great Lakes and the Hudson River in North America are well-known problem areas, and PCBs in fish, poultry, pork, and beef have caused scares in Europe and other sites globally.

Dioxins are a family of compounds produced as unintended by-products of pesticide manufacture, but they also occur naturally in trace amounts – humans have probably been producing dioxins ever since they could control fire, because combustion of wood or peat with trace amounts of natural Cl can lead to chlorinated dioxin production. For

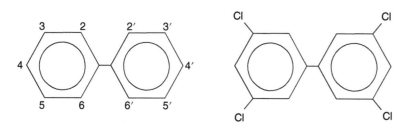

Fig. 3.13 Structural formula of the class of compounds known as polychlorinated biphenyls (PCBs). The numbered sites on the left correspond to potential bonding sites of functional groups (in PCBs, mainly H or Cl, sometimes OH). The example on the right is 3,3',5,5'-tetrachlorobiphenyl.

Focus Box 3.4

PCBs, Environmental Racism, and the Birth of the Environmental Justice Movement

In 1982, the state of North Carolina, USA, made the decision to locate a hazardous waste landfill in the predominately African-American community of *Warren County*. The landfill would contain PCB-contaminated soil collected from illegal dumping of toxic waste along roadways. The protests that resulted led to the arrests of > 500 protesters, including Dr. Benjamin F, Chavis, Jr. of the United Church of Christ, and Walter Fauntroy, at the time a member of the US House of Representatives. The Warren County protest did not prevent the development of the landfill, but it did provide the impetus for the 1987 study Toxic Waste and Race (by the United Church of Christ), which concluded that people of color are exposed at unjust rates to hazardous waste facilities based on where they live and work. The Warren County case is often cited as the beginning of the environmental justice movement.

example, dioxin has been discovered in bog sediments in Canada presumably derived from combustion well before the pre-industrial era (Silk et al. 1997). A similar occurrence has been noted in England, where the pre-industrial era dioxin is a different isomer than modern industrially associated versions (Green et al. 2001). Anthropogenic sources of dioxins include pulp mills that use chlorine to bleach paper – during this process, Cl from the bleaching agents combines with organic matter from wood pulp to produce dioxins, and for this reason there is a move to limit the amount of chlorine used and to capture dioxins and related compounds produced during this process. Polychlorinated dibenzo-*p*-dioxin (PCDD) is a common by-product of pesticide production, and similar to PCBs, chlorine atoms can occur at any of the sites numbered (in the case of dioxin) 1 through 4 and 6 through 9.

In the example in Figure 3.14, chlorine atoms occupy positions 2, 3, 7, and 8 and the compound is given the name 2,3,7,8-tetrachlorodibenzo-*p*-dioxin (TCDD), a common form of dioxin in the environment.

The herbicide *2,4-dichlorophenoxyacetic acid* (2,4-D) is a broadleaf-specific herbicide that usually does not kill grasses. For this reason, it is used to kill unwanted flowering and leafy plants, including noxious weeds, in grasslands and other areas. It is also an illustrative example of a synthetic organic compound because it contains numerous functional groups, including carboxyl, phenyl, and chlorine.

In the absence of dioxin, 2,4-D (Figure 3.15) has relatively low toxicity according to the USEPA, UKACP, and other organizations. It is relatively soluble due to the presence of the carboxyl, chlorine, and phenoxy groups and is prone to seepage to groundwater. Yet at the same time, its high solubility means that it is available to microbes and hence it decomposes relatively rapidly in well-aerated soils. Adding a third Cl atom to 2,4-D

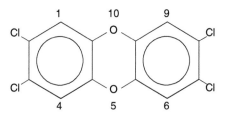

Fig. 3.14 2,3,7,8-tetrachlorodibenzo-*p*-dioxin (TCDD), a member of the dioxins. Sites labeled with 1, 4, 6, and 9 are occupied by H.

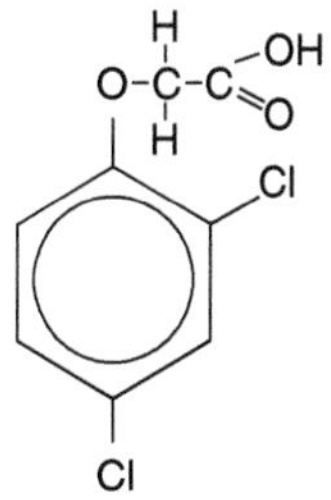

Fig. 3.15 Structural formula of 2,4-dichlorophenoxyacetic acid. Note the carboxyl (COOH) functional group in upper right of diagram and two Cl functional groups.

produces 2,4,5-trichloroethylene (2,4,5-T), the defoliant known as *Agent Orange*.

3.1.7 Nitrogen, phosphorus, and sulfur in organic compounds

Dimethyl sulfide (DMS) is a natural compound produced by organic decay in marine waters that plays a role in global climate. Released to the atmosphere as DMS gas, it oxidizes to produce sulfate (SO_4^{2-}) particles that scatter sunlight, limiting the amount of solar radiation that reaches the earth's surface.

Focus Box 3.5

Dioxins and Textbook Cases of Environmental Exposure to Toxins

The presence of dioxin in *Agent Orange* (2,4,5-T), the defoliant used to kill forest vegetation during the Vietnam–US war in the 1960s–1970s, has been attributed to *neurological* problems in exposed populations. Other selected incidents of human illness and ecological contamination include well-documented exposures at *Times Beach*, Missouri, where dioxin-contaminated oil was used to control dust, and at *Love Canal*, New York, where chemical waste containing dioxin was buried under soil under what was to become a subdivision. Both of these incidents occurred in

the late 1970s. An accident at an Agent Orange manufacturing plant in *Spolana*, Czechoslovakia, in 1968, caused short-term problems (human poisoning) and long-term ecological problems from wastes washed into rivers and soils by subsequent floods. In 2004, the then-candidate for the presidency of the Ukraine, Viktor Yushchenko, was poisoned by very high doses of dioxin (allegedly added to a drink or food) that produced *chloracne*, a characteristic symptom of chlorinated hydrocarbon poisoning.

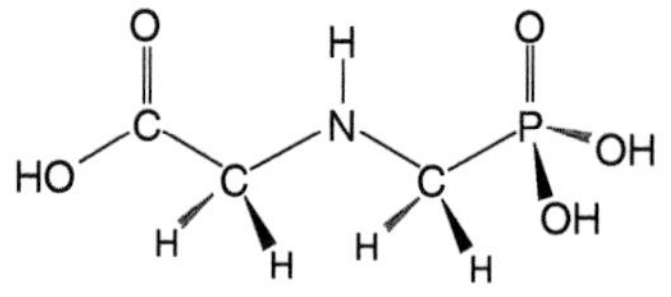

Fig. 3.16 Structural formula of glyphosate, known as N-(phosphonomethyl) glycine in IUPAC nomenclature. Note carboxyl group (far left), amine group (center), and phosphate (far right).

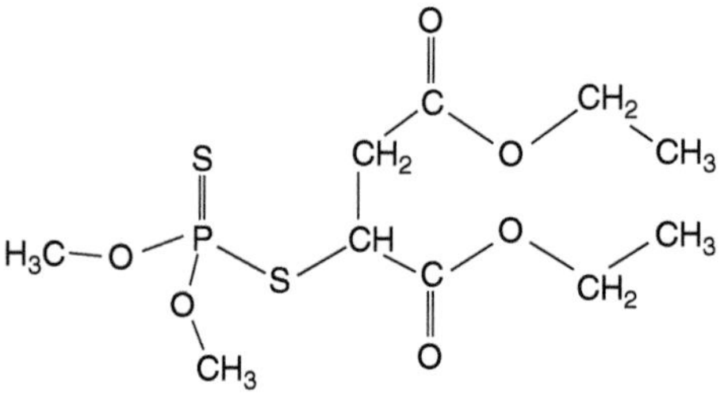

Glyphosate (Figure 3.16; IUPAC = N-(Phosphonomethyl) glycine) is an *organophosphate* herbicide that, as of 2017, is the most widely used in the world (Tarazona et al. 2017). It is the active ingredient in commercially available weed-killers like Roundup and Rodeo, and is a useful example because of its widespread use and the fact that it contains both N and P functional groups. The N occurs in a NH ketimine group (analogous to an amine group lacking one H atom) while the P occurs in a phosphono group, which is basically a phosphate group with a C atom in place of an O atom (note tetrahedral configuration). Note that hydroxyl groups are written here as HO rather than OH to illustrate bonding geometry; for example, in Figure 3.16 the hydroxyl on the left contains an O atom bonded to C, hence the HO on the left side of the molecule.

The presence of carboxyl and hydroxyl groups enhances the solubility of glyphosate (these two functional groups can lose a proton, especially in neutral to alkaline waters), and it tends to decompose more rapidly than most herbicides. It is also desirable because is believed to be less toxic than other herbicides (e.g. those in the organochlorine family), yet it still presents short-term respiratory risk and long-term risk to human kidneys and reproductive systems, according to the USEPA. Some genetically modified (GM) crops have been engineered to be resistant to glyphosate, so fields are sprayed with glyphosate in the hope that the glyphosate kills all plants but the GM crop.

Malathion is an organophosphate insecticide that has been used to spray wide areas for mosquito eradication, as was the case in New York City during the West Nile Virus scare of 1999–2000. It is also used as a topical remedy for killing head lice. Its relatively complex structure is illustrated in Figure 3.17. Note that the geometry is in a more abbreviated form than is shown for glyphosate above.

It is appreciably less toxic than parathion, diazinon, and other related organophosphates, yet still it is a potential carcinogen (the US EPA is still weighing evidence), and in the human body and in chlorinated water supplies it

Fig. 3.17 Structural formula of malathion, which is known as O,O-dimethyl dithiophosphate of diethyl mercaptosuccinate in IUPAC nomenclature. Note the numerous functional groups, including methyl (CH_3), phosphate with sulfide in place of an O atom (thiophosphate), and esters (region with O double-bonded to C—O).

transforms to the *metabolite* malaoxon, which is toxic in elevated amounts.

Picloram (IUPAC name: 4-amino-3,5,6-trichloro-2-pyridinecarboxylic acid) and *atrazine* (1-chloro-3-ethylamino-5-isopropylamino-2,4,6-triazine) are also common herbicides. Picloram is somewhat selective in that it is toxic to woody and broadleaf plants but not to most grasses, so is applied to rangelands to control noxious weeds. It is relatively nontoxic according to the USEPA. Atrazine is less selective and used for control of early-emergent woody, broad-leafed, and grassy plants, and it is believed to be toxic to wildlife (e.g. causing hermaphroditism in amphibians and reptiles; Hayes et al. 2002) and potentially carcinogenic to humans. It is not approved for use in the EU but is used widely elsewhere, including the US, where it is applied to combat the spread of weeds in cornfields.

The *carbamate* insecticide carbaryl (Sevin) is manufactured by reacting methyl isocyanate (MIC) with 1-napthol (Figure 3.18). Carbamates contain an NC(=O)O functional group (shown above the two phenyl groups in carbaryl in Figure 3.18) that typically is bonded to other alkyl or phenyl groups. The worst industrial disaster in human history occurred in December 1984 at a Union Carbide Chemical plant in Bhopal, India when the MIC used in carbaryl production was released to air by an explosion. At least 3800 people died in the minutes and hours following the explosion and thousands more have died premature deaths due to chronic disease (Broughton 2005). Methyl isocyanate is extremely toxic and has a high vapor pressure, so safe storage and use is imperative. The acute and chronic consequences of this disaster emphasize risk associated with manufacture of toxic organic compounds, and Section 8.11 provides additional details on the Bhopal tragedy.

Methyl tert-butyl ether (MTBE) is a gasoline additive – the formula is $(CH_3)_3COCH_3$ – containing oxygen

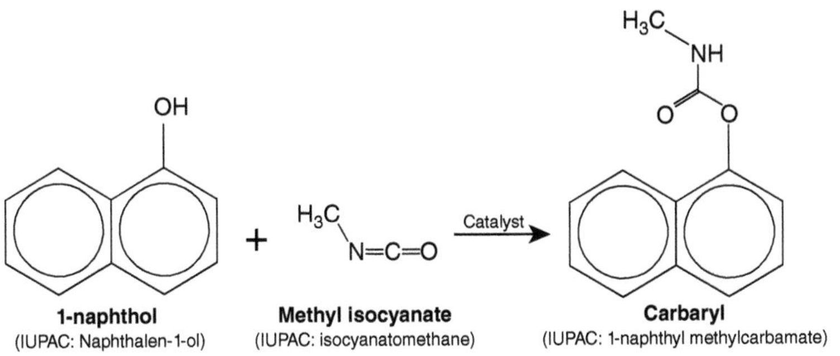

Fig. 3.18 Reaction of 1-naphthol with methyl isocyanate in the presence of a catalyst to produce the carbamate insecticide carbaryl (also known as Sevin and α-Naphthyl N-methylcarbamate).

used to minimize formation of volatile organic compounds (VOCs) produced by incomplete combustion. The 1990 amendments to the US Clean Air Act required the addition of oxygenates like MTBE in regions with smog problems, including large cities like LA, Houston, and Denver, as well as smaller cities in the intermountain American west like Missoula, Montana, and Boise, Idaho. While MTBE contributed to improved air quality, it may be carcinogenic and its high solubility makes it a groundwater contaminant that imparts an unpleasant, turpentine-like taste to water at concentrations as low as 20 ppb.

3.1.8 Pharmaceutical compounds

Antibiotics, synthetic hormones, painkillers, and antidepressants have emerged in the past two decades as threats to drinking water and aquatic ecosystems. If they are incompletely adsorbed or metabolized in the bodies of humans or livestock, these compounds enter aquatic systems. A more obvious pathway is haphazard disposal of pharmaceuticals as waste. Many are compounds that degrade slowly as they migrate through municipal wastewater systems to sewage treatment plants, or leak out of landfills into groundwater systems, so they tend

---

Focus Box 3.6

Organic Compounds and the Endocrine System

Many synthetic organic compounds are *endocrine disruptors* that interfere with the synthesis and cycling of natural hormones in the bodies of humans and other animals. At a molecular level, synthesized organic compounds are often compositionally and structurally similar to compounds in the endocrine system (Figure 3.19). As a result, endocrine disruptors can mimic behavior of natural hormones and hormonal triggers, thereby negatively affecting reproduction, in utero development, the neurological system, metabolism, and behavior. The effect of DDT on development of bird eggs is a clear result of endocrine disruption, and there are many other examples, including feminization of fish, mammals, amphibians, and reptiles, decreased sperm counts in human males, cancers in wildlife and humans, and much more. The list of compounds is very extensive, and ranges from pharmaceuticals, dioxin and dioxin-like compounds, PCBs, DDT and other pesticides, PFCs, and plasticizers such as bisphenol A (which may occur in food and bottled water packaging).

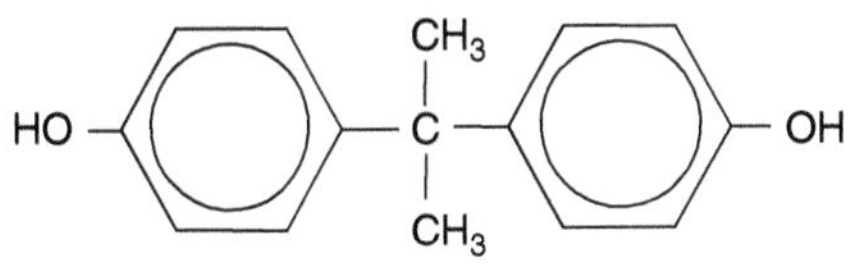

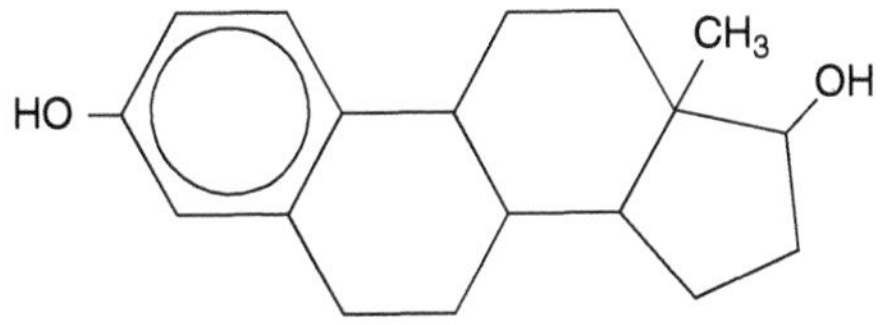

Fig. 3.19 Top = structural formula of bisphenol A, known as BPA and also as 4,4'-(propane-2,2-diyl)diphenol in IUPAC nomenclature. Bottom = structural formula of estradiol, also known in IUPAC nomenclature as (17β)-estra-1,3,5(10)-triene-3,17-diol, a hormone in the estrogen family of female sex hormones. Similarities in structure and composition cause BPA to bond to sites where estradiol normally would, and this is a mechanism for triggering endocrine disruption.

to travel through channels and by overland flow into streams, lakes, estuaries, and marine waters.

A classic case of environmental trade-offs is the case of the common birth control drug ethinylestradiol (17α-ethynylestradiol), which is a potent synthetic estrogen-based endocrine disruptor present in sewage effluent, but also cited as one approach to control human population growth. As a synthetic hormone, it is not hard to envision how it affects the endocrine system — it is intended to do so. Bisphenol A (Figure 3.19), on the other hand, is not designed to affect the endocrine system, but by comparing the similarities in composition and structure between natural sex hormones like estradiol (Figure 3.19) and the many synthetic organic compounds presented in the previous pages, it is not difficult to envision how the endocrine systems of humans and other animals can be affected, often during fetal development (e.g. Colborn and Thayer 2000).

3.1.9 Emerging contaminants – PFCs

Long-chain perfluorinated chemicals (PFCs) – also known as perfluorocarbons, poly- or perfluoroalkylacids (PFAAs) and polyfluoroalkyl substances (PFAS) – have recently emerged as a significant concern to human and environmental health. The most common PFCs in the

environment are long-chain compounds such as perfluorooctanoic acid (PFOA) and perfluorooctanesulfonate (PFOS). PFCs have been used since the 1950s in the production of synthetic performance products (Teflon, Scotchgard, GoreTex, airport firefighting foams [AFFs], and much more) and are now universal in surface and groundwaters around the world. Society has certainly benefited is some ways from PFCs – they are soluble in water and have the capacity to bond to insoluble (hydrophobic) compounds like waterproofing or nonstick compounds, so they have been used extensively to apply hydrophobic materials to fabrics and cookware. Problems arise from their resistance to degradation (due to the strength of the carbon–fluorine bond) and toxicity aggravated by their tendency to bioaccumulate (they bond to fatty acids in blood).

PFOA and PFOS consist of eight hydrophobic carbon–fluorine bonds (the eight-carbon chain gives them the common name C8); both compounds have hydrophilic tails that make them soluble in water (Figure 3.20). The hydrophilic tail in the case of PFOA is a carboxylic acid group (COOH) that can lose a proton to result in a soluble anion (i.e. a conjugate base), and PFOS has a sulfonic acid group that makes it very soluble in water. The anions are perfluorooctanoate (PFO) and PFOS. PFOS (often available with an alkali salt such as Na$^+$ or

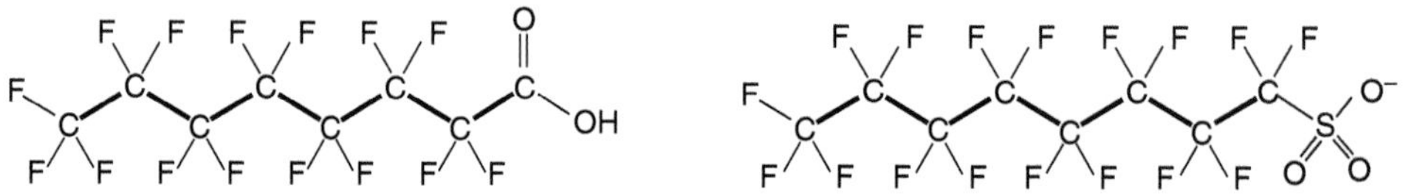

Fig. 3.20 Structural formulas for PFOA (left) and PFOS (right). Note proton on carboxylic acid group for PFOA and deprotonated sulfonic acid group for PFOS. The hydrophilic tails on the right side of each of these molecules greatly enhances solubility.

Focus Box 3.7

Toxicity and Regulation: PFCs and Drinking Water

The toxicity of a substance – from chemical elements like lead (Pb) or arsenic (As), or compounds like nitrate (NO_3^-) or benzene (C_6H_6), or synthetic chemicals such as 2,4-D or PFOA – is generally inversely correlated to what researchers and agencies deem as safe levels in drinking water; i.e. the more toxic a substance, the lower the safe level. The guidelines are often difficult to determine because studies are often done on lab animals and then extrapolated to humans, and also because effects on humans differ as a function of age (infants are typically most susceptible), gender, behavior, and much more. The technologies available to remove the contaminant also

influence guidelines and enforceable levels, to the extent that in the US, the maximum contaminant level goal for arsenic is 0.000 mg l^{-1}, but the legally enforceable concentration is 0.010 mg l^{-1} (0.01 ppm = 10 ppb). In the case of PFOA, some US states (New Jersey and Vermont) have enacted stringent action levels as low as 14 and 20 ng l^{-1} (parts per trillion!), respectively, whereas others have levels as high as 290 and 2000 ng l^{-1} (Texas and North Carolina) (Post et al. 2017). The US EPA action level is 70 ng l^{-1}, but this limit for drinking water is not enforceable given the relatively early stage of research on PFCs, health, and water treatment.

K^+) is more common than the undissociated compound perfluorooctanesulfonic acid (PFOSA).

A recent study indicates that drinking water supplies for 6 million US residents exceed the US EPA lifetime health advisory level of $70\,ng\,l^{-1}$ for PFOS + PFOA (Hu et al. 2016). Included among negative health impacts are low birth weight, attention deficit-hyperactivity disorder, testicular cancer, bladder cancer, liver cancer, early menopause, lowered sperm counts, and thyroid disease (e.g. Xiao et al. 2015). C8 compounds have been phased out in favor of shorter-chain C6 compounds (still PFCs, just shorter chains) which are similarly soluble and resistant to breakdown, and may have similar toxicities, although less is known about this compared to C8. Appendix II presents context and data from an ongoing study of PFOA contamination in Vermont, USA.

3.2 NATURAL ORGANIC COMPOUNDS AT THE EARTH SURFACE

Soils, surface water, and groundwater all contain natural organic compounds, some in dissolved forms (dissolved organic carbon or DOC) and others in solid forms, commonly termed particulate organic carbon (POC). Concentrations vary widely, from $< 1\,mg\,l^{-1}$ (ppm) in oceans and rain to ~100 ppm in some marsh and soil waters. The distinctive brownish tea-like color of streams in temperate and humid regions is due to DOC.

3.2.1 Humic and fulvic material

Humic and fulvic substances are among the most common forms of natural organic carbon in surface environments; combined, humic and fulvic substances are known as humus (Brady and Weil 2002). Examples include:

1 Humin, the insoluble component of humus, is derived from decomposition of the more resistant plant-derived organic components including carbohydrates, aromatic and alkane groups, proteins, and organic acids (sugars and starches are rapidly consumed by organisms).

2 Humic acid, the soluble form of humin that occurs abundantly in soil waters.

3 Fulvic acid, the most soluble form of humus that tends to occur in surface waters along with humic acid leached from soils. Plate 4 shows tea-colored river water with DOC in the form of humic and/or fulvic acids.

Humin, humic acid, and fulvic acid do not have fixed chemical compositions, but rather consist of a varied array of alkane chains and rings, aromatic, carboxyl, and hydroxyl groups, and some S- and N-based functional groups. Humic acid is richer in C and contains less O than fulvic acid (McBride 1994).

Carbon cycling, climate forcing, chemical weathering, storage and release of plant nutrients, complexation of Al, Fe, and trace metals (which enhances their mobility), and contaminant fate and transport (i.e. the behavior in soils, air, and water of contaminants) are all strongly influenced by the varied roles and forms of natural organic matter. The flux of large quantities of iron from tropical forests to the oceans is enhanced by complexation of Fe to organic matter, and this relationship becomes important when considering that iron is a limiting nutrient in the oceans which plays an important role in phytoplankton productivity and hence fixation of CO_2 from atmosphere to ocean. Similarly, formation and degradation of POC and DOC in soils is linked to C cycling and climate, in that flux of C to and from soil organic matter can be either a source or sink of atmospheric CO_2. Fine-grained particulate organic matter in soils strongly adsorbs many non-polar organic contaminants, limiting their mobility but also their degradation rates. Humus exerts strong controls on developing and preserving soil aggregates, and thus exerts controls on water movement through soils as well as resistance to erosion.

3.2.2 Origins and compositions of fossil fuels

Organic matter tends to oxidize to CO_2 in oxidizing/aerobic environments like aerated forest floors or grasslands, but in reducing environments like swamp bottoms or sediments on the deep ocean floor, organic matter does not tend to oxidize to CO_2. When remains of algae, bacteria, plants, and other forms of dead organisms are preserved in reducing environments and buried in sedimentary sequences, they may be transformed to the fossil fuels *coal* and *petroleum*: "petro" comes from the Greek word for rock and "oleum" is derived from the Latin word for oil. When buried under layers of sediment, dead marine organisms (mostly algae and zooplankton) avoid oxidation to CO_2 and instead transform to hydrocarbons; plant-derived terrestrial organic carbon transforms to coal. These post-burial transformations are forms of *diagenesis* (as are mineralogical transformations in sediments and sedimentary rocks such as smectite → illite). Kinetic and thermodynamic factors dictate that deep, long-term burial will transform immature organic matter into coal and petroleum hydrocarbons. (The origin of fossil fuels is covered in greater detail in the section on the carbon cycle in Chapter 8, but this section presents some basic concepts

---

Focus Box 3.8

Coal and the Relationship of Organic Matter to Metamorphism in Earth's Crust

Swamp-derived humic organic matter transforms to coal through the maturation sequence shown below:

peat → lignite (soft, low-grade coal) → bituminous

→ anthracite (hard, high-grade coal), and eventually

→ graphite (in metamorphic environments).

Lignite occurs in shallow crustal sedimentary sequences (e.g. $\leq$ 2 km, $\leq$100 °C); bituminous coal is associated with burial diagenesis typical of many sedimentary basins (2–10 km burial depth, ~100–250 °C), and anthracite

with low-grade metamorphism at temperatures > 250 °C. Somewhere in the range of 300–400 °C, anthracite begins to transform to graphite (Plate 2). In terms of organic chemical composition, coal consists of a mix of alkane chains and rings and aromatic groups (and hence is compositionally C and H). With increasing grade, coal becomes progressively more C-rich, from low-carbon lignite (70–80 wt% C) to high-carbon anthracite (> 90% C) and eventually to graphite, which is nearly purely C. One of the unfortunate impurities in coal is sulfur, which comprises 1–3% of most coals and is the source of sulfuric acid in acid rain.

---

to facilitate subsequent discussion of fate and transport of organic compounds).

Soon after burial in anoxic deep-ocean sediments, remains of microscopic marine organisms begin to decompose, and the lack of O_2 gas prevents oxidation of organic matter to CO_2 gas. Kerogen, a waxy compound with long-chain alkanes and various other immature hydrocarbons, forms in the early stages of maturation. The immature high-molecular-weight alkane chains undergo catagenesis, a process whereby long alkane chains are transformed to progressively shorter chains (this is also known as cracking), transforming waxy kerogen into tar-like hydrocarbons and eventually into liquid and gaseous hydrocarbons. The end-products of catagenesis are the gases methane and ethane. Liquid and gaseous hydrocarbons typically undergo maturation in organic-rich shales (known as source rocks) but often are forced out of the high-pressure pores in deeply buried shales and into permeable sandstones and limestones (reservoir rocks) with lower pore pressures. From there extracting liquid hydrocarbons is somewhat analogous to pumping water out of a groundwater well, not accounting for complications produced by geological structures, clogging of pores by clays, destruction of drilling rigs by high pore pressures, and various other issues.

The C stored in sedimentary rocks as coal and hydrocarbons was first fixed into plants through photosynthesis (the E represents solar energy).

$$nCO_2 + nH_2O \ (+E) = C_nH_{2n}O_n + nO_2 \qquad (3.2)$$

The oxygen found in $C_nH_{2n}O_n$ is likely to be consumed by mineralogical or biochemical reactions in the O_2-poor conditions of anoxic sediments, leaving C and H to form hydrocarbons or coal. So, when we burn fossil fuels,

we effectively are transferring carbon from a long-term geological repository – one where it is stored in reduced form – to the atmosphere in the oxidized form CO_2.

$$C_8H_{18} + {}^{25}/_2O_2 = 8CO_2 + 9H_2O \ (+E) \qquad (3.3)$$

One way to think of combustion (of octane in this case) is as the reverse of photosynthesis, where the reactants (instead of products) are O_2 and reduced C, and products (instead of reactants) are CO_2, H_2O, and energy (heat and light energy).

3.3 FATE AND TRANSPORT OF ORGANIC POLLUTANTS, CONTROLS ON BIOAVAILABILITY, BEHAVIOR OF DNAPLS AND LNAPLS, BIODEGRADATION, REMEDIATION

In the past five to six decades, many of the thousands of refined and synthetic organic compounds that have been introduced to the environment have emerged as some of the most problematic sources of pollution in air, soil, and water. Common contaminants include pesticides (including chlorinated hydrocarbons and organophosphates), organic solvents and pesticides, and fossil fuels and associated aromatic hydrocarbons – sources are varied and include industry, agriculture, golf courses, the military, and residential lawns. It is difficult to generalize the behavior of such a large group of compounds, but in general, organic contaminants tend to possess low solubility in water and high vapor pressures (volatility). The following factors (which, to some extent, are interdependent) help to frame controls on their fate and transport:

1 Partitioning of the compound into solid, liquid, or gaseous phases;

2 The solubility of the compound, a property that is generally inversely related to adsorption;

3 The degree to which the organic contaminant in question will adsorb to minerals or organic matter in soils and sediments;

4 The density of the compound (if it is a liquid) relative to water; and

5 The extent to which the compound will undergo degradation (decomposition into CO_2, O_2, and other products), a process that depends on factors including biological controls (the composition of the natural microbial community, for example), physical properties (e.g. temperature, moisture content, porosity, and permeability of soil or sediment), chemical characteristics (e.g. reduction-oxidation potential, pH, salinity), and kinetic controls.

3.3.1 Solid–liquid–gas phase considerations

Most organic contaminants are liquids and they often have high vapor pressures (to conceptualize this, think of the vapors that you can smell escaping from gasoline), so we must consider the potential of such compounds to vaporize when spilled or applied to the land surface or subsurface. The definition of vapor pressure is the pressure of the gas phase (vapor) in equilibrium with a liquid at a given temperature in a closed system (where the vapor cannot continuously evaporate out of the system). The higher the vapor pressure of a liquid, the lower the boiling point of the liquid, and the more likely it is to vaporize. The term for organic compounds with high vapor pressures at normal temperatures is VOCs. VOCs are usually insoluble in water and have boiling temperatures < 200 °C. This group includes the alkanes, aromatic hydrocarbons, some chlorinated hydrocarbons (e.g. TCE, PCE, pentachlorophenol), and aldehydes (e.g. formaldehyde).

3.3.2 Solubility considerations

Organic compounds are generally hydrophobic (that is, they are insoluble in water), but the presence of certain functional groups, particularly nitrogen-based (e.g. amine), oxygen-based (e.g. carboxylic and hydroxyl), and sometimes halide groups (e.g. Cl), enhances solubility. The term commonly used to express the solubility of organic compounds is the *octanol-water partition coefficient*:

$$K_{OW} = C_{octanol}/C_{water} \qquad (3.4)$$

It is a simple expression defined by the solubility of an organic compound in *n*-octanol ($C_8H_{17}OH$) divided by its solubility in water. (In many cases, octanol is used as a surrogate for fatty tissues in organisms.) Nonpolar organic compounds tend to be soluble in other organic compounds (like octanol) and insoluble in water, so K_{OW} values for organic compounds tend to be greater than 1, commonly by orders of magnitude, and for this reason, values of K_{OW} are often reported as log $[K_{OW}]$. Table 3.4 presents values of representative compounds. Any compound with $K_{OW} > 1$ will preferentially partition into organic compounds; this is the case for basically all potential organic pollutants, so the real difference in assessing mobility is the extent of the K_{OW}. The greater the K_{OW}, the less soluble in water and more likely to dissolve in (or adsorb onto) organic compounds, including natural organic matter in soils (see Section 3.3.4). Compounds in the upper rows of Table 3.4, especially those with log $[K_{OW}] > 4$, are very strongly sorbed onto/into POC (e.g. humus, decaying vegetation). Compounds with log $[K_{OW}]$ values < 4 are generally more polar and their fates in the environment become increasingly controlled by the higher aqueous solubility and potential to be sorbed onto surfaces of clays or other minerals.

3.3.3 Interactions of organic compounds and organisms

Considering human health and ecological impacts, the K_{OW} is generally proportional to the potential of an organic compound to *bioaccumulate* (or *bioconcentrate*), provided that the compound does not rapidly degrade. Bioaccumulation is the process by which a substance is incorporated into an organism and, over time, increases in concentration because rate of removal is lower than rate of intake. Bioconcentration is largely the same process but does not include ingestion, instead including only absorption through skin or via respiration. Stable compounds with high K_{OW} values are soluble in fatty tissues of organisms and are only sparingly removed from organisms in urine, so they tend to bioaccumulate. Figure 3.21 plots K_{OW} versus bioaccumulation factor (BAF), the ratio of the concentration of the organic compound in the organism (or fatty tissue of the organism) divided by the concentration of the organic compound in the environment (e.g. water or sediments).

While bioaccumulation (or bioconcentration) pertains to increasing concentration of a substance (with time) in an organism, *biomagnification* occurs across trophic levels. The classic case is when secondary and tertiary consumers consume lower trophic level organisms with relatively low concentrations of hydrophobic substances, and because the rate of ingestion is greater than the rate of excretion (resulting in bioaccumulation at the organismal

Table 3.4 Data on various properties of organic compounds.

Compound	Composition	Log Kow	Solubility (mg l^{-1})[a]	Specific gravity	Vapor pressure (Pa)
Aldrin	$C_{12}H_8Cl_6$	6.5	0.03	1.6	0.00008
p,p'-DDT	$C_{14}H_9Cl_5$	6.2	<0.1	1.6	0.000025
2,4,2',4'-tetrachlorobiphenyl	$C_{12}H_6Cl_4$	6.0	0.1	1.2	0.13
Hexachlorobenzene	C_6Cl_6	5.6	0.006	1.2	0.001
Chlordane	$C_{10}H_6Cl_8$	5.5	<0.1	1.6	0.0013
Heptachlor	$C_{10}H_5Cl_7$	5.3	<0.1	1.7	0.04
Octane	C_8H_{18}	5.2	0.7	0.7	1500
Pentachlorophenol	C_6Cl_5OH	5.0	1.0	1.3	2.7
Endrin	$C_{12}H_5Cl_6O$	4.6	0.2	1.7	0.000027
Heptane	C_7H_{16}	4.5	3.0	0.70	5300
Phenanthrene	$C_{14}H_{10}$	4.5	1.2	1.2	0.13
Dieldrin	$C_{12}H_8Cl_6O$	4.3	0.2	1.7	0.000024
1,2,4-trichlorobenzene	$C_6H_3Cl_3$	4.2	35	1.5	40
1,4-dichlorobenzene	$C_6H_4Cl_2$	3.6	120	1.3	160
Ethylbenzene	C_8H_{11}	3.2	170	0.87	950
Xylene	C_8H_{10}	3.1	180	0.88	1070
Perchloroethylene (PCE)	C_2Cl_4	3.0	200	1.6	1870
Chlorobenzene	C_6H_5Cl	2.8	450	1.1	1170
Toluene	C_7H_8	2.7	500	0.87	2930
Trichloroethylene (TCE)	C_2HCl_3	2.4	1200	1.5	7800
Benzene	C_6H_6	2.1	1800	0.88	10000
Phenol	C_6H_5OH	1.5	8400	1.05	48

[a] In water at 25 °C.

Source: Drever (1997), Girard (2005), and US Centers for Disease Control (http://www.cdc.gov/niosh).

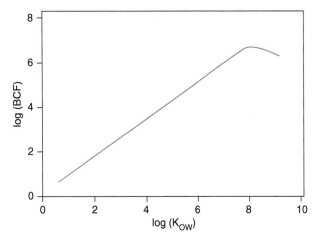

Fig. 3.21 Plot of log (K_{OW}) against log (bioconcentration factor, BCF), showing trend for a range of organic compounds in middle trophic level species (BCF generally increases with trophic level). Note the linear relationship for most of the plot (decreasing solubility in water = increased BCF), and the curious dip at high values of K_{OW}. Source: Created from data in Banerjee and Baughman (1991) and Arnot and Gobas (2006).

level), the result to the ecosystem in biomagification. Biomagnification tends to occur with insoluble organic compounds, and the classic example is concentration of

DDT up the food chain in aquatic ecosystems, where DDT concentrations commonly increase by orders of magnitude from low trophic level primary producers (plants) to primary consumers (herbivores) and eventually to high trophic level tertiary consumers (carnivores) according to the following scenario:

Tertiary Consumer = 50 ppm

↑

Secondary Consumer = 5 ppm

↑

Primary Consumer = 0.5 ppm

↑

Primary Producer = 0.05 ppm

Probably the most well-known consequences of this phenomenon were the precipitous drops in populations of eagles, peregrine falcons, and other predatory birds in the latter part of the twentieth century due to biomagnification of synthesized organochlorine compounds such as DDT and PCBs. While predatory bird populations have begun to rebound in response to restrictions on such compounds, biomagnification of dimethylmercury ($[CH_3]_2Hg$) is a problem that appears to be increasing in scope and magnitude. Hg in the environment is derived from numerous sources, from *non-point sources* like coal combustion and waste incineration that are broadcast over wide areas from multiple sources, to *point sources* of contamination such as industrial and mining operations. (More information on Hg is presented in Chapter 8 in the context of air quality.)

Carbon filters are generally effective at removal of organic compounds from drinking water – the nonpolar compounds are attracted to carbon, whether in designed filters or in natural organic matter in soils or sediments.

3.3.4 Adsorption of organic compounds

Sorption of organic pollutants to the surfaces of particles tends to limit pollutant mobility while also diminishing availability to decomposing microbes. *Adsorption* is a surface effect characterized by attachment of compounds or ions to particle surfaces by relatively strong covalent or electrostatic forces or relatively weak van der Waals forces. Viewed in this way, adsorption can play a positive role in

an environmental sense by limiting bioavailability of toxic organic compounds, but at the same time, adsorption can make a compound less accessible to microbes and hence potentially minimize degradation rates, causing the contaminant to persist for longer. At this point, it is worth making a couple of comments about terminology. The first is that adsorption is bonding of a compound or ion to a particle surface (refer to the description in Chapter 2.5), whereas absorption is incorporation of a compound into another substance.

Adsorption and *absorption* are types of *sorption*. Because organic molecules often diffuse into the structure of humus, rather than merely become attached to surfaces of humus, and also because it is difficult to discern whether organic molecules undergo adsorption, absorption, or a combination of the two, the more general term sorption is often applied. A sorbed organic compound may be held by van der Waals on the surface of a particle of humus, or it may have diffused into the structure of the particle. Lastly, the term desorption describes release of a compound back into the aqueous (or gaseous) phase from a particle, i.e. it the reverse of adsorption (or absorption).

Adsorption of organic compounds is controlled by properties of soil solids (the adsorbents) and the organic molecules themselves (the adsorbates).

Focus Box 3.9

Two Mechanisms for Sorption: Chemical vs. Physical

Adsorption produced by an electrostatic or covalent attraction between adsorbent and adsorbate is termed *chemisorption*. This is the process that governs ion exchange with surface of minerals or humus particles, and it is much stronger than *physisorption (physical adsorption)* that occurs when e.g. insoluble organic compounds form relatively weak van der Waals bonds with particles of generally any type. In chemisorption, the chemical properties of the solid absorbent (e.g. layer charge in clay minerals, surface charge in hydroxides, functional groups in humus) and adsorbate (is it a cation? anion? a polar compound?) play an important role in the adsorption process and often selectively partition certain phases out of the fluid and into adsorbed form. The proportion of a chemisorbed compound (on a particle surface) relative to other phases in the aqueous system may be significantly greater than its proportion in the fluid (often dissolved) phase. Physisorption generally does not result in any change in proportions of adsorbed species vs. those that remain in the fluid

phase – it does not involve changes to electronic structure of the atom or molecule, and is relatively indiscriminate and independent of solution chemistry and solid phases. Refer to Figure 3.22 and adsorption isotherms for more information.

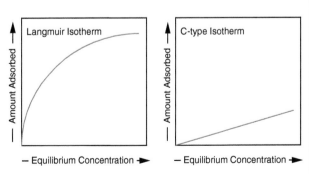

Fig. 3.22 Isotherms describing chemisorption (left) and physisorption (right) or organic compounds.

3.3.4.1 Adsorbates

Nonpolar compounds like alkanes, DDT, and PCBs have greater tendency to sorb strongly to natural organic matter and less tendency to sorb to clays, hydroxides, or other minerals. Conversely, the relatively soluble organic compounds, particularly ionic forms of organic compounds and, to a lesser extent, polar organic compounds, are attracted to ion exchange sites in minerals. Figure 3.23 presents solubility and K_{OW} data for selected organic compounds – recall that the higher the K_{OW}, the more likely it will be that an organic molecule will sorb to organic material like soil humus.

3.3.4.2 Adsorbents

Adsorption of organic contaminants by soil organic matter (humus) is generally greater than adsorption by inorganic clays, especially in organic-rich soils or silty-to-sandy soils with low clay content. In clay-rich soils with typical organic matter content (0.5–5%), however, clay minerals play an important role in organic matter adsorption, sometimes exerting greater control than humus by virtue of the abundance of clays. Clearly, minerals with high cation exchange capacity (CEC) such as smectites and vermiculites will more readily adsorb cationic (e.g. diquat, paraquat) and polar (e.g. atrazine, TCE) organic compounds than will those with low CEC (e.g. clays with no layer charge like kaolinite and coarse-grained neutral minerals such as quartz and feldspars).

Water content also is an important consideration when predicting adsorption of organic species. In general the moister the soil, the lower the extent of organic contaminant adsorption due to strong attraction of the polar water molecule to mineral surfaces (e.g. Sawhney and Singh 1997). To extend this further, saturated groundwaters with low organic matter content tend to exhibit poor retention of organic contaminants by adsorption, resulting in high mobility, whether as a dissolved (aqueous) phase, as a liquid or gas, or as fine-grained mobile particulate.

Adsorption isotherms are a common graphical means of representing adsorption data, containing information about the type of interaction between adsorbate and adsorbent. They are called isotherms because adsorption is temperature-dependent (sorption decreases with increasing temperature), and thus isotherms present adsorption features at fixed temperature. In these plots, the x-axis is the concentration of an adsorbate in a fluid (e.g. water), and the y-axis in the amount absorbed to a solid, when the system is at equilibrium. Two examples are presented: the Langmuir isotherm, which describes chemisorption, and a linear C-type (constant partitioning) isotherm that describes physisorption. Isotherms are presented here as they pertain to organic compounds and are referred to again in greater detail in Section 4.12.

Chemisorption is a monolayer process where the number of potential adsorption sites exerts a strong control on the amount of adsorbate that can bond to the particle surface, so as the equilibrium concentration (of absorbate) increases, the amount adsorbed increases logarithmically (the rate of increase diminishes). In physisorption, there is no monolayer restriction like that associated with chemisorption, and the number of available adsorption

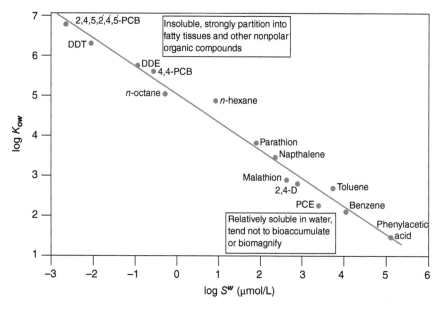

Fig. 3.23 Correlation between the solubility in water (presented as log S^w, the aqueous solubility) and octanol-water coefficient (presented as log K_{OW}) of organic compounds. Low-solubility compounds such as DDT and PCBs plot in the upper left of the line and relatively soluble compounds such as benzene, PCE, and 2,4-D plot in the lower right of the line. Source: Data are from Chiou et al. (1977, 2005).

Focus Box 3.10

Organic Compound Sorption and PZC

In acidic to neutral soils (pH < 8), iron hydroxides possess positive surface charge (Section 2.5.1.3, Table 2.4) and are capable of attracting anionic compounds, notably the acidic herbicides such as 2,4-D and picloram, and the insecticide pentachlorophenol. These compounds dissociate to produce H^+ and an anionic conjugate base in solution, with 2,4-D provided as an example (Figure 3.24).

In Eq. 3.5, R^- is any anion that might be dissolved in water. On the reactants side of Eq. 3.5, a chloride (Cl^-) anion is sorbed to ferrihydrite; on the products side, the anion symbolically represented as R^- (e.g. the 2,4-D anion in Figure 3.24) has undergone an exchange reaction with chloride and is now sorbed to the iron hydroxide.

$$R^- + [FERRIHYDRITE]\ Cl^- = Cl^- + [FERRIHYDRITE]\cdot R^- \tag{3.5}$$

Examples of anions that may sorb to hydroxides below the PZC include the 2,4-D anion as shown and

other organic anions; inorganic species such as nitrate (NO_3^-); the arsenite (AsO_3^{3-}) or arsenate anions (AsO_4^{3-}); phosphate anions (PO_4^{3-}); and many more.

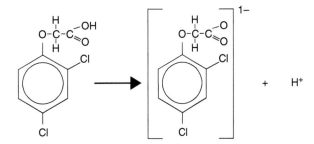

Fig. 3.24 Formation of 2,4-D anion by deprotonation, which occurs in nearly all aqueous systems regardless of pH. The resulting anion may sorb to anion exchange sites on humus or hydroxides if pH is < PZC.

sites on the particle surface is no longer a controlling variable. The result is a linear correlation between increasing concentration (e.g. of an organic compound in water) like that depicted in the C-type isotherm. At low concentrations, chemisorption can also produce a linear isotherm. This is because at very low concentrations of adsorbate, the number of available adsorption sites will not be a limiting factor on adsorption; only at higher concentrations where molecules must compete for adsorption sites will the isotherm become nonlinear (logarithmic). You can conduct experiments using adsorbents and adsorbates, measuring sorption as a function of adsorbate concentration, and the resulting curve will suggest (but not prove) potential sorption mechanisms. Proof only comes with direct measurement using molecular-level methods such as infrared spectroscopy, often combined with indirect techniques such as chromatography (e.g. Johnston et al. 2002).

The Langmuir isotherm can be derived from the Law of Mass Action (Chapter 1.7.2), presented here as a summarized version of the derivation given by McBride (1994). If A represents an absorbate, and S represents potential absorption sites, we can envision adsorption as follows:

$$S + A = S^*A \tag{3.6}$$

where S^*A indicates sites (S) that have adsorbed the compound (absorbate A). (To be clear, S^*A does not represent

the mathematical expression S times A.) Recall that at equilibrium, adsorption–desorption is a dynamic process and this reaction proceeds in both directions at equal rates. Thus this equation can be expressed as an equilibrium constant:

$$K_{ads} = [S^*A]/([S]\cdot) \tag{3.7}$$

As usual, brackets indicate activities (effective concentrations) of species in solution or on surfaces, where $[S]$ and $[S^*]$ are assumed to represent mole fractions of vacant adsorption sites and occupied sites, respectively.

The proportion of sites occupied by A (θ_A) can be written as:

$$\theta_A = [S^*A]/([S^*A] + [S]) \tag{3.8}$$

where $[S^*A] + [S]$ is the total number of possible sites, occupied (S^*A) and vacant (S). The proportion of vacant sites can be expressed as:

$$1 - \theta_A = [S]/([S^*A] + [S]) \tag{3.9}$$

It then follows that the number of occupied sites relative to unoccupied sites can be expressed as this ratio:

$$\theta_A/(1 - \theta_A) = [S^*A]/[S] = K_{ads}(A) \tag{3.10}$$

$K_{ads}(A)$ is the equilibrium constant that governs adsorption of A. Rearranging produces the standard Langmuir equation:

$$\theta_A = K_{abs}(A)/(1 + K_{abs}(A)) \qquad (3.11)$$

This equation produces the Langmuir isotherm and describes chemisorption.

For more detail on organic matter adsorption, please refer to Chiou (2002), McBride (1994), Langmuir (1997), and many others.

3.3.5 Non-aqueous phase liquids (NAPLs) in the environment

Nonpolar organic liquids are commonly referred to as non-aqueous phase liquids (NAPLs) because they can occur in various environments as liquid phases that are not dissolved in the aqueous phase. For example, octane and water are immiscible fluids; the log K_{OW} for octane is approximately 10^5, indicating that it is 100 000 times more soluble in octanol than water. Clearly, octane is quite insoluble in water, and with a density of $0.7\,g\,cm^{-3}$, octane will float on top of water ($\rho = 1.0\,g\,cm^{-3}$), whether the water occurs in a surface body like a lake or ocean or whether it occurs in an aquifer, where octane will saturate pores atop the water table.

Light non-aqueous phase liquids (LNAPLs) are water-insoluble liquid compounds like octane and other fuels with densities less than $1.0\,g\,cm^{-3}$ – these generally are the alkanes and alkenes (aliphatic hydrocarbons) and aromatic hydrocarbons (including BTEX and PAHs). Water-insoluble liquids with density $> 1\,g\,cm^{-3}$ are known as *DNAPLs*, dense non-aqueous phase liquids. DNAPLs generally include chlorinated hydrocarbons like the solvent TCE, the insecticide pentachlorophenol, PCB oils, and PAHs like those in creosote distilled from coal tar.

Behavior of NAPLs (i.e. fate and transport) will be dictated by the following possible pathways or processes (Figure 3.25):

1 Migration downward through soil toward the water table, where their behavior is dictated by solubility, density:

  A LNAPLs tend to fill pores between sediment grains and pool up atop the unconfined aquifer, and depending on their solubility, a certain amount will become dissolved in the aquifer below the water table and be transported laterally with groundwater flow. The dissolved phase plume is often rich in BTEX,

  B DNAPLs sink to the bottom of the aquifer and forms pools atop clay or rock confining layers. Any fractures in the confining layer permit further seepage into the deeper confined aquifer. Of course, depending on

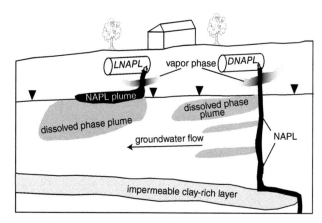

Fig. 3.25 Schematic diagram of behavior of NAPLs in the subsurface. Shown are two leaking underground storage tanks (USTs). Triangles indicate water table. The scale of the diagram does not permit depiction of sorption in the soil (refer to Section 9.8).

solubility, some DNAPL will dissolve in aquifer water and migrate with groundwater flow;

2 Sorption to soil particles (particularly soil organic matter and reactive clays like smectites);

3 Vaporization (~evaporation), a process that is most important for compounds with high vapor pressures (Table 3.4);

4 Biochemical degradation, to the extent permitted by natural microbial communities and physicochemical conditions in the soil or aquifer (Section 3.3.6).

The multiple pathways of NAPLs in soils and aquifers makes remediation (and modeling of behavior) difficult. Many questions must be addressed, for example: What percent of the spilled NAPL will sorb to soil particles? (Many factors must be considered, including solubility of the compound and composition of soil solids.) How much can be expected to vaporize? How much will be decomposed by the natural soil microbial community before it reaches the water table? How much will dissolve in the groundwater and how much will remain as a non-aqueous liquid (i.e. how much will be either pooled atop or settled at the base of the aquifer)? Appendix I presents a case study that examines leakage of PCE into an alluvial aquifer and attempts to remediate.

3.3.6 Biodegradation

Because they are composed largely of C and H, organic compounds are potential energy sources for microbial organisms, particularly bacteria. The C atoms in organic compounds occur in reduced form (C^{4-}), so the oxidation of C to C^{4+} releases electrons, the basic energy source for microbial life. In fact, the transition from reduced to oxidized carbon releases eight electrons per atom (or 8 mol of electrons per mole of C), making C one of the most energy-rich sources for microbial metabolism (reduced nitrogen is also very energy-rich when oxidizing from N^{5-} to N^{3+}).

Respiration is a common example of oxidation of organic matter that releases electrons to fuel life. The reactants are organic matter and O_2; the products are CO_2 and H_2O. The carbohydrates consumed by animals contain reduced carbon that reacts with oxygen to release energy (electron flow) to fuel life. Decomposition of organic compounds in soils and sediments is similar, but by virtue of their abundance relative to vertebrates, the microbial community is the main agent of decomposition (a cubic centimeter of humus-rich soil typically contains somewhere in the range of 100 million to 1 billion bacteria). A generic reaction for decomposition of organic matter (OM) is shown here:

$$OM + O_2 \rightarrow CO_2 + H_2O \qquad (3.12)$$

Note that this reaction typically is bacterially mediated, but organic compounds can also decompose abiotically by oxidation, a process that generally occurs at rates which are orders of magnitude slower than their microbially mediated counterparts, but abiotic decomposition can be an important process in systems with relatively low bacterial populations. One such example is the atmosphere, where strong oxidizers such as the hydroxyl (OH*) and hydroperoxy (HO$_2$*) radicals are very effective at oxidizing reduced carbon in organic compounds.

Reaction (3.12) is a simplified example that likely skips many steps of intermediate decomposition products known as *metabolites*. The toxicity and the behavior in the environment of a metabolite may differ appreciably from the original compound; for example, the metabolite may be more soluble or less soluble, or more likely (or less likely) to vaporize out of soil, than the original organic compound. The metabolite may decompose much more rapidly in the presence of soil microorganisms, or it may be more persistent than its parent compound (such as DDE and DDD compared to parent DDT).

Focus Box 3.11

Vinyl Chloride as an Example of Decomposition Pathways

Vinyl chloride (IUPAC name: chloroethene) is useful for understanding decomposition and metabolism because it can occur as a metabolite (e.g. of TCE) and it also decomposes to its own metabolites (Figure 3.26). It is carcinogenic and is a very common pollutant, occurring in 35% of the US EPA's National Priorities List (NPL) of contaminated sites. It is synthesized and used to make products such as polyvinyl chloride (PVC), so it may occur in the environment as a primary pollutant or as a metabolite (e.g. as a metabolite from TCE decomposition via DCE (dichloroethylene); Figure 3.26).

In shallow soil, vinyl chloride tends to vaporize due to its very high vapor pressure (340 000 Pa); if it does not vaporize, its high solubility (2700 mg l^{-1}) means that it may leach into groundwater, and if the groundwater is anaerobic (reducing), decomposition to the metabolite chloroethylene oxide tends to occur very slowly (note that oxygen is a reactant as shown in Figure 3.26), meaning that it will resist degradation or remediation in deeper groundwater or anaerobic shallow groundwater systems.

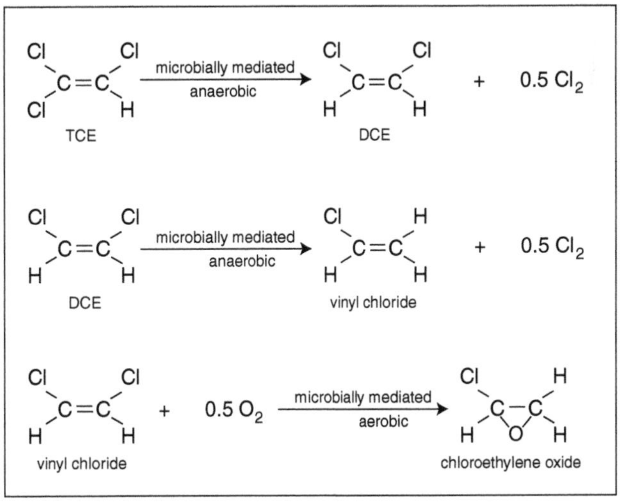

Fig. 3.26 Decomposition of trichloroethylene (TCE) to the metabolite dichloroethylene (DCE), then to the metabolite vinyl chloride, under (most likely) microbially mediated reducing conditions. Decomposition of vinyl chloride to the metabolite chloroethylene oxide tends to occur rapidly via microbial mediation in aerobic (oxidizing) waters but very slowly in anaerobic waters, and thus may persist in groundwater for years or decades.

3.3.7 Remediation

Common approaches to remediation of organic pollutants in soils, groundwater, and surface water often take advantage of the biochemical and physical properties of the substances, and are summarized as follows:

Bioremediation typically involves augmenting natural, in situ microbial communities to enhance microbial decay, from adding nutrients to foster increased populations and effectiveness of indigenous microbial populations to pumping O_2 into soil or shallow groundwater to enhance aerobic activity (a good example being the case of vinyl chloride, Figure 3.26).

Soil vapor extraction (SVE) is an in situ approach that takes advantage of the high vapor pressure of many organic pollutants. Vacuums are inserted into soil and the goal is to create a negative pressure gradient that causes desorption of VOCs and transport of vapors toward extraction wells, where VOCs can be extracted and treated.

Air sparging involves injection of air into soil or groundwater to cause desorption of VOCs from solids or transfer from dissolved state to a vapor phase. SVE is then used to remove VOC-containing air.

Chemical oxidation involves addition of strong oxidizers such as hydrogen peroxide (H_2O_2) to enhance oxidation of the organic contaminants (to CO_2 plus other products such as H_2O or Cl_2 depending on composition of the target compound).

Reductive dechlorination of organochlorine compounds may involve addition of a reducing agent (e.g. molasses, zero-valent iron) that aids in reduction of Cl that eventually (ideally) produces less-toxic metabolites (where $\sim Cl^+$ in organochlorine compounds is reduced to Cl^- in solution). Figure 3.26 depicts this approach and Appendix I presents a case study involving reductive dechlorination of PCE.

Pump and treat technology is often applied to groundwater where the above-listed approaches are ineffective – contaminated groundwater is pumped to the surface, treated to remove contaminants, and then (often) reinjected.

Land farming involves extraction of contaminated soil and transport (by truckloads) to rural regions where soils are spread out at the surface and natural vaporization, oxidation, and microbial decay remove or transform contaminants.

3.4 SUMMARY

Given the wide range in composition and structure of synthetic and natural organic compounds that occur as environmental contaminants, there is no simple rule of thumb to fully understand behavior. They are generally insoluble and tend to sorb to organic matter, but as Table 3.4 shows, solubility varies widely. PFCs, for example, are very soluble and tend not to sorb (case study in Appendix II), so the rule of thumb is useless for that family of contaminants. Furthermore, behavior is often site-specific, dependent on presence and composition of natural soil organic matter, mineralogy, pH, microbial community, temperature, and moisture content. Studies of organic compounds in field-based and experimental environments continue to expand our knowledge in this critical area of environmental science.

QUESTIONS

3.1 What volume of groundwater containing $2\,mg\,l^{-1}$ dissolved oxygen (DO) would be required to oxidize $23\,l$ of octane that leaked out of an underground storage tank (UST) and into an unconfined aquifer? List five factors that could affect reaction rate, and create a schematic diagram to show potential pathways of octane as it leaks out of the UST. List and briefly describe three potential remediation strategies.

3.2 If humus is transported by erosion from a field into a lake, will this result in increased DO, or decreased DO, in the lake? Support your response with a chemical equation.

3.3 Which is more likely to be sorbed by ferrihydrite in a forest soil at pH = 5, benzene or 2,4-D? Create a sketch to demonstrate. Also consider the potential for ferrihydrite to sorb 2,4-D at pH = 4 (e.g. tropical soil like Oxisol) relative to pH = 9 (e.g. arid soil like Aridisol); e.g. considering only ferrihydrite and 2,4-D, what factor related to pH might enhance (or limit) 2,4-D adsorption to a hydroxide like ferrihydrite (or goethite)? How might this allow you to predict sorption potential of 2,4-D as a function of soil type (in humid vs. arid climates)?

3.4 Which is more likely to respond to air sparging as a remediation tool, octane or pentachlorophenol? Explain using pertinent data.

3.5 Reductive dechlorination of PCE or TCE in groundwater may result in breakdown to less-chlorinated metabolites, including vinyl chloride, which is not responsive to reductive dechlorination (Figure 3.26). Describe potential strategies for remediation of groundwater with elevated vinyl chloride.

3.6 Which compound is more prone to bioaccumulation, toluene or the PCB termed 2,4,2′,4′-tetrachlorobiphenyl? Cite pertinent data in your response.

REFERENCES

Arnot, J.A. and Gobas, F.A.P.C. (2006). A review of bioconcentration factor (BCF) and bioaccumulation factor (BAF) assessments for organic chemicals in aquatic organisms. *Environmental Reviews* 14: 257–297.

Banerjee, S. and Baughman, G.L. (1991). Bioconcentration factors and lipid solubility. *Environmental Science and Technology* 25: 536–539.

Brady, N.C. and Weil, R.R. (2002). *The Nature and Properties of Soils*, 13e. Upper Saddle River, New Jersey, USA: Prentice-Hall.

Broughton, E. (2005). *The Bhopal disaster and its aftermath: A review*. Environmental Health: A Global Access Science Source https://doi.org/10.1186/1476-069X-4-6.

Chiou, C.T. (2002). *Partition and Adsorption of Organic Contaminants in Environmental Systems*, vol. 257. Hoboken, NJ: Wiley.

Chiou, C.T., Freed, V.H., Schmedding, D.W., and Kohnert, R.L. (1977). Partition coefficient and bioaccumulation of selected organic chemicals. *Environmental Science and Technology* 11: 475–478.

Chiou, C.T., Schmedding, D.W., and Manes, M. (2005). Improved prediction of octanol-water partition coefficients from liquid-solute water solubilities and molar volumes. *Environmental Science and Technology* 39: 8840–8846.

Colborn, T. and Thayer, K. (2000). Aquatic ecosystems: harbingers of endocrine disruption. *Ecological Applications* 10: 949–957.

Drever, J.I. (1997). *The Geochemistry of Natural Water Surface and Groundwater Environments*, 3e. Upper Saddle River, New Jersey, USA: Prentice-Hall.

Girard, J.E. (2005). *Principles of Environmental Chemistry*. Sudbury, Massachusetts, USA: Jones and Bartlett Publishers.

Green, N.J.L., Jones, J.L., and Jones, K.C. (2001). PCDD/F deposition time trend to Esthwaite water, UK, and its relevance to sources. *Environmental Science and Technology* 35: 2882–2888.

Hayes, T.B., Haston, K., Tsui, M. et al. (2002). Hermaphrodites beyond the corn field: atrazine-induced testicular oogenesis in leopard frogs (*Rana pipiens*). *Nature* 419: 895–896.

Hu, X.C., Andrews, D.Q., Lindstrom, A.B. et al. (2016). Detection of poly- and perfluoroalkyl substances (PFASs) in U.S. drinking water linked to industrial sites, military fire training areas, and wastewater treatment plants. *Environmental Science and Technology* 3: 344–350.

Johnston, C.T., Agnew, S.F., Schoonover, J.R. et al. (2002). Raman study of aluminum speciation in simulated alkaline nuclear waste. *Environmental Science and Technology* 36: 2451–2458.

Langmuir, D. (1997). *Aqueous Environmental Geochemistry*. Upper Saddle River, New Jersey, USA: Prentice-Hall.

McBride, M.B. (1994). *Environmental Chemistry of Soils*, 406. Oxford University Press.

Post, G.B., Gleason, J.A., and Cooper, K. (2017). Key scientific issues in developing drinking water guidelines for perfluoroalkyl acids: contaminants of emerging concern. *PLoS Biology* https://doi.org/10.1371/journal.pbio.2002855.

Sawhney, B.L. and Singh, S.S. (1997). Sorption of atrazine by Al- and Ca-saturated smectite. *Clays and Clay Minerals* 45: 333–338.

Silk, P., Lonergan, G.C., Arsenault, T.L., and Boyle, D.C. (1997). Evidence of natural organochlorine formation in peat bogs. *Chemosphere* 35: 2865–2880.

Tarazona, J.V., Court-Marques, D., Tiramani, M. et al. (2017). Glyphosate toxicity and carcinogenicity: a review of the scientific basis of the European Union assessment and its differences with IARC. *Archives of Toxicology* 91: 2723–2743.

Xiao, F., Simcik, M.F., Halbach, T.R., and Gulliver, J.S. (2015). Perfluoroctane sulfonate (PFOS) and perfluorooctanoate (PFOA) in soils and groundwater of a U.S. metropolitan area: migration and implications for human exposure. *Water Resources* 72: 64–74.

Aqueous Systems and Water Chemistry

4.1 INTRODUCTION TO THE GEOCHEMISTRY OF NATURAL WATERS

The composition of natural waters is largely controlled by two main factors: changes of state and biogeochemical processes. An example of a common change of state is evaporation, which is responsible for the low amounts of dissolved substances in atmospheric water; biogeochemical factors include exchange reactions, mineral crystallization, mineral dissolution, and decomposition of organic matter that occur in water in all reservoirs within the hydrologic cycle (Figure 4.1). Section 4.1.1 examines changes of state, beginning with evaporation and its effect on atmospheric water and surface water, followed by the effect of precipitation and infiltrated water on chemical weathering and the composition of surface water and groundwater.

4.1.1 Geochemistry and the hydrologic cycle

4.1.1.1 Evaporation and precipitation

Evaporation is effectively a distillation process in which dissolved ions and particles are left behind in the surface water body, and the water droplets that form in the atmosphere (and fall as precipitation) are the closest thing to pure H_2O on earth. In order to understand this distillation process, consider that water molecules (except for those rare ones with the heavier isotopes of oxygen or hydrogen, i.e. ^{17}O, ^{18}O, or ^{2}H) have a mass of 18 Da (or $18\,g\,mol^{-1}$), and typical dissolved ions such as Na^+, Ca^{2+}, Cl^-, and HCO_3^- have masses on the order of 23–61 Da. Thus,

the probability of a water molecule evaporating with an attached, electrostatically bonded ion or molecule tagging along is quite low. Their masses are just too high, and the result is that these solutes are left behind in surface water. As a result, atmospheric water is very dilute, typically with total dissolved solids (TDS) on the order of $<10\,mg\,l^{-1}$ (Table 4.1).

While evaporation tends to result in extremely dilute water vapor, it has the opposite effect on surface water bodies; in extreme scenarios (hot and arid), surface water tends to become progressively more concentrated in dissolved ions with progressive evaporation, which helps to explain why lakes or seas in isolated arid basins become so rich in salts – good examples include the Great Salt Lake in Utah, USA, the Aral Sea (or what is left of it) in central Asia, Lake Magadi in east Africa, and the shrinking Dead Sea in the Middle East, all water bodies with some of the highest TDS in the world. Evaporation of many such arid region lakes and seas is often so pronounced that it produces water that is oversaturated with respect to various dissolved ions, resulting in crystallization of evaporite salts that occur as crusty mineral deposits on shorelines (Figure 4.2) or as chemical sedimentary deposits in lake beds or the sea floor.

4.1.1.2 Infiltration, soils, chemical weathering

Once atmospheric water condenses and falls to the surface as *precipitation*, *infiltration* into soils and rocks tends to foster chemical weathering of minerals (a topic which is covered in detail in Chapter 9). Precipitation is generally slightly acidic (natural pH is generally 5.0–5.5) and this

Environmental and Low-Temperature Geochemistry, Second Edition. Peter Ryan.
© 2020 John Wiley & Sons Ltd. Published 2020 by John Wiley & Sons Ltd.

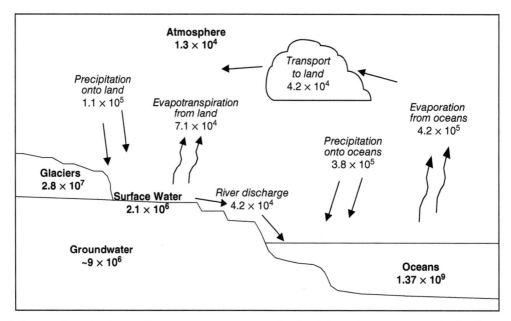

Fig. 4.1   Hydrologic cycle with volumes of reservoirs given in units of km^3 and *fluxes* shown in units of km^3 yr^{-1} . Surface water as shown above refers to fresh water on the continents, and the overwhelming majority of this reservoir is accounted for in a few large lakes. Also, less than one half of groundwater is accessible for use by humans; some is too deep and salty. Source: Modified from Winter et al. (1998).

Table 4.1   Representative values for natural waters (in mg/l) from various reservoirs within the hydrologic cycle.

Ion	Precipitation		Surface water			Groundwater	
	Continental	Coastal	Tropical river	Arid river	Seawater	Carbonate	Granitic
H$^+$ (as pH)	5.0–6.0[a]	5.0–6.0	6.5–7.0	7.0–8.5	8.1[b]	7.3	7.1
Na$^+$	0.2–1.0	1.0–5.0	1.5	117	10 770	6.1	0.8
Mg^{2+}	0.05–0.5	0.4–1.5	1.0	24	1 290	33.5	6.0
K$^+$	0.02–0.3	0.2–0.6	0.8	6.7	399	2.5	0.7
Ca^{2+}	0.02–3.0[a]	0.2–1.5	5.2	109	412	70.6	25.6
Si	0.1	0.02	7.2	30	2.8	9.6	8.0
HCO$_3^-$	0.1	0.1	20	183	122	268	113
Cl$^-$	0.2–2.0	1.0–10	1.1	171	19 500	9.0	1.7
SO$_4^{2-}$	1.0–3.0	1.0–3.0	1.7	238	2712	22.6	5.7
NO$_3^-$	0.4–1.3	0.1–0.5	<1	<1	<1	3.3	0.8

Nitrate values are highly variable due to cycling of organic matter and climate factors.

The example of tropical river water is the Amazon, the arid river is the Rio Grande, and seawater is global ocean mean.

[a]Higher values correspond to windblown dust in arid regions. Areas with acid rain have pH <5.

[b]Increasing CO$_2$ in ocean water is causing pH to decrease.

Source: Data are from Livingstone (1963), Holland (1978), Meybeck (1979), Mayo and Loucks (1995), Berner and Berner (1996), and other sources cited therein.

factor, combined with low concentrations of dissolved ions typical of precipitation water (Table 4.1), facilitates the chemical decomposition of minerals. The following reaction depicts the dissolution of albite (sodium plagioclase feldspar) in the presence of water and H$^+$:

$$NaAlSi_3O_{8(s)} + 4H^+_{(aq)} + 4H_2O_{(l)}$$
$$\rightarrow Na^+_{(aq)} + Al^{3+}_{(aq)} + 3H_4SiO_{4(aq)} \qquad (4.1)$$

In this example, chemical weathering releases dissolved Na, Al, and Si into solution.

Dissolution of calcite can be depicted as:

$$CaCO_3 + H^+ \rightarrow Ca^{2+}_{(aq)} + HCO_3^-_{(aq)} \qquad (4.2)$$

A quick glance at Table 4.1 reveals that the most abundant species in natural waters include Na$^+$, Ca^{2+}, Si (often

Fig. 4.2 The white deposits surrounding the shallow pond in the foreground are evaporite salts that crystallized because evaporation of spring water raised dissolved ion concentrations to saturation. Evaporite minerals in this area include halite, gypsum, calcite, and borax. Source: Death Valley, USA. US Geological Survey Photo Library. Photo ID: Hunt, C.B. 941.

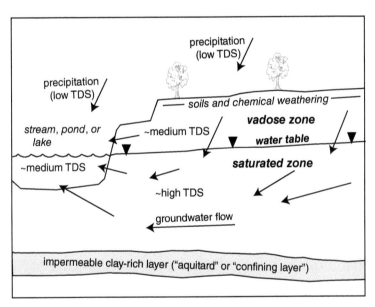

Fig. 4.3 Sketch showing, from top to bottom, the vadose zone (unsaturated zone), which includes soil and any sediment or rock above the water table and the shallow part of an aquifer (the saturated zone). Connections between groundwater and surface water are also shown by arrows indicating, typical groundwater flow paths. Relative total dissolved solids (TDS) of precipitation, soil water, groundwater, and surface water for a temperate region are shown.

as H_4SiO_4), and HCO_3^- (but not Al: its low solubility means that Al tends to become incorporated into minerals or disordered solids rather than remaining in solution). In addition to Na^+, Ca^{2+}, and Si (H_4SiO_4), Mg^{2+}, K^+, and SO_4^{2-} are also mainly derived from weathering of bedrock, while NO_3^- is mostly derived from decomposition of organic matter. Bicarbonate (HCO_3^-) comes from carbonate (e.g. calcite) dissolution, organic matter decomposition, or is delivered in rain as a product of CO_2 dissolution into atmospheric water.

4.1.1.3 Surface water and groundwater

The composition of *surface waters* and *groundwater* is influenced by local rock and soil composition as well as

climate – note the compositional differences in groundwater in carbonate regions as compared to granitic regions (Table 4.1), as well as the difference in surface water composition in tropical compared to arid climates.

The relatively high TDS of groundwater compared to surface water bodies (except where surface waters are concentrated by evaporation) is caused by high residence time of water in the saturated zone (Figure 4.3) and the lack of dilution by precipitation. Return flow of water from the subsurface (vadose or saturated zone) into surface water bodies (Figure 4.3) is the main source of dissolved species in surface water. In humid tropical or temperate regions, abundant precipitation tends to result in lakes and streams which are dilute relative to arid streams and lakes.

The saturated zone occurs below the water table and includes groundwater aquifers as well as saturated clays and bedrock units that do not produce water at sufficient rates to qualify as aquifers ($4\,l\,min^{-1}$ is typically the minimum production rate required for domestic use, and much more is required for agricultural, municipal, or industrial applications). The main gases that occur within pores in the saturated zone occur as small amounts of dissolved gases such as O_2, N_2, H_2S, or CO_2. The saturated zone is less oxidizing than most surface water, ranging from chemically reducing to intermediate between oxidizing and reducing.

Groundwater contains relatively low dissolved O_2 concentrations except in near-surface levels or in cases where surface water rapidly infiltrates, resulting in pores that contain high amounts of dissolved O_2. In contrast to the saturated zone, the vadose zone usually is chemically oxidizing except in cases where abnormally high amounts of organic matter (e.g. below landfills or wetlands) consume O_2 to produce reducing conditions. Ultimately, controls on the composition of surface water and groundwater are diverse and include climate, bedrock and soil type, topography, residence time, crystallization–dissolution of minerals, redox, acid–base reactions, ion exchange, biotic factors, and anthropogenic factors (e.g. land use and contamination).

4.1.1.4 Graphical analysis of climate and surface water composition

The effect of evaporation on surface water is often depicted by plotting the ratio of Na/(Na + Ca), where concentrations are expressed in mg l^{-1} on the x-axis against the log of the concentration of total dissolved solids (TDS, mg l^{-1}) on the y-axis (Gibbs 1970). Halite (NaCl) is much more soluble than calcite (CaCO$_3$), and when the concentration of dissolved solids is driven up by evaporation, the remaining surface water eventually becomes saturated with respect to calcite, causing calcite to crystallize and taking Ca out of solution, thus resulting in increased ratio of Na/Ca in solution. Consequently, streams and lakes from arid regions are characterized by high Na/Ca ratios and high TDS. So too is ocean water. These waters plot in the upper right of Figure 4.4. One environmental consequence of the high Na : Ca ratio and high TDS of surface waters (and therefore also often groundwaters) in many arid regions is sodification of agricultural soils when high-Na waters are used for irrigation. High Na destroys soil structure,

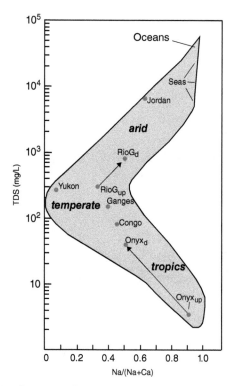

Fig. 4.4 Changes in the ratio of Na/(Na + Ca), based on units of mass (e.g. mg l^{-1}), plotted against total dissolved solids (TDS, in mg l^{-1}), for streams from different climate zones, and also for seas and oceans. Arrows indicate changes in stream composition that occur progressively downstream; for example, with increasing contributions from chemical weathering, tropical soils increase in TDS and the Na/(Na + Ca) ratio decreases. Source: Modified from Gibbs (1970) and Andrews et al. (2004).

reduces permeability, and alters nutrient availability in negative ways.

The points in the lower right of Figure 4.4 mostly represent headwater tropical streams, mountain streams, and meltwater streams with low TDS caused by dilution by rainfall and snowmelt. Streams in temperate climates where water composition is dominated by weathering of bedrock plot in the center left of the diagram. This is because limestones are the most rapidly weathering rocks in most regions of the world, and their dissolution releases Ca ions into solution. The Onyx River of Antarctica is a great example of a stream whose initial (headwater) composition has very low TDS and a high Na/(Na + Ca) ratio; it originates from glacial meltwaters. As it flows downstream, it becomes progressively more influenced by chemical weathering, which increases TDS and lowers the

Na/(Na + Ca) ratio. Another stream that illustrates these concepts well is the Rio Grande of the southwestern USA, which originates in the Rocky Mountains, where chemical weathering initially dominates stream composition, and then flows into arid basins where evaporation eventually is the dominant control on water composition. Conceptually, streams will trend from lower right to center left to upper right as high mountain streams fed by snowmelt (lower right) transition into more-evolved streams with a Ca-rich bedrock weathering signature (center left) and eventually, further downstream, toward the Na-rich, high TDS signature (upper right) produced by evaporation in arid environments.

This background on processes and reservoirs within the hydrologic cycle provides a framework for understanding some of the large-scale natural controls on water chemistry. Section 4.2 steps down to the atomic scale to explore some fundamental aspects of water molecules as they pertain to geochemistry.

4.2 THE STRUCTURE OF WATER – IMPLICATIONS OF GEOMETRY AND POLARITY

The structure and geometry of the water molecule explain many of the chemical and physical aspects of water that influence its behavior in the environment. Water is a covalently bonded molecule with a single oxygen atom in a −2 oxidation state and two +1-charged hydrogen atoms; it is not a linear molecule, and this is crucial to its physical and chemical behavior. The valence electrons of the oxygen atom are arranged in a tetrahedral configuration, where two points of the tetrahedron contain paired electrons of the oxygen atom, whereas the other two points are unpaired electrons that form bonds with H atoms. The result is a dipolar molecule with a negative pole on the side with the O atom and a positive pole on the side with the 2H atoms. The angle between the H atoms is approximately 105°, and the O–H bond distance is 0.97 Å (0.097 nm) (Figure 4.5).

The dipolar nature of H_2O makes it a very effective solvent that can form electrostatic attractions between anions and the H atoms of the positive pole, and between cations and the O atom of the negative pole. Figure 4.6 depicts a schematic example of H_2O dipoles oriented to facilitate solution of Na^+ and Cl^- ions – the dipolar character of water molecules, particularly their orientation

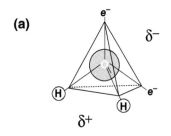

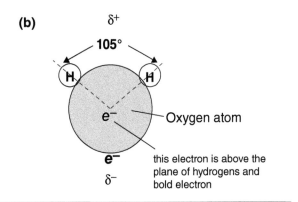

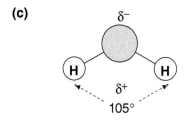

Fig. 4.5 The structure of the H_2O molecule (a) emphasizes the tetrahedral nature of the molecule, and (b, c) shows the structure and dipolar nature in two dimensions.

with respect to dissolved ions, helps to increase distance between cations (e.g. Na^+) and anions (e.g. Cl^-), thus decreasing the force of attraction between cation and anion, enhancing solubility and decreasing the likelihood of crystallizing a solid such as NaCl.

The geometry of the H_2O molecule, combined with hydrogen bonding between adjacent molecules, is responsible for the hexagonal open structure of H_2O_s (ice), resulting in the unusual situation that produces a solid phase ($\rho[H_2O_s] = 0.92\,g\,ml^{-1}$ at 1 atm) with lower density than its liquid phase ($\rho[H_2O_l] = 1.0\,g\,ml^{-1}$ at 1 atm). For this reason, ice floats on water and lakes do not freeze solid in cold winters, thus enabling complex aquatic ecosystems to exist in cold climates.

Hydrogen bonds are a type of van der Waals bond that form as the result of dipole–dipole attraction between a

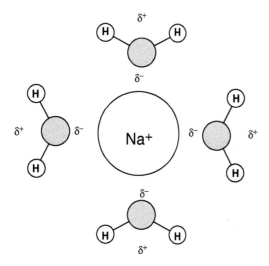

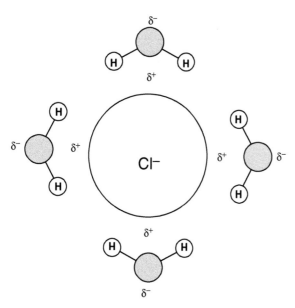

Fig. 4.6 Schematic sketch of the relationship of polar water molecules to a sodium cation (above) and chloride anion (below) in solution.

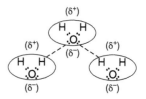

Fig. 4.7 Hydrogen bonds (dashed lines) between adjacent water molecules.

dissolve certain compounds of N and C, and high surface tension that causes capillary rise of water above the water table (Figure 4.3) and into the vadose zone.

4.3 DISSOLVED VERSUS PARTICULATE, SOLUTIONS, AND SUSPENSIONS

Aqueous systems are mixtures that contain a liquid phase (H_2O) with dissolved and suspended constituents. *Dissolved constituents* generally occur as ions that are electrostatically attracted to one of the positive or negative poles of water molecules, and are uniformly distributed throughout the liquid solvent (in an aqueous solution, the solvent is water). They are mobile and tend to diffuse readily and reach relatively constant concentrations within a water body, at least over reasonable spatial and temporal scales. Dissolved constituents will not settle out of water. The suspended (particulate) fraction consists of particles that can be physically separated from liquid, e.g. sand grains that settle out of suspension and form deltas where rivers flow into lakes or seas.

4.3.1 Dissolved constituents and the nature of solutions

Dissolution of the ionic solid halite by water into Na^+ and Cl^- ions produces a simple aqueous solution, which can be represented as:

$$NaCl_{(s)} \rightarrow Na^+_{(aq)} + Cl^-_{(aq)} \qquad (4.3)$$

Water-soluble substances like halite are often referred to as "hydrophilic" (water-loving) – they dissolve because the attraction between the H atom (or O atom) in a water molecule and an anion (or cation) in the solid substance is greater than the attraction between the cation and anion in the solid compound. This is especially true for ionic bonds, which are notoriously unstable in the presence of water. Most salts of Cl^-, SO_4^{2-}, and NO_3^- are easily dissolved (e.g. NaCl, KNO_3); carbonate minerals contain a mixture of bonds that range from covalent (C–O in CO_3^{2-}

hydrogen atom and unshared electron pairs from anions with high electronegativity (e.g. O, N, F, and C, to a lesser extent) in polar molecules. The attraction between an O atom in one H_2O molecule and an H atom in an adjacent H_2O molecule is an example of a hydrogen bond (Figure 4.7). Although intermolecular hydrogen bonds are only ~5% as strong as the intramolecular O–H covalent bonds (O–H hydrogen bond length is 1.97 Å), they are responsible for many of the unique qualities of water, including its anomalously high melting and boiling points compared to other liquids with similar mass, its ability to

anion) to ionic (the bond between Ca and the CO_3^{2-} anion) and are moderately soluble over time spans of months to thousands of years or more, depending on many factors (e.g. climate, composition and acidity of aqueous solution, bacterial activity, and flow rate).

4.3.2 Particulate (suspended) fraction

Minerals or other solids with covalent bonds are far less soluble than ionic solids – a good example is quartz (SiO_2). The polar covalent bond between Si and O atoms in quartz is far greater than the forces that might break those bonds; in an aqueous solution, the likely process by which quartz might dissolve would be the potential bond between an H atom (in H_2O) and an O atom in SiO_2. Quartz is too stable for this to occur at a rate fast enough to be significant in this consideration, so quartz remains in the particulate fraction. Similarly, many organic compounds are effectively insoluble in water, including compounds (polychlorinated biphenyls [PCBs], octane, and many organochlorine pesticides and solvents (refer to Chapter 3)), for similar reasons. Thus, quartz and many other silicate minerals including feldspars and phyllosilicates (clays and micas) and some oxides and organic particles tend to exist in aqueous systems as *particulates*, sediments, or as *suspended material*, rather than as dissolved ions. Ions can also exist as part of the particulate load when they are adsorbed onto mineral surfaces. Grains of quartz sand and phyllosilicates are common components of the suspended load in lakes, rivers, and oceans, as are the ions adsorbed onto their surfaces.

4.3.3 Immiscible liquids

Octane and other liquid hydrocarbons tend to exist as immiscible liquids in aqueous systems, often floating atop surface water due to their lower density ($\leq 1\ mg\ l^{-1}$). Such insoluble compounds are commonly referred to as "hydrophobic" or in environmental science, as non-aqueous phase liquids (NAPLs), a topic covered in Chapter 3.

4.3.4 Dissolved vs. particulate vs. colloidal

Aqueous systems consist of a mixture of dissolved ions and suspended particles, and in some cases, immiscible liquids too. Understanding whether or not a certain element or compound exists in a dissolved state or as particulate matter, or whether an organic liquid has partially dissolved, is often crucial in understanding environmental availability, fate, and transport. Typically, concentrations of soluble ions (e.g. Ca^{2+} or NO_3^-) or soluble liquids (e.g. benzene) in a wetland or lake may be relatively well mixed throughout the water body, whereas particulates and immiscible liquids are much more heterogeneous in their distribution. Particulates tend to get deposited out of suspension in slow-moving currents or stationary water, meaning any elements associated with those particulates would be concentrated in a certain part of the bottom sediments – this is the typical behavior of phosphorus and lead, for example. The behavior of other elements strongly depends on redox conditions, dissolved oxygen (DO), the presence or absence of various other ions in solution, biotic constituents, the presence of sorptive solids, and more.

An additional factor to consider related to the behavior of dissolved and particulate matter is that some particulates are extremely fine-grained, with particle diameters (according to IUPAC nomenclature) ranging from 10 Å (the thickness of a single unit cell of smectite clay) to 10,000 Å (=1 μm = grain of clay). These ultrafine-grained particles are known as *colloids*, and in the realm of environmental geochemistry include members of the smectite clay mineral group, hydroxides of iron and aluminum,

---

Focus Box 4.1

Examples of Dissolved vs. Particulate vs. Immiscible Forms

Uranium is an example of an element that, depending on conditions, can exist in soluble ionic forms or as insoluble particulates. If U in groundwater occurs in a particulate form it is less likely to be transported by slow-moving groundwater flow than would be dissolved uranium, so we would expect less uptake of $U_{particulate}$ into water supply wells than U_{aq}. Furthermore, $U_{particulate}$ is much more likely to be removed by simple filters than dissolved U_{aq}. Low-density immiscible liquids like octane and other fuels are largely partitioned atop the water body, whereas high-density immiscible liquids like chlorinated solvents tend to sink to the bottom of lakes, rivers, and aquifers. However, understanding partitioning of elements and compounds into dissolved vs. particulate (or immiscible liquid) phases is often not a simple task, complicated by the need to quantify kinetic factors, aqueous system composition, mineralogical content, and microbial activity, for example.

and humic organic compounds. Particles of such fine grain size tend to remain suspended in water regardless of flow rate and hence are far more mobile than typical particulates, only settling out of solution when the colloidal particles flocculate under certain geochemical conditions (e.g. seawater can cause clays to flocculate and hence precipitate onto the sea floor). Furthermore, their surfaces are highly reactive exchange sites, so colloids are often responsible for enhanced mobility of otherwise immobile elements (e.g. Cd, Pb, or U sorbed onto mobile, colloidal iron hydroxide in soil or sediments).

4.4 SPECIATION: SIMPLE IONS, POLYATOMIC IONS, AND AQUEOUS COMPLEXES

Not all ions in solution exist as simple *monoatomic ions* like Ca^{2+} or Cl^- but rather occur as charged molecules known as *polyatomic anions*. You have already seen some examples of polyatomic anions – representative members of this group include:
- carbonate (CO_3^{2-})
- bicarbonate (HCO_3^-)
- nitrate (NO_3^-)
- hydroxide (OH^-)
- phosphate (PO_4^{3-}), hydrogen phosphate anion (HPO_4^{2-}), dihydrogen phosphate anion ($H_2PO_4^-$)
- sulfate (SO_4^{2-})
- cyanide (CN^-)
- arsenate (AsO_4^{3-}), arsenite (AsO_3^{3-})
- dichromate ($Cr_2O_7^{2-}$)
- aluminum tetrahydroxide ($Al[OH]_4^-$).

Polyatomic cations are less abundant than their anionic counterparts. Examples include:
- aluminum monohydroxide ($Al[OH]^{2+}$) and aluminum dihydroxide ($Al[OH]_2^+$)
- ammonium (NH_4^+)
- uranyl (UO_2^{2+}).

Complexes are species formed by bonding between two simpler substances that typically results in greater solubility of one of the substances; for example, relatively insoluble trace metals that normally don't occur in solution can be dissolved when they form complexes with species such as organic matter or certain anions. A good example is uranium (in the U^{6+} form), which forms soluble complexes with the hydroxyl and carbonate anions, enhancing its solubility in water: two common uranium complexes are $UO_2(OH)^+$ and $UO_2(CO_3)_2^{2-}$. Organic compounds such as formic acid, humic acid, and oxalate also often form complexes that

enhance trace metal solubility in soils, surface water, and groundwater.

It is also important to recognize that there are some dissolved species that are not ionic. For example, Si commonly exists in solution as silicic acid, H_4SiO_4, which can also be written $SiO_2 \cdot 2H_2O$, illustrating that dissolved silica is effectively hydrated SiO_2. In fact, the best way to envision dissolved silica is as a SiO_4^{4-} tetrahedron bonded to four H^+, a sketch of which is presented in Figure 9.2 in reference to products of chemical weathering of silicate minerals.

The *speciation* of a given element in an aqueous system refers to the form in which it exists – does it occur as a simple dissolved ion, or as a polyatomic ion, or does it occur as a particulate? According to IUPAC terminology, *chemical species* describe the specific form of an element defined with respect to its isotopic composition, electronic or oxidation state, and/or complex or molecular structure. Speciation of an element is the distribution of the element among dissolved and particulate phases in a system. In the study of arsenic speciation, for example, we might inquire whether As occurs in one of the soluble anionic forms (including $H_2AsO_4^-$, $HAsO_4^{2-}$, AsO_4^{3-}, $H_2AsO_3^-$, $HAsO_3^{2-}$), as a solid arsenic sulfide (e.g. As_2S_3, AsS) or iron-arsenic sulfide (e.g. FeAsS), as a complex formed between As and colloidal iron hydroxide or As and dissolved (or particulate) organic matter, or as an ion sorbed (i.e. exchanged) onto the surface of a hydroxide or clay mineral. Considered in this way, speciation analysis involves the determination of chemical species present in solution and in the solid state, and the association of elements with those chemical species. Given the abundance of minerals and related solids in most systems, precise speciation analysis commonly requires determination of the chemical composition of minerals as well as amorphous solids and organic matter. Yet given the greater availability of species in solution, in some cases speciation analysis is more focused on dissolved constituents compared to solid forms.

4.5 CONTROLS ON THE SOLUBILITY OF INORGANIC ELEMENTS AND IONS

Perhaps the simplest means by which ions become dissolved in aqueous solutions is through the congruous (complete) dissolution of a mineral like the ionic salt calcium chloride, $CaCl_2$, for which dissolution is represented as:

$$CaCl_2 = Ca^{2+}_{(aq)} + 2Cl^-_{(aq)} \qquad (4.4)$$

There are numerous factors that control the solubility of solids, such as the mineral $CaCl_2$ and aqueous species like the dissolved Ca and Cl ions shown above. We will first consider a pair of important physical controls: temperature and residence time.

4.5.1 Role of temperature

One important physical control on the solubility of a given compound or element is water temperature, where increasing temperature generally increases solubility of ions due to the increased kinetic energy of the water molecules. However, there are two important exceptions. One pertains to substances that release heat when they dissolve. In such exothermic dissolution reactions, heat is a product of dissolution, so if we consider dissolution in terms of Le Châtelier's principle, adding heat to the solution will drive the reaction in the direction of the reactants, thus inhibiting dissolution.

A good example of an *exothermic dissolution reaction* is HCl in water – the ΔH (change in enthalpy) for $HCl \rightarrow H^+ + Cl^-$ is $-74.8\,kJ\,mol^{-1}$. What this means is that 74.8 kJ of heat are released by each mole of hydrochloric acid that dissolves in water. The negative sign for ΔH signifies that heat is released to the surroundings (in this case, the aqueous solution):

$$HCl \rightarrow H^+ + Cl^- + 74.8\,kJ\,mol^{-1} \qquad (4.5)$$

In Eq. (4.5), the products side shows $74.8\,kJ\,mol^{-1}$ of heat being released upon dissolution of HCl, and Le Châtelier's principle tells us that the solubility of HCl decreases as water temperature increases. (Note: the exothermic nature of this reaction is tangible because a graduated cylinder will heat up when you add HCl to water, something you do when you need to dilute acid for some lab procedures.)

The enthalpy change associated with dissolution of NaCl is $+3.9\,kJ\,mol^{-1}$, i.e. 3.9 kJ of heat are consumed per mol of NaCl that dissolves; the reaction is endothermic (absorbs heat), and predictably, NaCl solubility increases with increasing temperature. The dissolution of KNO_3 in water absorbs 35 kJ of heat per mole ($\Delta H = +35\,kJ\,mol^{-1}$), so (i) KNO_3 solubility increases with increasing temperature, and (ii) the temperature dependence for KNO_3 dissolution is much greater than it is for NaCl (Figure 4.8).

The solubility of gases is inversely proportional to temperature, a phenomenon described by the temperature dependence of the constant in Henry's Law, and also evident through the lens of Le Châtelier's principle, where heat produced by exothermic gas dissolution means that *gas solubility decreases as water becomes warmer*. Examples of this phenomenon at work in earth surface systems

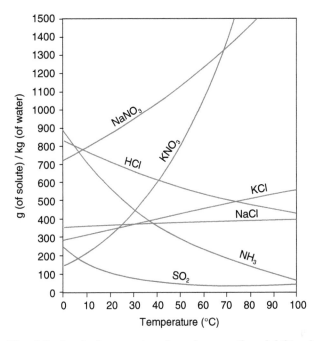

Fig. 4.8 Graph of temperature dependence on the solubility of various ionic compounds. Units are $g\,kg^{-1}$, which is effectively $g\,l^{-1}$. Dissolution of ammonia (NH_3) and sulfur dioxide (SO_2) in water is exothermic and these plots are typical for gases: higher T = lower solubility. The dissolution of the ionic salts is endothermic, but temperature is a much stronger factor for the nitrate salts than it is for the chloride salts. HCl is ionic but also can occur as a gas, so its behavior is intermediate to the salts and the gases.

include (i) thermal pollution of surface waters that causes decreases in DO that can impair aquatic ecosystems; and (ii) warming oceans that will have diminished capacity to retain dissolved CO_2. Examples of ΔH values for selected gases include $CO_2 = -19\,kJ\,mol^{-1}$, $O_2 = -12\,kJ\,mol^{-1}$, and $SO_2 = -39\,kJ\,mol^{-1}$.

4.5.2 Residence time

While it does not affect the inherent solubility of a given element or compound, residence time is an important control on the extent of dissolution because longer residence (exposure) times produce progressively more dissolution, although rate tends to decrease logarithmically with time as the system approaches saturation. Figures 1.13 and 1.18 depict this type of trajectory, showing rapid change initially, followed by progression toward a new state but at progressively slower rates, like a first-order kinetic reaction.

Residence time describes the amount of time a representative molecule of water will spend in a reservoir, where reservoirs can range from very small (e.g. pore spaces in soils or sediments) to large (e.g. lakes, the atmosphere, or the ocean). At steady state, i.e. where rate of inflow = rate of outflow, residence time is calculated as follows:

$$t_{res} = \frac{\text{volume (or mass) of reservoir (e.g. in l)}}{\text{flux into (or out of) reservoir (e.g. in l min}^{-1})} \tag{4.6}$$

Consider a small lake with a volume of $9.3\,km^3$ and a flux of $1.2\,km^3\,yr^{-1}$ (the flux here is river flow into the lake and out of the lake at equal rates). This scenario gives a residence time of 7.8 years. In a scenario where flux were reduced by 50%, residence time would double ($9.3\,km^3/0.6\,km^3\,yr^{-1} = 15.5$ years), and while the like might become more uniformly mixed (homogeneous), dissolved constituents (nutrients or contaminants e.g.) would have a longer residence time in the lake. The global ocean also serves to illustrate the concept of residence time. The approximate mass of water in the oceans of the world is $1.37 \times 10^{21}\,kg$ and flux is $3.9 \times 10^{17}\,kg$ of water per year. The flux assumes a steady state where the cumulative discharge of all of the world's rivers into the ocean plus precipitation into the ocean is $3.6 \times 10^4\,km^3\,yr^{-1}$, a value equaled by water lost by evaporation and seepage into subduction zones, sediments, and other sinks. This yields a mean residence time of water molecules in the oceans of $1.35 \times 10^{21}\,kg/3.9 \times 10^{17}\,kg\,yr^{-1} = 3.5 \times 10^3$ years (3500 years).

4.5.3 The ratio of ionic charge : ionic radius and its effect on solubility

The ratio of ionic charge : ionic radius ($|z|/r$) – also known as ionic potential – is useful in predicting solubility of ions because in general, the greater this ratio, the lower the solubility. So, we would predict that high-charge, low-radius ions like Al^{3+}, C^{4+}, Ti^{4+}, Si^{4+}, Zr^{4+}, N^{5+}, P^{5+}, and S^{6+} would be relatively insoluble and that relatively low-charge, high-radius ions like Na^+, K^+, Ca^{2+}, Pb^{2+}, and Br^- and Cl^- would be soluble, and we would be mostly correct. However, in some cases, the formation of polyatomic anions can greatly enhance the solubility of low-radius, high-charge ions like C^{4+} (HCO_3^-), N^{5+} (NO_3^-), S^{6+} (SO_4^{2-}), and, although not an anion, the formation of the polyatomic aqueous species H_4SiO_4 greatly enhances the solubility of Si^{4+}. Figure 4.9 depicts numerous elements plotted as a function of $|z|/r$ vs. the proportion present as particulate matter compared to the proportion present in solution (the higher this value, the less soluble). Notable features include the high solubility of low-charge, large-radius ions like Cl^-, Na^+, Ca^{2+}, and other halogens, alkalis, and alkaline earth metals; the intermediate solubilities of divalent transition metals (Zn^{2+}, Co^{2+}, etc.); low solubilities of high-charge, intermediate-size cations such as Al^{3+}, Fe^{3+}, and Ti^{4+}; and relatively high solubility of high-charge, small-radius cations like C^{4+}, N^{5+}, and S^{6+} which form the large-radius polyatomic anions carbonate, nitrate, and sulfate.

4.5.4 Reduction–oxidation reactions

Many elements are sensitive to reduction and oxidation; iron is a good example of a common element whose behavior in strongly controlled by redox conditions. Its oxidation state is controlled by redox environment, and in turn, the oxidation state largely dictates whether iron exists in dissolved or particulate form – ferrous iron (Fe^{2+})

Focus Box 4.2

Case Study of Aquifer Residence Time

In complex environments with multiple fluxes (inflows and outflows), a simple residence time calculation may not be sufficient. A good example is the carbonate aquifer of the Lakhssas Plateau in the Anti-Atlas Mountains of Morocco, where a study to understand groundwater age and residence time was carried out to inform water resource management (Ettayfi et al. 2012). Water in different compartments of the aquifer system were fingerprinted using isotopic tracers (oxygen, hydrogen, carbon, strontium) and radiocarbon dating. The radiocarbon dates indicate mean residence times of 10–15 ka for warm and saline waters in one aquifer compartment and ≤3 ka for fresher and cooler water in other parts of the system. The older water is isotopically distinct from the younger water, and the high residence time has at least two implications: (i) it has circulated deeper and dissolved carbonate and sulfate minerals along the way, producing water with high TDS; and (ii) groundwater in that compartment is not recharging, indicating that modern extraction effectively constitutes groundwater mining. Other compartments of the aquifer have isotopic signatures that indicate recent recharge from coastal moisture derived from the Atlantic Ocean, and the heterogeneity of the aquifer in terms of recharge and water quality requires informed management to sustain the water resource.

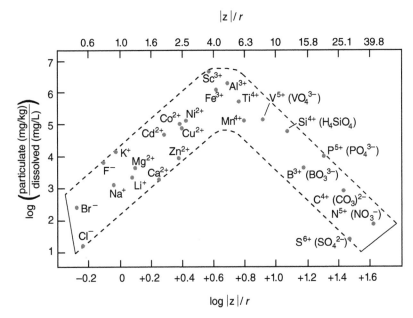

Fig. 4.9 Prediction of solubility of aqueous ions based on the ratio of ionic charge (z) to ionic radius (r, in Å), a ratio known as ionic potential. In the case of polyatomic molecules (e.g. H_4SiO_4), radius (r) is for the elemental ion shown (e.g. Si^{4+}), not for the entire molecule. The z values on both upper and lower x-axis are absolute values, and the y-axis data are based on concentrations in river water determined by Martin and Whitfield (1983). The plot shown is based on a graph in Andrews et al. (2004).

occurs in reducing (O_2-poor) aqueous systems and is relatively soluble, while the ferric iron (Fe^{3+}) that occurs in oxidizing (O_2-rich) systems is largely insoluble. Numerous elements have two or more common oxidation states, and changes in oxidation state produced by transfer of valence electrons is one of the most important concepts to consider when assessing elemental behavior in aqueous systems. In addition to Fe, numerous other elements exist in various oxidation states in nature, including the following:

- arsenic (As^{3-}, As^0, As^{3+}, As^{5+})
- carbon (C^{4-}, C^0, C^{2+}, C^{4+})
- chromium (Cr^{3+}, Cr^{6+})
- copper (Cu^0, Cu^{+1}, Cu^{2+})
- manganese (Mn^{2+}, Mn^{4+}, Mn^{7+})
- nitrogen (N^{3-}, N^0, N^{3+}, N^{5+})
- sulfur (S^{2-}, S^0, S^{4+}, S^{6+})
- uranium (U^{4+}, U^{6+}).

There are also many elements in nature that have only one oxidation state, including hydrogen (H^+), the alkali metals (Na^+, K^+), and alkaline earth metals (Mg^{2+}, Ca^{2+}).

When the highly electronegative elements oxygen and fluorine form chemical bonds in compounds, they have predictable charges – oxygen is O^{2-} in compounds (except in rare cases like peroxides, where it is partially reduced to O^- or a mixture of O^0 and O^{2-}) and fluorine is F^- in compounds. Redox reactions can also affect the elements that exist with only one valence state in nature. Examples include Ni^{2+} adsorbed to iron hydroxide that gets released to solution in conjunction with iron hydroxide dissolution that occurs when iron is reduced ($Fe^{3+} \rightarrow Fe^{2+}$), or Pb^{2+} that is released to solution when galena (PbS) dissolves via sulfide oxidation. Elements that occur in their native state, including metals like gold and silver (Au, Ag), the carbon minerals diamond and graphite (both are pure C), and the diatomic gases O_2, N_2, S_2, have oxidation states of zero.

Environments with abundant O_2 like the atmosphere, bubbly streams, aerated soils, and shallow groundwater are typical oxidizing environments, whereas O_2-poor

---

Focus Box 4.3

Chromium, Redox, Solubility, and Toxicity

As would be predicted based on $|z|/r$ rationale, the trivalent chromium ion (Cr^{3+}) is relatively insoluble (it is also nontoxic), yet when oxidized to the hexavalent state (Cr^{6+}), chromium is soluble as an oxyanion (soluble CrO_4^{2-} or $Cr_2O_7^{2-}$). This is important because in the hexavalent state, Cr is toxic. Predictably, if hexavalent chromium is reduced to Cr^{3+}, it becomes less soluble and less toxic. Thus, reduction and oxidation reactions, often referred to informally as redox reactions, are additional important controls on the solubility and bioavailability of many elements, particularly those that can exist in more than one oxidation state.

environments like anoxic wetlands, deep groundwater, and waterlogged soils are typical reducing environments.

4.5.4.1 Half-cell reactions

Half-cell reactions depict reduction or oxidation reactions for individual elements involved in a redox reaction involving two or more species, at least one undergoing oxidation and at least one undergoing reduction. The electron transfer associated with redox reactions is measurable (reduction–oxidation involves transfer of electrons and thus produces a current), and by convention, values are reported relative to the value for the oxidation of H_2 gas:

$$\tfrac{1}{2}H_2 = H^+ + e^- \qquad (4.7)$$

In Table 4.2, the values assigned to each half-cell reaction are defined relative to the standard hydrogen electrode (SHE), which is given the arbitrary standard electrode potential (E^o) value of zero. Note that E^o values in Table 4.2 are presented in order from strong oxidizing agents (e.g. F_2) to weak oxidizing agents (e.g. Li^+).

A positive E^o value indicates that an element is an electron recipient (oxidizing agent) likely to cause oxidation of less-electronegative elements (those lower in the table) – these elements (or ions) with positive E^o are the oxidizing agents that include the gases oxygen, fluorine, and chlorine, which respectively undergo reduction to O^{2-}, F^-, and Cl^-. Note that elements with negative E^o values are mostly alkali, alkaline earth, and trace metals, elements that are likely to undergo oxidation to forms ions such as Na^+, Fe^{2+}, Ni^{2+}, and Al^{3+}. Positive E^o values indicate that the reaction will occur spontaneously as written in Table 4.2, and the strongest oxidizing agents (those most likely to undergo reduction) have the most positive E^o values. By this convention, all reactions are written as reduction reactions, so keep in mind that half-cell reactions with negative E^o are not spontaneous as written and are likely to proceed in the reverse direction (the reaction of Na^+ to Na^0 will not occur in natural waters). You will also observe that some metals are intermingled with the strong oxidizers O_2, F_2, and Cl_2 – these are metals that can occur in nature in their native states (e.g. Au^0, Cu^0, Hg^0, Pt^0). (Note: in some tables you will find that the signs are opposite of those presented here, in which case negative values are associated with oxidizers like O_2 and F_2 and positive values with elements likely to be oxidized (e.g. alkalis, alkaline earths, trace metals.)

Standard electrode potential values are useful because they allow us to compare the tendency of one element

Table 4.2 Half-cell reactions and standard electrode potentials (25 °C, 1 atm) presented from strongest oxidizing agents (top) to weakest oxidizing agents (strongest reducers, bottom).

Half-reaction					Standard electrode potential, E^o (V)
Oxidizing agent			Reducing agent		
F_2	+	$2e^-$	→	$2F^-$	+2.88
Au^+	+	e^-	→	Au^0	+2.64
Cl_2	+	$2e^-$	→	$2Cl^-$	+1.36
$O_2 + 2e^-$	+	$4H^+$	→	$2H_2O$	+1.23
Br_2	+	$2e^-$	→	$2Br^-$	+1.08
Hg_2^{2+}	+	$2e^-$	→	$2Hg^0$	+0.85
I_2	+	$2e^-$	→	$2I^-$	+0.53
Cu^+	+	e^-	→	Cu^0	+0.52
Cu^{2+}	+	$2e^-$	→	Cu^0	+0.34
$2H^+$	+	$2e^-$	→	H_2	0.00
Pb^{2+}	+	$2e^-$	→	Pb^0	−0.13
Ni^{2+}	+	$2e^-$	→	Ni^0	−0.24
Cd^{2+}	+	$2e^-$	→	Cd^0	−0.40
Fe^{2+}	+	$2e^-$	→	Fe^0	−0.41
S^0	+	$2e^-$	→	S^{-2}	−0.44
Cr^{3+}	+	$3e^-$	→	Cr^0	−0.74
Zn^{2+}	+	$2e^-$	→	Zn^0	−0.76
U^{4+}	+	$4e^-$	→	U^0	−1.38
Al^{3+}	+	$3e^-$	→	Al^0	−1.70
La^{3+}	+	$3e^-$	→	La^0	−2.36
Mg^{2+}	+	$2e^-$	→	Mg^0	−2.36
Rb^+	+	e^-	→	Rb^0	−2.60
Na^+	+	e^-	→	Na^0	−2.71
Ca^{2+}	+	$2e^-$	→	Ca^0	−2.87
Sr^{2+}	+	$2e^-$	→	Sr^0	−2.90
Ba^{2+}	+	$2e^-$	→	Ba^0	−2.91
K^+	+	e^-	→	K^0	−2.94
Li^+	+	e^-	→	Li^0	−3.04

to undergo oxidation or reduction in the presence of another element. Consider the following two half-cell reactions:

$$Fe^{2+} + 2e^- \rightarrow Fe^0 \; (E^o = -0.41)$$

$$Cu^{2+} + 2e^- \rightarrow Cu^0 \rightarrow (E^o = +0.34)$$

These two half-cell reactions show reduction of divalent cations to native elements. For Fe, the negative E^o of this reaction indicates that it is not spontaneous as written, but rather is spontaneous in the opposite direction. The positive E^o of the Cu reduction reaction indicates that it is spontaneous in the direction written, so we can rewrite the equations in the directions of spontaneity:

$$Fe^0 \rightarrow Fe^{2+} + 2e^- \; (E^o = +0.41)$$

$$Cu^{2+} + 2e^- \rightarrow Cu^0 \rightarrow (E^o = +0.34)$$

Note that it is necessary to change the sign of E^o for the Fe half-cell reaction when the direction is reversed. The sum of the two reactions produces the following reduction–oxidation reaction (and E^o for the reaction):

$$Cu^{2+} + Fe^0 \rightarrow Cu^0 + Fe^{2+} \; (E^o = +0.75)$$

This indicates that in an aqueous system, iron will tend to be oxidized by copper to Fe^{2+} ions and the copper will occur as native copper metal, helping to explain why in some ore bodies, Cu exists as pure native copper (Cu^0) while iron exists as Fe^{2+} in pyrite (FeS_2).

4.5.4.2 Redox reactions in the environment

Organisms and organic matter often strongly influence Eh; for example, decomposition of organic matter typically consumes O_2 and results in reducing conditions. One classic example is eutrophication, where degradation of algae in lakes, ponds, and shallow marine waters reduces dissolved O_2 in the water. The microbial organisms responsible for degradation (consumption) of algae are mainly aerobes who consume O_2, and the decrease in O_2 ultimately alters community structure for the worse and in some cases increases mobility of redox-sensitive metals such as Cr, Fe, and Mn.

Another example of the effect of organic matter on Eh and trace element mobility is petroleum leaks into groundwater, where microbial degradation of petroleum compounds consumes O_2 and reduces Eh. This can result in reduction of ferric to ferrous iron and lead to dissolution of iron hydroxides, which in turn releases arsenic into solution – aquifers where dissolved arsenic was once well below drinking water standards (e.g. 10 ppb in the USA) can experience substantial increases of dissolved arsenic that exceed standards following petroleum leaks.

A few additional examples of how redox conditions control fate and transport of aqueous species include:

1 Nitrogen is less mobile in soils when it exists as N^{3-} in the reduced polyatomic ammonium cation (NH_4^+) as compared to when it exists as N^{5+} in the highly mobile, oxidized nitrate anion (NO_3^-) – this is an important consideration because migration of N compounds into surface water can lead to eutrophication (toxic algal blooms and hypoxia), a threat to many aquatic ecosystems;

2 Metal-bearing sulfide minerals such as pyrite (FeS_2), arsenopyrite (FeAsS), and galena (PbS) are stable in reducing environments but decompose in oxidizing environments driven by oxidation of the sulfide anion (S^{2-}) to S^{6+} (which occurs in the sulfate anion, SO_4^{2-}) – in this case, oxidation can result in release of trace metals and arsenic to soils and aquatic systems; and

3 Organic compounds, including both natural (e.g. humus in soil) and synthetic/refined (e.g. fuels, solvents, pesticides, and pharmaceuticals), are relatively stable in reducing environments but tend to decompose to products like CO_2 and H_2O in oxidizing environments.

The quantity of *DO* (as O_2 gas dissolved in water) strongly influences redox conditions in most aqueous systems. Highly oxygenated surface waters, e.g. in cool streams and lakes, contain approximately 7–9 mg l^{-1} of DO, and comprise the most oxidizing of aqueous systems at and near the Earth surface. In such high-DO environments, redox-dependent elements occur in oxidized states (e.g. Fe^{3+}, N^{5+} in NO_3^-, S^{6+} in SO_4^{2-}, Cr^{6+} in $Cr_2O_7^{2-}$). Wetlands and groundwaters contain naturally lower amounts of DO than do surface waters; for example, groundwaters generally contain DO from 1 to 6 mg l^{-1}, with concentrations decreasing with depth, and wetlands contain similarly low values (wetland sediments are often anoxic, with 0 mg l^{-1} DO). Accordingly, oxygen-poor environments of deep groundwater and stagnant wetlands (especially wetland sediments) contain reduced species such as S^{2-} (which occurs in sulfide minerals like pyrite or as H_2S, hydrogen sulfide gas), Fe^{2+} (in sulfide and silicate minerals), and N^{3-} (in NH_4^+, ammonium, or NH_3, ammonia gas). It is worth noting that values of DO below 5–6 mg l^{-1} generally do not foster healthy aquatic ecosystems (with the exception of natural swamps or deep marine realms that are adapted to these conditions).

The effects of reduction–oxidation on water chemistry appear again in Section 4.10 in terms of Eh–pH (pe–pH) diagrams, an approach which considers the paired role of redox and pH on the mobility of elements and compounds.

4.5.5 Acid–base considerations and pH

The pH of an aqueous solution exerts a strong control on the composition of aqueous systems, influencing the speciation and solubility of many elements and compounds. For example, metals are generally more soluble in acidic water than in neutral or basic water, and highly basic waters can enhance the solubility of compounds of elements such as aluminum, arsenic, and selenium.

The carbonate system strongly affects the pH of natural waters, and while this topic is covered in detail later in this chapter and Chapter 5, important points to consider at this point are: (i) the presence of dissolved CO_2 (as H_2CO_3, carbonic acid) in atmospheric water tends to result in pH values $\sim$5.0–5.5; and (ii) in soils, chemical weathering reactions tend to consume H^+ from carbonic H_2CO_3 (delivered in precipitation) and produce surface water and groundwater with pH values between 6.5–8.5. In some cases stronger acids may control pH, for example sulfuric acid (H_2SO_4) in acid mine drainage surface waters (often with pH = 2–3) and acid precipitation with pH $\sim$3.0–4.5. Evaporation in arid environments may produce briny sea or lake waters with pH values = 9–12.

Aluminum is an excellent example of an element whose solubility in solution is strongly influenced by pH. At neutral pH, aluminum is nearly insoluble and exists as the Al^{3+} ion in silicates such as kaolinite ($Al_2Si_2O_5[OH]_4$) and hydroxides such as gibbsite ($Al[OH]_3$). However, with increasing acidity, the affinity of hydrogen ions (H^+) for the hydroxyl anion (OH^-) in the mineral is greater than the bond between the OH^- and the Al^{3+} cation in the mineral lattice. This process, known as *hydrolysis*, is the dissolution of minerals and other solids caused by bonding of H^+ to an anion in the solid.

Two important consequences of hydrolysis are:

1 Consumption of H^+ ions that lowers solution acidity (i.e. buffers the system by consuming H^+, thus increasing pH); and
2 Dissolution of solids that releases ions into solution.

Aluminum and the mineral gibbsite allow us to examine how variations in pH of aqueous solutions control solution composition. In an acidic solution, hydrolysis of gibbsite can be depicted as follows:

$$Al(OH)_3 + H^+ = Al(OH)_2^+ + H_2O \qquad (4.8)$$

The products of this reaction are dissolved aluminum as a polyatomic aluminum hydroxide cation, $Al(OH)_2^+$, and water. Further increases in acidity can result in the following sequence of reactions:

$$Al(OH)_2^+ + H^+ = Al(OH)^{2+} + H_2O \qquad (4.9)$$

$$Al(OH)^{2+} + H^+ = Al^{3+} + H_2O \qquad (4.10)$$

where dissolved aluminum hydroxide anions are progressively stripped of OH^- by hydrolysis, ultimately resulting in dissolved aluminum as the simple Al^{3+} cation in highly acidic waters (Figure 4.10).

Speciation of dissolved aluminum as a function of pH – ranging from particulate $Al(OH)_3$ at neutral pH to

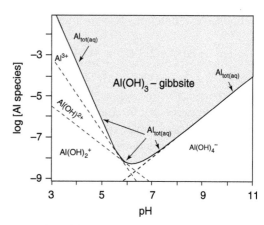

Fig. 4.10 The solid line forming the bottom of the shaded area is the sum of all dissolved Al species, indicating that dissolved Al in water reaches its lowest concentration at a pH of 6. Note the exponential increase in Al solubility with decreasing or increasing pH. The shaded area represents gibbsite – the mineral $Al(OH)_3$ – which will be most stable where Al solubility is lowest, i.e. pH = 6. Dashed lines correspond to concentrations of Al species indicated. At a pH of 3, Al speciation is predicted to be $Al(OH)_2^+ > Al(OH)^{2+} \geq Al^{3+}$, and gibbsite is predicted to be almost absent.

$Al(OH)_2^+$, $Al(OH)^{2+}$, and Al^{3+} with increasing acidity – is depicted in Figure 4.10.

Figure 4.10 also illustrates the principle that, in addition to increasing solubility at low pH, Al solubility also increases with increasing pH, an effect caused by reaction of $Al(OH)_3$ with H_2O in basic solutions:

$$Al(OH)_3 + H_2O = Al(OH)_4^- + H^+ \qquad (4.11)$$

In basic solutions where $[OH^-]$ is orders of magnitude greater than $[H^+]$, aluminum hydroxide (as gibbsite) reacts with water to produce hydrogen ions in order to bring the system closer to equilibrium, and the resulting $Al(OH)_4^-$ anions are soluble in alkaline water.

An alternative way to visualize $Al(OH)_3$ solubility in basic solutions is as follows, emphasizing the effect of high concentration of OH^- (i.e. high pH):

$$Al(OH)_3 + OH^- = Al(OH)_4^- \qquad (4.12)$$

Many other metals in addition to Al exhibit *amphoteric* behavior in solution, including arsenic (technically a metalloid rather than a metal), cadmium, iron, lead, and zinc – such amphoteric metals are able to react as either an acid or base, meaning that they can have higher solubilities at acidic and basic pH values as compared to neutral pH.

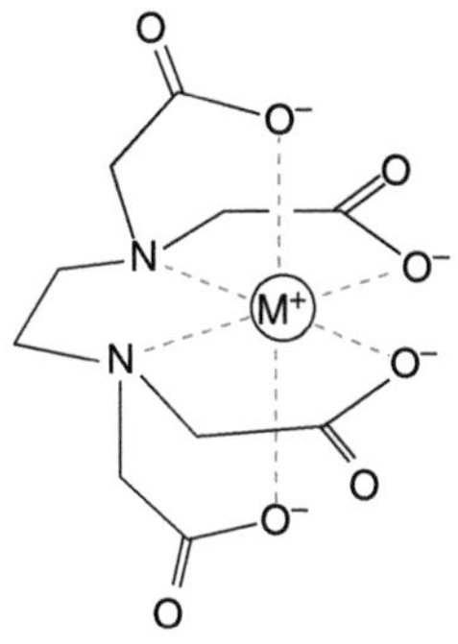

Fig. 4.11 Structural formula of ethylenediaminetetraacetate (EDTA) molecule, a good example of a chelating compound. Carbon atoms occur at all of the bends in the sketch. M^+ represents any metal cation.

4.5.6 Ligands and elemental mobility

Ligands are neutral molecules or anions (often organic) that can combine with a metal cation to form a *complex* that is usually more soluble than the un-complexed metal cation. Inorganic anion ligands include many of the examples presented in previous sections, including hydroxyl (OH^-), bicarbonate (HCO_3^-), sulfate (SO_4^{2-}), and phosphate (PO_4^{3-}). Common organic ligands include acetate (CH_3COO^-), oxalate ($C_2O_4^{2-}$), and the large molecular weight anion ethylenediaminetetraacetate (EDTA) (Figure 4.11).

EDTA is special type of ligand known as a *chelate*, a molecule that is able to form multiple bonds around a metal cation in a cage-like structure. Note the coordination of the metal cation (M^+) to multiple bonding sites within the chelate. In soils, the large molecular weight humic and fulvic compounds, products of decomposition of organic (~plant) matter, are important chelating agents that enhance migration of Al and Fe. In other cases, the mobility of trace elements ranging from arsenic to zinc may be enhanced by complexation to ligands.

4.6 ION ACTIVITIES, IONIC STRENGTH, TDS

4.6.1 Ion activity coefficients

In concentrated solutions (e.g. seawater, deep groundwater brines, acid mine drainage), the spacing between dissolved ions is low enough so that nearby cations and anions are electrostatically attracted to each other. These attractions diminish the ability of the ions to participate in chemical reactions (e.g. mineral growth, ion exchange), and because their effective concentration is lower than their actual concentration in solution, the concentration (of species i) is multiplied by an activity coefficient (γ) to produce a value known as the ion activity:

$$\mathbf{a}_i = [i] \times \gamma \qquad (4.13)$$

where $\mathbf{a}_i$ is activity of ion i, $[i]$ = concentration of ion i, and γ_i is the activity coefficient of ion i.

In dilute solutions, γ is approximately 1 and thus activity is nearly equal to concentration. In more concentrated solutions such as saline waters, many groundwaters, or acid mine waters, $\gamma < 1$ and the value of γ decreases with increasing concentration and temperature, reflecting the fact that higher temperatures and ion concentrations cause divergence of concentration and activity. In most fresh surface waters and groundwaters, γ values are ~0.8–1.0.

Ion activity coefficients can be calculated from this Debye-Huckel equation as follows:

$$\log \gamma_i = -A \times z_i^2 \times \sqrt{I} \qquad (4.14)$$

A is a constant that accounts for pressure and temperature. At 1 bar of pressure (acceptable for surface water and most groundwater), values of A range from 0.4913 at 0 °C to 0.5042 at 20 °C and 0.5221 at 40 °C. The term z_i is the ionic charge and I is ionic strength. This form of the Debye-Huckel equation is valid for ionic strengths (I) $< 10^{-2}$, and thus applies to most fresh waters, but not brines or sea water. For reference, arid streams with high dissolved solids are in the I = 10^{-2} range and most groundwater is similar or more dilute; temperate and tropical surface waters tend to have ionic strengths that are much lower (more dilute) than 10^{-2}. For briny solutions, a more detailed version of the Debye-Huckel equation that takes into account the finite size of ions is needed, but for the types of water present in most streams, lakes, and aquifers, the form presented here is sufficient.

A pair of calculations help to illustrate the importance of considering ion activity. In both calculations, temperature = 20 °C. The variables are the cations (Na^+ and Ca^{2+}) and the ionic strength (10^{-2} and 10^{-4}). In order to solve for γ_i the equation can be rearranged as:

$$\gamma_i = 10^{(-A \times z_i^2 \times \sqrt{I})} \qquad (4.15)$$

For Na^+ at I = 10^{-2}, γ_i = 0.89.
For Na^+ at I = 10^{-4}, γ_i = 0.99.
For Ca^{2+} at I = 10^{-2}, γ_i = 0.62.
For Ca^{2+} at I = 10^{-4}, γ_i = 0.95.

These examples help to illustrate that, in dilute solutions (e.g. $I = 10^{-4}$), activity coefficients are nearly 1.0 and the activity of a given ion is almost equal to its concentration. However, as ionic strength increases, ion activity decreases, and this effect is more pronounced for higher-charge ions. In effect, for $I = 10^{-2}$ at 20 °C, only ~62% of the calcium in solution is actually available to react. The remainder is likely bound in aqueous complexes such as $CaCO_3^0$ or $CaSO_4^0$ (where the superscript "0" indicates an uncharged aqueous complex that is not a mineral).

4.6.2 Ion activity product

Consider a general reaction at equilibrium where:

$$aA + bB = cC + dD \qquad (4.16)$$

A and B, and C and D are reactants and products, respectively, and the lowercase a, b, c, and d represent molar fractions of reactant or product. Recall from Chapter 1 the equation for the equilibrium constant:

$$K_{eq} = [C]^c \times [D]^d / ([A]^a \times [B]^b) \qquad (4.17)$$

where $[C]^c$ represents the concentration of reactant C raised to the c power, and so on. The term on the right side of the equal sign is the equilibrium product. We can also write this equation using ion activities rather than concentrations.

$$K_{eq} = a_C{}^c \times a_D{}^d / (a_A{}^a \times a_B{}^b) \qquad (4.18)$$

If the system is not at equilibrium, i.e. if the expression on the right side is greater to or less than the K_{eq}, then the expression on the right side is an *ion activity product (IAP)*, where:

$$IAP = a_C{}^c \times a_D{}^d / (a_A{}^a \times a_B{}^b) \qquad (4.19)$$

If the IAP is less than the K_{eq}, then the reaction will progress in the direction of products (i.e. to the right). Conversely, if the IAP > K_{eq}, the reaction will proceed to the left (forming A and B from C and D). The saturation state of the solution is represented by the symbol Ω (omega), where

$$\Omega = IAP / K_{sp} \qquad (4.20)$$

The following equations describe how the IAP is useful for examining saturation state:
- If $\Omega > 1$, the solution is oversaturated (or supersaturated);

- If $\Omega = 1$, the solution is at saturation;
- If $\Omega < 1$, the solution is undersaturated.

This should make sense intuitively, because if a system contains high concentrations (or activities) of reactants and low concentrations (or activities) of products, the reaction will proceed in the direction of products.

4.6.3 Ionic strength

The *ionic strength* of a solution is determined by:
1 concentrations of ions in solution; and
2 ionic charges of the ions in solution.
   Ionic strength is defined as follows:

$$I = 0.5 \, \Sigma \, [i] \times z_i{}^2 \qquad (4.21)$$

Units of ion concentration $[i]$ are mol/l and z_i is the charge on ion i. The ionic strength of a solution is thus the sum of the products of the concentrations of each ion in solution times their ionic charge squared, for all ions in solution. Typically, I for freshwater is approximately 10^{-3} to 10^{-4} mol l^{-1}, whereas seawater has a typical value of 0.7 mol l^{-1}. If we consider the dissolved major ions in a sample of groundwater from a bedrock aquifer in Vermont with the following composition, we can compute ionic strength (Table 4.3).

Given that the initial units of mg/l were converted to mmol units, it is necessary to convert the sum (in this case, 12.63) to molar units by multiplying by 10^{-3}. The resulting value is then multiplied by 0.5, and the ionic strength of this sample is 6.32×10^{-3}, typical of groundwater from deep aquifers of relatively high ionic strength (note that values are often presented unitless although technically they would be expressed as mol × charge^2 l^{-1}).

Table 4.3 Computation of ionic strength from dataset with concentrations in mg l^{-1}.

	mg l^{-1}	mmol l^{-1}	$z_i{}^2$	$[i] \times z_i{}^2$
pH = 7.5	—	—	—	—
Na$^+$	4.79	0.21	1	0.21
Mg^{2+}	21.8	0.90	4	3.59
K$^+$	0.61	0.02	1	0.02
Ca^{2+}	46.8	1.15	4	4.59
HCO$_3^-$	205.0	3.36	1	3.36
Cl$^-$	0.90	0.03	1	0.03
SO$_4^{2-}$	20.3	0.21	4	0.85
SUM	300.2	5.87		12.63
$I =$	6.32×10^{-3}			

Table 4.4 Ranges of total dissolved solids for different classifications of waters (after Gorrell 1958).

Type	TDS (mg l^{-1})
Fresh	$<10^3$
Brackish	10^3–10^4
Saline	10^4–10^5
Brine	$>10^5$

4.6.4 Total dissolved solids

The sum of the concentrations of all dissolved ions in solution is termed the TDS, and values are typically presented in terms of mg l^{-1} or mmol l^{-1} . For the groundwater example above, TDS is the sum of the concentrations of all cations (Na^+, Mg^{2+}, K^+, Ca^{2+}) and anions (HCO_3^-, Cl^-, SO_4^{2-}) in solution, which gives a value of 300.2 mg l^{-1} or 5.87 mmol l^{-1}. You will note that, unlike ionic strength, TDS does not account for ionic charge. TDS classifies waters as shown in Table 4.4.

Brackish waters occur in mixing zones (e.g. estuaries) between fresh water and marine water, while brines occur where hydrothermal processes or evaporation produce waters with extremely high TDS (> 35 g l^{-1}).

Interactions among solutes (mainly dissolved ions) and solids (mainly minerals) exert a strong control on water composition. Ions can be removed from solution by sorption to the surfaces of solids such as clay minerals, hydroxides, and organic matter, or by mineral crystallization, where mineral growth proceeds by precipitation of ions from solution.

4.7 SOLUBILITY PRODUCTS, SATURATION

The solubility of a compound can be determined by measuring the amount of the compound that dissolves to form a saturated solution; at saturation, no net dissolution occurs and the solid–aqueous solution system reaches a dynamic equilibrium. Solubility and saturation are two important concepts in the realm of aqueous geochemistry – highly soluble minerals will result in high concentrations of dissolved ions in solution, so clearly, considerations of solubility and saturation have clear applications to questions of contaminant behavior. Consider the case of three possible solid forms of lead, each presented in Table 4.5 with their respective K_{sp} values.

Lead chloride and lead sulfate are many orders of magnitude more soluble than lead carbonate, emphasizing the importance of mineralogical analysis in studies of aqueous

Table 4.5 Examples of solubility products for three lead (Pb) minerals (25 °C, 1 atm).

Formula	Mineral name	K_{sp}
$PbCO_3$	Cerussite	7.4×10^{-14}
$PbCl_2$	Cotunnite	1.7×10^{-5}
$PbSO_4$	Anglesite	1.6×10^{-8}

systems. A soil containing 100 ppm Pb in $PbCO_3$ may yield less dissolved Pb in solution water than would a soil with only 10 ppm Pb contained within $PbCl_2$. The speciation of trace metals in minerals or other solids can clearly exert a strong control over their mobility, and given enough data, geochemists can quantitatively predict equilibrium concentrations.

One way to determine solubility would be to immerse a crystal of a mineral such as anglesite into deionized water and measure the concentrations of Pb^{2+} and SO_4^{2-} until they no longer increase. At this point, a dynamic equilibrium will have been established where any new dissolution of the anglesite is matched by an equal amount of precipitation of calcite. The solution is saturated with respect to lead and sulfate ions, and no net dissolution or precipitation occurs. The reaction that describes this process is:

$$PbSO_4 = Pb^{2+} + SO_4^{2-} \qquad (4.22)$$

The reaction can be quantified by the equilibrium constant

$$K_{eq} = a_{Pb^{2+}} \times a_{SO_4^{2-}} / a_{PbSO_4} \qquad (4.23)$$

The solid crystal of anglesite (or what remains of it) cannot be represented in terms of activity, so by convention its activity is given the value of 1, and the equation now appears as

$$K_{eq} = a_{Pb^{2+}} \times a_{SO_4^{2-}} \qquad (4.24)$$

In reality, as long as there is solid anglesite, the amount of it will not affect the reaction between $Pb^{2+} + SO_4^{2-}$ ions and the solid anglesite – it is the concentrations of the ions that are most important, at equilibrium.

The K_{eq} for a solid and its associated dissolved ions is known as the *solubility product* (K_{sp}), which has the following formula and K_{sp} value for anglesite at 1 atm and 25 °C (Table 4.5):

$$K_{sp} = a_{Pb^{2+}} \times a_{SO_4^{2-}} = 1.6 \times 10^{-8} \text{ mol}^2 \text{ l}^{-2} \qquad (4.25)$$

Pb^{2+} activity is equal to SO_4^{2-} activity and the value of the activity of each is

$$\sqrt{1.6 \times 10^{-8} \text{ mol}^2 \text{ l}^{-2}} = 1.3 \times 10^{-4} \text{ mol l}^{-1}$$

Focus Box 4.4

Dissolved Pb: Comparison of Anglesite to Cerussite as Solid Control

The solubility product of anglesite can be used to determine the equilibrium concentration of Pb^{2+}, in a solution where anglesite is the solid phase controlling Pb concentration, is 1.3×10^{-4} mol l^{-1}. Now compare this to a situation where dissolution of cerussite into ions of Pb^{2+} and CO_3^{2-} is responsible for controlling the amount of dissolved Pb^{2+}. In this situation, predicted equilibrium concentration would be far lower than that for anglesite: based on the K_{eq} for cerussite from Table 4.5, the equilibrium concentration of

Pb is:

$$\sqrt{7.4 \times 10^{-14} \, mol^2 \, l^{-2}} = 2.7 \times 10^{-7} \, mol \, l^{-1} \, of \, Pb^{2+}$$

In other words, if dissolution (or precipitation) of cerussite is controlling Pb^{2+} in solution, concentrations of Pb^{2+} will be approximately 1000 times lower than if the more soluble anglesite is the corresponding mineral.

In this equilibrium situation, the concentrations of Pb^{2+} and SO_4^{2-} are each 1.3×10^{-4} mol l^{-1}, and these are the predicted concentrations at equilibrium that dissolution of anglesite would produce.

To reiterate a point made in Section 4.6.2, if the IAP is greater than the K_{sp} ($\Omega > 1$), the solution is oversaturated and anglesite will precipitate from solution. If the $\Omega < 1$ (the IAP $< K_{sp}$), the solution is undersaturated and anglesite will dissolve. If $\Omega = 1$ (IAP $= K_{sp}$), the system is saturated and at equilibrium, and rates of mineral dissolution and mineral crystallization will be equal. So a system with $[Pb^{2+}] = 0.7 \times 10^{-4}$ mol l^{-1} and $[SO_4^{2-}] = 1.7 \times 10^{-4}$ mol l^{-1} will be undersaturated with respect to anglesite because the IAP will be $< K_{sp}$.

However, a system with $[Pb^{2+}] = 0.9 \times 10^{-4}$ mol l^{-1} and $[SO_4^{2-}] = 6 \times 10^{-4}$ mol l^{-1} will be oversaturated with respect to anglesite, because the IAP $= 5.4 \times 10^{-8}$ mol l^{-1}, which is greater than the K_{sp}. Lead and sulfate ions will precipitate out of solution to form anglesite.

Using solubility product data makes it possible to predict whether a solution is saturated, oversaturated, or undersaturated with respect to any mineral. A detailed example pertaining to Ca^{2+}, HCO_3^-, and the mineral calcite is presented in Chapter 5.

4.8 COPRECIPITATION

The process by which a mineral crystallizes from solution and incorporates ions into its structure in addition to the dominant ions is coprecipitation. One of the most important examples of coprecipitation is the uptake of trace metals and metalloids (e.g. arsenic) into the structure of iron oxy-hydroxides that range from the oxides hematite (Fe_2O_3) and magnetite (Fe_3O_4) to less-crystalline hydroxides such as ferrihydrite (written as $5Fe_2O_3 \cdot 9H_2O$ or $Fe_5HO_8 \cdot 4H_2O$) and the polymorphs goethite (α-FeO[OH])

and lepidocrocite (γ-FeO[OH]). Incorporation of trace metals into Fe-hydroxides is often responsible for minimizing their concentration in solution or in exchange sites, making them less environmentally available. Elevated arsenic in the regional aquifer system of the Bengal Fan of the Indian subcontinent is caused – at least in part – by release of arsenic from Fe-hydroxides that dissolve in reducing zones (rich in organic matter) within the aquifer (Zheng et al. 2005).

Once an ion is incorporated into the structure of a mineral by coprecipitation, its fate will be dictated by the fate of the mineral – if the mineral is transported as a particulate, so too will be the coprecipitated constituent. If the mineral dissolves, the coprecipitated ion will be released back into solution, after which the ion may be adsorbed onto exchange sites, or remain in solution, or become incorporated into a new solid. As an example, consider an iron hydroxide that incorporated arsenic into its structure when it crystallized in an oxidizing soil, but then was eroded into a stagnant wetland. The reducing conditions of the wetland sediment likely will result in dissolution of the hydroxide and release of iron and arsenic into solution. If sufficient sulfur is available in the wetland sediment, it will probably exist as the sulfide anion (S^{2-}) and if it reacts with the liberated iron, could result in crystallization of pyrite (FeS_2) or a related sulfide (potentially with coprecipitated arsenic).

The uptake of radium into relatively insoluble barium sulfates (which can range from low-order hydrated forms to the mineral barite, $BaSO_4$) is an illustrative example of coprecipitation. Ra^{2+} is a relatively soluble cation, but it fits quite well into the $BaSO_4$ structure (substituted for Ba^{2+}), and thus can be removed from solution by coprecipitation when $BaSO_4$ crystallizes. This mechanism has been attributed to lowering concentrations of the radioactive Ra in aquifers. Many trioctahedral clay minerals contain metals such as Cu, Ni, and Zn incorporated into octahedral sheets and there is emerging evidence that As^{3+} and As^{5+} can substitute for Si^{4+} in the tetrahedral sheets of clay minerals. There is no question that in order to understand controls on aqueous systems, geochemists must identify the composition, abundance, mineralogical form, potential mobility (as particulates), and solubility of solids in the aqueous system.

4.9 BEHAVIOR OF SELECTED ELEMENTS IN AQUEOUS SYSTEMS

Many elements are inherently soluble whereas others are not, a point made in Section 4.5.3 and Figure 4.9, but even elements that are inherently soluble may be precipitated into insoluble minerals. Alkali and alkaline earth elements, including those commonly referred to as the *base cations* (Na, K, Mg, Ca), are among the most soluble metals in aqueous systems. Their high solubility is enhanced by low ionic charge (e.g. Na^+, Ca^{2+}), relatively large ionic radii, and the fact that, often, the minerals in which these elements reside are soluble (e.g. halite, $NaCl$, sylvite, KCl) or moderately soluble (e.g. calcite, $CaCO_3$, or other carbonates). In some cases, however, the solubility of the base cations will be limited by their occurrence in minerals of relatively low solubility, including albite ($NaAlSi_3O_8$), microcline ($KAlSi_3O_8$), and other feldspars; micas such as muscovite ($KAl_3Si_3O_{10}(OH)_2$) and biotite ($KAlMg_2Si_3O_{10}(OH)_2$); and other silicate minerals such as the amphibole hornblende ($Ca_2(Mg,Fe,Al)_5(Si,Al)_8O_{22}(OH)_2$). Again, it is clear that understanding the solubility of any element requires knowledge of the minerals or other solids (e.g. organic matter or amorphous inorganic solids) present in the system. Chapter 9 presents additional information on mineral solubility and dissolution rates in the framework of soils and chemical weathering.

4.9.1 Examples of heavy metals and metalloids

4.9.1.1 Heavy metals

The heavy metals are commonly defined as metallic elements with atomic number greater than 20, including the transition group metals (comprising the central rows of the periodic table), the lanthanides (La to Yb), and the actinides (elements heavier than Fr). These elements are far less abundant than the alkali and alkaline earth metals, and so are often referred to as *trace metals*. They occur as trace constituents of various rock types, and thus can occur naturally in soils and water, but often they are present at higher concentrations due to human activities like mining, metal refining, industrial activities, and waste disposal.

An environment where heavy metals are highly mobile is acid mine drainage settings (Plate 1) (or their natural acid drainage analogs; Plate 5), where (i) surface waters in contact with the atmosphere produce oxidizing conditions, and (ii) the chemical weathering oxidation of sulfides produces solutions of low pH that enhances metal solubility. At neutral and basic pH, heavy metal solubilities are limited by coprecipitation with solid phases such as oxides, hydroxides, and carbonates, and under reducing conditions they tend to bond with the sulfide anion (S^{2-}) to form sulfides that are insoluble in O_2-poor waters (e.g. PbS). Heavy metals that exhibit this typical "soluble in acidic, oxidizing water" behavior include cadmium, cobalt, copper, lead, nickel, and zinc. These metals commonly occur together in base metal sulfide ore deposits in minerals such as covellite (CuS), chalcopyrite ($CuFeS_2$), galena (PbS), and sphalerite (ZnS), along with pyrite and other sulfides.

The high solubilities of heavy metals in acidic, oxidizing conditions leads to one type of remediation strategy that relies on constructing alkaline wetlands to remove metals from rivers and other aquatic systems – these constructed

---

Focus Box 4.5

Minerals vs. Inorganic Noncrystalline Solids

In some cases, it is incorrect to use the term "mineral" to describe the solid phases that form in soil, acid mine drainage, aquifer, volcanic, or other systems where the kinetics of rapid crystallization can result in precipitation of amorphous (or noncrystalline) solids that lack the internal atomic structure required to be minerals. The volcanic glass obsidian is a tangible example of an inorganic solid that crystallizes so rapidly that it lacks internal crystalline structure, and noncrystalline Al and Fe hydroxides occur in many soils and are known to sorb many trace elements. The noncrystalline aluminosilicate allophane occurs in many Al-rich soils and is known to sorb the tetrahedrally configured anions phosphate (HPO_4^{2-}) and arsenate ($HAsO_4^{2-}$) as inner-sphere complexes, thus exerting a strong influence on their mobility in the environment (Arai et al. 2005).

wetlands are characterized by high pH and low Eh (low DO), and these environments facilitate precipitation of metals as carbonates, sulfides, or other solid phases. The low kinetic energy of the wetland is beneficial because particulates settle out of suspension, so the wetland acts as both a chemical filter and physical filter, causing chemical precipitation of metal sulfides or carbonates and physical settling of most particles that might remain in suspension.

Not all heavy metals behave according to the pattern described above. As mentioned above with reference to redox reactions, chromium is an example of a trace metal that can exist either as metal cation (Cr^{3+}) or, in the Cr^{6+} state, as chromate (CrO_4^{2-}) or dichromate ($Cr_2O_7^{2-}$) oxyanions. As you might expect from the high cation charge, hexavalent chromium tends to occur in oxidized environments. It also is the main form of Cr in waters with pH > 7, meaning that unlike the base metals described above, Cr is predicted to be elevated in basic/alkaline waters.

In oxidizing environments, uranium occurs as a hexavalent cation (U^{6+}) that forms polyatomic anions with carbonate and hydroxide anions in neutral to alkaline waters that greatly enhance solubility. Examples include soluble $UO_2(CO_3)_2^{2-}$ and $(UO_2)_2CO_3(OH)_3^-$. At low pH (< 6), U primarily occurs as UO_2^{2+}, a relatively insoluble cation (commonly termed the uranyl ion) that is likely to be sorbed by minerals or organic matter. The absence of available carbonate and hydroxide anions at pH < 6 prohibits formation of soluble U–CO_3–OH anions and renders U relatively insoluble in slightly acidic to highly acidic waters.

4.9.1.2 Metalloids

Elements that possess metallic and nonmetallic properties are metalloids, and for this reason one of their most well-known uses is as semiconductors. They are located on the right side of the periodic table between the metals and gases, and include boron, silicon, germanium, arsenic, selenium, and antimony. The metalloid that generates the greatest amount of concern and attention in environmental geochemistry is *arsenic*. It often occurs in base metal deposits or gold deposits, typically as arsenic sulfides such as realgar (AsS) or orpiment (As_2S_3), as the iron-arsenic sulfide arsenopyrite (FeAsS), or as arsenic co-precipitated in other sulfides (e.g. in pyrite).

Often, arsenic is released to surface waters or soils by oxidation of arsenic-bearing sulfides. The example below is for arsenopyrite:

$$FeAsS + 13Fe^{3+} + 8H_2O$$
$$\rightarrow 14Fe^{2+} + SO_4^{2-} + 13H^+ + H_3AsO_{4(aq)} \quad (4.26)$$

Note that in this reaction, the oxidation of sulfur is driven by the reduction of ferric iron (Fe^{3+}) to ferrous iron (Fe^{2+}). This is a typical process in iron-rich sulfide deposits, where the oxidation of pyrite (here shown as a single reaction that summarizes two or more steps) produces high quantities of available ferric iron (note: this equation describes formation of sulfuric acid associated with acid mine drainage; Chapter 9):

$$4FeS_2 + 15O_2 + 2H_2O \rightarrow 4Fe^{3+} + 8SO_4^{2-} + 4H^+ \quad (4.27)$$

Arsenic in natural environments occurs in three main oxidation states that span a wide redox range, from As^{3-} and As^0 to As^{3+} and As^{5+}, the latter two often written As(III) and As(V). Arsenic in the trivalent state can substitute for ferric iron given their identical charge and similarity in radii in octahedral coordination ($As^{3+} = 0.58$ Å; $Fe^{3+} = 0.65$ Å) (Table 2.1); As(V) is also sorbed into Fe-hydroxides, so As mobility tends to be lower in oxidizing waters where Fe-hydroxides are stable.

In solution, arsenic primarily exists in two oxidation states, As^{3+} and As^{5+}, both of which form oxyanions, arsenite in the case of As^{3+} and arsenate in the case of As^{5+}. The general form of arsenite is $H_nAsO_3^{-(3-n)}$, and arsenate is $H_nAsO_4^{-(3-n)}$. Over pH ranges in most natural settings, arsenite occurs as arsenious acid (H_3AsO_3) and the arsenite anion $H_2AsO_3^-$, whereas arsenate occurs as two arsenate anions, $H_2AsO_4^-$ and $HAsO_4^{2-}$. Arsenic acid (H_3AsO_4) occurs in oxidizing conditions at a pH < 2 and thus might be found in very acidic mine drainage situations. Lower pH means greater activity of H^+, so acidic waters tend to contain the more H-rich species H_3AsO_3 and $H_2AsO_4^-$. In fact, high concentrations of H^+ in solution drive arsenite (shown below) and arsenate anions toward more protonated states:

$$H_2AsO_3^- + H^+ \rightarrow H_3AsO_3 \quad (4.28)$$

In high-pH solutions, arsenic species tend to lose H^+ to solution as is illustrated for arsenate anion:

$$H_2AsO_4^- + OH^- \rightarrow HAsO_4^{2-} + H_2O \quad (4.29)$$

Arsenite predominates in reducing conditions whereas arsenate is more common in oxidizing conditions, a predictable relationship when considering that arsenite contains the more reduced As(III) and that arsenate contains the most oxidized form of arsenic, As(V). In this sense, arsenic would be termed a *redox-sensitive element* that commonly occurs as oxyanions.

Focus Box 4.6

The Problem of Remediating Sites Contaminated with Heavy Metals and Arsenic

Arsenic does not behave in the same way as most heavy metals, yet is often present in mining, landfill, or industrial sites contaminated with heavy metals. Arsenic is generally soluble under all conditions except for acidic, reducing environments that contain available sulfide anions which foster formation of arsenic sulfides or coprecipitation of As into iron sulfide. This can be problematic when trying to remediate sites contaminated with heavy metals and arsenic. Employing the strategy of creating alkaline, reducing conditions will immobilize base metals such as Cu, Ni, Pb, and Zn, but will likely create high dissolved

As concentrations. In addition, because of its anionic character in solution, As (e.g. as $H_2AsO_4^-$) is not attracted to cation exchange sites on surfaces of clay minerals or organic matter. At low pH, when the surfaces of iron oxides and hydroxides are positively charged (by H^+), arsenate is sorbed, forming relatively strong inner-sphere complexes. However, at higher pH (>8), sorption is inhibited by the strong negative charges on the surfaces of iron oxyhydroxides (caused by deprotonation) that results in repulsion of arsenate and arsenite anions.

The metalloid selenium takes on two oxidation states, Se^{4+} and Se^{6+}, and like arsenic, also occurs as oxyanions. The selenite anion, SeO_3^{2-}, is the dominant species in mildly reducing waters, whereas the selenate anion, SeO_4^{2-}, is the dominant form under oxidizing conditions. In acidic conditions, the selenite anion reacts with protons to form $HSeO_3^-$. While selenium in general is soluble, the more reduced form, selenite, is strongly sorbed to iron oxyhydroxides at circum-neutral pH, whereas the more oxidized selenate is only very weakly sorbed. In this way its behavior is the opposite of arsenic, where the more oxidized arsenate anion is strongly sorbed but arsenite is not.

4.10 EH–PH DIAGRAMS

4.10.1 Principles of Eh–pH

Trace metal speciation is commonly illustrated using *Eh–pH diagrams* (also known as Pourbaix diagrams after pioneering work in the 1940s by Belgian chemist M. Pourbaix), which are sometimes plotted as *pe–pH diagrams*. These diagrams depict the combined influences of redox conditions and pH on speciation. The flux of electrons associated with redox reactions can be measured, and while the tendency to gain (or lose) electrons does not correspond to a concentration of electrons in the same way that acidity corresponds to a concentration of H^+ ions, redox potential nonetheless can be envisioned as an activity of electrons in solution (a_{e^-}), where a_{e^-} really represents the tendency of a system to provide (or consume) electrons. From this reasoning, the unit p*e* can be defined in a manner analogous to pH (recall that $pH = -\log_{10} a_H+$):

$$pe = -\log_{10} a_{e^-} \qquad (4.30)$$

The p*e* scale in natural earth systems ranges from -15 to $+20$, where the lower limit corresponds to reduction of H_2O to H_2 gas and O^{2-}, and where the upper limit corresponds to oxidation of $2H_2O$ molecules to $4H^+$ ions and O_2 gas.

The activity of electrons in solution is often expressed relative to the SHE by the unit Eh, where the conversion from p*e* to Eh is:

$$Eh = pe \times 2.303RT/F \qquad (4.31)$$

This indicates that $Eh = pe/16.9$. In the above equation, the gas constant $R = 1.987 \times 10^{-3}$ kcal deg^{-1}, $T = 298.15$ K (i.e. $25\,°C$), $F = 23.06$ kcal/volt-gram equivalent, and 2.303 is the conversion factor for $\log_e$ to $\log_{10}$.

Eh–pH diagrams usually depict systems at $25\,°C$ and 1 atm with components that commonly include H_2O, CO_2, O, S, and the metal or metalloid in question. They also include information on concentrations or activities of the various components in the system. Alternatively, a user may specify these variables if using software such as Geochemist's Workbench or similar software (e.g. PHREEQE) to produce an Eh–pH diagram. Examples of where selected natural environments plot in Eh (p*e*) − pH space are shown in Figure 4.12.

The upper limit on the stability of water is expressed by this equation:

$$2H_2O = O_2 + 4H_+ + 4e^- \qquad (4.32)$$

Eh can be calculated from the Nernst equation expressed as follows:

$$Eh = E^0 + (RT/nF) \times \ln(K_e) \qquad (4.33)$$

where E^0 is the standard potential for the half-cell reaction, R is the gas constant, T is temperature (Kelvins), n is

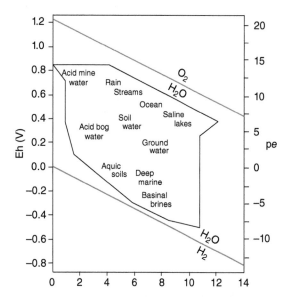

Fig. 4.12 Approximate locations of selected natural systems as a function of reduction–oxidation potential and pH. Note that redox units are given in terms of Eh (left) and p*e*. The thin line bounding the natural environments indicates the limits of nearly all natural waters. Source: After Baas Becking et al. (1960).

the number of moles of electrons transferred, and K_e is the equilibrium constant for the reaction.

Factoring in constants R, T (298.15 K, i.e. 25 °C), F, and 2.303 (to convert ln to $\log_{10}$), the equation becomes:

$$Eh = E^0 + 0.0592/n \times \log([C]^c \times [D]^d/[A]^a \times [B]^b)$$
(4.34)

So for the oxidation of H_2O shown in Eq. (4.32),

$$Eh = E^0 + 0.0592/4 \times \log(pO_2 \times [H^+]^4/[H_2O]^2) \quad (4.35)$$

However, given that pO_2 and $[H_2O]$ are both 1, the equation can be simplified to:

$$Eh = E^0 + 0.0592/4 \times \log([H^+]^4) \quad (4.36)$$

And considering that pH $= -\log[H^+]$ and that $\log([H^+]^4) = 4\log[H^+]$,

$$Eh = E^0 - 0.0592 \times pH \quad (4.37)$$

What this indicates is that the upper limit of the stability of water is a straight line with a slope of -0.0592. The y-intercept of E^0, which applies to Eq. (4.32), is $+1.23$ V, the same as the value listed in Table 4.2 for this reaction relative to the SHE.

A similar calculation for the lower limit on the stability of water gives us the same equation as (4.37):

$$Eh = E^0 - 0.0592 \times pH$$

In other words, the lower limit of the stability of water is also a straight line with a slope of -0.592. The difference is in the value of E^0, which for the SHE (Eq. (4.7)) is 0.0 V – this is by definition, given that the SHE is the reference for all other redox reactions and is assigned the value of zero. Therefore, the y-intercept for the lower stability limit of water is zero, as it appears on the Eh y-axis in typical Eh-pH diagrams (Figure 4.12).

Although very useful in depicting potential speciation, the disadvantage of Eh–pH diagrams is that they do not illustrate is the tendency of certain assemblages to exist in various states of disequilibrium. They also do not reflect the fact that kinetics (or reaction rates) often dictate mineral assemblages and dissolved species in low-temperature geological environments such as soils, surface waters, and groundwaters. We might not consider the potential for As uptake by Fe oxyhydroxides in an anoxic wetland based on equilibrium considerations alone. $Fe(OH)_3$ and FeOOH are not stable in anoxic settings; however, it might take years or decades or longer for Fe oxyhydroxides to dissolve in this type of setting, meaning that they could exert a control

---

Focus Box 4.7

Measuring Eh (or p*e*) in the Field: ORP Meters

Redox conditions of surface water, groundwater, etc. can be determined using oxidation–reduction potential (ORP) meters that report in units of volts (V) or millivolts (mV), the same units of measure as Eh. The difference between ORP and Eh is that ORP is often determined relative to an Ag-AgCl reference electrode, whereas Eh is relative to the SHE. The SHE is not convenient for field use, so field-based ORP values must be converted to Eh by converting the field-based ORP readings (e.g. relative to Ag-AgCl) to a SHE-normalized value – this accounts for the difference between the reference electrode and the SHE. Measuring the redox potential of the ZoBell solution used to calibrate the ORP meter (with Ag-AgCl reference electrode) gives a value of ~+230 mV, whereas the same ZoBell solution measured against the SHE will read ~+430 mV. Therefore, to convert ORP readings to Eh, 200 mV must be added to the ORP voltage, and a field ORP measurement of −53 mV (using Ag-AgCl as reference) would convert to an Eh value of +147 mV.

over arsenic mobility in spite of the fact that they are out of equilibrium with their environment.

Eh–pH diagrams also do not provide information on the potential of a given ion to be sorbed to mineral surfaces, nor do they consider coprecipitation reactions that might remove ions from solution, a good example being the tendency of Ba to remove Ra from solution upon crystallization of $BaSO_4$. It is rare that Eh–pH diagrams include Si, and even when they do, they do not portray incorporation of trace metals into silicate clays or other minerals as trace or minor components. Nonetheless, they are very useful at providing simplified graphic representations of elemental speciation, and they are widely used in environmental geochemistry.

4.10.2 Eh–pH diagrams for Cu, Pb, As, U, Fe, Al

Eh–pH diagrams of copper and lead are presented in Figure 4.13. The diagrams for these two metals depict the main patterns and types of species that occur for many other heavy metals, including Cd, Co, Ni, and Zn. In both diagrams, the system includes the heavy metal (Cu or Pb) plus oxygen, sulfur, water, and carbon dioxide (meaning that we can assess the role of carbonates). It is important to be aware of the system composition presented in such diagrams because it will affect speciation.

Eh–pH diagrams for arsenic and uranium are presented in Figure 4.14 to illustrate patterns for two of the elements that tend to form oxyanions, and also to provide data for elements that are inherently toxic. Compared to Cu, Pb,

and other trace metals that are effectively insoluble at neutral or higher pH, As and U possess broad aqueous fields at neutral and higher pH. In fact, they are soluble over most Eh and pH ranges, merely changing forms and oxidation states with changing redox and pH values.

Al and Fe are useful to consider in conjunction because they behave similarly under many natural geochemical conditions. Both are relatively insoluble in neutral waters and become progressively (and exponentially) more soluble with increasing acidity as well as increasing alkalinity. Both are abundant in silicate minerals, and they commonly precipitate out of soil solutions, surface waters, and groundwaters as hydroxides, e.g. $Fe(OH)_3$ and $Al(OH)_3$. However, there are important differences.

- First, as an ion, aluminum exists only in the Al^{3+} state and is not redox sensitive, whereas iron (as Fe^{2+} and Fe^{3+}) is very redox sensitive. This difference is apparent in the Eh–pH diagrams, where species boundaries for aluminum are vertical, meaning that changing Eh conditions do not result in a change in Al speciation. In a system controlled by Eh and pH, only changing pH will affect Al speciation. Conversely, changing Eh can result in a change in iron speciation, reflecting the fact that iron is sensitive to redox conditions. (Although in reality, if Al is coprecipitated in a redox-sensitive mineral like $Fe(OH)_3$ or Fe_2O_3, its behavior will be in part controlled by redox reactions affecting the iron hydroxide or oxide.) For example, decreasing Eh (reflecting transition to more-reducing conditions) at constant pH of 6 from Eh = 0.6 to 0.2 V will cause reduction of Fe^{3+} to Fe^{2+}, resulting in dissolution of $Fe(OH)_3$ and

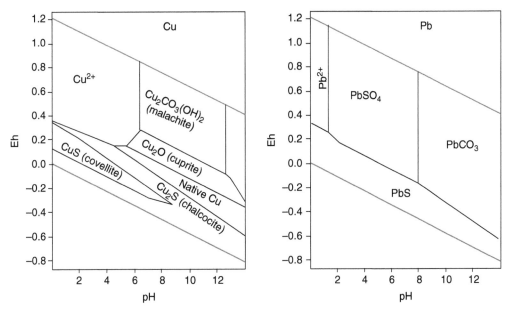

Fig. 4.13   Copper and lead Eh–pH diagrams at 25 °C and 1 atm pressure. Activities of Cu and Pb are 10^{-6} mol l^{-1}, pCO$_2$ = 400 ppmv ($10^{-3.4}$ atm) and total S = 10^{-2} mol l^{-1}.

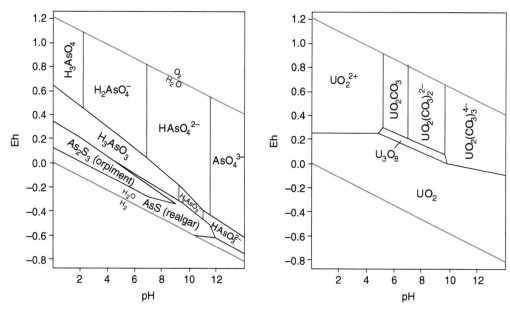

Fig. 4.14 Arsenic and uranium Eh–pH diagrams for systems at 25 °C and 1 atm pressure. For the arsenic diagram, activity of $As_{total} = 10^{-6}$ and S activity $= 10^{-2}$ mol l^{-1}. For the uranium diagram, activity of U $= 10^{-8}$ mol l^{-1} and C (as CO_2) is $10^{-3.4}$ atm.

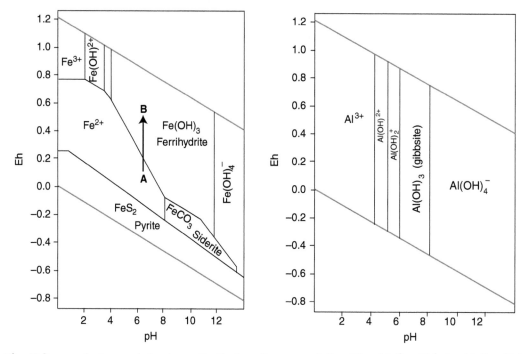

Fig. 4.15 Eh–pH diagram for iron and aluminum. For the iron diagram, activity of Fe $= 10^{-6}$ mol l^{-1}, S $= 10^{-2}$ mol l^{-1}, and $CO_2 = 10^{-3.4}$ atm. For aluminum diagram, Al $= 10^{-9}$ mol l^{-1}. Note that iron speciation is influenced by both Eh and pH; conversely, aluminum speciation is controlled by pH but not Eh. The arrow from A to B represents oxidation as shown in Plate 5.

leaving Fe^{2+} in solution. Note that at low to intermediate Eh, iron exists as the Fe^{2+} cation in solution or as Fe^{2+} in minerals such as pyrite and siderite. At high Eh values, iron occurs either as the Fe^{3+} simple cation in solution (in highly acidic, oxidizing water) or as Fe^{3+} in ferrihydrite and other iron oxides and hydroxides.

- Second, Fe forms compounds with S and CO_3^{2-}, Al effectively does not (note presence of pyrite and siderite in Eh–pH diagram, Figure 4.15).
- Third, based on the sizes of the stability fields at high pH, it appears that Al is more soluble in high-pH solutions than is Fe.

4.11 SILICON IN SOLUTION

The behavior of silicon is notable because it reaches its greatest solubility in aqueous solutions at high pH. At pH values < 9.8, the dominant silicon species is H_4SiO_4, alternately referred to as silica, hydrated silica, or silicic acid. This species is produced in a simple sense by dissolution of quartz by hydration:

$$SiO_2 + 2H_2O = SiO_2 \cdot 2H_2O = H_4SiO_4 \qquad (4.38)$$

If dissolved silica concentrations (as H_4SiO_4, but measured with respect to SiO_2) at values of pH less than 9.8 are 10^{-4} M, the solution is in equilibrium with quartz. If dissolved silica concentrations are $> 10^{-4}$ M ($6\,mg\,l^{-1}$) at pH < 9.8, then the solution is oversaturated with respect to quartz and we would expect quartz to precipitate out of solution. Alternatively, if the solution were in equilibrium with respect to amorphous silica, which is far more soluble than quartz, the equilibrium concentration of H_4SiO_4 (as SiO_2) would be approximately 2×10^{-3} M ($120\,mg\,l^{-1}$). These values compare well with concentrations of dissolved silica in ocean water, streams, groundwaters, and soil waters, which generally range from 10 to $100\,mg\,l^{-1}$. Values $< 1\,mg\,l^{-1}$ can be produced in lakes and marine environments where diatoms (siliceous phytoplankton) sequester Si from solution to form cell wells (tests), and values $> 100\,mg\,l^{-1}$ occur in hydrothermal systems where high temperatures enhance silica solubility.

Above a pH of 9.8, high concentrations of OH^- cause deprotonation of H_4SiO_4 as follows:

$$H_4SiO_4 + OH^- = H_3SiO_4^- + H_2O \qquad (4.39)$$

This reaction accompanies exponential increases in silica solubility at pH > 9 (Figure 4.16), and with further increases in pH, deprotonation produces $H_2SiO_4^{2-}$ to the extent that at the extremely high pH of 13.1, the concentration of $H_2SiO_4^{2-}$ in solution is greater than $H_3SiO_4^-$:

$$H_3SiO_4^- + OH^- = H_2SiO_4^{2-} + H_2O \qquad (4.40)$$

The total amount of dissolved silica will be the sum of the three hydrated silica species. At pH < 9, nearly all dissolved silica occurs as H_4SiO_4. Between pH = 9.8 and 13.1, the main dissolved species is $H_3SiO_4^-$, with lesser amounts of H_4SiO_4 and $H_2SiO_4^{2-}$.

4.12 EFFECT OF ADSORPTION AND ION EXCHANGE ON WATER CHEMISTRY

The topic of sorption and ion exchange was introduced in Chapter 2 where the focus was the properties of minerals as they influence ion exchange, as well as the role of pH (e.g. point of zero charge [PZC] and isolectric point [IEP], Table 2.4); here, the focus is equilibrium exchange and its effects on the composition of aqueous solutions.

The exchange reaction of Ca^{2+} from aqueous solutions into an exchange site (e.g. on a mineral or organic matter) first occupied by Na^+ can be represented as:

$$2Na^+ \cdot [clay] + Ca^{2+}_{aq} = Ca^{+2} \cdot [clay] + 2Na^+_{aq} \quad (4.41)$$

Why might such a reaction occur, i.e. why might Ca^{2+} replace Na^+ on a mineral surface exchange site?

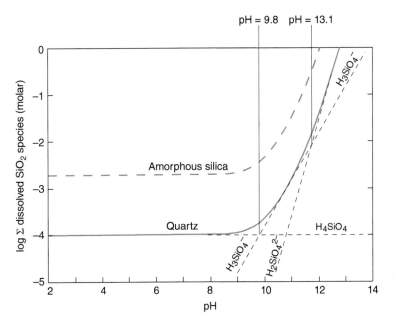

Fig. 4.16 Solubility of silica as a function of pH in equilibrium with amorphous silica and with quartz. Note that if amorphous silica is the solid phase in equilibrium with silica, dissolved silica concentrations are approximately an order of magnitude greater than when quartz controls silica solubility.

(Figures 2.25 and 2.26). One possible reason is that the attraction of divalent Ca^{2+} to the mineral surface is greater than the attraction of Na^+ to these sites, which would result in release of Na^+ to solution and adsorption of Ca^{2+}. (The issue of charge : radius is covered at the end of Section 2.5, including Eq. (2.6).)

4.12.1 Ionic potential, hydration radius, and adsorption

The ionic potential of an element, i.e. the ratio of ionic charge : ionic radius, helps to determine how strongly it will be attracted to exchange sites – high ratios indicate strong attraction. Although ionic radius is often used, the *hydration* radius of a cation is a better predictor of its attraction to exchange sites – this is because, in solution, ions exist within a shell of water molecules that are closely bonded to the ion. The hydration shell causes the charge on the ion to be more diffuse (in an extreme case, the hydration shell of Al^{3+} is almost 10 times greater than its ionic radius). Table 4.6 presents values of ionic radii as well as hydration radii for a group of cations.

Additional factors – such as the composition of the aqueous solution – are also important, but in a general sense, order of decreasing affinity for cation exchange sites (in ~dilute fresh water systems) among common cations, determined by ratio of hydration radius to ionic charge, is:

$$Al^{3+} > Ca^{2+} > Mg^{2+} > H_3O^+ (H^+) > K^+ > NH_4^+ > Na^+$$

4.12.2 Law of mass action and adsorption

An important control on exchange reactions and the composition of exchange sites is the chemistry of the aqueous solution, both in terms of concentrations of potential adsorbates (in this case cations like Na^+ and Ca^{2+}), as well

Table 4.6 Hydration radii relative to ionic radii and charges of selected cations (Volkov et al. 1997).

Cation	R_{ionic} (Å)	R_{ionic}:chg	R_{hydr} (Å)	R_{hydr}:chg
H_3O^+	1.15	1.15	2.80	2.80
NH_4^+	1.48	1.48	3.31	3.31
Li^+	0.94	0.94	3.82	3.82
Be^{2+}	0.31	0.16	4.59	2.30
Na^+	1.17	1.17	3.58	3.58
Mg^{2+}	0.72	0.36	4.28	2.14
Al^{3+}	0.53	0.18	4.80	1.60
K^+	1.49	1.49	3.31	3.31
Ca^{2+}	1.00	1.00	4.12	2.06
Rb^+	1.63	1.63	3.29	3.29
Cs^+	1.86	1.86	3.29	3.29
Zn^{2+}	0.74	0.37	4.30	2.15
Cd^{2+}	0.97	0.49	4.26	2.13
Pb^{2+}	1.32	0.66	4.01	2.01

R_{ionic} is ionic radius and R_{hydr} is hydration radius. Lower ratios of R_{hydr}:chg (chg = ionic charge) correspond to stronger adsorption to cation exchange sites. H_3O^+ is the hydronium ion form of H^+.

as overall ionic strength of the solution. In a general sense, the quantity of Ca^{2+} removed from solution will occur in approximate proportion to its abundance in solution accounting for competition with other ions (e.g. Mg^{2+}) and the composition of ions occupying exchange sites (e.g. Na^+).

Exchange of Ca^{2+} for Na^+ as shown in Eq. (4.41) is quantified by this formulation of the law of mass action:

$$K = a^2_{Na^+} * a_{Ca^{2+} \cdot [Clay]} / a_{Ca^{2+}} * a^2_{Na^+ \cdot [Clay]} \qquad (4.42)$$

K in this equation is not a constant, but rather is a coefficient (often referred to as "mass action coefficient" or "selectivity coefficient") that varies, especially as a function of changing ionic strength. As written here, with sorbed Ca^{2+} in the numerator, the higher the value of K, the more

---

Focus Box 4.8

Environmentally Important Soil Exchange Reactions

The high ionic potential of Al^{3+}, whether in ionic form or hydrated (Table 4.6), means that it will out-compete Na^+ and Ca^{2+} for exchange sites in acidic soils (e.g. at pH <5). This is a problem in acid rain-impacted soils prone to loss of Ca, a crucial limiting nutrient in many forest ecosystems. The strong attraction of certain smectite interlayers for K^+ make it more strongly sorbed than would be predicted based on ionic potential alone. In acidic soils, H^+ (as H_3O^+) is strongly attracted to mineral surfaces, behavior which is strongly controlled by the abundance of H^+ in acidic solutions. In alkaline soils, H^+ is likely to undergo desorption to neutralize pH. Also, as is demonstrated in Section 4.12.2, Na^+ is more strongly adsorbed than Ca^{2+} in water with high ionic strength, so exchange sites on particles in desert soils and marine sediments will likely be dominated by Na^+. In temperate soils and fresh waters, Ca^{2+} is likely to be the dominant adsorbed cation.

likely that Ca^{2+} will be more strongly adsorbed to the particle (clay in this case) than Na^+.

Given that exchange often involves ions of different charge (e.g. Na^+ vs. Ca^{2+}), a term known as the equivalent fraction (X_{M^+}) is introduced, where M^+ is a generic cation. X_{M^+} is the fraction of the cation exchange capacity occupied by cation M^+, where a divalent cation will occupy twice the capacity per mole as a monovalent cation, and $\Sigma X_{M^+} = 1.0$. Equation (4.42) can now be expressed as:

$$K = a^2_{Na^+} * X_{Ca^{+2} \cdot [Clay]} / (a_{Ca^{+2}} * X^2_{Na^+ \cdot [Clay]}) \quad (4.43)$$

This is the Kerr (1928) approach to expressing the selectivity coefficient. There are a few other approaches, including the Gaines-Thomas, Gapon, and Vanselow, but the relatively subtle differences among them do not merit detailed analysis given the scope of this book.

The relative abundances of Na^+ and Ca^{2+} in exchange sites can be predicted by assuming that $K = 1$ (i.e. that there is no strong preference by the clay for either Na^+ or Ca^{+2}, which is clearly a simplification) and by realizing that, in a simplified system of only two cations (Na^+ and Ca^{2+}), $X_{Na^+ [Clay]} + X_{Ca^{2+} [Clay]} = 1.0$ (i.e. that Na^+ and Ca^{2+} occupy 100% of exchange capacity). This being the case, we can examine two scenarios, one in high ionic strength water and one in low ionic strength water.

Setting activities of Na^+ and Ca^{2+} in solution to a very high end-member value of $1.0 M$ each, the equation becomes:

$$K = 1.0^2 * X_{Ca^{2+} \cdot [Clay]} / 1.0 * X^2_{Na^+ \cdot [Clay]} = 1$$

Rearranging gives:

$$X^2_{Na^+ \cdot [Clay]} - X_{Ca^{2+} \cdot [Clay]} = 0$$

Given that $X_{Na^+ \cdot [Clay]} + X_{Ca^{2+} \cdot [Clay]} = 1$, we can set:

$$X_{Na^+ \cdot [Clay]} = x$$

$$X_{Ca^{2+} \cdot [Clay]} = 1 - x$$

This results in an equation $x^2 + x - 1 = 0$, and using the quadratic formula, we find that $x = 0.618$, so of the total exchange sites, 61.8% are occupied by Na^+ ($X_{Na^+ \cdot [Clay]} = 0.618$), and the remaining 38.2% of exchange sites are occupied by Ca^{2+}.

If activities of Na^+ and Ca^{+2} in solution are set to a low end-member value of $0.001 M$ each, the equation becomes:

$$K = 0.001^2 * X_{Ca^{+2} \cdot [Clay]} / 0.001 * X^2_{Na^+ \cdot [Clay]} = 1$$

Solving as above, ($X_{Na^+ \cdot [Clay]} = 0.031$), indicating that only 3.1% of exchange sites are occupied by Na^+, whereas 96.9% are occupied by Ca^{2+}. This emphasizes the fact that, at low ionic strength (e.g. temperate river, lake and soil water, or groundwater), particles have a strong preference to adsorb Ca^{2+} over Na^+; conversely, particles in sea water, salty lake water, and certain hydrothermal waters of high ionic strength will have a preference to incorporate Na^+ rather than Ca^{2+} into exchange sites. Performing these calculations for other monovalent–divalent pairs will produce similar results. For example, in fresh waters, toxic metals such as Pb^{2+} and Cd^{2+} will sorb more strongly to clays than will Na^+, but if eroded and transported into the marine environment we would predict exchange reactions involving adsorption of Na^+ and desorption of Pb^{2+} or Cd^{2+}.

4.12.3 Adsorption edges

The charge on the particle surface is one of the main controls on adsorption of ionic species to surfaces of minerals and organic matter, and solution pH exerts an important influence on surface charge. At low pH, the high concentrations of H^+ in solution mean that in acidic solutions, particle surfaces will be dominated by H^+, producing positively charged particle surfaces. The result is that (i) the positively charged surfaces do not attract metal cations, and (ii) the positive charge imparted by the H^+-rich surface anions. At high pH, desorption of H^+ leaves negatively charged particle surfaces which can be illustrated by a simple equation.

$$[particle]^- \cdot H^+ + OH^- = [particle]^- + H_2O \quad (4.44)$$

The negatively charged, deprotonated surface favors adsorption of cations:

$$[particle]^- + M^{2+}_{(aq)} = [particle]^- \cdot M^{2+} \quad (4.45)$$

In this case, a divalent metal cation in solution is adsorbed to the surface of a negatively charged particle above the adsorption edge for that particle and that metal cation.

As a result, adsorption of heavy metal cations (Cu^{2+}, Pb^{2+}, Zn^{2+}, etc.) or base cations (Na^+, K^+, Ca^{2+}, Mg^{2+}) decreases with decreasing pH, and for most metal cations, adsorption will shift from nearly complete to nearly zero adsorption over a narrow pH range, typically 1–2 units of pH (Figure 4.17). This threshold is the *absorption edge*, or in more general terms, the *sorption edge*. The opposite

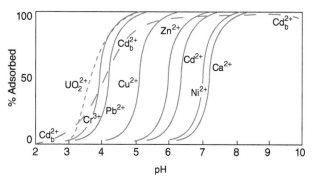

Fig. 4.17 Adsorption of metal cations onto ferrihydrite as a function of pH. Source: Data are from Dzombak and Morel (1990) except for uranyl, which is from Hsi and Langmuir (1985), and Cd_b^{2+}, which is the adsorption edge for Cd^{2+} onto a range of bacterial species (Yee and Fein 2001). The curvature in the adsorption edges indicate that, at low pH, adsorption is relatively weak; with increasing pH, cations are more strongly adsorbed. For example, at pH = 4, Cu^{2+} is weakly adsorbed (i.e. only a small proportion of Cu^{2+} ions in solution will be adsorbed at pH = 4), but at pH = 5.5, nearly all Cu^{2+} in solution will be adsorbed onto ferrihydrite. The boundary between pH 4 and 5.5 is the adsorption edge for Cu^{2+}.

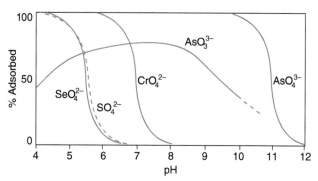

Fig. 4.18 Adsorption of the polyatomic anions selenate, sulfate, chromate, arsenite (AsO_3^{3-}), and arsenate (AsO_4^{3-}) onto ferrihydrite as a function of pH. Source: Data are from Dzombak and Morel (1990). Note that for all anions shown but arsenite, the shape of the curve is the inverse of the curves for metal cations shown in Figure 4.19, reflecting the fact that, as pH decreases, surfaces of ferrihydrite become progressively more occupied by H^+, creating positive surface charge needed for adsorption of anions. Thus, adsorption increases to the left of this diagram. Although not shown, decreasing pH causes protonation of the anions (e.g. to CrO_4^{2-} to $HCrO_4^-$, AsO_4^{3-} to $HAsO_4^{2-}$ to $H_2AsO_4^-$, etc.).

is true for adsorption of anions – adsorption will decrease with increasing pH as H^+ is released from surfaces to produce negatively charged surfaces that repel anions. The adsorption edge is controlled by pH, characteristics of the metal undergoing adsorption (or desorption), and the structure and charge distribution of the adsorbing particulate. The majority of adsorption edge data have been determined for iron hydroxides given their near ubiquity in surficial systems.

Absorption of Cr^{3+} and Pb^{2+} occurs at the lowest pH of the cations indicated in Figure 4.17, reflecting their low solubilities in solution and corresponding stronger likelihood of adsorption. The more soluble metal cations (Ni^{2+}, Zn^{2+}) exhibit adsorption edges at higher pH values, with the highly soluble Ca^{2+} shown for reference (Ca^{2+} is not adsorbed to ferrihydrite at pH <6). So, for example, at a pH of 5, we would predict that all Cr^{3+} and Pb^{2+} ions in an aqueous system would be adsorbed to surfaces of iron hydroxides (assuming sufficient $Fe(OH)_3$ is present in the system), and that Ni^{2+} and Zn^{2+} ions would be present in solution. Assuming that the $Fe(OH)_3$ occurs as fine-grained colloidal particles suspended in the water, filtration of this water using a standard 0.45 μm filter (Drever 1973) before chemical analysis should reveal the presence of Ni^{2+} and Zn^{2+} in solution, but not Cr^{3+} and

Pb^{2+}, which should be retained with the particles on the filter.

Adsorption of representative anions by $Fe(OH)_3$ in Figure 4.18 illustrates decreasing adsorption with increasing pH.

The adsorption edge for arsenate, AsO_4^{3-}, occurs at a pH of approximately 11, meaning that > 50% of arsenate in solution will be adsorbed at pH < 11, and nearly all arsenate will be adsorbed at pH < 9 (provided sufficient amounts of iron hydroxide occur in the soil or aquifer). It forms strong inner-sphere complexes with iron hydroxide over a broad range of pH values, from slightly basic (pH = 9) through neutral and very acidic. Arsenite (AsO_3^{3-}) on the other hand is never completely adsorbed. Predictably, it is adsorbed in progressively lower amounts at pH >8 due to increasing negative charge of the $Fe(OH)_3$ surface, yet it also exhibits progressively lower adsorption at pH values < 8. The differing behaviors of arsenate versus arsenite are related to speciation, specifically the extent of protonation. Recall that arsenate mainly occurs as two anionic species, $H_2AsO_4^-$ and $HAsO_4^{2-}$ – these will both be attracted to positively charged particle surfaces at pH < 9. However, as pH decreases toward more acidic values, anionic arsenite ($H_2AsO_3^-$) reacts with H^+ to form the uncharged H_3AsO_3 species (arsenious acid), which is

---

Adsorption, Adsorption Edges, and the Role of Bacteria

Microbial organisms – in addition to minerals (and non-crystalline inorganic solids) – also have the potential to control behavior of elements and compounds by adsorption, and like minerals, bacteria exhibit adsorption edges. Yee and Fein (2001) studied adsorption of cadmium (Cd^{2+}) in the presence of a wide range of bacteria, including Gram-negative (*Escherichia coli* and *Pseudomonas aeruginosa*) and Gram-positive (*Bacillus megaterium*, *Streptococcus faecalis*, *Staphylococcus aureus*, *Sporosarcina ureae*, and *Bacillus cereus*) species.

Their findings indicate that the adsorption edge is the same for all seven species, and that the adsorption edge occurs at lower pH than that of ferrihydrite (Figure 4.17). Subsequent studies have confirmed the ~ species-independent sorption behavior of microbes, but what is less certain is the role of microbial extracellular polymeric substances (EPSs) (Tourney and Ngwenya 2014). Nevertheless, any study assessing sorption in soils or other systems where microbes are present should consider their potential effects on sorption and mineralization.

---

not electrostatically attracted to particle surfaces and thus largely remains in solution.

One closing thought regarding adsorption edges involves the relationship between the adsorbing particle (i.e. the sorbent) and the potentially adsorbed ion (i.e. the sorbate). Table 2.4 shows that different minerals possess different PZCs and isoelectric points (IEPs), both of which pertain to pH thresholds at which the charge on a mineral surface (i.e. the sorbent) changes. Below this value, anions are likely to be adsorbed, and above this value, cations are likely to be adsorbed. However, properties of individual ions in combination with the PZC or IEP of the sorbent ultimately determine the pH of the adsorption edge. While PZCs and IEPs of iron oxides and hydroxides generally occur in the pH range of 6–8, adsorption edges of cations to the iron hydroxide ferrihydrite occur over a pH range from 3 to 8 (Figure 4.17). Particles with higher PZCs or IEPs (e.g. iron and aluminum hydroxides) are more suited to anion adsorption, especially at neutral to acidic pH, whereas particles with low PZCs or IEP (e.g. smectites and other silicate clays) are more suited to cation adsorption (except at extremely low pH). The factors that control adsorption are multiple and include mineral content (e.g. is ferrihydrite present? or is smectite the main sorbent in the system?), pH, and aqueous composition as well as kinetics, residence time, temperature, and competition among ions.

4.12.4 Adsorption isotherms

Adsorption isotherms quantify the extent of adsorption of a given substance as a function of its abundance in solution, at a constant temperature (explaining the term "isotherm"), accounting for variability in adsorption capacity of different solids (e.g. smectite vs. kaolinite). All adsorption isotherms show increasing adsorption with increasing concentration (or activity) or a substance in solution – the more of a given substance (e.g. simple or complex ion, organic liquid, gas) there is in solution, the more of it will adsorb to the surface of a solid, based on basic equilibrium considerations. The term "equilibrium adsorption isotherms" is often used, recognizing that kinetic factors are not inherently considered in this treatment.

The *linear* adsorption isotherm (or linear distribution coefficient), the simplest isotherm, is described by this equation:

$$C_{i(ads)} = K_d \times a_{i(aq)} \qquad (4.46)$$

$a_{i(aq)}$ is the activity of a given species (*i*) in solution, K_d is an empirically derived equilibrium constant that quantifies the tendency of species *i* to adsorb onto a given solid (relative to the amount that remains in solution), and $C_{i(ads)}$ is the concentration of species *i* adsorbed to the solid (typically in moles of *i* per kg of solid). The linear relationship shown in Figure 4.19a reflects a couple of simplistic and unrealistic assumptions, that (i) the solid has an unlimited number of adsorption sites, and (ii) that the rate of adsorption with increasing concentration of species *i* in solution does not decrease as adsorption sites become occupied by *i*. Two other models that deal with these overly simplistic assumptions are the Langmuir and Freundlich models for adsorption.

The need to account for the reality that there exist a finite number of adsorption sites – and that they eventually

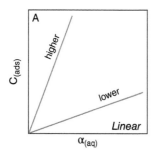
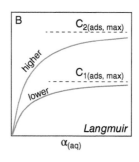
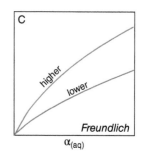

Fig. 4.19 Adsorption isotherms. A = linear isotherm, B = Langmuir isotherm, and C = Freundlich isotherm. X-axis is activity of sorbate (e.g. ion) and y-axis is extent of adsorption. Scenarios for lower and higher sorption potential are shown in each case, where controls are either characteristics of adsorbent (i.e. solid, e.g. CEC) or adsorbate (e.g. ionic charge), or both.

become saturated – led to development of the *Langmuir isotherm* (Figure 4.19b). This quantification (Eq. (4.47)) recognizes that as extent of adsorption increases and sites become saturated, the curve flattens and takes on a logarithmic form. It is expressed as:

$$C_{i(ads)} = C_{i(ads\ max)} \times K_L \times a_{i(aq)} / (1 + K_L + a_{i(aq)}) \quad (4.47)$$

$C_{i(ads\ max)}$ is the maximum quantity of species i that can be adsorbed to the solid surface and K_L is the Langmuir equilibrium adsorption coefficient. No matter how high the activity of i becomes in solution, the mass of i adsorbed ($C_{i(ads)}$) cannot exceed available adsorption sites ($C_{i(ads\ max)}$).

The *Freundlich isotherm* (Figure 4.19c) also accounts for the fact that adsorption is not linear and thus also reflects the effect of decreasing adsorption rate with increasing activity of i in solution; however, the Freundlich isotherm does not account for a finite number of adsorption sites, and instead reflects the potential that species can become progressively more weakly adsorbed in outer-sphere sites. It is expressed as:

$$C_{i(ads)} = K_L \times C_{i(ads)}{}^n \quad (4.48)$$

When the term n is <1, the curve becomes logarithmic.

4.13 OTHER GRAPHICAL REPRESENTATIONS OF AQUEOUS SYSTEMS: PIPER AND STIFF DIAGRAMS

Many different graphical representations of aqueous geochemistry are given in this text, including Eh–pH diagrams, adsorption edge diagrams, the Bjerrum plot of

H_2CO_3-$HCO_3{}^-$-$CO_3{}^{2-}$ (Figure 5.1), pH vs. activity, and more. Two additional diagrams that are commonly used to present aqueous chemistry data are Piper and Stiff diagrams. Piper diagrams plot the major aqueous cations Ca^{2+}, Mg^{2+}, and ($Na^+ + K^+$) on one ternary diagram (Figure 4.20, lower left), and the main aqueous anions chloride (Cl^-), sulfate ($SO_4{}^{2-}$), and alkalinity (mainly $HCO_3{}^-$; refer to carbonate geochemistry in Chapter 5 for definition of alkalinity) on a second ternary diagram (Figure 4.20, lower right). Hydrochemical facies as defined by Back (1966) are presented in the upper right and are used to delineate water compositions into different classifications.

For Piper plots, data are plotted in units of milliequivalents per liter (meq l^{-1}), which can be obtained from the more common mg l^{-1} unit as follows:

$$[C]\,(meq\,l^{-1}) = [C]\,(mg\,l^{-1}) \times mmol/mg$$
$$\times\ ionic\ charge\ on\ C\ (meq/mmol) \quad (4.49)$$

where [C] is concentration of ion C. Note that if data are already present in units of mmol l^{-1}, the conversion to meq l^{-1} is much simpler:

$$[C]\,(meq\,l^{-1}) = [C]\,(mmol\,l^{-1})$$
$$\times\ ionic\ charge\ on\ C\ (meq/mmol) \quad (4.50)$$

Data must then be normalized so that the sum of the three components, e.g. Ca, Mg, and (Na + K), equals 100%. The data presented in Figure 4.20 are from an aquifer system that consists of carbonate and clastic metasedimentary bedrock units.

Groundwater from the carbonate aquifer predominantly plots as calcium bicarbonate facies, whereas the

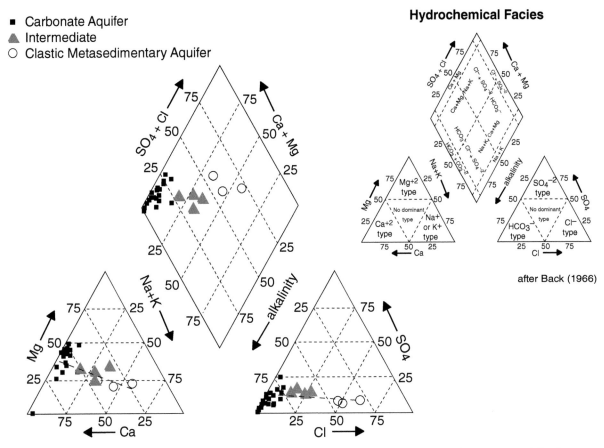

■ Carbonate Aquifer
▲ Intermediate
○ Clastic Metasedimentary Aquifer

Hydrochemical Facies

after Back (1966)

Fig. 4.20 Piper plots of ground water from aquifers of different bedrock type. Note the Na + K-Cl composition of the clastic metasedimentary aquifer (metamorphosed shales and sandstones with micas and feldspars) compared to the Ca–HCO₃ composition of the carbonate aquifer (mainly limestone and dolostone). In most ground water systems, alkalinity is the bicarbonate anion (HCO₃⁻).

clastic metasedimentary aquifer waters plot within or near to the field for (Na + K) chloride facies. Furthermore, mixing between these two end-members is indicated by the points (representing wells producing from both rock types) that fall along the mixing line (dashed line). As a reminder, the abundance of bicarbonate (HCO_3^-) versus carbonate (CO_3^{2-}), the two dominant species that comprise alkalinity, is controlled by pH. In waters of pH <8, HCO_3^- is the dominant species, whereas in waters of pH > 10, the carbonate anion (CO_3^{2-}) is the dominant carbonate species (Drever 1997).

Stiff diagrams are polygons that plot concentrations (in meq l⁻¹) of the cations Na (or Na + K), Ca, and Mg to the left of a center line, and the anions Cl⁻, ($HCO_3^- + CO_3^{2-}$), and SO_4^{2-} to the right of the center line. The farther from the center line, the greater the concentration of the anion or cation (Figure 4.21). Table 4.7 presents calculations for a Stiff plot using values presented earlier in Table 4.3.

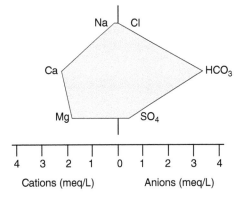

Fig. 4.21 Stiff diagram of water dominated by dissolved Ca, Mg, and bicarbonate.

Stiff polygons are often used to show how water chemistry varies spatially by plotting them on a map or schematically, as is shown in Figure 4.22. In this way, variability in the composition of streams, aquifers, and other water bodies can be summarized visually.

Table 4.7 Determination of values plotted on a Stiff diagram (Figure 4.21).

Element	mg l^{-1}	Conversion	meq l^{-1}
Mg^{2+}	21.8	×2 meq mmol^{-1} ÷ 24.31 mg mmol^{-1}	= 1.79
Ca^{2+}	46.8	×2 meq mmol^{-1} ÷ 40.08 mg mmol^{-1}	= 2.33
Na$^+$	4.79	×1 meq mmol^{-1} ÷ 22.98 mg mmol^{-1}	= 0.21
SO$_4$$^{2-}$	20.3	×2 meq mmol^{-1} ÷ 96.06 mg mmol^{-1}	= 0.42
HCO$_3$$^-$	205.0	×1 meq mmol^{-1} ÷ 61.02 mg mmol^{-1}	= 3.35
Cl$^-$	0.90	×1 meq mmol^{-1} ÷ 35.45 mg mmol^{-1}	= 0.025

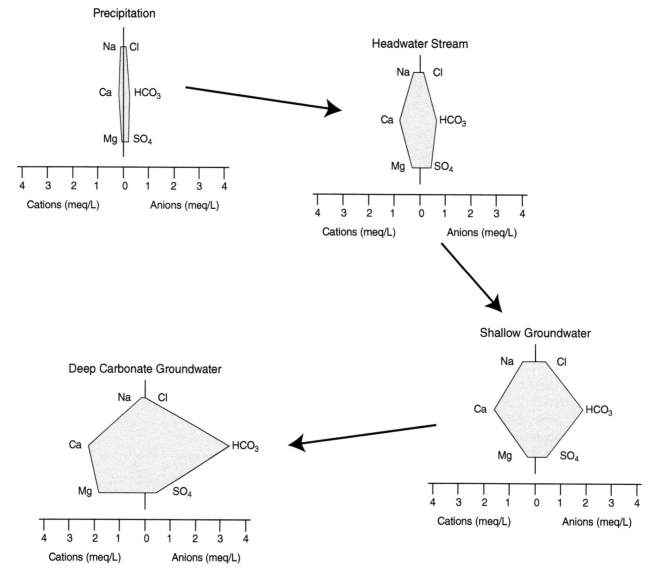

Fig. 4.22 Stiff plots showing water composition at different stages in the hydrologic cycle, emphasizing increasing TDS and alkalinity from atmosphere to surface water and eventually to deep (e.g. 150 m below surface) carbonate aquifer groundwater.

4.14 SUMMARY

Myriad factors control the chemistry of natural waters and contamination-affected waters. Evaporation effectively delivers distilled water to the atmosphere with very low TDS, and solution of CO_2 into atmospheric water results in naturally acidic precipitation. These factors help to produce water that has the capacity to dissolve minerals and organic matter at or below the surface. The composition of bedrock and biotic communities combined with climate strongly influence the natural composition of surface water and groundwater. In addition to pH (and the related important role of carbonate chemistry), reduction–oxidation reactions impart a strong control on mobility and stability of inorganic ions, organic compounds and minerals; for example, oxidizing waters foster breakdown of sulfide minerals and organic compounds, whereas reducing environments foster breakdown of iron hydroxides. Sorption is strongly influenced by minerals, organic matter, and microbial organisms present in the aqueous environment, so understanding the interrelationship of these factors allows for prediction and understanding of the fate and transport of natural compounds and contaminants in the environment.

Appendices I and II presents case studies that provide the opportunity to examine integrated concepts across aqueous geochemistry. Appendix III presents information on instrumental analysis in environmental geochemistry, including approaches used to analyze the composition of water, and Appendix IV contains thermodynamic data that allows quantification and comparison of stabilities.

QUESTIONS

4.1 Convert values listed for the arid river in Table 4.1 to mmol l^{-1} and cmol$_c$ l^{-1} (mmol of charge l^{-1}). Is this water body electrically neutral (i.e. do + and − charges balance?). Should it be electrically neutral? If it is not, give a possible explanation.

4.2 Determine the ionic strength and TDS (in mg l^{-1}) of the two river waters in Table 4.1, and classify the water based on values in Table 4.4. Plot these water body on the graph in Figure 4.4 – does it correctly classify these waters?

4.3 Predict the relative solubilities of Na^+, K^+, Mg^{2+}, Ca^{2+}, Co^{2+}, Fe^{2+}, Fe^{3+}, and Al^{3+} considering ionic charge and ionic radius, then compare your findings to the graph in Figure 4.9.

4.4 Provide a chemical reaction that shows the change in speciation of Al from a dissolved form to a mineral form as pH increases from 4 to 7.

4.5 Use the Debye-Huckel equation to determine ion activity coefficients (20 °C) for K^+ and Al^{3+} at ionic strengths of 10^{-2} and 10^{-4}. Why is this type of calculation important? List two examples of non-ionic aqueous complexes that might form involving K^+ and Al^{3+} (two each). Are these types of complexes more common or less common when ion activity coefficients are low?

4.6 Predict concentrations of Pb^{2+} and CO_3^{2-} in solution that would be produced by dissolution of cerussite.

4.7 Is uranium more soluble in oxidizing or reducing water? What are examples of species of U in solution? How does pH and presence of carbonate species affect U solubility in water?

4.8 Use the Fe and Al Eh–pH diagrams to explain controls on solubility and speciation of those two elements in aqueous systems.

4.9 Use the Law of Mass Action to predict relative abundances of Ca^{2+} and Al^{3+} in exchange sites for two waters, one of ~high ionic strength (10^{-1} M) and one more dilute (10^{-3} M). List pertinent assumptions and simplifications.

4.10 Compare linear vs. Langmuir vs. Freundlich adsorption models, including descriptions of their assumptions, graphical examples of adsorption isotherms, applicable equations and general applicability, i.e. which do you think best models adsorption in aqueous systems?

REFERENCES

Andrews, J.E., Brimblecombe, P., Jickells, T.D. et al. (2004). *An Introduction to Environmental Chemistry*, 2e. Blackwell Publishing.

Arai, Y., Sparks, D.L., and Davis, J.A. (2005). Arsenate adsorption mechanisms at the allophane-water interface. *Environmental Science and Technology* 39: 2537–2544.

Baas Becking, L.G.M., Kaplan, I.R., and Moore, D. (1960). Limits of the natural environment in terms of pH and oxidation-reduction potentials. *Journal of Geology* 68: 243–284.

Back, W. (1966). *Hydrochemical Facies and Ground-Water Flow Patterns in Northern Part of Atlantic Coastal Plain*. US Geol Surv Professional Paper 498-A. Washington, DC: US Government Printing Office.

Berner, E.K. and Berner, R.A. (1996). *Global Environment: Water, Air and Geochemical Cycles*. Upper Saddle River, New Jersey, USA: Prentice Hall.

Drever, J.I. (1973). The preparation of oriented clay mineral specimens for X-ray diffraction analysis by a filter membrane peel technique. *American Mineralogist* 58: 553–554.

Drever, J.I. (1997). *The Geochemistry of Natural Waters: Surface and Groundwater Environments*, 3e. Upper Saddle River, NJ, USA: Simon and Schuster.

Dzombak, D.A. and Morel, F.M.M. (1990). *Surface Complexation Modeling: Hydrous Ferric Oxide*. New York: Wiley-Interscience, 393 pp.

Ettayfi, N., Bouchaou, L., Michelot, J.L. et al. (2012). Geochemical and isotopic (oxygen, hydrogen, carbon, strontium) constraints for the origin, salinity, and residence time of groundwater from a carbonate aquifer in the Western Anti-Atlas Mountains, Morocco. *Journal of Hydrology* 438–439: 97–111. https://doi.org/10.1016/j.jhydrol.2012.03.003.

Gibbs, R.J. (1970). Mechanisms controlling world water chemistry. *Science* 170: 1088–1090.

Gorrell, H.A. (1958). Classification of formation waters based on sodium chloride content. *American Association of Petroleum Geologists Bulletin* 42: 2513.

Holland, H.D. (1978). *The Chemistry of the Atmosphere and Oceans*. New York: Wiley.

Hsi, C.D. and Langmuir, D. (1985). Adsorption of uranyl onto ferric oxyhydroxides; application of the surface complexation site-binding model. *Geochimica et Cosmochimica Acta* 49 (9): 1931–1941.

Kerr, H.W. (1928). The identification and composition of the soil alumino-silicate active in base exchange and soil acidity. *Soil Science* 26: 385–398.

Livingstone, D.A., 1963. *Chemical Composition of Rivers and Lakes*. United States Geological Survey Professional Paper 440-G.

Washington, D.C: U.S. Government Printing Office. Accessed 25 March 2019 online: http://pubs.er.usgs.gov/publication/pp440G.

Martin, J.M. and Whitfield, M. (1983). The significance of the river input of chemical elements to the ocean. In: *Trace Metals in Seawater* (ed. C.S. Wong, E. Boyle, K.W. Bruland, et al.), 265–296. New York: Plenum.

Mayo, A.L. and Loucks, D.L. (1995). Solute and isotopic geochemistry and ground water flow in the central Wasatch Range, Utah. *Journal of Hydrology* 172: 31–59.

Meybeck, M. (1979). Concentration des eaux fluviales en éléments majeurs et apports en solution aux oceans. *Review de Géologie Dynamique et de Géographie Physique* 21: 215–246.

Tourney, J. and Ngwenya, B.T. (2014). The role of bacterial extracellular polymeric substances in geomicrobiology. *Chemical Geology* 386: 115–132.

Volkov, A.G., Paula, S., and Deamer, D.W. (1997). Two mechanisms of permeation of small neutral molecules and hydrated ions across phospholipid bilayers. *Bioelectrochemistry and Bioenergetics* 42: 153–160.

Winter, T.C., Harvey, J.W., Franke, O.L., and Alley, W.M. (1998). *Ground Water and Surface Water, a Single Resource: The Hydrologic Cycle and Interactions of Ground Water and Surface Water*, vol. 1139. USGS Circular 79 p.

Yee, N. and Fein, J. (2001). Cd adsorption onto bacterial surfaces: a universal adsorption edge? *Geochimica et Cosmochimica Acta* 65: 2037–2042.

Zheng, Y., van Geen, A., Stute, M. et al. (2005). Geochemical and hydrogeological contrasts between shallow and deeper aquifers in two villages of Araihazar, Bangladesh: implications for deeper aquifers as drinking water sources. *Geochimica et Cosmochimica Acta* 69: 5203–5218.

5

Carbonate Geochemistry and the Carbon Cycle

The abundance, solubility, and reactivity of the carbonate species H_2CO_3, HCO_3^-, and CO_3^{2-} – as well as CO_2 – in air, surface water, soils, and groundwater means that the carbonate system exerts a very strong control on aqueous chemistry. Solution of CO_2 in H_2O creates carbonic acid (H_2CO_3) and reactions among H_2CO_3, HCO_3^- and CO_3^{2-} that absorb or release H^+ are important in controlling pH; furthermore, the capacity of carbonate species to form soluble complexes enhances the solubility of certain elements (e.g. U as $UO_2(CO_3)_2^{2-}$; Chapter 4) but may limit others (e.g. Pb as $PbCO_3$). Cycling of carbon among the terrestrial biosphere, the atmosphere, the oceans (as biotic C and inorganic C), and the crust (carbonate rocks, fossil fuel, and disseminated organic carbon) has profound impacts on climate, marine ecosystems, and energy sources. Accordingly, this chapter focuses on essential topics in carbonate geochemistry and the carbon cycle, first focusing of carbonate geochemistry and its impact on aqueous chemistry, then, in the second half, focusing on the cycling of carbon among reservoirs in the atmosphere, hydrosphere, and lithosphere.

5.1 INORGANIC CARBON IN THE ATMOSPHERE AND HYDROSPHERE

CO_2 dissolves in water (e.g. in the atmosphere and soils) to produce carbonic acid, a relatively weak acid that is responsible for many important processes at/near the Earth surface. It is the most abundant acid in natural waters, and it:

1 controls the pH of natural rain (at pH = 5.6 in equilibrium with CO_2) as well as many other natural waters via the action of carbonic acid plus the bicarbonate and carbonate anions; and
2 is the driving force behind the chemical weathering reactions that (i) release dissolved ions into streams and soils, and (ii) produce the minerals that form the solid component of soils.

Here is an example of the crucial role that carbonate can exert on water chemistry – the chemistry of aquifers or soils with only 1% calcite will be dominated by a "carbonate signature" due to the high rate of calcite dissolution and the production of dissolved bicarbonate (Langmuir 1997). These waters will be close to neutral pH and have high buffering capacity. Because the bicarbonate anion (HCO_3^-) can lose H^+ to buffer basic solutions and absorb H^+ to buffer acidic solutions, it is responsible for buffering most natural aqueous systems, and it is the main contributor to alkalinity in aqueous systems.

5.1.1 Atmospheric CO_2, carbonate species, and the pH of rain

CO_2 gas will dissolve into water (e.g. droplets in air, surface water, soil water, groundwater) until equilibrium is reached. Most of the $CO_{2(aq)}$ that diffuses into water remains as discrete CO_2 molecules dissolved in water, but by convention, all dissolved CO_2 in water, whether it is discrete $CO_{2(aq)}$ or whether it is truly carbonic acid (H_2CO_3), is typically referred to as H_2CO_3.

Environmental and Low-Temperature Geochemistry, Second Edition. Peter Ryan.
© 2020 John Wiley & Sons Ltd. Published 2020 by John Wiley & Sons Ltd.

The pH of rainfall is predominantly controlled by the reaction of water with CO_2 unless there are contributions from nitrogen oxides (NO_x) or sulfur oxides (potentially anthropogenic or locally derived from volcanoes and other isolated natural sources). When molecules of CO_2 interact with water (e.g. droplets in the atmosphere), the CO_2 dissolves in the water to an extent that is controlled by the equilibrium established between the two reactants, H_2O and CO_2, and the product, H_2CO_3.

$$H_2O + CO_{2(g)} = H_2CO_3 \qquad (5.1)$$

To know the pH of natural rain controlled by equilibrium with CO_2, we must know the following parameters:

1 the amount of CO_2 in the atmosphere – this is because the greater the amount of CO_2 in air, the greater the extent of contact between carbon dioxide and water to produce carbonic acid (to take an extreme example, if these molecules never interact, carbonic acid will not form);

2 the equilibrium constant for the reaction of CO_2 and H_2O to H_2CO_3 – the mere presence of CO_2 and H_2O does not guarantee that they will react to form H_2CO_3, so we need a term that quantifies the extent to which CO_2 will dissolve in H_2O (i.e. the probability that these two components will react with each other to form H_2CO_3 given a certain concentration of CO_2 and assuming abundant H_2O); and

3 the extent to which carbonic acid will dissociate into H^+ and HCO_3^- (bicarbonate) once CO_2 has dissolved in H_2O. This term will be the K_a for carbonic acid.

The calculations that follow assume a CO_2 concentration of 410 ppm, which is the expected mean atmospheric concentration in the year 2020. 410 ppm is equal to 410×10^{-6} atm.

First, the equilibrium constant for the reaction of CO_2 and H_2O to H_2CO_3 is:

$$K_{CO2} = [H_2CO_3]/[H_2O] \times pCO_2 \qquad (5.2)$$

Given that the concentration (or activity) of water is assumed to be 1, the equation simplifies to:

$$K_{CO2} = [H_2CO_3]/pCO_2 \qquad (5.3)$$

Rearranging to solve for the concentration of carbonic acid in water droplets in the atmosphere:

$$[H_2CO_3] = pCO_2 \times K_{CO2} \qquad (5.4)$$

Intuitively, this should make sense because this equation is saying that the amount of carbonic acid in atmospheric water is controlled by the amount of CO_2 in air (pCO_2) and

by the probability that CO_2 and H_2O will react to form H_2CO_3 (K_{CO2}).

The partial pressure of CO_2 at 410 ppm is 410×10^{-6} atm (or $10^{-3.387}$) and the K_{CO2} at 20 °C is $10^{-1.41}$ (or 0.0389). The K_{CO2} term is the Henry's gas constant, and the value of this constant is temperature dependent. At 25 °C, the value is $10^{-1.47}$, whereas at 0 °C the K_{CO2} is $10^{-1.11}$ and at 60 °C it is $10^{-1.78}$ (Plummer and Busenberg 1982). The variability in the K_{CO2} reflects the fact that the solubility of CO_2 in a liquid increases with decreasing temperature, a phenomenon common to gases.

Thus, $[H_2CO_3] = 410 \times 10^{-6} \times 3.89 \times 10^{-2} = 1.59 \times 10^{-5}$ (or $10^{-4.81}$)

Carbonic acid is a weak acid that only dissociates to a small extent, and its dissociation:

$$H_2CO_3 = H^+ + HCO_3^- \qquad (5.5)$$

is represented by the equilibrium constant K_1 (a K_a that is the first acid dissociation constant for H_2CO_3):

$$K_1 = [H^+] \times [HCO_3^-]/[H_2CO_3] \qquad (5.6)$$

and rearranging gives us this equation:

$$[H^+] \times [HCO_3^-] = K_1 \times [H_2CO_3] \qquad (5.7)$$

Given that dissociation of H_2CO_3 produces equal amounts of H^+ and HCO_3^-, the left side of the equation above is mathematically identical to $[H^+]^2$.

$$[H^+] \times [HCO_3^-] = [H^+]^2 \qquad (5.8)$$

In order to determine the pH of atmospheric water we first need the concentration of hydrogen ions $[H^+]$, which is influenced by the propensity of H_2CO_3 to dissociate, so we can write the equation containing the K_1 term as follows:

$$[H^+]^2 = K_1 \times [H_2CO_3] \qquad (5.9)$$

The K_1 is a constant value of $10^{-6.38}$ (or 4.17×10^{-7}) at 20 °C – for comparison, the value at 10 °C is $10^{-6.46}$ and at 30 °C it is $10^{-6.33}$. We determined above that $[H_2CO_3]$ in equilibrium with 410 ppm $CO_2 = 1.59 \times 10^{-5}$, so we can now determine the concentration of H^+ squared as follows:

$$[H^+]^2 = (4.17 \times 10^{-7}) \times (1.59 \times 10^{-5}) = 6.65 \times 10^{-12}$$

Taking the square root of both sides gives:

$$[H^+] = 2.58 \times 10^{-6} = 10^{-5.59}$$

And pH $= -\log [H^+] = -\log [10^{-5.59}] = 5.59$, which is the predicted pH of rain water in equilibrium with CO_2 at

20 °C at 410 ppm CO_2. This gives you the skills needed to repeat this calculation under different scenarios, including varied temperature and pCO_2, enabling projection of how CO_2 concentrations of 500, 750, or 1000 ppm might affect the pH of rain.

5.1.2 Speciation in the carbonate system as a function of pH

To assess carbonate speciation as a function of pH in a more quantitative manner (i.e. which of the following species is the dominant dissolved form of carbonate at a given pH: H_2CO_3, HCO_3^-, or CO_3^{2-}?), the equilibrium constants can be rearranged as follows. First, we will work with K_1, and – assuming an aqueous solution that is not exchanging CO_2 with the atmosphere – we can determine the pH at which H_2CO_3 and HCO_3^- will exist in equal concentrations. At 25 °C,

$$K_1 = [H^+] \times [HCO_3^-]/[H_2CO_3] = 10^{-6.35} \quad (5.10)$$

(at 20 °C, $K_1 = 10^{-6.38}$). This low K value is typical of weak acids that only partially dissociate in solution, especially when compared to the K values for the strong acids HCl and H_2SO_4, each of which are approximately 10^3, indicating nearly complete dissociation. (In the case of strong acids like these, the product of the concentration of H^+ and the conjugate base is one thousand times greater than the concentration of undissociated acid, i.e. HCl, H_2SO_4.)

The above equation can be rearranged as follows:

$$[H^+]/10^{-6.35} = [H_2CO_3]/[HCO_3^-] \quad (5.11)$$

and when the concentrations of H_2CO_3 and HCO_3^- are equal, the right side of the equation will equal 1:

$$[H^+]/10^{-6.35} = 1$$

(this is true when $[H_2CO_3] = [HCO_3^-]$).

Thus, in the case where $[H_2CO_3] = [HCO_3^-]$, the concentration of hydrogen ions $[H^+]$ is $10^{-6.35}$, and pH = 6.35. When pH is <6.35, H_2CO_3 is the dominant species of carbonate in solution, and at pH > 6.35, HCO_3^- is the dominant dissolved species.

However, as pH rises above 6.35 toward more basic conditions, HCO_3^- is likely to lose a hydrogen ion to form CO_3^{2-}, and we can also determine the pH at which they are equal in concentration, i.e. $[HCO_3^-] = [CO_3^{2-}]$, where K_2 is the second acid dissociation constant for H_2CO_3:

$$K_2 = [H^+] \times [CO_3^{2-}]/[HCO_3^-] = 10^{-10.33} \quad (5.12)$$

And when $[CO_3^{2-}] = [HCO_3^-]$:

$$[CO_3^{2-}]/[HCO_3^-] = 1 = 10^{-10.33}/[H^+] \quad (5.13)$$

In order to satisfy this equation, $[H^+] = 10^{-10.33}$, and the pH at which aqueous bicarbonate is equal to aqueous carbonate ion is 10.33 (at 25 °C).

It is important to emphasize that, in the K_1 and K_2 equations, $[H^+]$ is a variable, and this means that the ratios of H_2CO_3/HCO_3^- and HCO_3^-/CO_3^{2-} will be controlled by pH. So in a qualitative sense, H_2CO_3 will dominate under acidic conditions because the high concentrations of H^+ will drive the system to produce carbonic acid from bicarbonate.

$$HCO_3^- + H^+ \rightarrow H_2CO_3 \quad (5.5a)$$

In basic solutions, bicarbonate will release H^+ to neutralize OH^-, producing a carbonate anion and a water molecule.

$$HCO_3^- + OH^- \rightarrow CO_3^{2-} + H_2O \quad (5.14)$$

You can begin to see the ways in which bicarbonate tends to buffer solutions.

5.1.3 Alkalinity

The ability of a water body to neutralize acids is *alkalinity*, and in most waters, alkalinity is determined by the amount of dissolved bicarbonate and carbonate anion in solution as depicted in Eq. (5.15) (where concentrations are represented as molar values):

$$Carbonate\ alkalinity = [HCO_3^-] + 2 \times [CO_3^{2-}] \quad (5.15)$$

The concentration of the carbonate anion is multiplied by 2 to account for charge, which means that alkalinity is presented as equivalents per liter (typically meq l^{-1} or µeq l^{-1}). The reason for the doubling is that the carbonate anion can neutralize twice as much H^+ as can the bicarbonate anion. As suggested above, in the pH range of 6.4–10.3, alkalinity will be dominated by bicarbonate, and above pH = 10.3, alkalinity will be dominated by the carbonate anion.

Technically the above expression is the *carbonate alkalinity*, but only in rare cases do ions other than HCO_3^- or CO_3^{2-} contribute appreciably to alkalinity. Examples of other dissolved species that can contribute to alkalinity include dissolved silica (as $H_3SiO_4^-$), dissolved boron (as $B(OH)_4^-$), the hydroxyl anion (OH^-), and various organic anions, but typically the concentrations of HCO_3^- or

Focus Box 5.1

The Bjerrum Plot, Summarizing Carbonate Species and pH

In the system H_2CO_3–HCO_3^-–CO_3^{2-}, bicarbonate is the dominant species in solution between pH = 6.4 and pH = 10.3. Below a pH of approximately 6.4, carbonic acid is the dominant carbonate species in solution, and the carbonate anion CO_3^{2-} dominates above pH = 10.3. When different values of [H^+] are substituted into the equations in Section 5.1.2, the ratios of [H_2CO_3] : [HCO_3^-] and [HCO_3^-] : [CO_3^{2-}] can be determined over a range of pH values. The Bjerrum plot in Figure 5.1 displays these relationships graphically.

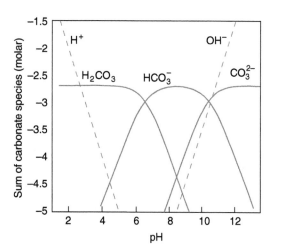

Fig. 5.1 Carbonate species as a function of pH (at 25 °C) in mol/l. Concentrations of H^+ and OH^- are also shown in mol/L. This type of figure is a Bjerrum plot.

Shaded areas represents regions within the conterminous United States of America where surface water alkalinity is < 200 µeq/L.

Fig. 5.2   This map of alkalinity is based on field-based measurements from 39 000 lakes and streams and summarized by the United States Geological Survey from numerous prior reports. This particular map is redrawn to show areas where alkalinity is < 200 µeq l^{-1}; the original USGS map provides intervals that show greater detail (e.g. 0–50, 50–100, 100–200, 200–400 and > 400 µeq l^{-1}). It provides a general idea of areas sensitive to acid precipitation due to naturally low buffering capacities. There is a strong relationship to underlying bedrock, notably the Sierra Nevada (SN) granitic batholith and intermediate to felsic Cascade Range (CR) volcanoes in the west, isolated felsic rocks or quartzite-dominated areas (e.g. Uinta Mountains, UM) in the interior, and the mixed lithologies of the Appalachian Mountains and related rocks in the eastern USA (Omernik et al. 2010, published online). https://water.usgs.gov/edu/graphics/alkalinity-map-large.gif.

$CO_3{}^{2-}$ are far higher than these other species, especially in fresh waters.

Because alkalinity reflects the ability of a water body to buffer acids, it plays a crucial role in determining how aquatic ecosystems will respond to acidification from acid rain, acid mine drainage, and other acidic wastes. Regions that lack carbonate rocks or carbonate minerals in soils and are exposed to acid rain are far more prone to soil and lake acidification than are regions containing carbonates (Figure 5.2).

Focus Box 5.2

How Alkalinity is Determined

Alkalinity determination by titration starts with a filtered water sample of a precisely measured water volume (e.g. 100 ml) and an acid of known concentration (as a titrant). Common titrants are H_2SO_4 and HCl at strengths of 0.02–1 N; for reference, 0.1600 N H_2SO_4 is a titrant commonly used by the US Geological Survey to determine alkalinity in natural waters with normal alkalinity (e.g. 50–200 mg $CaCO_3$ l^{-1}).

1. Determine initial pH of the water sample and record the value.
2. Slowly add small amount of titrant (e.g. 0.5 ml if using 100 ml of water) and allow to equilibrate by gentle, continuous stirring (a magnetic stirrer is commonly used). It is better to add titrant too slowly or in aliquots that are too small rather than too rapidly or too voluminous. Record the pH and volume of titrant added frequently during this process e.g. every 0.5 ml of titrant.
3. Continue to slowly add titrant and allow the sample to equilibrate at each step until a pH of 4.0 is reached. Be sure to continually record data. One way to do this is to set up a spreadsheet with volume of titrant added in one column and pH in another column.
4. As the solution approaches pH = 4.0, be sure to add progressively smaller volumes of titrant and verify that pH reaches a steady, nonfluctuating value (i.e. equilibrium) before adding the next aliquot of titrant. Avoid overshooting this pH range – pH will rapidly drop between pH = 5 and 3 as all remaining $HCO_3{}^-$ is converted to H_2CO_3.
5. Prepare a titration curve that plots volume of titrant added (in ml) on the x-axis and pH on the y-axis, or vice versa. For neutral to slightly alkaline waters the curve will typically have the form shown in Figure 5.3.
6. Determine the values of pH that correspond to the inflection points. The first inflection should occur at a pH of 8.3 which, for most terrestrial waters (streams, lakes, groundwater), corresponds to conversion of $CO_3{}^{2-}$ to $HCO_3{}^-$. This inflection point is sometimes referred to as "P alkalinity." The second point, which will occur around a pH of 4.3, corresponds to the conversion of $HCO_3{}^-$ to H_2CO_3. This inflection point is sometimes referred to as "T alkalinity."
7. Determine the volume of titrant (in ml) consumed in order to reach the P and T alkalinities, respectively.

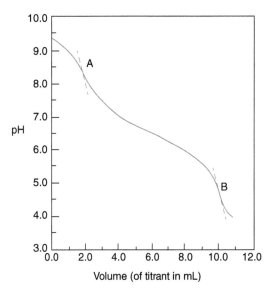

Fig. 5.3 Representative example of a titration curve. The steep slope at point A corresponds to conversion of $CO_3{}^{2-}$ to $HCO_3{}^-$, and the steep slope at point B corresponds to conversion of $HCO_3{}^-$ to H_2CO_3. At point B, all carbonate alkalinity has been converted to H_2CO_3.

Alkalinity of samples with an alkalinity >20 mg l^{-1} (i.e. most natural waters) is calculated using the following formula, where alkalinity is expressed in terms of meq l^{-1} :

$$Alkalinity = (A_{mL} \times A_N \times 1000 \, ml \, l^{-1})/H_2O_{mL}$$

- A_{mL} = volume of acid added
- A_N = normality of acid used as titrant
- H_2O_{mL} = initial volume of water to be titrated.
- 1000 ml = conversion from ml to l

To report results in units of mg $CaCO_3$ l^{-1}, convert the meq l^{-1} result as follows:

$$X \, meq \, l^{-1} \times 1 \, mmol/2 \, meq \times 100.1 \, mg \, mmol^{-1} \, (CaCO_3) \tag{5.16}$$

The rationale for this conversion is that Ca^{2+} contains 2 meq of charge per millimole and the mass of 1 mmol of $CaCO_3$ is 100.1 mg.

Given that alkalinity is effectively a measure of the buffering capacity of an aqueous solution, alkalinity is typically measured by titration using an acid like HCl, where HCl is added until the pH drops to 4 – greater amounts of bicarbonate and carbonate anions will require greater amounts of acid to lower the solution to pH = 4. At a pH of 4, effectively all HCO_3^- and CO_3^{2-} will have been converted to H_2CO_3 (or $CO_{2(aq)}$) (Figure 5.2).

A concept somewhat related to carbonate alkalinity is *dissolved inorganic carbon (DIC)*, which is the sum of all dissolved carbonate species:

$$DIC = [H_2CO_3] + [HCO_3^-] + [CO_3^{2-}] \quad (5.17)$$

Given that CO_2 in solution is effectively synonymous with H_2CO_3, the above equation is sometimes written with $CO_{2(aq)}$ instead of H_2CO_3. The designation DIC differentiates inorganic carbon from dissolved organic carbon (DOC) which is typically comprised of soluble fractions of humic and fulvic acids and potentially also soluble hydrocarbons such as benzene and toluene.

5.1.4 Carbonate solubility and saturation

With some understanding of carbonate speciation and controls on calcite dissolution, we can quantitatively address questions that allow us to predict behavior in the carbonate system in an aquifer, e.g. given the groundwater composition, is calcite stable? Will it precipitate from solution or will it dissolve? The answers depend on whether or not Ca^{2+} and HCO_3^- are saturated with respect to calcite, assuming that calcite is controlling equilibrium concentrations of Ca^{2+} and HCO_3^-. Such questions may be important to address, because (i) precipitation of calcite might be a means of reducing concentration of Pb or U or other trace elements by isomorphic substitution; or (ii) the opposite scenario, i.e. whether or not dissolution of calcite might release trace elements to groundwater.

For carbonate groundwater with pH = 7.5 (a typical value for an aquifer whose pH is controlled by carbonate equilibria), we first recognize that the concentration of bicarbonate is orders of magnitude greater than carbonic acid and the carbonate anion (Figure 5.1); > 99% of dissolved carbonate in a system at 7.5 pH will be HCO_3^-. At this pH, we can ignore the presence of HCO_3^-.

The dissolution of calcite can thus be represented as:

$$CaCO_3 + H_2O + CO_2 = Ca^{2+} + 2HCO_3^- \quad (5.18)$$

This reaction can be quantified by the following equilibrium constant (note that the concentration of bicarbonate

Table 5.1 Equilibrium reactions and constants for the carbonate system in fresh water at 10 °C (K_{10}) and 25 °C (K_{25}).

Reaction	K_{10}	K_{25}
$CO_{2(g)} + H_2O \rightarrow H_2CO_3$	$K_{CO2} = 10^{-1.11}$	$K_{CO2} = 10^{-1.47}$
$H_2CO_3 \rightarrow H^+ + HCO_3^-$	$K_1 = 10^{-6.46}$	$K_1 = 10^{-6.35}$
$HCO_3^- \rightarrow H^+ + CO_3^{2-}$	$K_2 = 10^{-10.49}$	$K_2 = 10^{-10.33}$
$CaCO_3 (cc) \rightarrow Ca^{2+} + CO_3^{2-}$	$K_{sp} = 10^{-8.41}$	$K_{sp} = 10^{-8.48}$
$H_2O \rightarrow OH^- + H^+$	$K_w = 10^{-14.53}$	$K_w = 10^{-14.00}$

Note that (cc) refers to calcite (rather than $CaCO_3$ polymorphs aragonite or vaterite).

is squared to reflect its molar abundance):

$$K_{eq} = [Ca^{2+}] \times [HCO_3^-]^2 / [CaCO_3] \times [H_2O] \times [CO_2] \quad (5.19)$$

The K_{eq} for reaction (5.19) is not typically listed in tables of thermodynamic data, so it is necessary to combine K_{eq} values of reactions listed in Table 5.1. The solid crystal of calcite in equilibrium with dissolved species cannot be represented in terms of concentration, so by convention its concentration is given the value of 1. This is because, in reality, as long as there is solid calcite, the amount will not affect the equilibrium between Ca^{2+} and HCO_3^- ions and the reactants carbonic acid (or $H_2O + CO_2$) and solid calcite – it is the constituents dissolved in water that are most important, at equilibrium. In dilute solutions (e.g. fresh water, groundwater), $[H_2O]$ is also assigned a value of 1, and the concentration of CO_2 is represented by the partial pressure of CO_2 (pCO_2), so the equation for the equilibrium constant is now:

$$K_{eq} = [Ca^{2+}] \times [HCO_3^-]^2 / pCO_2 \quad (5.20)$$

The reactions in Table 5.1 for a system at 10 °C (K_{10}) are suitable for temperate climates where groundwater is typically in the range of 8–12 °C. The net sum of exponents of these four reactions produces the equation for calcite dissolution in the presence of carbonic acid.

$CaCO_3(cc) \rightarrow Ca^{2+} + CO_3^{2-}$	$K_{sp} = 10^{-8.41}$
$CO_{2(g)} + H_2O \rightarrow H_2CO_3$	$K_{CO2} = 10^{-1.14}$
$H_2CO_3 \rightarrow H^+ + HCO_3^-$	$K_1 = 10^{-6.46}$
$H^+ + CO_3^{2-} \rightarrow HCO_3^-$	$K_2' = 10^{+10.49}$
$CaCO_3(cc) + CO_{2(g)} + H_2O$ $\rightarrow Ca^{2+} + 2HCO_3^-$	$K_{sp} = 10^{-5.49}$

Combining these equations as follows:

$$K_{eq} = K_{CO2} \times K_1 \times K_2' \times K_{sp}$$
$$= 10^{-1.41} \times 10^{-6.46} \times 10^{+10.49} \times 10^{-8.41} = 10^{-5.49}$$

results in Eq. (5.21):

$$10^{-5.49} = K_{eq} = [Ca^{2+}] \times [HCO_3^-]^2/pCO_2 \quad (5.21)$$

Note that because the reaction for the dissociation of bicarbonate needs to be expressed in the reverse direction of what is typically presented in tables of thermodynamic data (e.g. Table 5.1), its value in the calculation is the inverse of what is presented in Table 5.1.

The concentration of dissolved CO_2 in groundwater typically occurs at levels greater than what is in equilibrium with the atmosphere. This is because as surface water percolates downward through soils and the vadose zone toward the saturated zone, aerobic decomposition of organic matter produces dissolved CO_2 that drives partial pressures of CO_2 in groundwater up into the range of $\sim 10^{-2}$.

$$C_{org} + O_2 \rightarrow CO_2 \quad (5.22)$$

Note that in the absence of CO_2 produced by decomposition of soil organic matter, natural waters in equilibrium with the atmosphere correspond to a $pCO_2 = 10^{-3.4}$ (using an atmospheric CO_2 concentration of 410 ppm).

In groundwater where organic matter decay (Eq. (5.22)) has produced $pCO_2 = 10^{-2}$, we can express the equilibrium constant as:

$$K_{cc} = [Ca^{2+}] \times [HCO_3^-]^2 = 10^{-7.49} \quad (5.23)$$

Given the stoichiometry of calcite dissolution as expressed in Eq. (5.18), i.e. 2 mol of HCO_3^- are produced per mol of Ca^{2+}, i.e. $[HCO_3^-] = [2 \times Ca^{2+}]$, Eq. (5.23) can be expressed as:

$$[Ca^{2+}] \times [2 \times Ca^{2+}]^2 = 10^{-7.49} \rightarrow 4 \times [Ca^{2+}]^3$$
$$= 10^{-7.49} \rightarrow [Ca^{2+}]^3 = 10^{-7.49}/4$$

Assuming that activities (or concentrations) of Ca^{2+} and HCO_3^- are controlled by calcite dissolution, and also assuming that in these relatively dilute fresh waters that activities are $\sim$ equal to concentrations, we can determine $[Ca^{2+}]$:

$$[Ca^{2+}] = (10^{-7.49}/4)^{0.33} = 3.19 \times 10^{-3} \, mol \, l^{-1}$$

Then, given that calcite decomposition in the presence of carbonic acid yields 2 mol of bicarbonate per mole of calcium,

$$2 \times [Ca^{2+}] = [HCO_3^-] = 6.38 \times 10^{-3} \, mol \, l^{-1}.$$

Converting molar values to mg l^{-1}, $[Ca^{2+}] = 3.19 \times 10^{-3} \, mol \, l^{-1} \times 40.08 \, g \, mol^{-1} \times 1000 \, mg \, g^{-1} = 128 \, mg \, l^{-1}$.

By the same approach, $[HCO_3^-] = 389 \, mg \, l^{-1}$. In other words, at 10 °C with $pCO_2 = 10^{-2}$ atm, the equilibrium concentrations for groundwater right at saturation are $[Ca^{2+}] = 128$ ppm and $[HCO_3^-] = 389$ ppm Stated in simple terms, above these levels, calcite will precipitate from solution, and below these levels, calcite will dissolve.

To precisely determine if groundwater is saturated with respect to carbonate, recall that if the ion activity product (IAP) (Section 4.6.2) is greater than the K_{sp} ($\Omega > 1$), the solution is oversaturated and calcite will precipitate from solution. If the $\Omega < 1$ (the IAP $< K_{sp}$), the solution is undersaturated and calcite will dissolve. If $\Omega = 1$ (IAP $= K_{sp}$), the system is saturated and at equilibrium, and rates of mineral dissolution and mineral crystallization will be equal.

5.1.5 The effect of CO_2 partial pressure on stability of carbonate minerals

Based on the K_{cc} reaction derived above (i.e. the K_{eq} or K_{sp} for calcite dissolution, Eq. (5.21)), it is clear that the amount of carbon dioxide dissolved in water will exert some control over the solubility (or stability) of calcite (and other carbonate minerals too). The K_{cc} is restated here for emphasis:

$$K_{CC} = [Ca^{2+}] \times [HCO_3^-]^2/pCO_2 = 10^{-5.49} \quad (5.24)$$

The K_{cc} is a constant, so increasing pCO_2 must cause an increase in the concentrations of dissolved calcium and bicarbonate, meaning that increasing CO_2 means decreased stability of calcite.

As demonstrated in Eq. (5.23), in soils or groundwater where CO_2 is elevated by organic matter decay, a $pCO_2 = 10^{-2}$ would correspond to a K_{CC} value (at 10 °C) $= 10^{-7.49}$. If the CO_2 content of surface water or groundwater is solely controlled by equilibrium with the atmosphere, $pCO_2 = 10^{-3.4}$ and the value of K_{CC} (at 10 °C) $= 10^{-8.89}$.

When $pCO_2 = 10^{-2}$ atm (and the K_{cc} is $10^{-7.49}$), equilibrium concentrations of calcium and bicarbonate in equilibrium with calcite at 10 °C are:

$$[Ca^{2+}] = 3.19 \times 10^{-3} \, mol \, l^{-1} \text{ or } 128 \, mg \, l^{-1}$$

$$[HCO_3^-] = 6.38 \times 10^{-3} \, mol \, l^{-1} \text{ or } 389 \, mg \, l^{-1}$$

By comparison, when $pCO_2 = 10^{-3.4}$ atm (and the Kcc is $10^{-8.89}$), equilibrium concentrations of calcium and bicarbonate in equilibrium with calcite at 10 °C are much lower:

$$[Ca^{2+}] = 1.09 \times 10^{-3} \, mol \, l^{-1} \text{ or } 43.6 \, mg \, l^{-1}$$

$$[HCO_3^-] = 2.18 \times 10^{-3} \, mol \, l^{-1} \text{ or } 133 \, mg \, l^{-1}$$

Focus Box 5.3

Two Examples of Carbonate Equilibria in Aquifers

First consider a system at 10 °C like the one shown in Table 5.1, with 46.8 mg l^{-1} Ca^{2+} (1.17×10^{-3} mol l^{-1}), 205.0 mg l^{-1} HCO$_3^-$ (3.36×10^{-3} mol l^{-1}), and pCO$_2$ = 10^{-2} atm. In this case, the IAP = 1.17×10^{-3} mol l^{-1} × (1.68×10^{-3} mol l^{-1})^2 = 3.30×10^{-9} (or 10$^{-8.48}$). The value of Ω is:

$$\Omega = 10^{-8.48}/10^{-7.49} = 0.10.$$

The fact that $\Omega < 1$ indicates that these groundwaters are undersaturated with respect to calcite. Under these conditions, calcite will dissolve, and if, as is the case in some limestone aquifers, the calcite contains uranium, we would predict that U will also be released to solution, and likely would be soluble as a complex anion e.g. UO$_2$(CO$_3$)$_2^{2-}$.

Now consider a system at 10 °C and pH = 7.2 containing higher concentrations of dissolved C and bicarbonate: [Ca^{2+}] = 123 mg l^{-1} = 3.07×10^{-3} mol l^{-1} and [HCO$_3^-$] = 216 mg l^{-1} = 3.54×10^{-3} mol l^{-1}. Calculations indicate that this groundwater will be oversaturated with respect to calcite because the IAP = 3.85× 10^{-8} mol^2 l^{-2} = 10$^{-7.41}$; in this case,

$$\Omega = 10^{-7.41}/10^{-7.49} = 1.20$$

Given that the IAP is greater than the equilibrium constant, i.e. $\Omega > 1$, this solution is slightly oversaturated and we predict that calcium and bicarbonate ions will precipitate out of solution to form calcite.

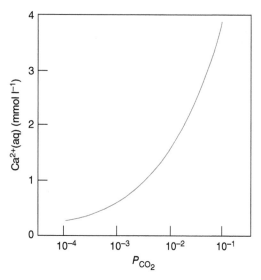

Fig. 5.4 Relationship between CO$_2$ dissolved in water and the stability of calcite (as indicated by the concentration of Ca^{2+}$_{(aq)}$ in equilibrium with calcite). Increasing Ca^{2+} indicates decreasing calcite stability, so this graph illustrates that increasing dissolved CO$_2$ enhances the solubility of calcite.

These relatively low concentrations reflect greater stability (lower solubility) of calcite at low dissolved CO$_2$ (e.g. pCO$_2$ = 10$^{-3.4}$ atm) compared to lower calcite stability at higher pCO$_2$ (e.g. 10^{-2} atm). Figure 5.4 shows this relationship graphically. As a qualitative reminder, when CO$_2$ dissolves in water it produces H$_2$CO$_3$, carbonic acid that enhances calcite dissolution.

Calculations like these throughout Section 5.1 help to understand and predict the effect of dissolved CO$_2$ on the stability of carbonate minerals in soils, surface waters

and groundwater. They also can help to provide insight into the devastating effect that increasing atmospheric CO$_2$ will have on the stability of marine CaCO$_3$ (including skeletons and shells of organisms, polyp secretions in coral reefs, and other parts of the marine ecosystem). Based on these principles and the plot in Figure 5.4, it is clear that increasing atmospheric CO$_2$ will diminish the stability of CaCO$_3$ in the marine environment, and this in turn is bound to have negative consequences for marine ecosystems forward into the twenty-first century.

5.1.6 The effect of mineral composition on stability of carbonate minerals

The often-interrelated variables of pH, dissolved CO$_2$, and temperature are important controls on the stability of calcite (and therefore also on water chemistry), but examining other carbonate minerals it is clear that carbonate solubility is also controlled by mineral composition and, in the case of the CaCO$_3$ polymorphs aragonite and calcite, by mineral structure. Aragonite is less stable than calcite at standard temperature and pressure (Table 5.2), and the least stable of the carbonate minerals presented in Table 5.2 is magnesite. The lower K$_{sp}$ of dolomite (vs. CaCO$_3$) is evident in weathering patterns of rocks that contain layers of calcite and dolomite. The more rapidly weathering calcite is recessed and layers of dolomite form positive relief reflecting the lower chemical weathering rate of dolomite relative to calcite (Plate 7).

Table 5.2 Equilibrium constants of common carbonate minerals at 20 °C and 1 atm.

Mineral	Formula	Ksp
Aragonite	$CaCO_3$	$10^{-8.30}$
Calcite	$CaCO_3$	$10^{-8.48}$
Dolomite	$Ca_{0.5}Mg_{0.5}(CO_3)$	$10^{-9.08}$
Magnesite	$MgCO_3$	$10^{-8.05}$
Rhodochrosite	$MnCO_3$	$10^{-10.09}$
Siderite	$FeCO_3$	$10^{-10.53}$

Dolomite has an asterisk because often when the formula of dolomite is given as $CaMg(CO_3)_2$, the Ksp is given as $10^{-18.16}$. The difference in the K_{sp} values reflects moles of CO_3^{2-} per unit formula.

Source: Morse et al. 2007 and sources cited therein.

5.2 THE CARBON CYCLE

The carbon cycle may be the most important elemental cycle to understand, at least in terms of environmental analysis. Carbon forms the basis of life, plays a major role in Earth's climate, drives chemical weathering (carbonic acid), and buffers aquatic ecosystems (bicarbonate anion). Carbon occurs in organic and inorganic forms and is an important constituent in all reservoirs, including the lithosphere (as carbonate minerals and organic matter), atmosphere (mainly as CO_2 and to a lesser extent as CH_4 and CO), oceans (as dissolved CO_2, bicarbonate, $CaCO_3$ in shells, and as organic carbon in plankton and other organisms), and terrestrial ecosystems (e.g. in living organisms, sediments, and soils). For these reasons and more, the carbon cycle is relatively complex, yet understanding the carbon cycle is crucial to the practice of environmental geochemistry, climate analysis, ecosystem analysis, and much more in the twenty-first century.

Principles of biogeochemical cycling of elements are presented at the beginning of Section 6.1, so depending on your background it may be worthwhile reading that section before reading this section.

5.2.1 Oxidation states of carbon

Carbon is a redox-sensitive element with oxidation states that range from C^{4-} to C^{4+} (keeping in mind that carbon tends to form covalent or polar covalent bonds). Arranged in order from most-reduced to most-oxidized, carbon occurs in the following oxidation states and compounds:

1 The most-reduced form of carbon, C^{4-}, is the oxidation state of carbon in methane, CH_4 (recall that hydrogen is H^+ except in certain cases when it forms metal hydrides).

2 Carbon atoms in an alkane like octane (C_8H_{18}) are slightly less reduced, with an average oxidation state of $C^{2.22-}$ – this non-integer oxidation state reflects covalent bonds where carbon atoms can be thought of as a mixture of C^0 and C^{4-}, with C^{4-} slightly dominant over C^0.

3 An approximate molecular formula for humus is $C_{28}H_{26}O_{19}$ (Manahan 2000) and in this compound the average carbon atom is $C^{0.4-}$ (again, reflecting a mix of C^0 and C^{4-}, in this case dominated by C^0).

4 In some types of organic matter, the average oxidation state of carbon atoms is zero. Consider glucose, $C_6H_{12}O_6$, where the charges on H^+ and O^{2-} balance each other out, leaving carbon atoms as C^0. The carbon atoms in graphite and diamond, both forms of pure elemental carbon, also contain C^0 and the atoms are held together by covalent bonds (as well as van der Waals bonds in the case of graphite sheets).

5 *The scenarios presented above represent compounds with chemically reduced carbon.*

6 C^{2+} occurs in carbon monoxide (CO), a compound that represents a more-oxidized form of carbon than those above, yet less oxidized than the compounds presented below. In CO, the C^{2+} atom balances the O^{2-} atom.

7 C^{4+}, the most-oxidized state of carbon, occurs in CO_2 as well as in carbonate minerals ($CaCO_3$, $MgCO_3$, etc), the bicarbonate anion (HCO_3^-) and the carbonate anion (CO_3^{2-}).

Thus, reduced carbon – including C^{4-} and C^0 and mixtures of these oxidation states – occurs in organic matter and anaerobically decomposed organic matter, and oxidized carbon occurs in CO_2 and carbonates. Not surprisingly, organic matter and other forms of reduced carbon (inorganic CH_4, for example) are stable in the reducing/anaerobic/anoxic to suboxic environments of the Earth's crust, mantle, wetlands, marine sediments, and other low-O_2 environments; conversely, carbon dioxide and carbonates are stable in O_2-rich environments of oxidized/aerobic sea water, the atmosphere, soils, and fresh water.

Much of the cycling of carbon is controlled by redox reactions, one of which can be summarized by the oxidation of methane and C^{4-} to C^{4+} (note: this reaction in the atmosphere is more complicated than this summation; see Chapter 8):

$$CH_4 + 2O_2 = CO_2 + 2H_2O \qquad (5.25)$$

If viewed in reverse, this reaction would represent the reduction of carbon from C^{4+} to C^{4-} associated with formation of CH_4. In addition to inorganic processes like oxidation (or combustion) of CH_4, biological processes also play a significant role in carbon cycling, from photosynthesis and respiration to microbially mediated oxidation of soil organic matter to CO_2.

The global carbon cycle (Figure 5.5) includes all plants, animal, and microorganisms, all decomposing plant remains, water bodies large (oceans) and small (lakes, ponds), soils (including various forms of organic and inorganic carbon), the crust and mantle of the Earth, products of volcanic eruptions, and more. In this section, carbon cycling is mainly presented through fluxes and reservoirs that exist on a global scale.

5.2.2 Global-scale reservoirs and fluxes of carbon

The masses of reservoirs at global scales are measured in units of petagrams (10^{15} g) or gigatons (Gt). The units are the same, i.e. 1 Gt = 1 Pg, but the Pg is the accepted SI unit so this text refers to these massive quantities (10^{15} g) in terms of Pg. Also, unless otherwise indicated, values indicate the mass of carbon rather than the mass of the molecule. So if this text were to report a value of 1.5 Pg of carbon per year (1.5 Pg C yr^{-1}), the equivalent amount of carbon dioxide would be 5.5 Pg CO_2 yr^{-1} (where the conversion is 44 g of CO_2 per 12 g of C). Commonly masses (or moles) of carbon are given rather than masses of molecules (e.g. CO_2, CH_4, C_6H_6, etc.) to facilitate tracking of carbon, in part because of oxidation reactions that transform reduced gases like CH_4 to the oxidized form, CO_2.

The globally important C reservoir with the greatest amount of carbon by far is the *mantle* of the Earth, with *340 000 000* Pg of carbon (most of which likely occurs as graphite, diamonds, carbonate, and possibly methane). Yet in spite of the enormous size of the mantle carbon reservoir, the flux between the mantle and other reservoirs (crust, atmosphere, terrestrial ecosystems, and oceans) is very low – ~0.02 Pg yr^{-1} (Cartigny et al. 2008), a quantity that is effectively negligible for the conversation that follows – and so will not be considered in detail.

After the mantle, the next largest reservoir of carbon on Earth is the *crust*, with *75 000 000 Pg C*. Of this amount, a relatively small quantity (5000 Pg) occurs as hydrocarbons, and the remainder, the vast majority, occurs either in carbonate minerals in rocks (e.g. limestone, dolostone) or as dispersed organic carbon (as kerogen, humin, and other forms). The long-term non-anthropogenic flux of carbon to and from the crust at steady state is ~0.2 Pg yr^{-1}, and

while this too is a low rate of exchange, it (i) is ten times the mantle flux, and (ii) will serve as an important reference point in the discussion of carbon cycling, fossil fuels, and climate.

The *oceans* contain ~*40 000 Pg C*, the vast majority of which exists as dissolved bicarbonate in the deep ocean (*39 000 Pg C*). Lesser amounts are found as dissolved CO_2, DOC, and carbon in living organisms, mainly in the shallow ocean. The shallow ocean (depths <100 m) is characterized by photosynthesis and greater mixing with the atmosphere; the deep ocean is more isolated from the atmosphere and at these depths carbon primarily occurs in inorganic forms (e.g. dissolved CO_2, HCO_3^-). While the long-term steady-state flux between atmosphere and oceans is ~90 Pg yr^{-1}, this balance has shifted to a net flux into the oceans since the onset of the industrial revolution (Figure 5.5).

Terrestrial ecosystems contain *2200 Pg C* in the sub-reservoirs of biota (600 Pg C) and soils (1600 Pg C). The steady-state flux between terrestrial ecosystems and atmosphere is ~125 Pg yr^{-1}, but this value is strongly dependent on factors such as land use, soil temperature, and other parameters that are very susceptible to short-term change.

The smallest of the global reservoirs is the *atmosphere* with ~*870 Pg C*; however, the current flux of carbon from the crust of the Earth to the atmosphere (~10 Pg yr^{-1}) is larger than the flux of C from the atmosphere back into the crust (~0.2 Pg yr^{-1}) or into marine and terrestrial reservoirs (combined = ~3 Pg yr^{-1}), an imbalance that is one way to conceptualize increasing atmospheric CO_2. For reference, the amount of CO_2 (quantified as C) in the year 1750 was ~600 Pg.

The following sections explore in detail the carbon cycle from the perspective of each of the main global-scale reservoirs.

5.2.3 Processes that transfer carbon into the crust

The main processes that fix carbon into the crustal reservoir occur over time spans of millions of years, and include the following two main pathways: (i) chemical weathering that facilitates transport of inorganic carbon to the ocean, where it can become buried and compacted in sediments (and form limestone, for example); and (ii) photosynthesis followed by burial and compaction of organic carbon in marine sediments.

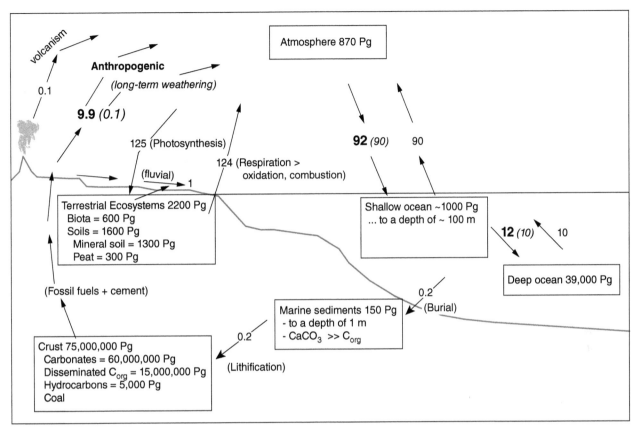

Fig. 5.5 Schematic representation of the carbon cycle with values for the year 2016, with reservoirs given in units of Pg C (see text) and fluxes in Pg C yr^{-1}. Note that current fluxes to atmosphere and ocean (in bold) are a net positive, causing increasing concentrations of carbon in these reservoirs. Fossil fuel combustion and cement production are depleting the crust (lithospheric reservoir) of carbon. Sources: Sundquist and Visser (2004); Solomon et al. (2007); Friedlingstein et al. (2010); Ruddiman (2013); Boden et al. (2017); and Olivier et al. (2017). Values in parentheses are pre-industrial estimates, which draws attention to increasing fluxes of crust → atmosphere and atmosphere → oceans.

5.2.3.1 Carbonate rocks

Chemical weathering of silicate rocks, using the mineral wollastonite ($CaSiO_3$) as a simplified example, can be represented as follows:

$$CaSiO_3 + 2H_2CO_3 + H_2O \rightarrow Ca^{2+} + 2HCO_3^{-}$$

$$+ H_4SiO_{4(aq)} \tag{5.26}$$

Section 5.1 presents formation of carbonic acid (H_2CO_3) in the atmosphere from $CO_2 + H_2O$, so chemical weathering is one of the steps involved in the transfer of carbon from the atmospheric reservoir to the crustal reservoir. Another important step is the fluvial (i.e. by rivers) transport of dissolved Ca^{2+} and HCO_3^{-} to the marine environment, where dissolved calcium and bicarbonate react to form calcium carbonate (often in the form of aragonite, a polymorph of calcite), a process that is

often biotically mediated when marine organisms form their shells:

$$Ca^{2+} + HCO_3^{-} \rightarrow CaCO_3 + H^+$$

or, if the bicarbonate has been transformed to carbonate anions in the alkaline marine environment:

$$Ca^{2+} + CO_3^{2-} \rightarrow CaCO_3$$

Burial of $CaCO_3$ in marine sediments and eventual compaction and cementation into carbonate-bearing rocks (e.g. as $CaCO_3$) provides long-term storage of carbon in the crust of the Earth. By convention, C in carbonate minerals (including calcite in fossilized shells of marine organisms) is considered *inorganic carbon*. For this reason, when calcite, dolomite, or other carbonate minerals undergo dissolution, the resulting $HCO_3^{-}{}_{(aq)}$ or $CO_3^{2-}{}_{(aq)}$

anions are considered *DIC*. To reiterate, CO_2 dissolved in water is also DIC, whether represented as $CO_{2(aq)}$ or H_2CO_3. Conversely, dissolved organic matter (e.g. humic acid) is characterized as *DOC*. In most cases, surface water, soil water, and groundwater typically contain both DIC and DOC.

5.2.3.2 Fixation of organic carbon into organisms

Fixation of *organic carbon* into Earth's crust begins with photosynthesis, much of which takes place in the photic zone of the oceans. Alfred Redfield determined in the 1930s that the average ratio (or stoichiometry) of C : N : P in marine plankton is 106 : 16 : 1, and the photosynthetic reaction can be represented as:

$$106CO_2 + 16NO_3^- + PO_4^{3-} + 122H_2O + 19H^+ + E$$

$$\rightarrow C_{106}H_{263}O_{110}N_{16}P + 138O_2 \qquad (5.27)$$

The marine plankton shown in Eq. (5.27) as a product contains carbon that has been photochemically transformed from atmospheric CO_2 to carbon fixed into the biological realm of the oceanic reservoir. The chemical formula $C_{106}H_{263}O_{110}N_{16}P$ represents a variety of proteins, carbohydrates, lipids, lignins, and other components of biotic matter. In this compound, the oxidation states are H^+, O^{2-}, N^{3-}, and P^{5+}; carbon occurs as C^0.

When plankton die and undergo decay in oxidizing (aerobic or oxic) environments like those typified in shallow, well-circulated, oxygenated marine water or sediments, the reaction shown above will proceed in reverse and carbon will shortly be returned to the atmosphere as CO_2 – this oxidation reaction is respiration ($CH_2O + O_2 \rightarrow CO_2 + H_2O + E$). This may occur when zooplankton consume phytoplankton; another mechanism is aerobic decay mediated by microbes. Note also that aerobic decay also returns the vital nutrients N and P to sea water or sediments (i.e. the nutrients are "mineralized") in the form of nitrate and phosphate. In order to be transferred into the crustal reservoir in significant amounts, remains of organisms usually require burial in reducing environments of the sea floor or terrestrial swamps.

5.2.3.3 Formation of hydrocarbons

When dead marine plankton settle through the water column to depths below the photic zone, they are likely to end up deposited as particulate organic matter (POM) on the sea floor, where decay is likely to occur via *anaerobic* pathways in the suboxic or anoxic (chemically reducing)

environment common to many marine and swamp sediments. The absence or limited amount of free O_2 in these environments limits conversion of organic C (C^0) to CO_2 (C^{4+}); instead, the residues of POM in anaerobic/reducing/anoxic environments are subsequently broken down by anaerobic bacteria into progressively more hydrocarbon-like compounds.

During the first stages of transformation at burial depths of a few hundred meters below the sea floor, anaerobic decomposition causes the O : C, H : C, and N : C ratios of the remaining organic matter to decrease. This produces the insoluble, viscous organic compound *humin* (~similar to humic matter in soils), and later, waxy *kerogen* (pronounced kair-o-jen). Much of the lost O, H, and N are converted to mobile liquids or gases, e.g. H_2O, CH_4, CO_2, and NH_3. The CH_4 that forms in these early stages ($<50\,°C$) of organic maturation is known as *biogenic* methane, which has a different origin than the *inorganic* methane that forms at temperatures of 150–200 °C at deeper crustal levels approaching the end stage of petroleum maturation. Figure 5.6 is a Van Krevelen diagram showing (i) changing ratios of O : C and H : C with progressive organic maturation, and (ii) differences in the types of organic matter from which hydrocarbons can

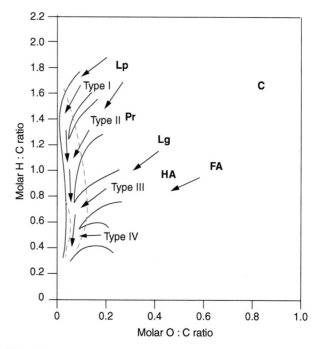

Fig. 5.6 van Krevelen diagram showing approximate composition of components of organic matter and trajectory of organic matter maturation from immature organic matter through the petroleum maturation window. Lp, lipids; Pr, proteins; Lg, lignite; C, carbohydrate; HA, humic acid; FA, fulvic acid. Types I–IV are described in the text.

Focus Box 5.4

Kerogen and Bitumen in the Crust

Kerogen is a waxy mixture of organic compounds with an approximate stoichiometry of $C_{200}H_{300}O_{11}N_5S$. Type I kerogen is mainly derived from algae and they commonly mature into tar-like hydrocarbons found in oil shales. Type II kerogen is the main source of liquid hydrocarbons (their source is marine plankton), while type III kerogen is derived from higher-order plants and are the common source of coal. In addition to their position in the transformation of marine organic carbon (plankton) to hydrocarbons, kerogen is the main form of organic matter in the crustal/geological reservoir, comprising ~90% of organic matter in sedimentary rocks. Most organic matter is very dispersed in the crust, and only rarely does organic matter occur in the concentrations, stoichiometries, and temperature range required to be considered a hydrocarbon resource. *Bitumen* (pronounced "bit-you-men") occurs in conjunction with kerogen and also is a product of organic matter maturation in sediments. Whereas kerogens are waxy and insoluble in organic compounds, bitumens are black and tarry and soluble in organic compounds. Both store carbon for millions of years in sediments and sedimentary rocks, and given the right circumstances can transform into hydrocarbon fuel resources. At burial depths of ~1 km (~60 °C), long-chain alkanes and related compounds derived from kerogen and bitumen begin to undergo catagenesis (also termed "cracking") that results in the formation of shorter-chain hydrocarbons. Once sediments have been buried to depths of ~2–3 km (~75–125 °C), organic matter maturation reaches the point where alkanes with ~4–20 carbons occur – this is the crude oil that humans pump from the ground as a fuel source.

form. Note the low O : C ratio of lipids compared to other components of living organic matter – it is this attribute that makes lipids most suitable for conversion to kerogen and ultimately to hydrocarbons.

5.2.3.4 Formation of coal

The decomposition of terrestrial plants in the reducing (anaerobic) environments of swampy environments may result in coal deposits if these organic-rich sediments undergo sufficient burial and heating. The oldest coal deposits occur in sedimentary rocks that date back to the Devonian Period (~400 Ma); there can be none older because land plants, the primary source of carbon for coal deposits, had not sufficiently evolved and colonized the continents prior to this time. The majority of economic coal deposits are from the following time periods:
- Carboniferous and Permian periods (~355–250 Ma);
- Jurassic and Cretaceous periods (~202–65 Ma);
- Paleogene Period (Early Cenozoic) (~65–23 Ma).

High global sea level during these periods led to the development of broad, gently sloping coastlines with vast forested wetlands. Modern-day examples of similar swampy environments include the Pantanal in Brazil and the Okefenokee Swamp in the southeastern USA. Decomposition of fallen trees and other organic matter in anaerobic bottom waters and sediments limits loss of carbon (as CO_2) to the atmosphere and instead fosters a decay sequence somewhat similar to hydrocarbon maturation.

This sequence occurs with burial, compaction, and increasing temperature and follows the maturation trend (temperatures are approximate):

$$peat \rightarrow lignite\ (<100°C)$$
$$\rightarrow bituminous\ coal\ (100–250°C)$$
$$\rightarrow anthracite\ (>250°C)$$

As with hydrocarbons, the carbon content of coal increases with increasing grade, from ~50% by mass in lignite to 90% or 95% in anthracite (Section 3.2.2).

5.2.4 Rates of organic carbon flux to and from the crust

Transformation of decaying organic matter to hydrocarbons or coal occurs slowly and requires tens of millions of years. The flux into the crust is slow ($0.2\ Pg\ C\ yr^{-1}$) and most petroleum is produced from rocks whose ages range from Pennsylvanian (~300 million years old [Ma]) to Miocene (~15 Ma); most coals are similarly old. The molecules comprising the gasoline, natural gas, diesel, and coal that we are burning today began their journey into sedimentary rocks as plankton or other plant remains long, long ago.

Cement production is another anthropogenic cause of net carbon transfer from Earth's crust to atmosphere. Cement is derived from limestone in a high-temperature process

($\sim$900 °C) that results in conversion of inorganic carbon stored in the crust (as calcite) to CO_2:

$$CaCO_3 + heat \rightarrow CaO + CO_2 \qquad (5.28)$$

CaO is "lime," a component of cement, and CO_2 is released to the atmosphere as a byproduct. Approximately 10% of anthropogenic CO_2 emissions are derived from cement production.

Like the carbon in fossil fuels (hydrocarbons and coal), the inorganic carbon (e.g. $CaCO_3$) currently in carbonate rocks was fixed into the crust tens of millions to billions of years ago. The pure limestone used to make cement in the St. Lawrence River valley of southern Quebec is Ordovician ($\sim$460 Ma) in age; the carbon stored in the ancient calcite is being extracted from the crustal reservoir at rates many orders of magnitude more rapidly than it would be transferred via the long-term rates of natural uplift and chemical weathering.

As a whole, the flux of carbon with respect to the crust is as follows. The transfer of carbon from atmosphere into sediments (mainly marine sediments), where they may ultimately be fixed back into the crustal reservoir, is 0.2 Pg yr^{-1}, a value that represents the combined fixation of organic and inorganic carbon. Humans are removing carbon from oil fields, coal mines, and cement

quarries and converting it to CO_2 at a rate of $\sim$10 Pg yr^{-1} (Table 5.3). This value is reasonably well known because society keeps track of the amount of fossil fuels burned and cement manufactured each year. Thus, the simple mass balance with respect to the crust is:

$$Flux_{in} = 0.2 \text{ Pg yr}^{-1}$$

$$Flux_{out} = 10 \text{ Pg yr}^{-1}$$

The low rate of natural fixation of C into the crust, and the need to round because of uncertainty, results in a net flux of C from crust to atmosphere of $\sim$10 Pg yr^{-1}, ignoring uptake into other reservoirs. Some of the C transferred from crust to atmosphere is absorbed by the oceans but proportionally this quantity is decreasing (Table 5.3).

Another factor to address quantitatively is the long-term flux of C from the crust (or lithosphere) to the atmosphere. Uplift and exposure of carbonates and other C-bearing rocks to the atmosphere result in oxidation of organic C and chemical weathering of minerals, as illustrated with the weathering of wollastonite (Eq. [5.26]). The long-term rate prior to major human alteration was $\sim$0.2 Pg C yr^{-1}, representing a pre-industrial steady state where $Flux_{out}$ and $Flux_{in}$ were both 0.2 Pg C yr^{-1}. The current non-steady-state situation is resulting in transfer of C from crust to atmosphere at a rate $\sim$50 times greater

Table 5.3 Changing fluxes of carbon among the main reservoirs for three time periods[a,b].

(a) Fluxes of Carbon (Pg yr^{-1}). Negative values indicate reservoirs which are losing C.

Time Period	Fossil fuel and cement	Land-use source	Land sink	Terrestrial mass balance	Ocean sink	Atmospheric increase
1960–1980	−3.8	−1.4	1.4	0.0	1.6	2.3
1980–2000	−5.9	−1.5	2.0	0.6	2.1	3.3
2000–2015	−8.5	−1.0	2.4	1.4	2.4	4.7

(b) Sources and Sinks of C Relative to the Atmosphere (Pg yr^{-1} for first three columns, and % released or absorbed for last three columns).

Time Period	Gross flux in (FF/cem + LUC)	Sinks (Ocean, Land)	Net atm increase	FF/cem C % remain in atm	FF/cem C % into ocean	% FF/cem C sequestered
1960–1980	5.3	3.0	2.3	42	44	58
1980–2000	7.4	4.1	3.3	44	36	56
2000–2015	9.5	4.8	4.7	49	25	51

[a]In Table 5.3a, negative values indicate that combustion of fossil fuel, production of cement, and land-use practices (e.g. deforestation) are causing loss of C from those three reservoirs; reforestation and other changes to land are also a sink for C ("land sink"), and the net change in C in the terrestrial reservoir (mainly biota and soil) is the "terrestrial mass balance" – from 1980 to 2015, the terrestrial reservoir has experienced a net gain in C. So too has the ocean reservoir and the atmosphere.

[b]Table 5.3b depicts gross transfer of C from the lithosphere (FF/cem = fossil fuels and cement) and land-use change (LUC) to the atmosphere (combined = "FF/cem + LUC"). Total C removed from the atmosphere is "Sinks (Ocean, Land)." Also shown are percentages of anthropogenic (fossil fuel + cement + land-use change) C that remains in the atmosphere (increasing from 1960 to 2010), anthropogenic C that is sequestered in the ocean (decreasing from 1960 to 2010), and total fossil fuels and cement sequestered (oceans + land; also decreasing with time). For sources see Figure 5.5.

(based on ~ 10 Pg yr^{-1} ÷ 0.2 Pg yr^{-1}) than the long-term geological (natural) flux; the current rate is also ~ 50 times greater than the flux of C to the crust. We are depleting carbon from the crust and adding it to the atmosphere at a rate 50 times greater than it is being returned to the crust.

The fate of the excess carbon transferred from the geological to the atmospheric reservoir depends in large part on reservoirs (e.g. oceans, forests) that are able to fix carbon over much shorter time spans compared to the slow fixation rate into the crust. Table 5.3 delineates the main fluxes of carbon among reservoirs. In addition to the average 2000–2015 flux of 8.5 Pg yr^{-1} of carbon to the atmosphere from the crustal/lithospheric reservoir, average 2000–2015 uptake of carbon by the oceans was ~ 2.4 Pg C yr^{-1} and *net* transfer from the atmosphere to terrestrial reservoirs (via photosynthetic uptake into plants plus fixation into soils) was ~ 1.4 Pg yr^{-1} of carbon. This complex system is difficult to precisely quantify, but the mass balance of carbon fluxes, and measurements of atmospheric carbon concentrations, indicates that from 2000 to 2015, the atmosphere gained an average net amount of ~ 4.7 Pg of carbon per year (in 2017 it gained about 6 Pg C). This means that $\sim 50\%$ (from 4.8 ÷ 9.5 Pg C yr^{-1}) of total fossil fuel and cement-derived carbon emissions annually between 2000 and 2015 has been fixed into the ocean and terrestrial sinks. This is less than ocean-terrestrial uptake rate from 1960 through 1980, where oceanic and terrestrial reservoirs absorbed $\sim 58\%$ of annual carbon emissions.

5.2.5 The ocean reservoir

The largest of the Earth surface reservoirs is the global ocean with 40 000 Pg C. Most carbon in the oceans is derived from, and cycled back to, the atmosphere due to the vast permeable boundary between these two reservoirs. CO_2 from the atmosphere is fixed into the oceanic reservoir in two main ways. The first is via photosynthesis by phytoplankton as described above; the second is solution of CO_2 into sea water. Carbon also enters the oceans via fluvial transport, mainly as dissolved bicarbonate with lesser amounts of organic carbon, but this flux (1.2 Pg C yr^{-1}) is small compared to the amount fixed into the oceans from the atmosphere (90 Pg C yr^{-1}).

5.2.5.1 Fixation of C into oceans

Incorporation of carbon into the oceans by photosynthesis is generally a very short-term fixation (in other words, this carbon generally has a low residence time), especially if the phytoplankton are consumed by higher-trophic level organisms and their carbon is returned to the atmosphere as CO_2 via respiration. However, if after the plankton die they sink to great depths, the carbon will be removed from the shallow ocean, facilitating fixation of organic carbon into marine sediments. Provided that the C stored in these sediments is not oxidized to CO_2, this process can contribute to long-term C storage.

The solution of CO_2 gas into sea water is the dominant means by which carbon is transferred from atmosphere to oceans, and it exists as dissolved CO_2 gas molecules, as well as carbonic acid (H_2CO_3) and the bicarbonate (HCO_3^-) and carbonate (CO_3^{2-}) anions. The difference between these two constituents, CO_2, and H_2CO_3, is difficult to determine analytically so they are often grouped together as CO_2 (aq). Figure 5.5 shows that the oceans are currently fixing 92 Pg yr^{-1} of carbon while they are emitting 90 Pg C yr^{-1}, indicating a net increase of ~ 2 Pg yr^{-1} of carbon (the average for 2000–2015 is 2.4 Pg C yr^{-1}).

Approximately one-third of anthropogenic CO_2 added to the atmosphere has been absorbed by the oceans. This is a predictable consequence of the increasing CO_2 in air considered in terms of *Henry's Law*, which states that, at constant temperature, the concentration of a gas dissolved in a liquid is directly proportional to the concentration (or partial pressure) of the gas in air in contact with, and in equilibrium with, the liquid.

$$k_H = c_{aq}/c_{atm} \qquad (5.29)$$

In this equation, k_H is the Henry's Law constant for CO_2, c_{aq} is the concentration of CO_2 in the aqueous solution (e.g. sea water), and c_{atm} is the concentration of CO_2 in the atmosphere. The eloquently simple mathematics here indicate that when c_{atm} increases, and in order for k_H to remain constant (which it must for a given temperature), c_{aq} must also increase. So this is another aspect of the complex climate system that is not so complex. Simple first principles of chemistry dictate that increasing CO_2 in the atmosphere will increase dissolved CO_2 in sea water (note: this ignores for now decreasing solubility of CO_2 with increasing temperature of sea water).

5.2.5.2 Ocean acidification

A consequence of incorporation of CO_2 into sea water is the formation of carbonic acid, as we have seen previously:

$$CO_2 + H_2O \rightarrow H_2CO_3$$

From this reaction it is evident that increased solution of CO_2 in sea water will drive the system toward increased

H_2CO_3, dissociation of which will yield H^+ and bicarbonate ions in solution:

$$H_2CO_3 \rightarrow H^+ + HCO_3^-$$

Further dissociation yields another H^+ and a carbonate anion:

$$HCO_3^- \rightarrow H^+ + CO_3^{2-}$$

Though carbonic acid is not a strong acid, its dissociation in the marine environment drives the carbonate system toward lower pH, and ocean acidification (or decreased ocean alkalinity, depending on point of view) caused by increased solution of CO_2 in sea water is predictable and is occurring (Raven et al. 2005; Lauvset et al. 2015). Decreasing pH of sea water raises concerns about marine ecosystems, particularly pertaining to organisms that manufacture shells and other body parts from $CaCO_3$. Calcium carbonate solubility is highly sensitive to pH, and ocean acidification is likely to diminish the stability of calcium carbonate-based organisms and ecosystems; calcareous plankton, an important component of the marine food chain, and coral reefs, are expected to be particularly sensitive to lowered pH.

The impact of increasing CO_2 on marine $CaCO_3$-based organisms can be viewed in terms of the equilibrium between dissolved CO_2 in sea water, H_2O molecules, and carbonic acid. Increasing CO_2 increases the amount of carbonic acid, and subsequent H_2CO_3 dissociation yields H^+, and the fate of calcareous organisms in sea water is clearly pH-dependent:

$$CaCO_3 + H^+ = Ca^{2+} + HCO_3^-$$

Increasing H^+ due to increased CO_2 destabilizes $CaCO_3$ and the organisms that rely on it.

We can also examine the relationship between shell formation from dissolved carbonate anions and the role of increased CO_2 in sea water:

$$Ca^{2+} + CO_3^{2-} = CaCO_3$$

However, excess CO_2 fosters this reaction:

$$CO_2 + H_2O + CO_3^{2-} = 2HCO_3^-$$

which results in decreased carbonate anion concentration, which in turn inhibits production of $CaCO_3$ by organisms (Orr et al. 2005).

A 2004 report by the Royal Society of London (Raven et al. 2005) indicates that global sea water pH has decreased from 8.15 to 8.05 over the past 200 years due to anthropogenic CO_2 emissions (which, due to the log scale by which acidity is measured, corresponds to a ~25% increase in H^+ in sea water). The consensus of climate scientists and oceanographers summarized in the 2007 IPCC report (Solomon et al. 2007) projects significant decreases in ocean pH with progressive increases in CO_2. Current data from the Pacific Ocean north of Hawaii (Station Aloha) depict some predictable features, including increasing inorganic carbon (i.e. dissolved CO_2) and decreasing pH (from mean annual value of 8.12 in 1990 to 8.06 in 2016). As noted above, a decrease in pH of 0.05 or 0.1 units, while it might appear small, is not – it represents a 12% or 25% increase in $[H^+]$.

Focus Box 5.5

Le Châtelier's Principle, Henry's Law, and Ocean Acidification

In sea water saturated with respect to calcium carbonate (this is the situation now and the one that has prevailed during most of Phanerozoic time, i.e. for the past 540 Ma), the following reaction occurs (Eq. [5.18] in reverse):

$$Ca^{2+} + 2HCO_3^- = CaCO_3 + CO_2 + H_2O \qquad (5.30)$$

The $CaCO_3$ is often the foundation of shells or other hard parts of marine organisms. The CO_2 produced by reaction (5.30) is in equilibrium with atmospheric CO_2. Viewed in this way, it is possible to predict the effect of increased $CO_{2(atm)}$ on the stability of $CaCO_3$ in sea water, and by simple extension, viability of marine ecosystems with increasing atmospheric CO_2. Considered in terms of *Le Châtelier's*

principle, increasing the concentration of any of the products (in this case, CO_2) will cause a shift in the equilibrium toward the direction of reactants. *Henry's Law* dictates that the amount of a dissolved gas in a liquid – in this case atmospheric CO_2 in contact with sea water – is proportional to its partial pressure (which is increasing as we convert fossil fuels and other C sources to CO_2 in air). Both of these fundamental principles of chemistry indicate that increasing atmospheric CO_2 will shift the equilibrium position toward reactants, in this case dissolved calcium and bicarbonate, and this will occur at the expense of calcium carbonate and marine organisms who rely on the stability of $CaCO_3$ in sea water. Basic chemical concepts demonstrate the risk faced by twenty-first century marine ecosystems.

Focus Box 5.6

A Geological Perspective on Ocean Acidification

The geological record often serves as a useful tool in assessing the potential response of oceans and marine ecosystems to increasing CO_2. Volcanically derived CO_2 increased rapidly in the atmosphere during the Aptian Age of the Cretaceous Period 120 million years ago – the volcanic activity was located at and resulted in production of the Ontong-Java plateau, and this event is particularly useful to compare to the modern situation because, like now, the rate of CO_2 flux to the atmosphere was rapid, not gradual. Erba et al. (2010) examined the effect of this event on marine calcifiers, in this case studying their response to increasing solution of CO_2 into sea water. In summary, results indicate that calcareous organisms responded in predictable ways, exhibiting malformations, dwarfism, and decreases in abundance associated with sea water acidification. Overall flux of $CaCO_3$ to marine sediments also decreased, further evidence for decreased marine ecosystem health, and also a symptom of lowered saturation state of $CaCO_3$ in sea water.

5.2.5.3 Long-term viability of oceans as C sink

Yet another concern related to the potential for CO_2 to dissolve in sea water is the effect of increasing ocean temperature on the solubility of CO_2 gas. If we are concerned about the effect of too much CO_2 on sea water chemistry, we also need to be concerned about the effect of rising ocean temperature on CO_2 solubility. For a given partial pressure of CO_2 in air, the amount of CO_2 in solution in a liquid (e.g. sea water) in contact with the air decreases as temperature increases (Figure 5.7). This is a concern because the ocean has been a major sink for anthropogenic CO_2 over the past 150 years. From 1960 to 1980, 44% of fossil fuel plus cement emissions were fixed into the oceans as dissolved CO_2; the same value for the 2000–2010 time period is 30%. (It is important to note that while the proportion of CO_2 emissions sequestered into the oceans has decreased, the actual amount of CO_2 sequestered has increased, from $2.1\,\mathrm{Pg\,C\,yr^{-1}}$ in the 1960–1980 period to $2.3\,\mathrm{Pg\,C\,yr^{-1}}$ in the 2000–2010 period.) The inverse correlation between the solubility of CO_2 in sea water and temperature (Figure 5.7) indicates that the capacity of the oceans to fix CO_2 from air will diminish as the oceans warm; however, the myriad controls on ocean chemistry and dissolved CO_2, from ocean circulation patterns to biological activity, serve to complicate precise prediction of CO_2 fixation in the oceans.

CO_2 concentrations in sea water range from 4.4×10^{-3} g $CO_2/100\,g$ (of sea water) up to 3.5×10^{-2} g/100 g depending on temperature, where higher values correspond to cooler, deeper waters (Berner and Berner 1996). Figure 5.7 emphasizes that (i) CO_2 is most soluble in cool sea water at great depths, (ii) CO_2 is least soluble in warm surface water, and (iii) at atmospheric pressure, the solubility of CO_2 in sea water decreases with increasing temperature.

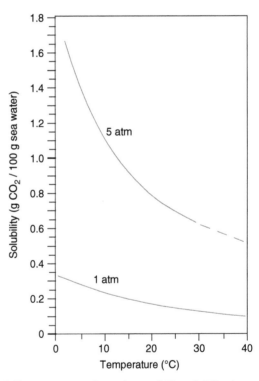

Fig. 5.7 Temperature dependence of CO_2 solubility in sea water at two different pressures. At a given pressure, increasing temperature decreases the solubility of CO_2 in sea water. Source: Data are from Stewart and Munjal (1970) and sources cited therein.

5.2.5.4 Methane hydrates

An additional important and interesting reservoir of carbon that occurs primarily in marine sediments, but also in arctic soils and sediments, are the slushy-to-icy compounds known as methane hydrates (synonymous terms include gas hydrates and methane clathrates). These compounds consist of methane encased in a crystalline dodecahedral structure formed by H_2O

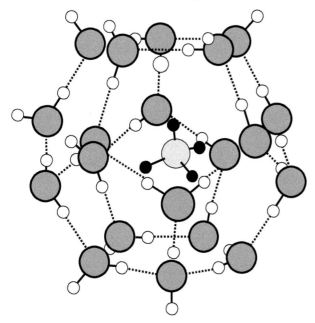

Fig. 5.8 Structure of methane clathrate (methane hydrate). Note the dodecahedral (12-sided) crystal comprised of H_2O molecules encasing a methane molecule. Oxygen atoms in water are dark blue with black outlines and the hydrogen atoms in water are white. The carbon atom in methane is light blue and the four hydrogen atoms in the methane molecule are black. Covalent bonds within the water and methane molecules are indicated by black lines while the hydrogen bonds between adjacent water molecules are indicated by dashed lines.

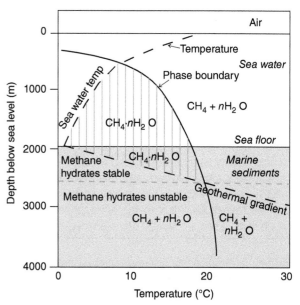

Fig. 5.9 Stability diagram for methane hydrates in marine environments. The high pressure of deep sea water is not directly shown here, but rather is indirectly indicated by water depth. The combination of high pressure and low temperature in the deep sea are the main forces that lead to formation and preservation of clathrates in marine sediments. In this example, methane hydrates are stable below 500 m in sea water and down to a depth of 500 m in marine sediments (or 2500 m below the sea floor) – the stability field is indicated by vertical lines. Warming of deep sea water would shrink the size of the stability field. Source: Modified from data in Caldeira et al. (2005).

molecules coordinated around methane molecules (Figure 5.8); the stoichiometry is $(CH_4)_8 (H_2O)_{46}$ (Spiro and Stigliani 2003).

The term for any such crystal where an H_2O cage-like structure taps a gas molecule is clathrate, and when methane is the gas in the structure, they are methane hydrates or methane clathrates. The global reservoir of methane hydrates is approximately 10 000 Pg C, which is twice global hydrocarbon reserves. For this reason, energy agencies and companies are interested in them as a potential fuel source (yet even with modern technology, extraction is difficult). The ability of methane to act as a strong greenhouse gas in the atmosphere raises concerns about release of methane from clathrates to the air.

In marine sediments, methane hydrates typically occur within the upper 500 m of sediments (i.e. within 500 m of the sea floor); at greater depths, heat emanating from within the earth melts the clathrate and releases methane and water. The temperature of deep sea water near the sea floor is ~2 °C, and the combination of low temperature and high pressure of the deep seas stabilizes these structures as lenses or pods within sediments (Figure 5.9). In some areas, methane hydrates even outcrop at the sea floor.

Most occur on the continental shelf, and the origin of the carbon in the methane may be terrestrial organic matter transported to oceans via streams, or marine organic matter that settles through the sea water column toward the sea floor. In either case, microbial degradation in anaerobic deep sea water facilitates formation of methane which then becomes incorporated into clathrates in cold, high-pressure, deep sea water and sediment. Methane in clathrates may also be derived via outgassing from the lithosphere (e.g. gas escaping from hydrocarbons), where it can migrate upward through faults or fractures toward the sea floor to the shallow depths within marine sediments where it becomes stable in a clathrate structure.

5.2.6 Carbon in cold region soils

Soils and sediments of the vast Arctic region (e.g. Alaska, Siberia, Scandanavia, northern Canada) contain methane hydrates that are analogous to those in marine sediments. The quantity is estimated at approximately 1500–2000 Pg C in Arctic clathrates, an amount which is roughly double the amount of C currently in the atmosphere (Heimann 2010). In Arctic soils, low mean annual air temperature and the pressure of overlying sediments stabilize clathrates to depths of $\sim$500 m. Methane hydrates often occur in permafrost soils and sediments and the origin is similar to the marine environment – anaerobic microbes facilitate decomposition of organic matter into methane, and in the arctic environment, anaerobic conditions are fostered by vast wetlands.

Concern with methane hydrates in terms of air quality and climate arises from the unsettling thought that global warming will decrease their stability, i.e. cause them to melt, a prospect that could release very large amounts of methane to the atmosphere. Marine sediments contain a far greater content of methane in clathrates than do Arctic soils and sediments ($\sim$10 000 Pg C vs. $\sim$1500 Pg C); however, the great pressure of sea water and the isolation of deep sea water from the atmosphere suggests that deep marine methane hydrates may be relatively stable. Terrestrial hydrates, on the other hand, are likely much less stable. Air temperature is very closely linked to the stability (and depth below surface) of permafrost and associated clathrates, so many climate scientists are concerned about the potential for release of methane from clathrates buried at shallow depths in Arctic soils and sediments.

5.2.7 The atmospheric reservoir

Relationships between carbon in the atmosphere and the other carbon reservoirs have been explored in the previous sections, so this section focuses on aspects of the C cycle related to the atmosphere that have not yet been covered. It is important to emphasize that the concentration of CO_2 (plus CH_4 and other less-abundant forms of C) in the atmosphere is controlled by fluxes that range from very rapid – e.g. removal of CO_2 from air via photosynthesis, or via solution of CO_2 into sea water – to very slow, such as the transfer of CO_2 to the crust. At the same time, humans have greatly altered the natural cycling of C by accelerating the rate at which carbon is transferred from the crust to the atmosphere. Flux out of the atmosphere has not kept pace with anthropogenic increases, so the CO_2 concentration in the atmosphere is not in a steady state.

5.2.7.1 Changes to atmospheric carbon over geological time

Over Earth's history, atmospheric CO_2 has varied by orders of magnitude, due in large part to evolution of life and burgeoning of photosynthesis that provided a new mechanism to fix carbon from air onto marine and eventually terrestrial reservoirs. In particular, the appearance and increase in abundance of cyanobacteria in the Early Proterozoic Era heralded the onset of oxygenic[1] photosynthesis, facilitating the fixation of CO_2 from atmosphere to biota and the production of O_2; increasing O_2 transformed the atmosphere and shallow marine realms from reducing into oxidizing environments, a shift that occurred $\sim$2.2–2.1 billion years ago and is recorded in the appearance of widespread ironstones and redbeds (Fe_2O_3-bearing sandstones) and disappearance of pyrite as a type of clast deposited in stream sediments (a good summary of early-Earth C cycling is presented by Sundquist and Visser 2004).

Atmospheric carbon concentrations likely were equal to or greater than current concentrations during most of the Phanerozoic (540 Ma to now), when values often exceeded 1000 ppm (Royer et al. 2004). What kept the Earth from being extremely warm earlier in its history is that the sun was less luminous than it is now. The early Earth had a stronger greenhouse effect but lower insolation than the modern world, so in spite of much higher CO_2 concentrations, temperatures were within $\sim$8 °C of modern temperatures (Veizer et al. 1999), and were lower during periods of glaciations (notably parts of the late Proterozoic, Permian and, of course, Pleistocene). During most of the Pleistocene and Holocene (i.e. the past 2 Ma), atmospheric CO_2 has ranged from $\sim$180 to 280 ppm (as indicated by air bubbles trapped in glacial ice), and only in the past 150 years has it exceeded 300 ppm.

The main sinks for atmospheric carbon over Earth's history have been:

1 Chemical weathering of silicate minerals, particularly Ca- and Mg-silicates. The reaction of wollastonite was presented earlier in Section 5.2.3.1, and the weathering of forsterite (the magnesian variety of olivine) is shown below:

$$Mg_2SiO_4 + 4H_2O + 4CO_2$$
$$\rightarrow 2Mg^{2+} + 4HCO_3^- + H_4SiO_{4(aq)} \qquad (5.31)$$

This process transfers CO_2 from air into soils, streams, or ocean water, where the CO_2 here is transformed into the bicarbonate anion. The carbon may then be stored for long periods of time (e.g. time scales of 10^6 years and longer) if the carbonate reacts with Ca to form aragonite or calcite in marine sediments and sedimentary rocks.

---

[1] Note – prior to the appearance of cyanobacteria, more-primitive types of bacteria carried out anoxygenic photosynthesis that, as the name would suggest, did not produce O_2. C fixation may have been facilitated by reactions involving H, S, or some other element(s). Some of these bacteria still exist today (e.g. green bacteria).

2 Chemical weathering of carbonates, as shown in Eq. (5.18), which can result in transfer of C to oceans via fluvial transport; then, in the marine environment, if the bicarbonate anion reacts with Ca to form carbonate minerals it may become buried in sedimentary basins, where the carbon can be sequestered for millions of years.

3 Photosynthesis that facilitates transfer of carbon to organic matter followed by burial, as described above for hydrocarbons, coal, and methane hydrates. In each of these three scenarios, carbon can be stored for millions to billions of years in the crust of the Earth if buried in sedimentary rock sequences.

Carbon is released from long-term geological reservoirs to the atmosphere by the following processes:

1 Chemical weathering of carbonates that produces CO_2 gas (the equation shown below is a different way to think of this reaction than what is shown in Eq. (5.18)):

$$CaCO_3 + 2H^+ \rightarrow Ca^{2+} + H_2O + CO_2 \qquad (5.32)$$

In this example, we can assume that (i) tectonic forces have caused uplift and exhumation of once deeply buried sedimentary rocks, and (ii) the source of the acid driving the weathering reaction is HNO_3 or H_2SO_4 derived from volcanism, natural combustion, or other sources. If the H^+ source is carbonic acid, the result may be no net addition of carbon to the atmosphere (see Eq. (5.18)).

2 Oxidation of organic matter in the crust (e.g. hydrocarbons) and marine sediments (e.g. methane hydrates). This process too would be facilitated by uplift and exposure of sedimentary rocks. Whether or not uplift leads to a net release of carbon (via these two examples), or whether it leads to a net removal of carbon from the atmosphere, depends on the amount of silicate weathering as compared to oxidation of organic matter and loss of carbon from carbonate rocks (Ruddiman 2013).

3 Volcanic emissions of carbon dioxide.

4 Metamorphism of carbonate rocks at temperatures high enough to liberate CO_2 (Eq. (5.28)). You will note that this natural process is analogous to manufacture of cement as shown earlier in this chapter. CO_2 can escape from metamorphic rocks via fractures and other pathways.

5 Metamorphism of organic matter, e.g. in metasedimentary rocks, a process where reduced carbon is oxidized to CO_2 and allowed to escape via fractures and other pathways.

5.2.7.2 Feedback loops

The main long-term controls on atmospheric CO_2 involve transfer of carbon between the atmosphere and the main lithospheric reservoirs, namely carbonates and organic matter in the crust. The rates of these processes are relatively slow compared to modern anthropogenic fluxes (Figure 5.5; Table 5.3). The Phanerozoic record of atmospheric CO_2 concentration reflects the influence of feedbacks (or feedback loops), where *positive feedbacks* amplify processes already in progress (these processes or triggers are sometimes termed climate forcings), whereas *negative feedbacks* counteract processes already underway.

A geochemically related positive feedback loop is as follows:

1 increased insolation (due to sunspot activity or orbital variations) causes warming of the atmosphere;

2 the warming atmosphere heats the shallow ocean;

3 increased ocean temperature decreases the solubility of CO_2 in sea water;

4 loss of CO_2 from sea water to the atmosphere causes the atmosphere to undergo further warming;

5 the further increase in air temperature causes additional losses of CO_2 from sea water; and so on.

In a positive feedback loop, the initial forcing (in this cause warming of air due to increased solar energy reaching the Earth) sets a process in motion that can lead to greatly amplified results.

A negative feedback in the carbon-climate system is chemical weathering. Consider again an initial forcing (e.g. increased insolation) that causes temperature of the atmosphere to rise, liberating CO_2 from sea water and terrestrial sources. However, given that CO_2 is a reactant in chemical weathering reactions, increasing atmospheric CO_2 enhances rate of weathering reactions, transferring carbon from CO_2 to HCO_3^-, where it can be fixed into marine carbonates. This process decreases the concentration of CO_2 in the atmosphere, or at least it counters the increase caused by the initial trigger, and in doing so counteracts the initial temperature increase.

The pronounced global cooling during late Cenozoic time that lead to the Pleistocene glacial epoch has been attributed to chemical weathering associated with uplift in the Himalaya, Andes, and other mountain ranges that lowered atmospheric CO_2 concentrations (refer to sources in Ruddiman 2013) – this period of elevated tectonic activity, much of which was related to contraction and uplift from western Europe to east Asia, raised mountains to high elevations and supported a greater-than-normal abundance of alpine glaciers. The mechanical action of alpine glaciers produces fine-grained glacial flour with high surface area : volume ratio, thus increasing the rate of chemical weathering. Increased rates of chemical weathering result in decreased amounts of CO_2 in the atmosphere; in turn, lower CO_2 in the atmosphere leads to global cooling which at the time may have fostered an even greater abundance and extent of alpine glaciers, which in turn would have ground up even more rock, further enhancing chemical weathering, decreasing CO_2 in air to a greater extent, and setting in motion a positive feedback loop whose result was two million years of extensive alpine and continental glaciation.

It is the back-and-forth between the many mechanisms controlling fluxes between reservoirs, amplified at times by feedback loops in the climate system, that regulate carbon content in the atmosphere, crust, and terrestrial and marine reservoirs (the reader is referred to Ruddiman 2013 for a more extensive account of climate feedbacks, paleoclimate, and the climate system in general). A geological time scale perspective on CO_2 is presented in Figure 7.2, showing changes in atmospheric CO_2 over the past 550 Ma (i.e Phanerozoic time).

5.2.7.3 Anthropogenic C and the atmosphere

The mass of carbon in the atmosphere has been increasing progressively since the late eighteenth century as carbon from the crust (hydrocarbons, coal, and calcite) and, to a lesser extent, carbon from terrestrial ecosystems, is converted to CO_2. Understanding the behavior of the modern atmosphere considered as a reservoir within the carbon cycle requires the following assumptions and information:

1 First, in terms of understanding the twenty-first century carbon cycle, transfer of carbon from atmosphere to the crust (or mantle) by natural processes is very small compared to anthropogenic fluxes. Fixation of C into the crust occurs at rates $\sim$50 times slower than the human-enhanced flux of carbon from crust to atmosphere (Figure 5.5; Table 5.3). The only way that fixation of C into the lithosphere will become significant is if humans undertake large-scale carbon sequestration via mineral carbonation or injection of CO_2 into deep reservoirs.

Table 5.4 The abundance of carbon in major reservoirs globally, as of 2018.

Reservoir	Amount ($\times 10^{15}$ g C)
Crust	75 000 000
Carbonate rocks	60 000 000
Dispersed organic C	15 000 000
Hydrocarbons, coal	5 000
Oceans	40 000
Soils	1 600
Atmosphere	870
Land Plants	600

2 Second, the oceans are currently absorbing a net quantity of $\sim$2–3 Pg C yr^{-1} from the atmosphere, a process driven by the change in equilibrium between atmosphere and ocean. Increasing partial pressure of CO_2 in air will force greater amounts into sea water.

3 Third, given that CO_2 is a reactant in photosynthesis, the equilibrium of this reaction should be shifted to the products side – Le Châtelier's Principle indicates that photosynthesis should occur at a greater rate; however, this interpretation ignores complications such potential absence of limiting nutrients, soil erosion, and desertification, all of which would limit photosynthesis (and hence C uptake into the terrestrial reservoir) on a global scale.

4 Fourth, atmospheric CO_2 and global mean temperature have been closely linked throughout Earth history, particularly the Pleistocene (the past two million years) and Holocene (the past $\sim$10 000 years). Regardless of whether the initial trigger of warmings in the past have been temperature increase (e.g. due to increased insolation) or addition of CO_2 to the atmosphere (e.g. via uplift and oxidation of organic matter in sedimentary rocks), temperature and CO_2 have the potential to act in concert via various feedback loops to produce rapid warming. Geologic and ice core evidence tells us that when the Earth's atmosphere warms, it does so very rapidly, and the negative feedback loops that eventually kick in to counter warming and instigate cooling operate at much slower rates. The issues are covered in more detail in Chapter 7.

Exchange of carbon between atmosphere and the oceans, and between the atmosphere and terrestrial reservoirs (soils, forests), occurs on relatively short time scales, typically days to months to years to hundreds or, in the case of deep circulating marine water, thousands of years. Carbon that humans are extracting from the crustal reservoir will either become partitioned into terrestrial or marine reservoirs, or it will build up in the atmosphere. All of these are happening right now, and the residence time in these reservoirs is short compared to the crustal reservoir. For example, CO_2 in air is converted to organic carbon by photosynthesis over the life span of a plant, and the cycle is completed when the plant decays in the presence of water, oxygen, and microorganisms and the carbon is returned to the atmosphere as CO_2. This occurs in marine and terrestrial ecosystems. Similarly, solution of CO_2 gas from atmosphere into ocean water or groundwater can result in storage of C as HCO_3^- for tens to thousands of years, depending on how deeply the water circulates into the marine environment or aquifer.

The tenuous status of carbon in terrestrial and marine reservoirs is a concern. What happens if the oceans reach saturation and do not continue to absorb roughly one half of anthropogenic carbon emissions, as has been the case over the past 50 years? Similarly, if organic carbon and methane in warming permafrost soils are released to the atmosphere, how will society respond? Cutting emissions is clearly one strategy (progress has been made, but the entrained status of fossil fuels makes it a difficult one to achieve); intentional sequestration of carbon into stable reservoirs is another strategy currently being explored.

5.2.8 Carbon sequestration

With atmospheric and oceanic CO_2 steadily rising into the twenty-first century, researchers have begun to explore ways to intentionally remove CO_2 from air though a group of processes that fall under the umbrella term sequestration. Many of the approaches are biological and involve fixation via photosynthesis of carbon into ecosystems, particularly marine reservoirs, e.g. via iron fertilization of sea water to accelerate marine microbial fixation, or terrestrial reservoirs, including reforestation (to store C in biomass) and improved soil management to fix and store photosynthetically fixed carbon. However, some sequestration approaches are geochemical in nature; the two we will examine in this section are mineral carbonation and injection of CO_2 into permeable rock deep in sedimentary basins.

Chemical reaction between a magnesium silicate mineral and CO_2 (in the form of carbonic acid) is mineral carbonation. Many natural weathering reactions are carbonation reactions (Eqs. (5.18), (5.31)); shown below is the reaction of serpentine with carbonic acid:

$$Mg_3Si_2O_5(OH)_4 + 2H_2O + 3CO_2$$

$$\rightarrow 3MgCO_3 + 2\,SiO_2{\cdot}H_2O_{(am)} \qquad (5.33)$$

The magnesite ($MgCO_3$) produced in this reaction can be sequestered as white $MgCO_3$ powder. There are many abandoned mines and quarries with tailings piles rich in serpentine-group minerals, and if this reaction could be carried out on a large scale, we might have a legitimate means of removing carbon from the atmosphere via a process that mimics natural chemical weathering. Some hydrocarbon reservoir sandstones contain authigenic Mg-rich berthierine and chlorite minerals that also may be suitable for mineral carbonation. The US Department of Energy has funded studies on mineral carbonation, and while in principle this approach has potential, the main barrier is that the reaction occurs very slowly at ambient temperatures and requires heat to carry out at rates that would be effective. For this reason, the

process is not cost-effective with current technology and economics.

Another approach to carbon sequestration with connections to geochemistry is injection of CO_2 into deep porous aquifers with impermeable caps, the classic example being sandstones beneath shale confining layers (or aquitards). In some cases, the host reservoirs are sandstones from which petroleum has been produced – in fact, deep injection of CO_2 has already successfully taken place in hydrocarbon reservoirs of the North Sea, and the Society of Petroleum Engineers indicate that any remaining technical barriers to carbon capture and storage (CCS) on a large scale are surmountable (https://www.spe.org/industry/carbon-capture-sequestration.php). Note that CCS differs from mineral carbonation in that the mechanism for retaining the CO_2 in the reservoir is via solution of CO_2 into deep basin groundwater. One of the potential problems with this approach is that CO_2 produces carbonic acid, which has the potential to dissolve calcite in the reservoir rock.

Though beyond the scope of this text, it is also worth drawing attention to the fact that there are many other C sequestration schemes, ranging from intentionally fixing C into soil by conservation farming practices to fertilizing ocean water with Fe, where the rationale is that Fe is a limiting nutrient in the oceans and increased Fe will result in increased primary productivity, fixing C into phytoplankton.

The topic of carbon cycling is highlighted again in Chapter 7, where the focus is the atmosphere, first examining the geological topic of atmospheric change over the past ~3 billion years, and then focusing on C as a global atmospheric pollutant.

QUESTIONS

5.1 Calculate the pH of rain if CO_2 concentration in the atmosphere were to reach 1000 ppm

5.2 Predict the concentration of Ca^{2+} in solution (e.g. soil water) in equilibrium with calcite at 25 °C.

5.3 What is the oxidation state of C in CO_2? In CH_4? What does this difference imply about the stability of CH_4 in contact with the atmosphere?

5.4 Is the arid river water in Table 4.1 saturated with respect to Ca and HCO_3 (in equilibrium with calcite), or is it below saturation?

5.5 Determine the mean residence time of a carbon atom in the atmosphere. In your response, indicate what value you used for the size of the reservoir as well as the value used for flux.

5.6 Compare the rate of burial of C in marine sediments to the pre-industrial rate of exhumation of C (e.g. via weathering and volcanic emissions). How do these rates compare to each other, and to the modern rate of anthropogenic C emissions?

5.7 Describe the two main means by which C is transferred from atmosphere to ocean (i.e. organic vs. inorganic). Provide details, and indicate which tends to result in long-term storage. Describe two long-term environmental concerns related to transfer of C from rock to ocean, using chemical equations and graphs to support your answer.

5.8 Warming air and sea water are likely to destabilize C stored in methane hydrates – which is likely more vulnerable, CH_4 in permafrost soils or CH_4 in deep marine environments, and why?

5.9 Provide an example of a mineral carbonation chemical reaction, showing how C might be fixed from the atmosphere into long-term mineral storage. What is the main factor limiting practical application of mineral carbonation as a strategy for managing carbon?

REFERENCES

Berner, E.K. and Berner, R.A. (1996). *Global Environment: Water, Air and Geochemical Cycles*. Prentice-Hall.

Boden, T.A., Marland, G., and Andres, R.J. (2017). Global, Regional, and National Fossil-Fuel CO_2 Emissions. Carbon Dioxide Information Analysis Center, Oak Ridge National Laboratory, U.S. Department of Energy, Oak Ridge, Tenn., U.S.A. doi 10.3334/CDIAC/00001_V2017.

Caldeira, K., Akai, M., Brewer, P. et al. (2005). Ocean storage. In: *IPCC Special Report on Carbon Dioxide Capture and Storage* (ed. B. Metz, O. Davidson, H. de Coninck, et al.), 277–317. Cambridge: Cambridge University Press.

Cartigny, P., Pineau, F., Aubaud, C., and Javoy, M. (2008). Towards a consistent mantle carbon flux estimate: Insights from volatile systematics (H2O/Ce, δD, CO2/Nb) in the North Atlantic mantle (14 °N and 34 °N). *Earth and Planetary Science Letters* 265: 672–685.

Erba, E., Bottini, C., Weissert, H.J., and Keller, C.E. (2010). Calcareous Nannoplankton response to surface-water acidification around oceanic anoxic event 1a. *Science* 329: 428–432.

Friedlingstein, P., Houghton, R.A., Marland, G. et al. (2010). Update on CO_2 emissions. *Nature Geoscience* 3: 811–812.

Heimann, M. (2010). How stable is the methane cycle? *Science* 327 (5970): 1211–1212.

Langmuir, D. (1997). *Aqueous Environmental Geochemistry*. Upper Saddle River, New Jersey, USA: Prentice-Hall.

Lauvset, S.K., Gruber, N., Landschützer, P. et al. (2015). Trends and drivers in global surface ocean pH over the past 3 decades. *Biogeosciences* 12: 1285–1298. https://doi.org/10.5194/bg-12-1285-2015.

Manahan, S.E. (2000). *Environmental Chemistry*, 7e. CRC Press.

Morse, J.W., Arvidson, R.S., and Lüttge, A. (2007). Calcium carbonate formation and dissolution. *Chemical Reviews* 107: 342–381.

Olivier, J.G.J., Peters, J.A.H.W. and Schure, K.M. (2017). *Trends in global emissions of CO_2 and other greenhouse gases: 2017 Report*. PBL report no. 2674, PBL Netherlands Environmental Assessment Agency. http://www.pbl.nl/en/news/newsitems/2017/greenhouse-gas-emission-levels-continued-to-rise-in-2016.

Omernik, J.M., Griffith, G.E., Irish, J.T. and Johnson, C.B. (2010). Total Alkalinity of Surface Waters. Corvallis, Oregon, USA: Environmental Protection Agency. Accessed online 25 March 2019: https://water.usgs.gov/edu/graphics/alkalinity-map-large.gif.

Orr, J.C., Fabry, V.J., Aumont, O. et al. (2005). Anthropogenic ocean acidification over the twenty-first century and its impact on calcifying organisms. *Nature* 437: 681–686.

Plummer, L.N. and Busenberg, E. (1982). The solubilities of calcite, aragonite and vaterite in CO_2-H_2O solutions between 0 and 90 °C, and an evaluation of the aqueous model for the system $CaCO_3$-CO_2-H_2O. *Geochimica et Cosmochimica Acta* 46: 1011–1040.

Raven, J., Caldeira, K., Elderfield, H. et al. (2005). *Ocean Acidification Due to Increasing Atmospheric Carbon Dioxide*. Policy Document., 60. London: The Royal Society.

Royer, D.L., Berner, R.A., Montanez, I.P. et al. (2004). CO_2 as a primary driver of Phanerozoic climate change. *GSA Today* 14: 4–10.

Ruddiman, W.F. (2013). *Earth's Climate: Past and Future*, 3e. New York: W.H Freeman and Co.

Solomon, S., Qin, D., and Manning, M. (Eds.) (2007). *Climate Change 2007: The Physical Science Basis*. Contribution of Working Group I to the Fourth Assessment Report of the Intergovernmental Panel on Climate Change.

Spiro, T.G. and Stigliani, W.M. (2003). *Chemistry of the Environment*, 2e, 489. Upper Saddle River, NJ, USA: Prentice Hall.

Stewart, P.B. and Munjal, P. (1970). Solubility of carbon dioxide in pure water, synthetic sea water, and synthetic sea water concentrates at −50 to 250 °C and 10 to 45 atm pressure. *Journal of Chemical and Engineering Data* 15: 67–71.

Sundquist, E.T. and Visser, K. (2004). The geological history of the carbon cycle. In: *Biogeochemistry, Treatise on Geochemistry*, vol. 8 (ed. W.H. Schlesinger), 425–472. Elsevier, Ch. 9.

Veizer, J., Ala, D., Azmy, K. et al. (1999). Sr^{87}/Sr^{86}, C^{13}, and O^{18} evolution of Phanerozoic sea water. *Chemical Geology* 161: 59–88.

6

Biogeochemical Systems and Cycles (N, P, S)

Transfer of elements among various compounds and reservoirs that differ in form and scale occurs as a result of biogeochemical reactions and changes of state. Some elements have very strong affinities for certain reservoirs; for example, silicon is strongly partitioned into minerals in the crust and in sediments at the surface of the earth, whereas only small amounts of Si, mainly as wind-blown mineral particles, are found in the atmosphere. Conversely, nitrogen is strongly partitioned into the atmosphere and concentrations of nitrogen in the crust are very low compared to other elements, and to N abundance in air. In some cases, elemental cycling occurs primarily by inorganic processes (e.g. mineral reactions deep within sedimentary basins or hydrothermal systems), but often elemental cycling is influenced or strongly controlled by biological factors. Prominent examples of biogeochemical processes are the cycling of carbon, where photosynthesis fixes carbon into living organisms, and the cycling of nitrogen, where soil microbial activity can help to fix nitrogen from the atmosphere (as N_2) into soils (e.g. as NH_4^+ and NO_3^-).

6.1 SYSTEMS AND ELEMENTAL CYCLES

Reservoirs, *fluxes*, *residence times*, and *steady state* are all important concepts to understand when considering elemental cycling, so the following sections present some basic concepts and terms on these topics.

6.1.1 Reservoirs, fluxes, and systems

Reservoirs range in scale depending on the scope of the question under consideration. In general, any reservoir will contain a quantifiable amount of material with definable borders defined by distinct chemical, physical, or biological characteristics. Essentially, when seeking to quantify the amount of material in various parts of the lithosphere–hydrosphere–biosphere–atmosphere system, and when trying to examine the movement of material among these systems, the constituents of Earth need to be defined as containers (reservoirs) that gain and lose material. Some reservoirs are global in scale, e.g. the atmosphere, oceans, soils, and biota. Studies of carbon cycling often examine reservoirs at this scale (e.g. "what is the flux of carbon to the atmosphere from the crustal reservoir?"). However, if the scope of a study is cycling of phosphorus in a small lake, pertinent reservoirs might include the water column (which could contain dissolved P as well as P sorbed to inorganic or organic particulate matter) and lake sediments (which would likely include P in minerals and P in organic matter). Scale can go all the way down to reservoirs on the molecular scale, e.g. "is the selenium in this soil partitioned into mineral exchange sites, the aqueous solution, or organic matter?" – each of these would comprise a microscale reservoir in that soil system.

The Earth is effectively a *closed system* with respect to elemental cycling – elements are recycled within reservoirs

Environmental and Low-Temperature Geochemistry, Second Edition. Peter Ryan.
© 2020 John Wiley & Sons Ltd. Published 2020 by John Wiley & Sons Ltd.

from the atmosphere to down into the solid Earth, but little matter enters or leaves the Earth system. Clearly this ignores addition of matter by extraterrestrial impacts (proportionally very small inputs) and negligible losses from the top of the atmosphere to space, but the consideration of Earth as a closed system is the common approach because these minor fluxes are negligible at the scale of the whole Earth. Unlike the closed-system approach used to characterize the Earth as a whole, reservoirs in environmental science and geology often are *open systems* characterized by exchange of material; it would be hard to model the atmosphere as a closed system with respect to carbon, for example, given the constant exchange of that element among biosphere, atmosphere, hydrosphere, and lithosphere.

Flux is the rate of exchange of material between reservoirs. Units are typically mass/time, e.g. kg yr^{-1}, or molar values such as mol day^{-1}. In some cases fluxes are directly measured (e.g. transport of particulate iron to the oceans in rivers), whereas in other cases fluxes are inferred based on measurable changes in reservoirs. Controls on flux include:

1 geochemical processes such as the oxidation of CO to CO_2 in the atmosphere (which will at least partially control CO and CO_2 abundance), dissolution of minerals during chemical weathering, and sorption of elements from solution into (or onto) minerals;

2 biochemical and biotic processes such as photosynthesis, respiration, nutrient uptake by plants, and microbially mediated decomposition of dead biota; and

3 physical processes such as evaporation and condensation, ocean and air currents, and river flow.

The average amount of time an element resides in a reservoir is the *residence time*, a parameter which is controlled by the volume (or mass) of the reservoir and the flux into (and out of) the reservoir (Eq. (4.6), Chapter 4). Clearly, the larger the reservoir and the slower the flux, the longer the residence time; for example, residence time of an average water molecule in the ocean is 3500 years whereas in lakes or rivers, residence time may be more likely on the order of years (small lake) to days or weeks (river).

6.1.2 The concept of steady state vs. dynamic equilibrium

An important assumption associated with residence time calculations is the existence of *steady state*, a situation in which the flux of a given component into the reservoir is equal to the flux out. Thus, at steady state, the concentration of a given element or component C will not change with time, a statement that can be represented as:

$$dC/dt = 0 \qquad (6.1)$$

where t is time.

Steady-state Scenario

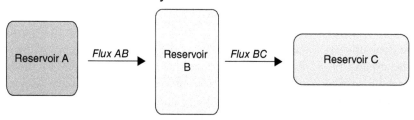

If *Flux AB = Flux BC*, then Reservoir B is in steady state with respect to the material (e.g. H_2O) or element (e.g. Ca) flowing into and out of the reservoir. If *Flux AB ≠ Flux BC*, the reservoir will either experience net gain or loss of the material or element in question.

Dynamic Equilibrium Scenario

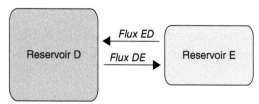

If *Flux DE = Flux ED*, then the system comprising Reservoirs D and E is in dynamic equilibrium. The net amount of the material will remain constant in each reservoir. If *Flux DE ≠ Flux ED*, then the system is out of equilibrium and one reservoir will experience net gain and the other net loss.

Fig. 6.1 Box model representations of the differences between steady state and dynamic equilibrium.

Steady state is not equivalent to dynamic chemical equilibrium, a more strictly defined situation requiring that processes be reversible and that the forward rate is equal to the reverse rate (see Chapter 1). To probe this in more depth, an illustrative example of steady state is the concentration of calcium in seawater. Over time spans of at least thousands of years, the flux of Ca into seawater, mainly via river transport of weathered bedrock constituents, is equal to the rate at which it is removed from seawater, mainly by precipitation of $CaCO_3$ and incorporation into marine sediments. However, this cannot be considered as an example of *dynamic chemical equilibrium* because the reactions, particularly the way by which Ca is added to the oceans (i.e. river flow), are not reversible, so the oceans are considered to exist "at steady state with respect to calcium concentration."

Geochemical cycles are often represented by box models, where boxes represent reservoirs and arrows between boxes represent fluxes or, in some cases, processes that are responsible for transfer between reservoirs. Simple box models help to illustrate the difference between steady state and chemical equilibrium (Figure 6.1). More complicated box plots are useful for illustrating reservoirs, fluxes, and processes in elemental cycling.

If we liken the steady-state scenario depicted in Figure 6.1 to the concentration of calcium in seawater, *Flux AB* primarily represents river transport of Ca into seawater and *Flux BC* represents chemical precipitation of $CaCO_3$ out of seawater. In this example, Reservoir A would be the terrestrial Ca reservoir, Reservoir B would be the seawater Ca reservoir, and Reservoir C would be the ocean sediment Ca reservoir. The concentration of Ca in seawater will be at a steady state if *Flux AB = Flux BC*.

6.2 ELEMENTAL CYCLES

The following sections on N, P, and S first present elemental cycling as controlled by natural processes and then consider human alterations to the cycle. Evidence from the geological record provides a reference for preindustrial reservoir sizes, fluxes, and processes, and this background serves as a reference to assess modern-day fluxes and understand anthropogenic impacts. The main geological records used in quantifying the cycling of N, P, and S are marine and lacustrine sediments and ice cores. Other records used include corals, paleosols, deep sea water, fossilized organisms, sedimentary rocks (especially marine carbonates and shales), and speleothems (cave deposits).

The main reservoirs for each of the cycles will be the lithosphere (mainly the crust but the mantle also plays a role in cycling of some elements), terrestrial ecosystems (mainly soils and land plants), the atmosphere, and the oceans (which consist of inorganic and organic forms of various elements in the water column and in ocean-bottom sediments). In some cases, we will also examine smaller reservoirs such as lakes (especially in the case of nitrogen and phosphorus cycling), rivers, soils, and wetlands. For the most part, terrestrial water bodies are insignificant reservoirs in terms of global-scale cycles, but can be important in terms of flux (e.g. fluvial transport of elements from terrestrial to marine reservoirs) and ecological ramifications (e.g. nitrogen and phosphorus in freshwater ecosystems).

6.2.1 The nitrogen cycle

Humans have profoundly altered the nitrogen cycle, perhaps more so than the cycle of any other element. By the late 1900s we had reached the point where humans

---

Focus Box 6.1

Steady State vs. Non-Steady State: The Example of CO_2 in Air

Steady-state conditions usually exist over fixed amounts of time, but over long periods, disturbances eventually disrupt steady-state situations. One example is the long-term concentration of CO_2 in the atmosphere – prior to the industrial era (or Anthropocene), when CO_2 concentration is viewed over time scales of hundreds to thousands of years, its atmospheric concentration may appear to be (or have been) at steady state. Over longer periods, changes in Earth's orbit (Milankovitch cycles), volcanism, mountain uplift, and other factors shift the balance of atmospheric CO_2 in one direction or the other. Certainly over time scales on the order of $> 10^4$ years, CO_2 varies notably and would not be characterized as steady state, yet during the ~10 000 years of the Holocene, atmospheric CO_2 remained at 270 ± 10 ppm, representing a relatively prolonged steady state. The fact that CO_2 began to climb in the nineteenth century, and has surpassed 400 ppm, indicates that the atmosphere is no longer at steady state with respect to CO_2.

had become responsible for a greater amount of reactive nitrogen in forests, lakes, farms, and streams than can be produced by natural processes (Vitousek et al. 1997) and the rate of anthropogenic nitrogen fixation continues to increase.

The real crux of the problem is that humans are capable of transforming unreactive N_2 (in air) to reactive species such as ammonium, nitrate, and NO_x, which then are free to infiltrate terrestrial and aquatic ecosystems. Very few organisms on Earth are capable of breaking the triple N:::N bond in N_2 gas (this bond is so strong it renders N_2 nearly inert). Those that do possess this ability include certain species of prokaryotes (bacteria and archaea), yet in many natural systems, nitrogen is the element that limits growth due to the unavailability of reactive nitrogen. What these prokaryotes are able to do is convert stable atmospheric N_2, a form that is unavailable as a nutrient to nearly all living organisms, to molecules of reactive nitrogen (Nr) that can be used by plants as nutrients.

Nitrogen is a limiting nutrient in most agricultural soils, so it is added as fertilizer to increase crop yields; reactive nitrogen is also enriched in animal waste, so it can be introduced to surface water and shallow groundwater from sewage treatment plants, septic systems, and animal feed operations. Furthermore, combustion of fossil fuels and biomass releases reactive nitrogen oxide (NO_x) species into the atmosphere that ultimately result in increased nitrogen in soils and surface waters.

Consequences of the anthropogenically triggered shift in nitrogen cycling include:

1 eutrophication of lacustrine and nearshore marine ecosystems and associated loss of plant and animal diversity;

2 acid rain (the nitric acid variety) and related acidification of soils and surface waters;

3 acid rain-related leaching of essential soil nutrients, particularly calcium, and potassium that are particularly important in forested ecosystems;

4 increased concentrations of the potent greenhouse gas nitrous oxide (N_2O) in the atmosphere; and

5 increased concentrations of nitrogen oxides that contribute to the formation of photochemical smog.

The following sections present some of the important concepts related to nitrogen species, nitrogen cycling, and the environmental impacts of human-caused alteration of the nitrogen cycle.

6.2.1.1 Nitrogen oxidation states, nitrogen species

Nitrogen occurs in many oxidation states, ranging from the most-reduced form, N^{3-}, to the most-oxidized form, N^{5+}. A neutral nitrogen atom contains seven electrons, five of which are valence electrons configured as $2s^2\,2p^3$, meaning that a nitrogen atom can achieve the octet rule by (i) gaining three electrons to produce N^{3-}; (ii) losing five electrons to become N^{5+}; and (iii) sharing three pairs of electrons with another nitrogen atom to form N^0 in N_2. As is shown in Table 6.1, nitrogen can occur in all oxidation states between N^{3-} and N^{5+}, although N^{2-} and N^- are rare or nonexistent in nature.

Reduced nitrogen as N^{3-} occurs in organic matter where it is typically chemically bonded to carbon, and in species that are stable in reducing environments, including the ammonium ion (NH_4^+) and ammonia gas (NH_3). Wetland sediments, crustal environments with low dissolved oxygen, deep marine waters, and decaying organic matter all contain ammonium and have the potential to produce ammonia gas. One example of the relationship between ammonium and ammonia occurs in alkaline soils, where NH_4^+ can lose a proton to solution, resulting in NH_3. Similarly, when urea (Table 6.1) is applied as a nitrogen fertilizer, it is prone to degassing as ammonia in alkaline, waterlogged soils.

N_2 gas, the most common form of nitrogen in the Earth surface/shallow crustal environment, is the least reactive of the nitrogen species, and this attribute tends to limit the amount of reactive/available nitrogen in terrestrial, aquatic, and marine ecosystems. Only in certain biochemical environments can N_2 be transformed by natural processes to reactive nitrogen species – these include soils and waters that contain certain bacteria and related microbes, and in situations where lightning or combustion produce NO_x.

Compounds containing N^+, N^{2+}, or N^{3+} tend to form in oxidized aqueous environments (e.g. streams, shallow

Table 6.1 Some common nitrogen species and the oxidation state of nitrogen in the compounds.

Species	Name	Oxidation state
NH_3	Ammonia	N^{3-}
NH_4^+	Ammonium ion	N^{3-}
$(NH_2)_2CO$	Urea	N^{3-}
N_2H_4	Hydrazine[a]	N^{2-}
NH_2OH	Hydroxylamine[a]	N^-
N_2	Molecular nitrogen (N_2 gas)	N^0
N_2O	Nitrous oxide	N^+
NO	Nitric oxide	N^{2+}
HNO_2	Nitrous acid	N^{3+}
NO_2^-	Nitrite anion	N^{3+}
NO_2	Nitrogen dioxide	N^{4+}
HNO_3	Nitric acid	N^{5+}
NO_3^-	Nitrate ion	N^{5+}

[a] These compounds are synthetic.

groundwater, aerated soils, oxygenated lakes, sea water, etc.), often via oxidation of organic matter or ammonium, or via microbially mediated fixation from N_2. Of the species presented in Table 6.1, some of the more common ones in the environment include:

1 Nitrous oxide (N_2O), a greenhouse gas that is ~300 times more potent on a molecule-by-molecule basis than CO_2; one important origin of this trace atmospheric gas is denitrification of fertilizers, notably in tropical and subtropical soils; it also forms via natural processes of denitrification and nitrification (see Section 6.2.1.2) in soils and marine waters.

2 Nitric monoxide (NO, also termed nitrogen oxide) and nitrogen dioxide (NO_2), molecules which often form during combustion and are often grouped together as NO_x.

3 Nitric acid (HNO_3), which forms in the atmosphere via the oxidation of NO and NO_2. It is also a product of volcanic eruptions.

4 Nitrate (NO_3^-), a highly soluble anion that is a product of natural oxidation of organic matter, yet also one whose concentration at the Earth surface is greatly enhanced by anthropogenic activities, especially addition of fertilizers and combustion, where nitrate can form via oxidation of NO_x by-products (more-detailed information on nitrogen gases as atmospheric pollutants is presented in Chapter 8). Nitrate is responsible for eutrophication in many aquatic environments and is regulated in public water supplies because it is linked to *methemoglobinemia*, a condition linked to "blue baby syndrome." Anaerobic bacteria in the stomach reduces nitrate to nitrite, which in turn can oxidize ferrous iron (Fe^{2+}) in hemoglobin to ferric iron (Fe^{3+}), preventing transport of O_2 in blood. This condition is most pronounced in children younger than six months of age.

Nitrogen concentrations are often reported with reference to N atoms in a molecule, rather than the mass of the entire molecule (akin to C in CO_2, CH_4, and other C compounds). A good example is nitrate, which is commonly reported as NO_3^--N; if a water sample contains $6.6\ mg\ l^{-1}$ of dissolved nitrate, it will contain $1.5\ mg\ l^{-1}\ NO_3^-$–N. The conversion factor is based on $14.01\ g$ of nitrogen in 1 mol of nitrate ($62.01\ g$). Therefore, to convert the mass of NO_3^- to NO_3^--N, multiply by $0.226\ [14.01\ g\ mol^{-1}\ (N)/62.01\ g\ mol^{-1}(NO_3^-)]$. The US EPA maximum contaminant level for nitrate in public water supplies is $10\ mg\ l^{-1}\ NO_3^-$-N, which corresponds to $44\ mg\ l^{-1}\ NO_3$.

6.2.1.2 Processes operating within the nitrogen cycle

The wide range in oxidation states means that reduction–oxidation reactions strongly control nitrogen cycling, and

microorganisms usually play a critical role. Figure 6.2 summarizes many of the transformations that occur in the N cycle.

A good starting point for examining biogeochemical transformations in the N cycle is atmospheric N_2, the largest reservoir of nitrogen on earth. This form of nitrogen is largely unavailable to living organisms; for example, it is not available for uptake by photosynthetic plants, and when humans and other animals inhale air containing N_2, the gas is eventually exhaled in the same, unaltered form. For this reason, animals must obtain nitrogen from proteins and other food sources. Fortunately, nitrogen-fixing microorganisms such as *Rhizobium*, immobile bacteria which live in root nodules of legumes (e.g. alfalfa, clover, lentils, peas, beans), and *Azotobacter*, a motile genus of soil bacteria, are capable of reducing N^0 (in N_2) to N^{3-}. Often the bacteria form a mutualistic relationship with a plant; by transforming N_2 to NH_3 or NH_4^+, *Rhizobium* bacteria provide an available form of nitrogen as a nutrient to the plant, and in return, the bacteria obtain carbohydrates from the plant.

Transformation from atmospheric N_2 to plant-available ammonium is *fixation* and it can be summarized in this reaction:

$$N_2 + 8H^+ + 8e^- + 16\ ATP = 2NH_3$$
$$+ H_2 + 16\ ADP + 16P_i \qquad (6.2)$$

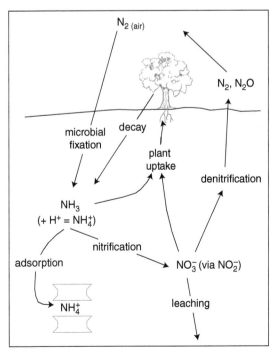

Fig. 6.2 Biogeochemical transformations in the N cycle.

ATP and ADP are adenosine triphosphate and adenosine diphosphate and P_i is inorganic phosphorous. Ammonia commonly reacts with water to form the ammonium ion:

$$NH_3 + H_2O = NH_4^+ + OH^- \qquad (6.3)$$

NH_4^+ is a plant-available form of nitrogen and it may be incorporated into plant roots in this state. If the ammonium cation persists in soil (the likely scenario in waterlogged neutral to acidic soils), it may become adsorbed (e.g. to clays or humus) or it may be transported in solution.

In aerated soils, NH_4^+ may undergo oxidation to *nitrite* via a process mediated by bacteria (commonly those of the *Nitrosomonas* species) as follows:

$$NH_4^+ + 1.5O_2 \rightarrow NO_2^- + 2H^+ + H_2O \qquad (6.4)$$

The change in Gibbs free energy for this reaction is $-66\,kJ\,mol^{-1}$ ($\Delta G^o_R = -66\,kJ\,mol^{-1}$), so it is spontaneous; however, *Nitrosomonas* bacteria act as a catalyst, increasing reaction rate. What do the bacteria get from this reaction? Electrons (recall that electron transfer is the main motivation for bacteria who mediate chemical reactions). The oxidation of 1 mol of ammonium to nitrite releases 6 mol of electrons ($N^{3-} \rightarrow N^{3+} + 6e^-$), and this helps fuel bacterial metabolism. Nitrite is unstable relative to *nitrate*, so nearly all NO_2^- is further oxidized to NO_3^- in

the presence of *Nitrobacter* bacteria:

$$NO_2^- + 0.5O_2 \rightarrow NO_3^- \qquad (6.5)$$

The oxidation of nitrogen in this case is of a smaller magnitude than for ammonium – only 2 mol of electrons are released per mole of nitrite oxidized to nitrate ($N^{3+} \rightarrow N^{5+} + 2e^-$). Nitrate is typically 10 times more abundant than nitrite in aerated soils and aquatic systems, and the term for the sequential transformation of $NH_4^+ \rightarrow NO_2^- \rightarrow NO_3^-$ is *nitrification*.

In chemically reducing environments such as waterlogged soils and wetland sediments, nitrates are reduced to N_2 or less commonly, N_2O, gases which seep back into the atmosphere, effectively completing the nitrogen cycle. This process of transforming mineralized or fixed nitrogen to N_2 or N_2O gas is known as *denitrification*. As with most transformations in the nitrogen cycle, this process is facilitated by bacteria (e.g. *Pseudomonas*) which, in this case, use the nitrate as an electron acceptor during respiration. This is necessary because O_2 gas, the main electron acceptor in oxidizing environments, is largely unavailable in reducing/anaerobic/anoxic environments.

The main oxidation state of nitrogen in organic matter is N^{3-} regardless of whether a plant incorporates nitrogen as NH_4^+, NO_3^-, or some other form. Thus, if nitrogen is incorporated into a plant as nitrate, the plant will reduce

---

Focus Box 6.2

Decomposition of N-Bearing Organic Matter: The Example of Urea

The following example summarizes the fate of organic matter undergoing decomposition in soils using the simple compound urea (Figure 6.3) to illustrate nitrogen in organic matter. The degradation of urea and eventual oxidation of ammonium released by urea breakdown helps to illustrate the types of reactions associated with release of nitrogen from organic matter in soils and waters.

The first step, which occurs in the presence of urease bacteria, is hydrolysis of urea to yield two ammonium atoms plus bicarbonate. Note that in this first step, the oxidation state of nitrogen is unchanged (it is N^{3-} in both urea and ammonium).

$$(NH_2)_2CO + 2H_2O + H^+ \rightarrow 2NH_4^+ + HCO_3^- \qquad (6.6)$$

The general term for any reaction, e.g. decay of plant matter, which releases available nitrogen (or any nutrient) to soil is *mineralization*; this specific case is known as *ammonification*.

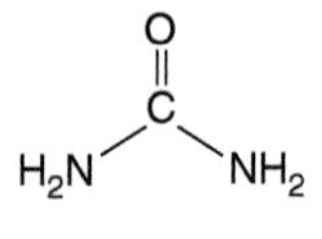

Fig. 6.3 Structural formula of urea, $(NH_2)_2CO$.

At this point, the ammonium ion could lose a proton to form ammonia, which could then be volatilized into the atmosphere. Given the abundance of OH^- in alkaline solutions, this process tends to occur in alkaline soils (reaction (6.3) in reverse):

$$NH_4^+ + OH^- \rightarrow NH_3 + H_2O \qquad (6.7)$$

In oxidizing environments, ammonia or ammonium is likely to be oxidized to nitrate as shown in Eqs. (6.4) and (6.5). Urea is often used as a N fertilizer, so this example helps to illustrate the likely pathways of that form of fertilizer in different soils.

the nitrogen from N^{5+} to N^{3-} so that it can become a constituent of organic matter, e.g. as amine functional groups.

Once nitrogen has been incorporated into organic matter, it can either be consumed as N^{3-} (e.g. in proteins) and metabolized by animals (a process often termed *assimilation*), or it can undergo a series of transformations to other oxidation states, all of which are higher (less negative) than N^{3-}. Thus, any reaction that changes the oxidation state of nitrogen in organic matter will be an oxidation reaction.

Combustion is another common process responsible for the transformation of nitrogen from reduced species to oxidized species. Reduced nitrogen in organic matter is oxidized to species such as NO and NO_2 during combustion, but this is only a small part of nitrogen emitted by combustion (e.g. of biomass, hydrocarbons or coal). The majority of N released to air comes from N_2 in the atmosphere that combines with atmospheric O_2 in the elevated temperatures of the internal combustion engine or forest fire to form nitrogen oxides. Two examples of nitrogen–oxygen reactions that occur during combustion are:

$$N_2 + O_2 = 2NO \tag{6.8}$$

$$N_2 + 2O_2 = 2NO_2 \tag{6.9}$$

In reaction (6.8), nitrogen is oxidized from N^0 to N^{2+}; in reaction (6.9), nitrogen is oxidized from N^0 to N^{4+} (Table 6.1). The species produced in these reactions are generically referred to as NO_x. High temperatures associated with lightning in the atmosphere also foster formation of NO_x.

Nitrogen species in solution are typically quantified and classified as either *dissolved inorganic nitrogen (DIN)* or as *dissolved organic nitrogen (DON)*. DIN typically takes the form of low-molecular weight compounds such as ammonium (NH_4^+), nitrite (NO_2^-), and nitrate (NO_3^-) whereas in DON, nitrogen is bonded to carbon in more-complex, high-molecular weight organic molecules (e.g. humic acid). DIN is typically more available for uptake by plants in terrestrial and aquatic ecosystems (and farms, lawns, and golf courses), and for this reason chemical fertilizers typically contain ammonium or nitrate, or both (ammonium nitrate is a common fertilizer). Organic fertilizers (e.g. compost, manure) commonly contain more DON than DIN, meaning that the nitrogen is released slowly over time.

6.2.1.3 Global-scale reservoirs and fluxes of nitrogen

The abundance of nitrogen in Earth surface reservoirs is measured in units of teragrams or Tg (10^{12} g). As was the

case for carbon (Chapter 5), values represent the mass of nitrogen rather than the mass of the molecule. Fluxes are given in teragrams of N per year ($TgN\ yr^{-1}$).

Of the globally important N reservoirs at or near the Earth's surface, the one with the overwhelming majority of nitrogen is the *atmosphere* (Figure 6.4), which contains 4×10^9 TgN (3×10^{20} mol of N), nearly all of it as N_2 gas. The remainder – approximately 10^3 Tg of atmospheric N ($< 0.01\%$ of total N in air) – exists as the more-reactive species NO, NO_2, N_2O, and NH_3. The residence time of N_2 in the atmosphere can be computed from reservoir size (4×10^9 Tg) and flux of N_2 to terrestrial and marine ecosystems via fixation ($\sim 3 \times 10^2$ Tg yr^{-1}), yielding a T_{res} value of $\sim 10^7$ years. Residence times of NO_x and NH_3 are far shorter and average from days to weeks; residence time of N_2O in the atmosphere is ~ 100 years.

Natural processes capable of transferring nitrogen out of the atmosphere to other reservoirs include biological fixation (mediated by bacteria and archaea) and lightning, where temperatures are sufficiently high to break N:::N and O::O bonds and foster formation of NO_x species that can dissolve in rain and be delivered to terrestrial or marine environments. Combustion (e.g. forest fires) is also a natural means by which N_2 can be transformed to reactive species and removed from the atmosphere. The net flux of nitrogen out of the atmosphere and into terrestrial and marine environments is causing major impacts to ecosystems, but the mass of N_2 in the atmosphere is so great that this transfer is effectively undetectable in the concentration of nitrogen in air.

The next largest reservoir of nitrogen in the Earth surface and near-surface system encompasses the sedimentary rocks of the shallow crust, with $\sim 8 \times 10^8$ TgN. The main forms of nitrogen in this reservoir are ammonium cations (e.g. NH_4^+ can substitute for K^+ in mica interlayers) and organic matter; coal, for example, typically contains ~ 0.1–1% N. Transfer from the crustal N reservoir to the more-surficial N reservoirs (atmosphere, soils, terrestrial, and marine ecosystems) occurs by way of chemical weathering of rocks and by volcanic emissions. Combustion of fossil fuels also releases reactive nitrogen to the atmosphere and, ultimately, ecosystems.

The *oceans* contain $\sim 6 \times 10^5$ TgN, the vast majority of which exists as inorganic nitrogen and as nitrogen fixed into organic matter in the deep ocean. Lesser amounts occur in organic matter in the shallow ocean. Fixation of nitrogen into marine environments occurs by bacterially mediated reduction of N_2 to NH_3, by atmospheric deposition of NH_4^+ and NO_x, and by fluvial inputs of reactive nitrogen (especially NO_3^-) from terrestrial environments. The oceans receive 50 TgN per year in the form of NO_3^-,

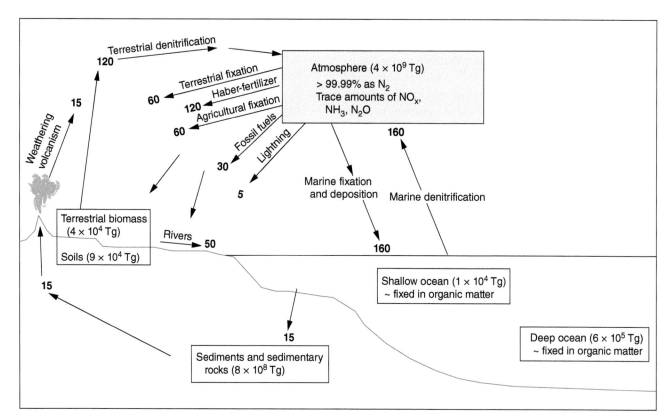

Fig. 6.4 Schematic sketch of the nitrogen cycle showing main reservoirs (in boxes) and fluxes (arrows). Numbers in bold type indicate flux between reservoirs in TgN yr^{-1}, and the mass of nitrogen in each reservoir is given in Tg. "Terrestrial fixation" represents fixation of N to natural ecosystems by *Rhizobium* and other species. "Agricultural fixation" applies to N fixed into crops by *Rhizobium* and other species (in mutualisms with legumes). "Haber-fertilizer" represents transformation of N$_2$ to NH$_3$ via the Haber-Bosch process – this flux is often effectively synonymous with industrial N fixation. "Marine fixation and deposition" represents nitrogen incorporated into oceans by biological fixation and rainfall (e.g. dissolved NH$_4^+$, NO$_x$, much of which is anthropogenic, including from fossil fuel combustion. Source: Data are from Galloway et al. (2004), Schlesinger (2009), Fowler et al. (2013), and Yang and Gruber (2016), and given large uncertainties, these are estimates, especially fluxes (notably marine denitrification). The net flux in the modern N cycle clearly reflects the enrichment of terrestrial and marine ecosystems in N via transformation of N$_2$ to reactive nitrogen species (from Haber-Bosch process/fertilizer manufacture and formation of NO$_x$ from combustion of fossil fuels).

NO$_x$, NH$_4^+$, and related reactive N species produced by fossil fuel combustion, fluvial transport of fertilizer runoff, and other anthropogenic sources; the preindustrial value was 40 TgN yr^{-1} (Yang and Gruber 2016).

Terrestrial ecosystems (including soils) contain $\sim$1.3 $\times$ 10^5 TgN in subreservoirs of terrestrial biota (4 $\times$ 10^4 TgN) and soils (9 $\times$ 10^4 TgN). The steady-state flux between terrestrial ecosystems and atmosphere is $\sim$100 TgN yr^{-1}, but addition of fertilizers, NO$_x$ from fossil fuel combustion, and other forms of Nr have increased the flux from terrestrial systems to the atmosphere. Leaching of nitrate and other Nr species from soils to rivers and ultimately lakes and shallow marine environments contributes to eutrophication, a topic covered in Section 6.2.1.4.

Although not shown as a reservoir given the ephemeral nature of fluvial systems, *rivers* are an important component of the nitrogen cycle. Estimates of preindustrial N flux via rivers from terrestrial to marine ecosystems indicate a value of $\sim$20 TgN yr^{-1}; current riverine N fluxes are approximately 50 TgN yr^{-1}. Some of the reactive N transported by rivers is denitrified in the nearshore marine environments, which returns N$_2$ to the atmosphere. If fertilizer use continues to increase, fluvial transport of nitrogen to lakes and shallow marine ecosystems are also likely to increase.

6.2.1.4 Human perturbation of the nitrogen cycle and resulting environmental impacts

The roots of environmental problems pertaining to alteration of the nitrogen cycle all stem from the same source – transformation of molecular nitrogen (N_2) to reactive nitrogen (Nr). Three main mechanisms by which N_2 is transformed to Nr include:

1 manufacture of nitrogen fertilizers by the Haber-Bosch process and application to fields;

2 fixation of reactive nitrogen into agricultural soils by cultivation of legumes; and

3 formation of nitrogen oxides as by-products of fossil fuel combustion.

Prior to the twentieth century, the main sources of nitrogen-bearing fertilizers were organic (as opposed to inorganic), including animal manure, bat and seabird guano, and composted organic matter. In fact, meters-thick deposits of N-rich guano were exported globally (especially from Chile) in the nineteenth century to satisfy fertilizer and explosive needs. By 1909, Fritz Haber had discovered a process for synthesizing ammonia from N_2 and H_2. Soon after, the German chemical company BASF purchased the rights to the process and by 1913 Carl Bosch was able to increase Haber's approach from benchtop scale to industrial scale. Initially, the ammonia developed by the Haber-Bosch method was used to manufacture fertilizer, but soon after was used to make explosives for World War I. The origin of the hydrogen used in the Haber-Bosch process is methane, which is purified to remove impurities (e.g. H_2S, SO_2) and reacted as follows:

$$CH_4 + 2H_2O \rightarrow CO_2 + 4H_2 \qquad (6.10)$$

or

$$CH_4 + H_2O \rightarrow CO + 3H_2 \qquad (6.11)$$

The hydrogen is then reacted with nitrogen gas under high temperature and pressure (~400 °C and ~200 bars) to form ammonia:

$$N_2 + 3H_2 \rightarrow 2NH_3 \qquad (6.12)$$

At this point, the ammonia can be converted to ammonium (by reaction with H^+) or nitrate (by oxidation) to form N-rich fertilizers such as NH_4NO_3, KNO_3, and urea.

Nitrogen and phosphorus are the two main limiting nutrients in soils and aquatic ecosystems – nitrogen due to the fact that the most abundant form occurs as relatively unreactive N_2 gas, and phosphorus due to its general insolubility in soils and waters. As a result, farm fields, residential lawns, golf courses, and gardens are commonly fertilized with N and P to increase productivity. Since the mid-1900s, the most common origin of the nitrogen used in fertilizers is fixation from N_2 by the Haber-Bosch process, and the N molecules added to fields are either

the highly mobile nitrate anion, or are molecules such as ammonium and urea, species which are readily transformed to nitrate in the oxidizing conditions of most soils.

Nitrate is easily leached from soils and transported into surface water bodies and shallow groundwater due to its high solubility and low sorption potential. (It also tends to leach out of sewage treatment plants, feedlots, and septic systems). Given that low nitrogen concentrations in surface waters limits algal growth, anthropogenic activities that increase concentrations of nitrate and other Nr species in surface waters tend to trigger increased algal growth. In many lakes and shallow marine waters, increasing Nr is directly proportional to algal growth (Figure 6.5), the result of which is a sequence of changes to the ecosystem termed eutrophication (from Greek: "eu" = enriched or sufficient, "trophic" = living or nutrient).

When the reactive nitrogen is anthropogenically derived, the term *cultural eutrophication* is applied to distinguish this phenomenon from natural eutrophication that occurs in some wetlands, shallow lakes, and coastal marine ecosystems, where deep, cold, oligotrophic ("oligo" = low) lakes can eventually evolve into eutrophic lakes or swamps as they fill in with sediment and organic matter over time.

Eutrophic lakes, bays, shallow marine waters (especially near river deltas), and rivers (especially slow-moving ones)

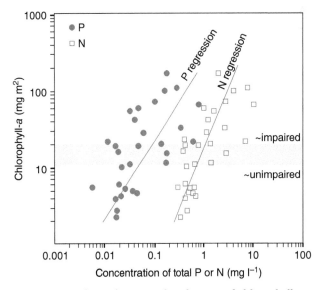

Fig. 6.5 Correlation between phosphorus and chlorophyll-*a*, and between nitrogen and chlorophyll-*a*, in streams of the northeastern United States. The shaded area represents the transition zone between unimpaired streams and impaired streams in the northeastern US. Note that the axes are logarithmic and that the x-axis is in units of milligrams per liter (ppm). Source: Data are from Riskin et al. (2003).

Focus Box 6.3

Case Study of the Nitrogen Fertilizers and Surface Water Quality

An example of the relationship between N fertilizer application and response was illustrated in a study done in Wisconsin (USA) from the 1960s through the 1990s (Portage County Groundwater Conditions 2000), where application of fertilizer was compared to stream and groundwater nitrate concentrations. The results of this study are summarized in Figure 6.6, which shows two graphs, one of nitrogen fertilizer sales in Wisconsin, and the other of nitrate concentrations in baseflow of the Little Plover River. The measurements of NO_3^- in the Little Plover River were made in January when the majority of stream flow is derived from soil and groundwater influx to the stream; thus, measuring the stream during baseflow conditions provides a proxy for soil and shallow groundwater nitrate flux. The increase in nitrate shown in Figure 6.6 is typical for streams from across the globe, and is illustrative of the changes in surface water chemistry that lead to eutrophication. It also reflects the fact that N fertilizers are being applied at rates far greater than the potential for N uptake by plants, leaving excess nitrate available for leaching into groundwater and for runoff into streams, lakes, estuaries, and other surface water bodies.

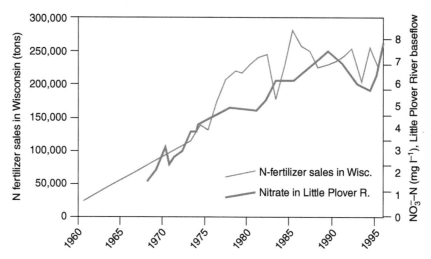

Fig. 6.6 Relationship between nitrogen fertilizer use and dissolved nitrate in the Little Plover River of central Wisconsin. Source: Based on data from Portage County Groundwater Conditions (2000). Note that nitrate data are in units of mg/L NO_3^--N, meaning that only the mass of nitrogen in nitrate is included in the calculation. Sales of nitrogen fertilizers in Wisconsin parallel national and global trends.

are characterized by abundant growth of blue-green algae known as cyanobacteria which often form extensive mats on the water surface (see Plate 8). The algae blooms themselves can be toxic due to the presence of various cyanotoxins including microcystins, anatoxin-a, and saxitoxins (Graham et al. 2010), but in an ecological sense, the greatest damage results when the algae, which have short life spans, die and decay in the water. Decomposition of the algae is carried out by aerobic bacteria which consume oxygen, and the greater the abundance of algae undergoing decay, the greater the amount of O_2 consumed from the water column by the aerobic decomposers. This cyclic process depletes the water of available oxygen and causes the death of other organisms, particularly fish and amphibians. In the oceans, these affected areas are known as *marine dead zones*.

Notable water bodies affected by cultural eutrophication are the delta of the Mississippi River in the Gulf of Mexico, Chesapeake Bay, and San Francisco Bay in the USA, Lake Erie in the USA, shallow nearshore waters of the Baltic Sea, the Black Sea (from nutrients transported by the Danube River), parts of the South China Sea, and far too many other water bodies to list here. It is also important to remember that phosphorus is also a key cause of eutrophication, and the sources and behavior of P is covered in Section 6.2.2 on the phosphorus cycle.

Changes in global nitrogen cycling presented in Figure 6.7 illustrate the following:

1 notable increases in nitrous oxide in the atmosphere (this record is derived from the composition of air bubbles trapped in glacial ice);

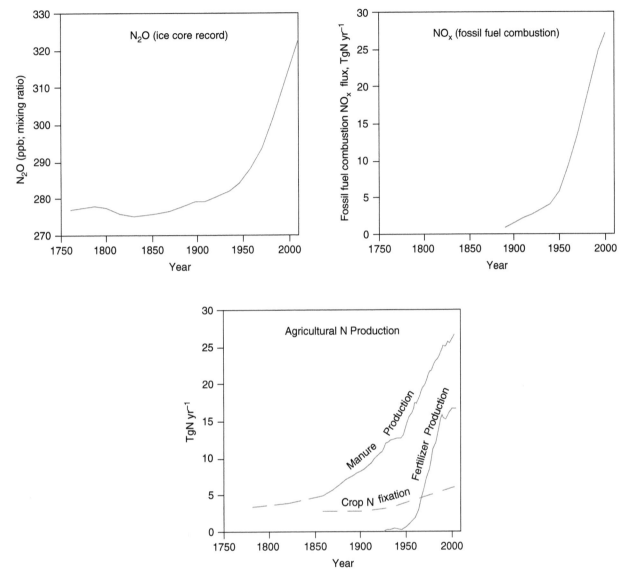

Fig. 6.7 Global trends in sources and cycling of reactive nitrogen. (a) increasing atmospheric nitrous oxide recorded in ice cores; (b) increasing emissions of nitrogen oxides from fossil fuel emissions (~2/3 of total NO_x emissions); (c) increases in production of fertilizers derived from Haber-Bosch process and manure, as well as uptake of N by crops. Note that (a) represents atmospheric *concentration* of N_2O while (b) and (c) present *fluxes* (Holland et al. 2005; MacFarling Meure et al. 2006).

2 increases in nitrogen oxides (NO_x) produced by fossil fuel combustion throughout the twentieth century, as well as a recent tapering of the curve due to emissions controls in automobiles, coal plants, and other sources; and

3 increased agricultural application of reactive N to the landscape, both from increased manure production from farm animals as well as the industrial transformation of N_2 into reactive nitrogen in fertilizer via the Haber-Bosch process. It is important to note that the curves for agricultural N are similar to population growth curves, and that additions of nitrogen to crops have far exceeded crop uptake of nitrogen since the early twentieth century. The excess nitrogen (i.e. that not fixed into crops) is largely available

for leaching into groundwater and surface water, where it can cause human health problems (i.e. methemoglobine-mia) and ecological problems (i.e. eutrophication).

Nitrous oxide emissions are mainly from agricultural soils where N-bearing fertilizers are converted to N_2O – in 2008 in the United States; 68% of N_2O emissions were related to "agricultural soil management" (US EPA; http://www.epa.gov/nitrousoxide/sources.html), so it is not surprising that the N_2O curve and the agricultural N production curves are similar. N_2O also forms naturally via marine nitrification and is produced in sewage treatment plant lagoons and natural wetlands. N_2O is not easily decomposed in soils or in the troposphere, and most decomposition takes place in the stratosphere; as a result, atmospheric concentrations of N_2O have increased ~20% since the mid-1800s (Figure 6.7). In the stratosphere, N_2O is decomposed by reaction with oxygen radicals that acts to lower concentrations of beneficial stratospheric ozone.

The NO_x emissions represented in Figure 6.7 precipitate onto the terrestrial landscape and nearshore marine ecosystems when NO_x is oxidized and dissolved in rain, and then delivered to the earth's surface (typically) as nitric acid. The high K_a of nitric acid (2.4×10^1) means that it nearly completely dissociates, resulting in delivery of dissolved H^+ and nitrate to soils and aquatic ecosystems. Details of N_2O and NO_x reactions in the atmosphere are presented in Chapter 8, and issues associated with acid rain are presented in Chapters 8 and 9.

6.2.2 The phosphorus cycle

The phosphorus cycle is much simpler than the carbon and nitrogen cycles for two reasons. First, in natural environments at or near the Earth's surface, phosphorus has no significant gas phase – it occurs in solid inorganic minerals and organic solids as well as sorbed and dissolved forms (it can also occur in a molten state in magma or lava). Second, phosphorus has only one valence state in nature, P^{5+}, so it is not prone to the wide array of reduction–oxidation reactions that occur in the C and N (and S) cycles. Like carbon and nitrogen, phosphorus is critical to living organisms; it is a crucial component of DNA and RNA and essential in the transmission of chemical energy via ATP and ADP molecules.

The primary mineral source of P is apatite, $Ca_5(PO_4)_3(OH,Cl,F)$, a calcium phosphate which occurs in nearly all rock and sediment types at concentrations typically in the range of 0.1–1%. The OH^-, Cl^-, and F^- anions substitute for each other, and the F-rich end-member, fluorapatite, is the hardest (most resistant to chemical breakdown) of the apatite varieties. Given that tooth enamel is essentially comprised of apatite, you can surmise why fluoride is added to toothpaste. A rare but P-rich rock type known as phosphorite, a sedimentary rock of marine origin, often contains 20–40% P_2O_5 (~10–20% apatite) and is mined as a source of P for fertilizers. Other phosphate minerals are far more rare than apatite and include turquoise, $CuAl_6(PO_4)_4(OH)_8 \cdot 5H_2O$; the uranium-bearing mineral autunite, $Ca(UO_2)_2(PO_4)_2 \cdot 10\,H_2O$; and monazite (Ce, La, Nd, Th, Y)PO_4, a mineral useful to geochronologists for dating of igneous rocks.

6.2.2.1 P cycling in soils

Chemical weathering of hydroxy-apatite in soils is a good starting point for the P cycle:

$$Ca_5(PO_4)_3OH + 4H_2CO_3$$
$$\rightarrow 5Ca^{2+} + 3HPO_4^{2-} + 4HCO_3^- + H_2O \qquad (6.13)$$

The speciation of phosphate produced in this reaction depends on pH (Figure 6.8). In typical pH ranges for soils, the dominant forms are $H_2PO_4^-$ in the pH range of 2–7 and HPO_4^{2-} in the pH range of 7–12. In very acidic situations (e.g. acid mine drainage), H_3PO_4 may occur, and under extremely alkaline conditions, the PO_4^{3-} ion is predicted to dominate.

P is nearly ubiquitous in rocks and soils (it typically comprises 200–500 mg kg^{-1} in rock or soil), but two factors combine to make it a limiting nutrient in soils and aquatic ecosystems:

1 Phosphate is insoluble or only sparingly soluble under most conditions, so in soils the phosphate anion (PO_4^{3-}) tends to form poorly crystalline hydrated solids by bonding with Ca^{2+}, Fe^{3+}, or Al^{3+} to produce species such as $Ca_3(PO_4)_2 \cdot 2H_2O$, $FePO_4 \cdot 2H_2O$, and $AlPO_4 \cdot 2H_2O$. Ca-phosphate tends to occur in alkaline soils whereas Fe- and Al-phosphate are more likely to form in acidic soils. Incorporation of phosphate anions into hydroxides of Fe and Al also limit P solubility.

2 Apatite is a relatively stable mineral, meaning that P is only slowly released to the soil solution during chemical weathering. The solubility product of apatite is ~10^{-58}, a value which would produce an equilibrium concentration of dissolved P of only ~10^{-8} mol l^{-1} (Schlesinger, 1997).

The consequence of these two factors is that P is dominantly bound in the solid phase, including well-crystalline

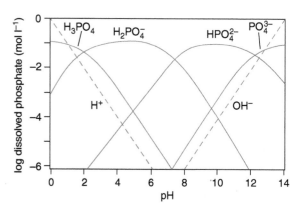

Fig. 6.8 Variation in speciation of phosphate as a function of pH. Concentrations of H^+ and OH^- are also shown for reference, emphasizing e.g. how increasing $[H^+]$ favors the more H-rich $H_2PO_4^-$ over HPO_4^{2-}.

apatite and poorly crystalline soil hydroxides and phosphates. A general guideline regarding P speciation is that in soils, terrestrial surface waters, and groundwater, the ratio of particulate P to dissolved P is ~10:1 (Howarth et al. 1995), with much of the dissolved P occurring in organic acids (e.g. fulvic acid), situations where the organic acid functions as a chelate. This ratio can vary, however, depending on source (e.g. P tends to be more easily dissolved from organic matter), pH (P is *least insoluble* between pH of ~6 and 8), and location. Open ocean water, where nearly all P is dissolved, is a major exception to this general set of rules (recall that Fe tends to be the limiting nutrient in open marine waters).

Formation of iron and calcium phosphates following weathering of apatite or release of P from organic matter or other solids illustrates the incorporation of P into the relatively insoluble minerals that limit P solubility:

$$H_2PO_4^- + Fe^{3+} + 2H_2O \rightarrow FePO_4 \cdot 2H_2O + 2H^+ \quad (6.14)$$

$$2HPO_4^{2-} + 3Ca^{2+} + 2H_2O \rightarrow Ca_3(PO_4)_2 \cdot 2H_2O + 2H^+ \quad (6.15)$$

P may also sorb onto surfaces of iron hydroxides or become incorporated into hydroxides as inner-sphere complexes, and this is demonstrated schematically in Figure 6.9. If the reduction–oxidation conditions at the sediment-water boundary are oxidizing, P is likely to remain sorbed to the iron hydroxides. If the conditions are reducing (anoxic), which is commonly the case in sediments at the bottoms of lakes and estuaries, iron will be reduced, the iron hydroxide will dissolve, and P will be released to solution and become available for biological uptake. $Fe(OH)_3$ that was stable in the oxygenated soil and fluvial environment will be prone to $Fe(III) \rightarrow Fe(II)$ reduction and release of P to solution in anoxic sediments. This mechanism is at least partly responsible for availability of P in lacustrine and estuarine sediments.

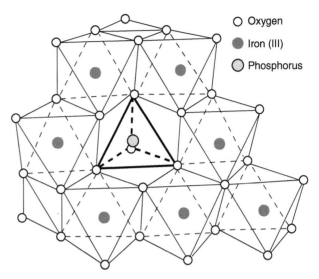

○ Oxygen
● Iron (III)
◉ Phosphorus

Fig. 6.9 Schematic sketch of showing tetrahedral PO_4^{3-} molecule incorporated into an iron hydroxide structure. This sketch is based on a ferrihydrite structure determined by Michel et al. (2007).

The P cycle in soils and terrestrial ecosystems is summarized in Figure 6.10. The largest reservoir in the near-surface and Earth surface system is sedimentary rocks and sediments, but this P is largely unavailable for biological use. Most of the P available for plant uptake is partitioned into the organic P reservoir, contained within decaying organic matter, humus, fulvic acids, microbial organisms, and related organic reservoirs. Any P not fixed immediately by plants is likely to be incorporated into inorganic minerals or noncrystalline solid phases such as Fe-, Al-, or Ca-phosphates and iron hydroxides. At the same time, many plants exude organic acids, chelating compounds, and enzymes that are capable of extracting P from inorganic and organic solid phases, so there certainly is some flux between secondary minerals and biota.

Focus Box 6.4

Redox Effects on the Mobility of a Non-redox-sensitive Element (P)

P itself does not undergo reduction or oxidation, but if incorporated into a redox-sensitive phase such as $Fe(OH)_3$ or $FePO_4 \cdot 2H_2O$, redox conditions will influence P fate and transport. Young and Ross (2001) studied agricultural fields that are flooded each spring and found that the reducing conditions imparted by flooding caused ~10-fold increases in dissolved Fe^{2+} and P in soil porewaters. Furthermore, given that the majority of phosphorus transport to lakes and estuaries occurs as P sorbed to iron hydroxides (Figure 6.9), the behavior of P in these water bodies will largely be controlled by biogeochemical conditions that affect the stability of iron hydroxides. This was first recognized in the 1930s and 1940s (e.g. Einsele 1936; Mortimer 1941) and has become the paradigm for understanding P mobility in aquatic systems.

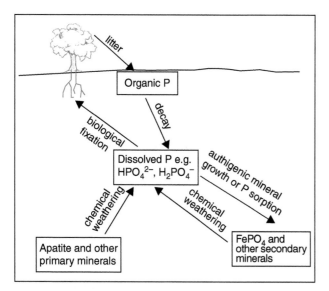

Fig. 6.10 Schematic sketch of the phosphorus cycle in soils. Due to the insolubility of phosphorus, most P cycling in terrestrial ecosystems takes place between organic P and biota.

6.2.2.2 The global phosphorus cycle

The largest reservoir of P at and near the Earth's surface is sedimentary rock and sediments, and the main means by which P is transported at the Earth's surface is by stream flow. Given that P is largely insoluble, the majority is transported as particulate material. Approximately $22 \, \text{TgP} \, \text{yr}^{-1}$ are transported by rivers to lakes and to the global ocean, and $\sim 90\%$ is carried as adsorbed P or as

P minerals (e.g. phosphates) – this component is mainly deposited in deltas and other types of marine sediments near continents. To reiterate a point made above, the main particulates responsible for P transport in the solid state are iron hydroxides and particulate organic matter (Ruttenberg 2003). The dissolved P carried by rivers is ultimately transported by currents into the open ocean, where it can be taken up as a nutrient or can circulate at the surface or to deeper levels.

Physical and chemical weathering of rock and sediment plus decomposition and erosion of organic matter yields $\sim 20 \, \text{TgP} \, \text{yr}^{-1}$ to biota and soils, and this flux is matched by erosion of soil P and biologically bound P into streams and, ultimately, the ocean. The cycle is closed by burial of marine P into marine sediments and the eventual transformation of these sediments in rock. In general, the flux of P between the main reservoirs (sediment/rock $\rightarrow$ soils/biota $\rightarrow$ streams $\rightarrow$ ocean $\rightarrow$ sediment/rock) is $\sim 20 \, \text{TgP} \, \text{yr}^{-1}$, although as is indicated on Figure 6.11, the riverine flux is about 10% greater, reflecting increased erosion due to anthropogenic activities, including road building, construction, logging, agriculture, and application of mined phosphorus as fertilizer. The increased fluvially transported P is responsible for much eutrophication.

In the open ocean, phosphorus is dominated by dissolved species – this is because the particulate P carried by rivers is mostly deposited in river deltas and on the continental shelf (Paytan and McLaughlin 2007). In the

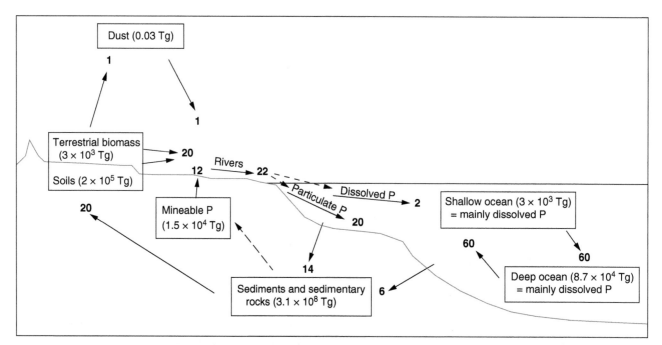

Fig. 6.11 The global P cycle. The unit for reservoir sizes is TgP and fluxes are in $\text{TgP} \, \text{yr}^{-1}$. Source: Data sources are from Schlesinger (1997) and Ruttenberg (2003).

shallow oceans, ~1–5% of total P is contained within biota, but the overwhelming majority of shallow marine P is dissolved (Ruttenberg 2003) The flux of P between dissolved and biotic P reservoirs in the shallow oceans is ~9×10^2 Tg yr^{-1} – this flux is not shown on Figure 6.11. Upwelling of deep marine dissolved P (~60 Tg yr^{-1}) is locally responsible for productive marine ecosystems (e.g. the northwest coast of South America), and this flux is ~in steady state with downwelling of dissolved P and biological P from the shallow ocean.

Two final points worth emphasizing are (i) unlike carbon and nitrogen, the atmospheric reservoir is very small and the only significant way that P is transported in air is as dust, and (ii) the reservoir of phosphate rock which can be mined for P comprises ~1.5×10^4 Tg; of this, ~12 TgP are mined each year, resulting in transfer of P from the sedimentary rock reservoir to soils, biota, river flux, and surface water bodies.

6.2.2.3 Phosphorus and eutrophication

Phosphorus is often the main limiting nutrient to growth in terrestrial and marine ecosystems at trophic levels ranging from phytoplankton to crops and trees. Phosphorus deficiencies can be problematic in agricultural soils, and thus P is one of the target elements applied in fertilizer, whether it is chemical fertilizer derived from phosphate mining, or P contained in manure or other organic forms. Erosion of P from stream banks, farm fields, construction sites, and numerous other sources has increased input of P to aquatic systems in rivers, lakes, estuaries, and the oceans. The role of P as a limiting nutrient with profound effects on lake ecology is illustrated in Figure 6.12, a graph plotting total P against chlorophyll-*a* for dozens of lakes studied globally. This graph indicates that, regardless of climate, geology, biota, and any other factors, P concentration controls primary productivity in aquatic ecosystems.

6.2.3 Comparison of N and P

Data from numerous streams in New England, USA, help to illustrate the effect of P on aquatic ecosystems, and also accentuate some of the main differences between phosphorus and nitrogen in aquatic systems. Figure 6.5 plots the concentrations of total P and total N separately against the abundance of chlorophyll-*a* in water, where chlorophyll-*a* is a useful proxy for abundance of cyanobacteria/algae commonly associated with eutrophication. Both elements clearly exhibit a positive correlation with chlorophyll-*a*,

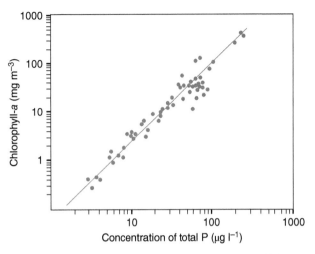

Fig. 6.12 Plot of total P vs chlorophyll-*a* for dozens of lakes from different ecosystems globally. In each case, total P is the amount of P present in lake water at the time of spring overturn; chlorophyll-*a* concentrations reflect peak summer abundances. Source: Data are from Sakamoto (1966), Dillon and Rigler (1974), and sources cited therein. Note that the x-axis in this diagram is in units of micrograms per liter (ppb).

but an important difference is apparent – growth of chlorophyll-*a* (i.e. bacteria and algae) occurs at P concentrations that are approximately an order of magnitude lower than what is required of N to produce equal amounts of bacteria and algae. Total P concentrations of 0.1 mg l^{-1} (ppm) are likely to produce an aquatic ecosystem of impaired quality, whereas the transition from unimpaired to impaired occurs at ~1 mg l^{-1} nitrogen.

Although both have potential to be limiting nutrients in terrestrial and aquatic ecosystems, the differences between N and P are striking. Included among these differences are:

1 phosphorus tends to be relatively insoluble and partitioned into the particulate fraction (i.e. silt, clay), whereas in surface waters nitrogen typically occurs as the highly soluble nitrate anion;

2 nitrogen-fixing microbes can introduce reactive N to soils and aquatic ecosystems from air; however, if P is lacking, it cannot be fixed from air and must be introduced to natural systems by rivers, chemical weathering, or windblown dust;

3 P has no significant gaseous phase, whereas N is the dominant gas in the atmosphere;

4 the primary source of N is the atmosphere, whereas the primary source of P is rock;

5 P only occurs in the pentavalent (P^{5+}) state and is prone to reduction–oxidation controls only if it is sorbed into iron hydroxides or other redox-sensitive phases, whereas N has numerous oxidation states and thus is prone to strong reduction–oxidation controls on speciation;

6 while both are limiting nutrients in aquatic ecosystems, P appears to be the truly limiting nutrient in most cases – this has been demonstrated experimentally by Schindler et al. (2008), who found that even in lakes with *excess P and limited N*, eutrophication occurred because the cyanobacteria were able to fix N from the atmosphere in the absence of abundant nitrate in the water. In lakes with *excess N but limited P*, eutrophication did not occur (P cannot be fixed from air via microbially mediated reactions); and

7 in many estuaries and in the ocean, N may be the true limiting nutrient to eutrophication (Carpenter 2008).

These differences are important when considering their common environmental impact, eutrophication. Because P is partitioned into the solid phase in terrestrial surface waters, its transport to surface waters is potentially easier to address than that of nitrate, whose high solubility and resistance to sorption (due to its negative charge) make it very mobile. Drainage ditches lined with grasses, hay bales, or small retention ponds will tend to foster settling of P-bearing particulates, while the highly soluble nitrate is largely unaffected – it will remain in solution and percolate through hay bales or dams and migrate down gradient in solution. Also, whereas the particulate forms of P are minimally soluble in soil and groundwater, nitrate is highly mobile in the subsurface and is capable of contaminating groundwater and of traveling along subsurface flow paths toward surface water bodies. For these reasons, P is often the focus of efforts to mitigate nutrient transport to eutrophication-sensitive water bodies.

6.2.4 The sulfur cycle

Sulfur is a relatively mobile element and, like C and N, can occur in a wide range of oxidation states (from S^{2-} to S^{6+}) and gas phases. Its cycling is strongly controlled by microbial processes. Unlike N or P, sulfur often controls fate and transport of heavy metals and arsenic in rocks, soils, and waters. Like C, N, and P, sulfur cycling has been altered by human activities, and one of the main consequences is acid precipitation in the form of sulfuric acid derived from fossil fuel combustion (especially coal) and in the refining of metals contained in sulfide ores.

6.2.4.1 Sulfur oxidation states, sulfur species

The location of sulfur in column VI of the periodic table and its valence electron structure of $3s^2\ 3p^4$ allow us to predict that S will occur in +6 and −2 oxidation states, i.e. it can satisfy the octet rule by losing six electrons or gaining two electrons. S^{4+}, S^{2+}, and S^0 oxidations states also occur. Table 6.2 presents the numerous oxidation states of sulfur and provides examples of many S-bearing compounds.

In most of the compounds shown in Table 6.2, sulfur occurs in a single oxidation state; for example, in galena (lead sulfide), sulfur occurs as the sulfide (S^{2-}) anion; in sulfur dioxide, sulfur is S^{4+}, and in sulfates, S^{6+} occurs. This is not the case with pyrite, where the oxidation state of sulfur is best characterized as $S_2{}^{2-}$. This is the disulfide ion, a molecule comprised of two covalently bonded S atoms whose overall charge is −2. The *average* charge on sulfur is −1. The same concept applies to arsenopyrite, where, in a unit cell, an arsenic atom substitutes for a sulfur atom to create As–S pairs whose sum charge is −2. Many of the intermediate oxidation states occur as metastable intermediary phases, notably SO_2 and SO_3 (combined, SO_x), compounds which form during combustion and oxidation of combustion products in air and volcanic emissions. These gases are oxidized in the atmosphere to H_2SO_4, a strong mineral acid that dissociates completely, or nearly so, to $2H^+ + SO_4{}^{2-}$. The first dissociation constant (K_1) for H_2SO_4 is very high (2.4×10^6), and the second dissociation constant (K_2) for $HSO_4{}^- = H^+ + SO_4{}^{2-}$ (1×10^{-2}) is high enough that $SO_4{}^{2-}$ is present in higher amounts than $HSO_4{}^-$ (the bisulfate anion) in natural waters.

Focus Box 6.5

The Sulfide Anion vs. the Sulfate Anion: Occurrences and Significance

The S^{2-} anion – the most reduced form of sulfur – and the compounds it forms are generally stable in reducing environments of wetland sediments, low-O_2 crustal environments, and deep marine environments. In addition to occurring in sulfide minerals such as pyrite and galena (Table 6.2), the sulfide anion is present in organic compounds and biological molecules. Any compound containing the sulfide anion is likely unstable in oxidizing conditions (Eq. (6.20)). The other extreme of S oxidation states is S^{6+}, which in most compounds combines with oxygen to form the sulfate anion. $SO_4{}^{2-}$-bearing compounds, including sulfate minerals (e.g. gypsum, jarosite) and sulfuric acid, are most stable in oxidizing environments such as rivers, the atmosphere, shallow ocean water, and aerated soils, but are unstable in reducing environments (Eqs. (6.16) and (6.17)).

Table 6.2 Examples of sulfur species and the oxidation states of sulfur in these compounds, presented from most-reduced, S^{2-}-bearing species to the most-oxidized, sulfate-bearing species.

Species	Name	Oxidation state(s) of sulfur
S^{2-}	Sulfide anion	S^{2-}
PbS	Galena (lead sulfide)	S^{2-}
ZnS	Sphalerite	S^{2-}
$CuFeS_2$	Chalcopyrite	S^{2-}
HgS	Cinnabar	S^{2-}
$(CH_3)_2S$	Dimethyl sulfide (DMS)	S^{2-}
$(CH3)_2SO$	Dimethylsulfoxide (DMSO)	S^{2-}
$(CH3)_2S^+CH_2CH_2COO^-$	Dimethylsulfoniopropionate (DMSP)	S^{2-}
$C_7H_{14}N_2O_2S$	Aldicarb[a]	S^{2-}
H_2S	Hydrogen sulfide	S^{2-}
COS or OCS	Carbonyl sulfide	S^{2-}
CS_2	Carbon disulfide	CS_2
S_2^{2-}	Disulfide anion	S_2^{2-}
FeS_2	Pyrite (iron sulfide)	S_2^{2-}
FeAsS	Arsenopyrite	$(AsS)^{2-}$
S_8	Elemental sulfur	S^0
SO	Sulfur monoxide	S^{2+}
H_2SO_3	Sulfurous acid	S^{4+}
SO_3^{2-}	Sulfite anion	S^{4+}
SO_2	Sulfur dioxide	S^{4+}
SF_6	Sulfur hexafluoride [s]	S^{6+}
SO_3	Sulfur trioxide	S^{6+}
$BaSO_4$	Barite (barium sulfate)	S^{6+}
$CaSO_4 \cdot 2H_2O$	Gypsum (calcium sulfate)	S^{6+}
$KFe_3^{3+}(OH)_6(SO4)_2$	Jarosite	S^{6+}
H_2SO_4	Sulfuric acid	S^{6+}
SO_4^{2-}	Sulfate anion	S^{6+}

[a]These compounds are synthetic.

6.2.4.2 The global S cycle

The dominant reservoir of S (Figure 6.13) is the lithosphere (2×10^{10} TgS), especially pyrite in sediments and in rocks; lesser amounts of S in the lithosphere occur in sulfate minerals (e.g. gypsum) in evaporite deposits. Most of the S in the shallow lithosphere is fixed from seawater into marine sediments as pyrite or other sulfide minerals. Cycling of S from the lithosphere to the next largest reservoir, terrestrial biota and soils, occurs via chemical weathering of sulfides and the associated transformation of S^{2-} to SO_4^{2-}, a form which is soluble and readily available for plant uptake and transport through soils and surface waters.

Volcanism is also responsible for transferring S (e.g. in the forms H_2S, SO_2, and H_2SO_4) from the lithosphere to soils, biota, and the atmosphere. The main natural terrestrial sources of S to the atmosphere are volcanism, dust, and wetland-derived reduced S gases (e.g. H_2S); however, anthropogenic activities currently add three to

five times as much S to the terrestrial atmosphere than do natural processes. Rivers transport dissolved sulfate to the oceans, at which point the sulfate may remain in solution, become fixed as a nutrient by marine biota, or undergo reduction to sulfide in anoxic sediments, which can result in the full-cycle transfer of S back to the lithosphere as pyrite. Within the ocean-oceanic atmosphere system, decaying phytoplankton release dimethyl sulfide (DMS) to the atmosphere, and most of it is rapidly returned to ocean water.

6.2.4.3 The marine S cycle

The starting point for any cycle is somewhat arbitrary, but for sulfur a good starting point is dissolved sulfate in sea water. Sulfate comprises $2712\,mg\,l^{-1}$ of seawater and the main source of this dissolved sulfate is river influx, current estimates of which are 220 teragrams of S per year ($TgS\,yr^{-1}$). At least one half of the current river flux is anthropogenic (Brimblecombe 2003). Atmospheric

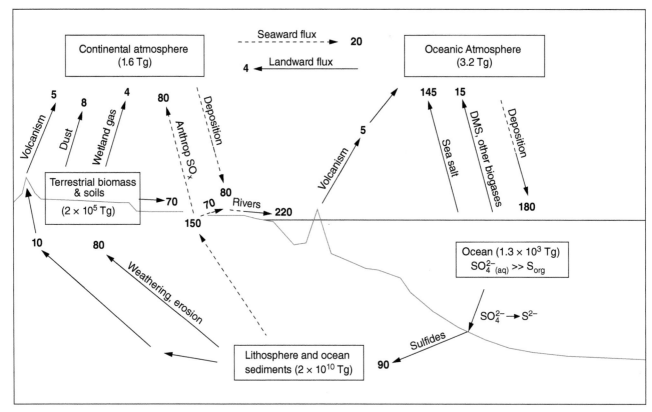

Fig. 6.13 Global sulfur cycle with reservoir masses given in TgS and fluxes in TgS yr^{-1}. Fluxes with strong anthropogenic inputs are shown with dashed lines. Of the 150 Tg of sulfur transferred from the crust (fossil fuels, metal mining and refining, fertilizers) to the Earth's surface, 80 TgS yr^{-1} are emitted to the atmosphere (mainly by fossil fuel combustion) and 70 TgS yr^{-1} are transferred to soils, lakes, rivers, and other terrestrial pools. Source: Data sources are Ivanov (1981), Schlesinger (1997), and Brimblecombe (2003).

deposition onto the oceans also delivers a large amount of sulfur to this reservoir, but most of this is sulfate sea spray blown from sea to air and then back to sea again, so it is not a net source of S to ocean water (Figure 6.13). The oceans also receive a net atmospheric flux of 16 TgS yr^{-1} of anthropogenic sulfur (mainly SO$_x$) carried by prevailing winds from land to sea.

The main net flux of S out of seawater is caused by sulfate reduction (SO$_4{}^{2-}$ to S^{2-}) that ultimately results in formation and burial of sulfide minerals (mainly pyrite or its disordered diagenetic precursor mackinawite). These reactions typically take place at the sediment–sea water interface in chemically reducing conditions in the presence of organic matter (approximated as CH$_2$O), or at hydrothermal vents. In the presence of organic matter and anaerobic microorganisms, reduction of sulfate happens as follows:

$$2CH_2O + SO_4{}^{2-} \rightarrow 2CO_2 + S^{2-} + 2H_2O \qquad (6.16)$$

or

$$2CH_2O + SO_4{}^{2-} + 2H^+ \rightarrow 2H_2CO_2 + H_2S \qquad (6.17)$$

The individual sulfur reduction reaction is:

$$S^{6+} + 8e^- \rightarrow S^{2-}$$

and the paired oxidation reaction (of carbon) is

$$C^0 \rightarrow C^{4+} + 4e^-$$

The reactions are balanced in terms of electrons by the fact that two moles of carbon undergo oxidation. Anaerobic bacteria of the genus *Desulfotomaculum* and genus *Desulfovibrio* use sulfur as an electron acceptor, allowing the organisms to obtain energy while also enabling them to act as catalysts for sulfate reduction.

Marine sediments that contain sufficient amounts of Fe^{2+} (i.e. most marine sediments) foster formation of disordered iron monosulfide as a metastable phase before recrystallizing into pyrite in the presence of hydrogen sulfide (Hurtgen et al. 1999):

$$Fe^{2+} + S^{2-} \rightarrow FeS \qquad (6.18)$$

then

$$FeS + H_2S \rightarrow FeS_2 + 2H^+ + 2e^- \qquad (6.19)$$

Formation of pyrite and hydrogen sulfide in anoxic marine sediments is shown schematically in Figure 10.14 (Chapter 10); fractionation of sulfur isotopes takes place during this process – isotopically heavy sulfur (^{34}S) is partitioned into pyrite in sediments whereas isotopically light sulfur (^{32}S) is partitioned into hydrogen sulfide, some of which migrates out of sea water and into the atmosphere. Some H_2S may also participate in the FeS to FeS_2 transformation. Pyrite exerts a strong control on cycling of trace metals such as Cu, Pb, and Zn because these metals can substitute for the Fe^{2+} atom; arsenic can substitute for sulfur atoms in pyrite, in some cases reaching thousands of $mg\ kg^{-1}$ As in the pyrite (Smedley and Kinniburgh 2002). When As and S are in equal abundance, the mineral is termed arsenopyrite.

If pyrite that forms in marine sediments (or hydrothermal vents) is buried and compressed into rock, the sulfur is removed from sea water and is sequestered into the crustal/lithospheric S reservoir.

Biological fixation of dissolved sulfate by phytoplankton (e.g. *Dinophyceae* and *Prymnesiophyceae*) is another important component of marine S cycling. The S^{6+} in dissolved sulfate is reduced to S^{2-} and fixed into dimethylsulfoniopropionate (DMSP), a compound that regulates osmosis and other biological functions. When higher trophic-level organisms (e.g. zooplankton) consume phytoplankton, or if they die and undergo decay, DMSP decomposes to DMS and methanethiol (CH_3SH). Of these two S-bearing degradation products, methanethiol rapidly disappears, perhaps incorporated into organic matter or degraded by oxidation; most DMS is microbially degraded or fixed into biota but a small amount seeps into the atmosphere where it comprises an important part of marine atmospheric chemistry. The flux of $15\ TgS\ yr^{-1}$ to the atmosphere is the largest single *natural* source of S to air. For reference, current (early twenty-first century) emissions of SO_x from fossil fuel burning and refining and sulfide ore metal refining (and atmospheric oxidation products) are approximately five times greater than the DMS flux.

Carbonyl sulfide and hydrogen sulfide are other trace biogases emitted to the atmosphere from the oceans, but their contributions are at least an order of magnitude less than DMS (Schlesinger 1997). DMS is important in the oceanic atmosphere because (i) in air it is oxidized to sulfate aerosols (in a multistage reaction series) which act as cloud condensation nuclei, and (ii) sulfate aerosols formed via oxidation of DMS also act to scatter incoming solar radiation, thus contributing a cooling force to the climate system. The residence time of DMS in the atmosphere is ~2 days, so the majority of DMS flux to the oceanic atmosphere is returned quickly to the oceans, commonly as oxidation products that are redeposited into the ocean or coastal regions in precipitation (~rain).

6.2.4.4 Sulfur, soils, and biota

Chemical weathering of pyrite-bearing rocks following uplift and erosion introduces sulfur to soils and biota, a process summarized by this weathering reaction (and shown in different formulations in chapters 1, 2, 4 and 9:

$$2FeS_2 + 7O_2 + 2H_2O \rightarrow 2Fe^{2+} + 4SO_4{}^{2-} + 4H^+ \quad (6.20)$$

Note that oxidation of sulfur, not ferrous iron, drives this weathering reaction. The bacteria *Thiobacillus thiooxidans* and *Thiobacillus ferrooxidans* accelerate weathering rates of pyrite but often their effects are most strongly pronounced in acid mine drainage situations. The -2 charge on sulfate ($SO_4{}^{2-}$) indicates that it tends to be more strongly sorbed to anion exchange sites (particularly sites associated with Fe hydroxides or humus) than are monovalent anions like Cl^- and $NO_3{}^-$, but less so than more strongly sorbed phosphate anions. This indicates that sulfate is moderately mobile in soils and available for leaching to surface waters or groundwaters, or for plant uptake.

In most soils, sulfur eventually is incorporated into biota or is bound to humus and other forms of particulate and dissolved organic matter. When incorporated into plants, most sulfate undergoes assimilatory reduction to S^{2-} so that the sulfur can be bound to carbon in, for example, the amino acids cysteine and methionine. Decomposition of organic matter in soils commonly results in sequestration of reduced S in humus or, in some cases, can result in oxidation of sulfide to sulfate. Much sulfur is cycled back and forth between reduced sulfur in organic matter in soil, sulfate in the soil solution, and reduced sulfur in plant tissue.

Transfer of sulfur from soils and terrestrial biota to the atmosphere occurs by two main mechanisms. The first is simply by physical erosion of particles into air by wind. While the annual flux of S from soils and terrestrial biota is relatively small at $8\ TgS\ yr^{-1}$, it is important climatically because sulfate particles act as cloud condensation nuclei and function as scatterers of incoming solar radiation (i.e. exert a cooling effect). Sulfur is also transferred from terrestrial ecosystems to atmosphere in the form of reduced gases such as hydrogen sulfide, carbonyl sulfide, and DMS which are produced in, and emanate from, wetland sediments and waterlogged soils.

6.2.4.5 Sulfur and the atmosphere

Due to short residence times of sulfur species in the lower troposphere, most models of the global S cycle treat the terrestrial and oceanic atmospheres as separate reservoirs with relatively small fluxes between the two. Flux from the oceanic atmosphere to the terrestrial atmosphere is 4 TgS yr^{-1}, and 20 TgS yr^{-1} migrate in the opposite direction; most of this flux is anthropogenic S derived from fossil fuel combustion and refining.

From Figure 6.13 it is clear that the greatest flux associated with the terrestrial atmosphere is addition of SO$_x$ to the atmosphere from *anthropogenic emissions* – approximately 80% of sulfur emissions to the terrestrial atmosphere are from anthropogenic activities, including:

1 coal combustion (typically S comprises 1–3% of coal);
2 petroleum refining and combustion; and
3 refining of metal sulfide ores.

Increased S emissions have negative consequences for human health (respiratory tract issues like asthma from SO$_2$) and ecosystems, especially acid rain in the form of sulfuric acid via transformation of precursor SO$_x$ and other oxidizable sulfur gases. These topics are covered in in Chapters 8 and 9.

After anthropogenic emissions, the next most abundant net source of S to the atmosphere is volcanic eruptions. Prior to the industrial era, volcanic gases were the main sources of S to the terrestrial troposphere, but now this natural source accounts for less than 10% of total S flux to the terrestrial atmosphere. Volcanic gases include a wide variety of SO$_x$ phases, hydrogen sulfide, and less commonly, carbonyl sulfide and carbon disulfide. These gases generally oxidize to sulfate and dissolve in precipitation, although very strong (Plinean) volcanic eruptions can eject sulfates to the stratosphere, where they can persist for two to three years, exerting a short-term cooling effect on mean global climate – the eruption of Mount Pinatubo in the Philippines in 1991 and the resulting three years of lowered mean global temperature is a good example of this effect.

Anthropogenic activities affect the oceanic atmosphere less than the terrestrial atmosphere. The majority of SO$_x$ pollution is deposited on land, much as H$_2$SO$_4$ acid rain. Within the oceanic atmosphere–sea water system, most biologically derived DMS is rapidly cycled from the atmosphere back into sea water, so little DMS reaches land. Given that DMS production is a function of marine primary productivity (i.e. phytoplankton abundance), and that warming oceans may alter primary productivity, there are concerns about future production rates of DMS.

6.2.4.6 Sulfur and river flux

The greatest flux in the modern global sulfur cycle is the fluvial transport of dissolved and particulate S from land to the oceans, and at least 50% of current river S flux is anthropogenic – preindustrial riverine S flux was likely ~100 TgS yr^{-1} (Brimblecombe 2003), compared to current global flux of ~220 TgS yr^{-1}. Natural sources of sulfate in rivers include deposition of volcanic gases and wetland-derived gases plus sulfate derived from pyrite weathering and various organic forms of S including humus, DMS, and more. The main anthropogenic species in river water are SO$_x$ and H$_2$SO$_4$ derived from fossil fuel combustion, mining and processing of sulfide ores, and to a lesser extent, organic and inorganic forms of S derived from fertilizers. The S compounds are delivered to streams in many ways, including overland flow, throughflow via soils and groundwater, erosion of stream banks and soils, and direct inputs of SO$_4{}^{2-}$ from acid mine drainage sites.

In summary, the greatest disturbances to the global sulfur cycle are related to fossil fuel combustion as well

---

Focus Box 6.6

Potential Consequences of Global Warming on DMS and Sulfate Production

Possible links between ocean warming and DMS are related to projected changes in ocean dynamics; in one scenario, warmer waters would lead to greater ocean stratification, decreased upwelling, and lower nutrient availability. This would lead to a reduction in marine net primary productivity, which ultimately would yield a decrease in the DMS production in the ocean and less DMS in air. And, given that DMS oxidizes to sulfate particles that scatter incoming solar radiation and foster formation of cloud condensation nuclei, less DMS would mean less reflection of incoming radiation and greater warming, potentially initiating the feedback loop: more warming → less DMS → more warming → less DMS etc. This is the CLAW hypothesis, named after the four authors who researched the topic (Charlson et al. 1987). Modeling by Cameron-Smith et al. (2011) indicates that a warming ocean will cause shifts in DMS productivity, with decreasing production in the tropical ocean and increasing production poleward.

as mining and processing of metal sulfide ores, and the greatest impacts have been to forested ecosystems and lakes affected by sulfuric acid deposition, a topic covered in Chapters 8 and 9. Concerns about the fate of DMS in the oceanic atmosphere rise from projections of warming oceans and changing oceanic dynamics. On the positive side, anthropogenic sulfur emissions are predicted to decrease into the twenty-first century (Figure 6.14) as the result of improved technologies and greater social, economic, and political will.

6.2.5 Integrating the C, N, P, and S cycles

Understanding the overlaps in cycling and behavior of C, N, P, and S is in important aspect of comprehending and predicting geochemical processes at and near the Earth's surface. Outlined below are some of the main concepts and mechanisms that connect these four critical elements:

1 C, N, P, and S are essential elements in living organisms.

2 All four elements are cycled abundantly between biota and organic and inorganic forms.

3 While all four are important components of terrestrial and marine ecosystems, only C, N, and S have significant atmospheric fluxes. Very little P is cycled through the atmosphere.

4 P is relatively immobile compared to most species of C, N, and S, and as a result it should be easier to control its movement from soils to aquatic systems.

5 All four elements play roles in global climate, from the obvious effects of infrared-active CO_2, CH_4, and N_2O in trapping heat, to solar radiation-scattering SO_4^{2-} in exerting a cooling influence, to the less-obvious role that P plays in controlling primary productivity and hence uptake of CO_2; the topic of global climate is covered in Chapter 7.

6 N and P are limiting nutrients that often control primary productivity in terrestrial and aquatic ecosystems, and hence affect the cycling of C (through controls on photosynthesis) and S (via plant uptake) as well.

7 Formation of carbonic acid in the atmosphere exerts a strong control on weathering rates and release of P and S to soil solution; formation of sulfuric acid (anthropogenic rate ≫ natural rate) can enhance cycling of all elements by speeding up chemical weathering and oxidation of organic matter.

8 Various species of C, N, and S are released to air during combustion of fossil fuels, so this one anthropogenic activity greatly affects the cycles of these three elements.

9 Improving environmental quality of terrestrial and aquatic ecosystems and the atmosphere will require better management of all four of these critical elements.

Other important biogeochemical cycles include those of O, H (sometimes cycling of CHONPS are considered together), Ca, As, Pb, Hg, various pesticides, fuels, and solvents, and many more. While not covered in this chapter, pertinent concepts related to the behavior and cycling of

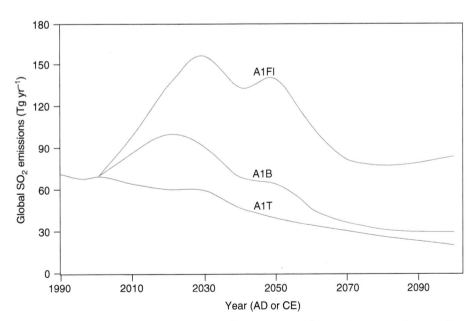

Fig. 6.14 Three scenarios for SO_2 emissions into the twenty-first century. Each scenario assumes increased population growth to 9 billion by 2050. The difference in the scenarios is that A1FI assumes continued intensive fossil fuel combustion, especially coal; A1B assumes a balance of fuel sources; and A1T assumes nearly complete introduction of non-fossil fuel energy sources. Source: IPCC Special Report on Emissions Scenarios (Nakicenovic and Swart 2000).

other elements and compounds are presented in various sections. A quick glance at the index at the end of the book will direct the reader to specific sections on these components, including their presence and speciation in minerals, behavior in weathering environments, their occurrence in organic compounds (synthesized or natural), their roles in atmospheric and aqueous chemistry, and more.

QUESTIONS

6.1 Describe the difference between steady state and dynamic equilibrium and provide an example of each.

6.2 What is the most-reduced oxidation state of N? Name a compound in which it occurs. Similarly, indicate which is the most-oxidized state of N, and name a compound in which it occurs. Describe why microbes exert such a strong control on N transformations and use a chemical equation to illustrate your point.

6.3 What is the most-reduced oxidation state of S? Name a compound in which it occurs. Similarly, indicate which is the most-oxidized state of S, and name a compound in which it occurs.

6.4 What are the main ways by which humans have altered the S cycle? In particular, indicate the reservoirs from which humans are extracting S, the chemical reactions responsible for its transfer to other reservoirs, and two environmental problems resulting from altered S cycling.

6.5 Compare the solubility of nitrate and phosphate in soil. Values of the K_{sp} for pertinent species at $25\,^{\circ}C$ include: $AlPO_4 = 6.3 \times 10^{-19}$, $CaHPO_4 = 1 \times 10^{-7}$, $FePO_4 = 1.3 \times 10^{-22}$, and, while not commonly determined for soluble salts, the K_{sp} for KNO_3 is $\sim 10^1$. Write the chemical equation associated with the K_{sp} for each of these four reactions. Given these data, in what form(s) is P most likely to occur and in what form(s) is N most likely to occur?

6.6 How do microbes facilitate nitrification and denitrification?

6.7 Create a box model to schematic show the relationship of marine primary productivity and dimethylsulfide formation to sulfate aerosol production in air above seawater. Explain the potential climate significance of decreased marine productivity of DMS.

6.8 Given that both P and N can cause eutrophication, indicate by what process(es) N and P commonly enter aquatic bodies. Show these pathways with a schematic sketch.

6.9 Humans are responsible for a greater amount of reactive nitrogen in forests, lakes, farms, and streams than is produced by natural processes – what are the main origins of reactive nitrogen (i.e. how is it fixed, both naturally and anthropogenically, into soils and aquatic bodies), and what are its main forms?

REFERENCES

Brimblecombe, P. (2003). The Global S Cycle. In: *Biogeochemistry, Treatise on Geochemistry* (ed. W.H. Schlesinger), 645–682. Amsterdam, The Netherlands: Elsevier.

Cameron-Smith, P., Elliott, S., Maltrud, M. et al. (2011). Changes in dimethyl sulfide oceanic distribution due to climate change. *Geophysical Research Letters* 38: L07704: 5. https://doi.org/10.1029/2011GL047069.

Carpenter, S.P. (2008). Phosphorus control is critical to mitigating eutrophication. *Proceedings of the National Academy of Sciences of the United States of America* 105: 11039–11040.

Charlson, R.J., Lovelock, J.E., Andreae, M.O., and Warren, S.G. (1987). Oceanic phytoplankton, atmospheric sulfur, cloud albedo and climate. *Nature* 326: 655–661.

Dillon, P.J. and Rigler, F.H. (1974). The phosphorus-chlorophyll relationship in lakes. *Limnology and Oceanography* 19: 767–773.

Einsele, W. (1936). Über die Beziehungen des Eisenkreislaufs zum Phosphorkreislauf im eutrophen See. *Archiv für Hydrobiologie* 29: 664–686.

Fowler, D., Coyle, M., Skiba, U. et al. (2013). The global nitrogen cycle in the twenty-first century. *Philosophical Transactions of the Royal Society of London, Series B: Biological Sciences* 368: https://doi.org/10.1098/rstb.2013.0164.

Galloway, J.N., Dentener, F.J., Capone, D.G. et al. (2004). Nitrogen cycles: past, present and future. *Biogeochemistry* 70: 153–226.

Graham, J.L., Loftin, K.A., Meyer, M.T., and Ziegler, A.C. (2010). Cyanotoxin mixtures and taste-and-odor compounds in cyanobacterial blooms from the Midwestern United States. *Environmental Science and Technology* https://doi.org/10.1021/es1008938.

Holland, E.A., Lee-Taylor, J., Nevison, C., and Sulzman, J. (2005). *Global N Cycle: Fluxes and N₂O Mixing Ratios Originating from Human Activity. Data set.* Available online [http://www.daac.ornl.gov] from Oak Ridge National Laboratory Distributed Active Archive Center, Oak Ridge, Tennessee, U.S.A. doi:https://doi.org/10.3334/ORNLDAAC/797.

Howarth, R.W., Jensen, H.S., Marino, R., and Postma, H. (1995). Transport to and processing of P in near-shore and oceanic waters. In: *Phosphorus in the Global Environment* (ed. H. Tiessen), 323–245. New York: Wiley.

Hurtgen, M.T., Lyons, T.W., Ingalls, E.D., and Cruse, A.M. (1999). Anomalous enrichments of iron monosulfide in euxinic marine sediments and the role of H₂S in iron sulfide transformations: examples from Effingham Inlet, Orca

Basin, and the Black Sea. *American Journal of Science* 299: 556–588.

Ivanov, M.V. (1981). The global biogeochemical sulphur cycle. In: *Some Perspectives of the Major Biogeochemical Cycles* (ed. G.E. Likens), 61–78. SCOPE Report 17. New York: Wiley.

MacFarling Meure, C., Etheridge, D., Trudinger, C. et al. (2006). The law dome CO_2, CH_4 and N_2O ice core records extended to 2000 years BP. *Geophysical Research Letters* 33 (14): L14810. https://doi.org/10.1029/2006GL026152.

Michel, F.M., Ehm, L., Antao, S.M. et al. (2007). The structure of ferrihydrite, a nanocrystalline material. *Science* 316: 1726–1729. https://doi.org/10.1126/science.1142525.

Mortimer, C.H. (1941). The exchange of dissolved substances between mud and water in lakes. *Journal of Ecology* 29: 280–329.

Nakicenovic, N. and Swart, R. (eds.) (2000). *Special Report on Emissions Scenarios: A Special Report of Working Group III of the Intergovernmental Panel on Climate Change*, 612. Cambridge, UK: Cambridge University Press. Available online at: http://www.grida.no/publications/other/ipcc_sr/?src=/climate/ipcc/emission.

Paytan, A. and McLaughlin, K. (2007). The oceanic phosphorus cycle. *Chemical Reviews* 107 (2): 563–576. https://doi.org/10.1021/cr0503613.

Portage County Groundwater Conditions (2000). https://www.co.portage.wi.us/Groundwater/undrstnd/pcstate2.htm. Accessed 13 June 2011.

Riskin, M.L., Deacon, J.R., Liebman, M.L. and Robinson, K.W. (2003). Nutrient and Chlorophyll Relations in Selected Streams of the New England Coastal Basins in Massachusetts and New Hampshire, June–September 2001. *U.S. Geological Survey Water-Resources Investigations Report* 03-4191, 15 pp.

Ruttenberg, K.C. (2003). The global phosphorus cycle. In: *Biogeochemistry, Treatise on Geochemistry* (ed. W.H. Schlesinger), 585–643. Amsterdam, the Netherlands: Elsevier.

Sakamoto, M. (1966). Primary production by the phytoplankton community in some Japanese lakes and its dependence on lake depth. *Archiv fuer Hydrobiologie* 62: 1–28.

Schindler, D.W., Hecky, R.W., Findlay, D.L. et al. (2008). Eutrophication of lakes cannot be controlled by reducing nitrogen input: results of a 37 year whole ecosystem experiment. *Proceedings of the National Academy of Science USA* 105: 11254–11258.

Schlesinger, W.H. (1997). *Biogeochemistry: An Analysis of Global Change*, 2e. San Diego, USA: Academic Press.

Schlesinger, W.H. (2009). On the fate of anthropogenic nitrogen. *Proceedings of the National Academy of Sciences of the United States of America* 106 (1): 203–208.

Smedley, P.L. and Kinniburgh, D.G. (2002). A review of the source, behavior and distribution of arsenic in natural waters. *Applied Geochemistry* 17: 517–568.

Vitousek, P.M., Aber, J., Howarth, R.W. et al. (1997). *Human Alteration of the Global Nitrogen Cycle: Causes and Consequences. Issues in Ecology* No. 1, 17. Washington, DC: Ecological Society of America.

Yang, S. and Gruber, N. (2016). The anthropogenic perturbation of the marine nitrogen cycle by atmospheric deposition: nitrogen cycle feedbacks and the ^{15}N Haber-Bosch effect. *Global Biogeochemical Cycles* https://doi.org/10.1002/2016GB005421.

Young, E.O. and Ross, D.S. (2001). Phosphate release from seasonally flooded soils: a laboratory microcosm study. *Journal of Environmental Quality* 30: 91–101.

7

The Global Atmosphere: Composition, Evolution, and Anthropogenic Change

This chapter is designed to cover background required for understanding the natural structure and composition of the global atmosphere, including how the atmosphere has varied in composition over time from billion-year and million-year time scales across geological time, to more recent changes that have occurred in the Pleistocene and Holocene epochs. Chapter 7 also presents some of the fundamental concepts that help us to predict and understand atmospheric circulation, which are applicable to contaminant transport as well as local and regional climate. The chapter concludes by presenting data and concepts related to greenhouse gases (GHGs), including molecular-level properties that impart lead to infrared (IR) trapping capability, as well as the role of GHGs in global climate, from preindustrial time to the twenty-first century.

7.1 ATMOSPHERIC STRUCTURE, CIRCULATION, AND COMPOSITION

The atmosphere differs from the other main components of the Earth system in at least two important ways:

1 It is the smallest reservoir at or near the Earth's surface in terms of mass, containing only 5×10^9 Tg of matter as compared to the hydrosphere, which contains 2.4×10^{12} Tg of matter (H_2O), and the crust of the Earth, which contains 2.4×10^{13} Tg of matter (rock). One important result of its small size is that the atmosphere can be very sensitive to small amounts of contamination.

2 The atmosphere has a high mixing rate relative to the hydrosphere, soils, sediments, and bedrock. This means that (i) contaminants can rapidly spread throughout the

atmosphere, and (ii) the atmosphere has a more homogeneous distribution of constituents than the hydrosphere or the lithosphere. Point-source contaminants in the hydrosphere are often confined to a particular zone within an aquifer or to a specific stream, whereas point-source contaminants in the atmosphere spread widely in time frames as short as a few days. Illustrative examples are sulfuric acid and mercury derived from coal-fired power plants – these contaminants are deposited into terrestrial and aquatic ecosystems hundreds of kilometers downwind of their sources. Radioactive fallout from the 1986 Chernobyl disaster in the Ukraine (part of the USSR at the time) spread to much of northern Europe within hours to days, and sulfate particles derived from very strong Plinian volcanic eruptions can reach the stratosphere and be dispersed globally.

7.1.1 Structure and layering of the atmosphere

The atmosphere consists of distinct layers characterized by differences in air pressure, temperature, and composition. Changes in atmospheric pressure as a function of altitude are explained by this derivation of Dalton's law of partial pressures:

$$p_z = p_0 \times e^{(-z/H)} \qquad (7.1)$$

where p_z is atmospheric pressure at altitude z, p_0 is atmospheric pressure at sea level, and H is a term known as the "scale height," a measure of the effective thickness of an atmospheric layer which is expressed mathematically as $H = k \times T/(m \times g)$ (where k is Boltzmann's constant,

Environmental and Low-Temperature Geochemistry, Second Edition. Peter Ryan.
© 2020 John Wiley & Sons Ltd. Published 2020 by John Wiley & Sons Ltd.

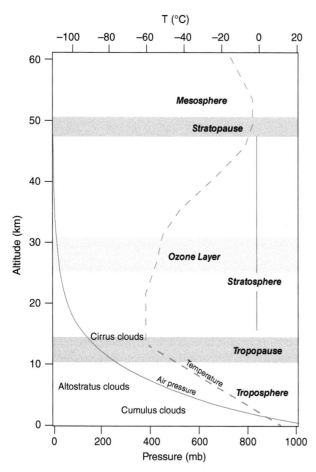

Fig. 7.1 Structure of the atmosphere to an altitude of 60 km. Exponentially decreasing air pressure is indicated by the solid line and air temperature is indicated by the dotted line. Not shown on this figure are the upper mesosphere, the mesopause (temperature minimum at ~80 to 100 km), thermosphere (~100 to 600 km altitude), and exosphere. The altitude shown for the tropopause is representative of low- to mid-latitudes – see text for explanation.

T is temperature in Kelvins, m is the molecular mass of the parcel of atmosphere being considered, and g is acceleration due to gravity).

Dalton's law of partial pressures dictates that increasing z causes an exponential drop in atmospheric pressure (Figure 7.1). Then, from the ideal gas law

$$pv = nRT \qquad (7.2)$$

it is evident that the pressure of a gas (or mixture of gases) is proportional to the number of gas molecules (assuming constant volume and temperature). This becomes apparent at the elevation of high mountainous regions (~6 to 9 km), where atmospheric pressure is $\leq 50\%$ of its value at sea level (1 atm = 1013.25 mb), and thus the amount of available oxygen to breathe is $\sim\leq 50\%$ less than at sea level.

For this reason, hikers become easily exhausted at high elevations and climbers of Mt. Everest require oxygen tanks. At an altitude of 50 km, air pressure is only 1 millibar, or 0.1% of air pressure at sea level.

Temperature, like air pressure, also decreases markedly with increasing altitude, but only to an altitude of ~15 km, at which point absorption of ultraviolet (UV) light by O_2 molecules produces heat and causes a *temperature inversion* (Figure 7.1). The lower part of the atmosphere below the temperature inversion is the *troposphere* – given that the tallest mountains reach elevations of ~8.5 to 9 km, and that transoceanic flights travel at altitudes of ~11 km, nearly all human activities occur within the troposphere. The height of the troposphere varies as a function of latitude. At the equator, the top of the troposphere occurs approximately 18–20 km above sea level; at 45° latitude, the troposphere extends to an altitude of 8–9 km, and at the poles its extent is only 6–7 km. Weather-related phenomena and nearly all clouds occur within the troposphere because the inversion at the top of the troposphere (warm stratospheric air resting atop cold upper tropospheric air) prevents air from rising through the *tropopause*, the boundary between the troposphere and the overlying *stratosphere* (Figure 7.1).

Considering atmospheric chemistry, the most important part of the stratosphere is the *ozone layer*, where incoming UV radiation causes dissociation of O_2 molecules and formation of O_3 (ozone) molecules (Section 8.9). Above the stratosphere are the *mesosphere, thermosphere,* and *exosphere* (not shown on Figure 7.1), layers which combined contain < 0.1% of the mass of the atmosphere. These layers of the upper atmosphere are characterized by intense UV bombardment and are richer in lighter gases such as H and He. Details of atmospheric chemistry in layers above the stratosphere are not considered other than in Chapter 11 when considering formation of cosmogenic radionuclides.

7.1.2 Geological record of atmospheric composition

Earth's earliest atmosphere formed by degassing of volatile compounds (e.g. H_2O, N_2, CO_2) from the mantle and crust, and from volatile compounds contained in meteorites, comets, and other extraterrestrial impact bodies that collided with Earth early in its history (mainly before 4 Ga). By 3.8 Ga the atmosphere was likely dominated by CO_2 and H_2O with additional constituents such as SO_2, HCl, Cl_2, N_2, H_2, CH_4 (methane), and NH_3 (ammonia). This early atmosphere likely had little or no O_2, an idea that is supported by the presence of pyrite cobbles in Archean stream deposits. Pyrite is unstable in the presence of O_2 gas, so these deposits are evidence for the O_2-poor character of the early atmosphere, and any O_2 in the early atmosphere would have only existed due to photochemical dissociation of small amounts of H_2O into O_2. However, by ~3.5 Ga, photosynthetic bacteria had appeared in the oceans, an evolutionary advance that would eventually result in profound changes to the composition of Earth's atmosphere.

Photosynthesis by cyanobacteria and other early prokaryotes converted CO_2 and H_2O into organic matter and oxygen, and from this point forward the atmosphere began to shift toward one of greater O_2 content at the expense of CO_2 (Chapter 3, Eq. (3.2)):

$$n\,CO_2 + n\,H_2O + E = C_nH_{2n}O_n + n\,O_2$$

Atmospheric O_2 content has likely varied between 15% and 30% during the ~550 Ma of Phanerozoic time (Figure 7.2) (Berner 1997), controlled by changes in photosynthesis related to evolution of new life forms (particularly the appearance of vascular plants at ~375 Ma) as well as other aspects of the carbon cycle and the sulfur cycle. For example, the bacterially mediated reduction of sulfate, as illustrated in reaction (7.3), is an additional means of producing atmospheric O_2. In reverse (i.e. right

to left), this would depict reaction of O_2 gas with pyrite, an example of an oxidation reaction that would remove O_2 from the atmosphere.

$$2Fe_2O_3 + 16Ca^{2+} + 16HCO_3^- + 8SO_4^{2-}$$
$$\leftrightarrow 4FeS_2 + 15O_2 + 16CaCO_3 + 8H_2O \qquad (7.3)$$

Relative to O_2, CO_2 has undergone more profound changes in atmospheric abundance during the Phanerozoic, particularly between 400 and 300 Ma (middle Devonian to middle Carboniferous time), a period during which evolution and appearance of land plants provided a major sink for CO_2. Since ~350 Ma, CO_2 and O_2 have broadly fluctuated in an inverse manner, likely reflecting the influence of photosynthesizing organisms on atmospheric composition (e.g. Lovelock and Margulis 1974).

The Quaternary (Pleistocene and Holocene) record of atmospheric composition is a useful framework because it is the time period in which humans have evolved and flourished. The onset of the Pleistocene coincides with cooling climate and expanding ice sheets. Over the past 2 Ma, humans and their predecessors have witnessed numerous cyclic periods of glacial advance and retreat. Atmospheric CO_2 has decreased through most of the Cenozoic (Figure 7.2), which is one of the main forcings responsible for Pleistocene ice ages.

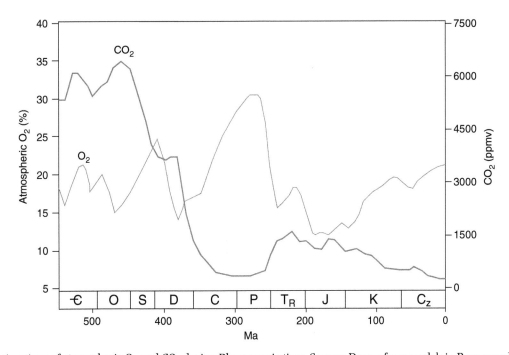

Fig. 7.2 Estimations of atmospheric O_2 and CO_2 during Phanerozoic time. Source: Drawn from models in Berner and Berner (1996) and Berner (2006). Uncertainty in CO_2 values is ~± 30% of plotted value, meaning that at the Cambrian–Ordovician boundary, CO_2 was ~6000 ± 1800 ppmv. From left to right and oldest to youngest, geological periods are Cambrian (Є), Ordovician (O), Silurian (S), Devonian (D), Carboniferous (C), Permian (P), Triassic (T_R), Jurassic (J), Cretaceous (K), and Cenozoic (C_z).

Focus Box 7.1

The Paleocene–Eocene Thermal Maximum (55 Ma)

Temperature of the global ocean increased 5–9 °C during a period of rapid global warming (over a period of 10 000–20 000 years) at the boundary of the Paleocene and Eocene epochs (55 Ma). (Although rapid in geologic time, this rate is still less than one-tenth of current warming rate.) Evidence from the geologic record includes a pronounced negative shift of −4 to −5 per mil in carbon isotope compositions of marine sediments, and similar shifts in the oxygen isotope composition of benthic and planktonic foraminifera. The oxygen isotope record informs temperature estimates, and the carbon isotope shift is consistent with flux of a very large amount (2000 Pg or 2000 Gt) of isotopically light carbon to the atmosphere (refer to Chapter 10 for stable isotopes). Carbon in methane hydrates (Section 5.2.5.4) is isotopically very light (negative) relative to carbon in other reservoirs, and many studies attribute the rapid warming to release of methane to the air from marine sediments. Global temperature had been rising slowly at a steady rate, and at 55 Ma may have crossed a threshold where slowly warming ocean sediments began to release methane that then set into effect pronounced feedback loops. Conversion of CH_4 to CO_2 in the atmosphere and solution of CO_2 into seawater then led to an extended period of ocean acidification – evidence of which is preserved in decreased $CaCO_3$ content of marine sediments – that coincides with mass extinctions of benthic foraminifera (e.g. Zachos et al. 2005).

7.1.3 Climate proxies

Climate scientists make use of many climate proxies to understand interactions between the composition of the atmosphere and average global temperature, and the most common ones are from ice cores. The usefulness of ice cores lies in the fact that they typically contain annual layers of snow which, over time, are compressed into ice. Bubbles of atmosphere become trapped in the snow and ice, allowing paleoclimatologists to directly measure paleoatmospheric composition. Variations in stable isotope compositions of hydrogen (δD) and oxygen ($\delta^{18}O$) in H_2O molecules that originally fell as snow provide a mechanism to determine paleotemperature (details are presented in Chapter 10) – these isotopic values are controlled by global temperature and local temperature, where warmer oceans and air foster heavier (or less negative) values during warm periods. Dating of ice layers is possible because at shallow depth within the ice, annual layers of snow and ice can be counted backwards in time to determine the age of a given layer; deeper in the core, where individual layers are no longer distinguishable due to compaction, volcanic ash deposits and carbon from large forest fires are common means for determining age.

Well-known ice core records of Pleistocene and Holocene paleotemperature and atmospheric gas concentrations are the Vostok core (Petit et al. 1999) and the EPICA (European Project for Ice Coring in Antarctica) dome core (Augustin et al. 2004), both from Antarctica, and the GISP (Greenland Ice Sheet Project) and GRIP (Greenland Ice Core Project) cores from Greenland (e.g. Steffensen et al. 2008 and references therein). Data from the Vostok core are presented in Figure 7.3 – this core was chosen to illustrate relationships among temperature (and, by definition, the stable isotope signatures δD and $\delta^{18}O$), CO_2, CH_4, and insolation because the Vostok record extends farther back in time than do the GISP/GRIP records, and because pre-420 ka data from the EPICA dome record are of relatively poor resolution compared to pre-420 ka data from Vostok and EPICA.

Atmospheric CO_2, CH_4, and temperature have shown a very strong coherence for many hundreds of thousands of years (Figure 7.3), where cooler temperatures correspond to relatively low amounts of carbon dioxide and methane in air, and warmer periods are correlated to higher amounts of these GHGs. The strong relationships among CO_2, CH_4, and temperature are best understood in the framework of one of the main feedback loops operating within the climate system.

7.1.4 Orbital control on C

Variations in the tilt of Earth's axis of rotation (obliquity), in the orientation of the rotation axis of Earth (precession), and in the eccentricity of Earth's orbit combine to produce periods when mean insolation (incident solar radation) is greater than average. During these times, increased insolation produces increased average global temperature, which in turn causes changes to the carbon cycle; one consequence is warming of the oceans which results in a net flux of CO_2 out of the oceans. Warming also causes melting of ice caps which, in combination with expansion of warming ocean water, raises sea level. This

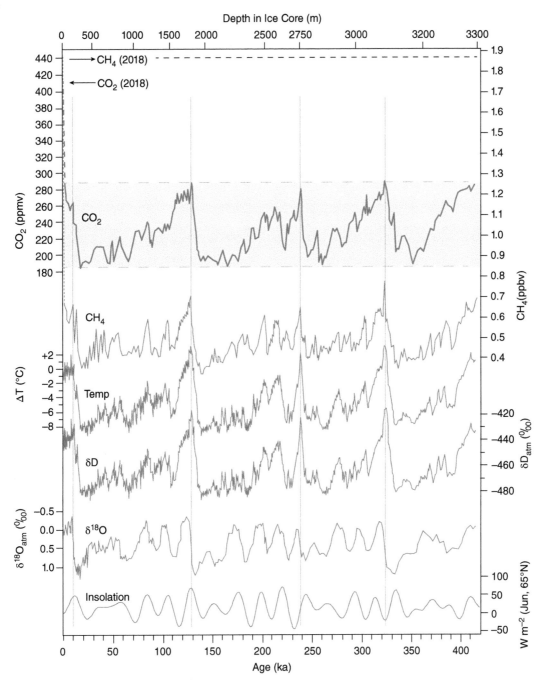

Fig. 7.3   Ice core record from Vostok, Antarctica, of atmospheric CO_2 and CH_4 concentrations (measured from bubbles trapped in ice) as well as changes in atmospheric temperature derived from D/H ratios (expressed as δD) of H_2O molecules, over the past 420 000 years (420 ka). Oxygen isotope data are also shown, as is insolation for the month of June at 65°S latitude. Note that the preindustrial range of CO_2 concentrations was ~180 to 290 ppmv, and that CH_4 ranged from ~0.4 to 0.8 ppbv. In June 2018, CO_2 was 411 ppmv and CH_4 was 1.85 ppmv (https://www.esrl.noaa.gov/gmd/ccgg/trends_ch4). Values for CO_2 and CH_4 concentrations above the shaded area represent values not seen in the ice cores (shaded box shows CO_2 range over past 420 ka). Source: Vostok data are from Petit et al. (1999) and references cited therein.

in turn produces a greater extent of methane-producing coastal swamps; these factors, and many others, combine to increase atmospheric concentrations of CO_2 and CH_4, which trap more heat in the atmosphere, causing further warming, which releases more CO_2 from oceans, and so on. This positive feedback loop causes rapid warming during transitions from ice ages to interglacial periods, a feature that is noticeable in Figure 7.3.

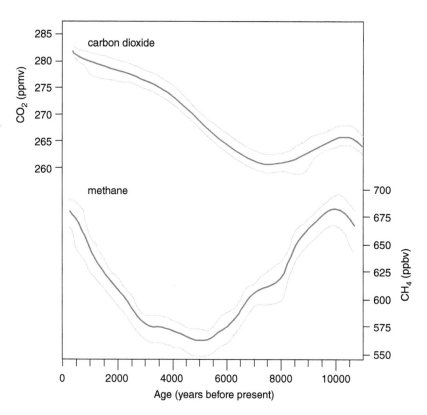

Fig. 7.4 Trends for carbon dioxide and methane during the Holocene epoch (plotted here as ~11 400 to 400 years before present) derived from air bubbles in glacial ice and firn (from the EPICA Dome C project). Dark blue curves are fitted trends and the light blue lines depict estimated uncertainty. Source: Data were obtained from the USA NOAA National Climate Data Center (http://www.ncdc.noaa .gov/paleo/data.html) and were presented in a similar manner by Flückiger et al. (2002).

During periods when orbital factors produce cooler than average temperatures, CO_2 is fixed more readily into oceans, methane-producing swamps shrink, organic matter decomposes more slowly, and the concentrations of CO_2 and CH_4 in the atmosphere decrease, trapping less heat and setting into motion a cooling feedback loop. Global cooling periods associated with the transitions from warm interglacial periods to ice ages occur far more slowly than do glacial to interglacial warming periods, differences which are attributable to the rates and magnitudes of the feedback loops. Also notable is that while the initial driving mechanism behind these shifts in climate appears to be related to Earth–Sun distance (i.e. changing temperature appears to be the initial trigger), the GHGs CO_2 and CH_4 play important roles in feedback loops that ultimately result in the magnitudes of warming (and cooling) observed in the δD and $\delta^{18}O$ ice core records.

The orbital mechanisms underlying these climate transitions have occurred in a more or less cyclic manner for hundreds of thousands of years, and this phenomenon was discovered by the Serbian mathematician and engineer Milutin Milanković and published in 1914. For that reason, sometimes these orbital variation and their controls on climate on 20 000–100 000-year time scales are known as Milankovitch cycles.

The Holocene epoch is often regarded as a time of stable climate and steady greenhouse gas concentrations, yet detailed analysis of bubbles trapped in glacial ice reveals some interesting trends, notably a slow but steady increase in CO_2 from 260 ppmv at 7000 years before present to 250 years ago, when CO_2 was at its preindustrial value of 280 ppmv. Methane also increased during the Holocene, although its rise did not begin until 5000 years before present (Figure 7.4). These trends are anomalous compared to the three previous interglacial periods (125, 240, 325 ka) recorded in the Vostok ice core record (Figure 7.3), where rapid warming and associated rapid increase in atmospheric CO_2 and CH_4 during deglaciation were followed by thousands of years of slow cooling and slowly decreasing CO_2 and CH_4 as Earth headed back toward glacial conditions.

7.1.5 Composition of the current atmosphere

Table 7.1 presents the composition of the modern atmosphere (to an altitude of 25 km). The concentrations of many of these gases, including nitrogen, oxygen, argon, neon, helium, and xenon, are relatively constant over time scales on the order of 10^3–10^4 years or longer. These gases do not vary appreciably over glacial–interglacial cycles, as do carbon dioxide and methane, and they do not vary over time scales of hours to days, as does water vapor.

Focus Box 7.2

When Did Humans Begin to Influence Global Climate?

The mid-Holocene increases in CO_2 and CH_4 may be anomalous compared to other post-glacial periods, and after examining insolation and other potential causes, Ruddiman (2003) speculated that the development and spread of agriculture 5000–7000 years ago (and its continuation to the onset of the industrial age) was responsible for these trends. In order to grow crops (from Southeast Asia to Mesopotamia to Mesoamerica), humans flooded land for rice cultivation (a source of CH_4) and they cut down trees and converted C_{org} into CO_2 in the atmosphere. Ruddiman argues that in the absence of human activities during the Holocene, atmospheric CO_2 and CH_4 would likely have continued to drop toward glacial maxima values of ~190 ppmv rather than recovering after their initial decreases in the earliest Holocene (Ruddiman 2003). Yet even if humans have been responsible for altering greenhouse gas concentrations for thousands of years, the magnitude of the preindustrial era increases and the corresponding rates of increase are mere fractions of the trends seen in GHGs over the past two centuries, and this is cause for concern.

Table 7.1 Concentrations of selected fixed gases in the atmosphere (sea level).

Gas	Concentration
Nitrogen (N_2)	78.09%
Oxygen (O_2)	20.94%
Argon (Ar)	0.93%
Carbon Dioxide (CO_2)	411 ppmv[a]
Neon (Ne)	18.2 ppmv
Helium (He)	5.2 ppmv
Methane (CH_4)	1.85 ppmv[a]
Krypton (Kr)	1.1 ppmv
Hydrogen (H_2)	0.55 ppmv
Nitrous oxide (N_2O)	0.32 ppmv[a]

[a]The concentrations of these gases are increasing due to anthropogenic emissions.

Source: Data are from Spiro and Stigliani (2003) and the Global Monitoring Division of the National Oceanic and Atmospheric Administration (USA) data on greenhouse gases. Data from Earth System Research Laboratory of US National Oceanic and Atmospheric Administration (http://www.esrl.noaa.gov/gmd/ccgg/trends).

The concentration of water vapor in the atmosphere ranges from < 1% in arid regions to 4% in hot humid tropical air, so when water vapor content is at its lowest extreme, the actual concentrations of the constant gases (N_2, O_2, Ar, etc.) are effectively the same as the values given in Table 7.1, and the sum of the main constant components (i.e. the fixed gases) $N_2 + O_2 + Ar$ is 99.96%. When water vapor is 4% on a humid, sticky day in the tropics, however, the actual concentrations of the fixed gases are proportionally lower. In this case, the amount of N_2 gas in humid air with 4% H_2O_g is $78.08\% \times 96/100 = 74.96\%$. Similarly, the amount of $O_2 = 20.95\% \times 96/100 = 20.11\%$, Ar = 0.89%, and the sum of the fixed gases $N_2 + O_2 + Ar$ is 95.96%.

The abundance of O_2 gas makes the atmosphere a strongly oxidizing environment and this helps to explain the presence of CO_2 and H_2O, gases that form by oxidation of, for example, organic matter and methane. These oxygen-bearing gases are relatively stable in the atmosphere and stand in contrast to gases that are more stable in reducing environments such as CH_4 and H_2S. These reduced gases (and others), however, do exist in the atmosphere at relatively constant concentrations, at least during the ~12 000 years of the Holocene prior to the industrial age (e.g. CH_4 concentration was ~0.8 ppmv throughout the Holocene; its current concentration is 1.85 ppmv).

If we consider the fate in the atmosphere of chemically reduced gases like CH_4, we would predict that they will react with O_2 as follows:

$$CH_4 + 2O_2 \rightarrow CO_2 + 2H_2O \qquad (7.4)$$

(Chapter 8 explores complexities of oxidation reactions which typically involve other species such as the hydroxyl and peroxy radicals, but this equation serves to emphasize the point being illustrated here.)

From tables of geochemical constants, the equilibrium constant for this reaction can be written as:

$$K_{eq} = [CO_2] \times [H_2O]^2 / [CH_4] \times [O_2]^2 = 10^{140} \qquad (7.5)$$

and from this equation, the concentrations of the gases can also be represented as partial pressures:

$$K_{eq} = pCO_2 \times pH_2O^2 / pCH_4 \times pO_2^2 = 10^{140} \qquad (7.6)$$

The K_{eq} of 10^{140} indicates that an equilibrium assemblage of CH_4, O_2, CO_2, and H_2O will be very strongly

---

Focus Box 7.3

Atmospheric Gas Concentrations

Concentrations of atmospheric gases are based on abundance of a gas on a volumetric basis, reflecting the fact that one mole of each gas occupies approximately equal volumes ($\sim$22.4 l mol^{-1}). In tangible terms, if a trace gas is present in the atmosphere at a concentration of 1 part per million volumetrically (1 ppmv), there will be 1 cm^3 of that gas in 10^6 cm^3 of air. It also implies that 1 ppmv of a given gas corresponds to 1 mol of that gas per 10^6 mol of air, and similarly, that this gas will effectively comprise 1 molecule out of every 10^6 molecules of air. So volumetrically and by mass, N_2 occupies 75–78% of air (depending on H_2O content), and CO_2 currently comprises $\sim$0.041% of air (and rising).

---

dominated by the products, which in this case are carbon dioxide and water. Using partial pressure values for O_2, CO_2, and H_2O of 0.21, 0.004, and 0.01 atm, respectively (if N_2 were part of the calculation its partial pressure value would be 0.78 atm, where the sum of all partial pressures = 1.0 atm), in the K_{eq} equation, as follows:

$$K_{eq} = 0.0004 \text{ atm} \times 0.01 \text{ atm}/p\text{CH}_4 \times 0.21 \text{ atm}$$

$$= 10^{140}$$

and rearranging to solve for $p\text{CH}_4$:

$$p\text{CH}_4 = 0.0004 \text{ atm} \times 0.01 \text{ atm}/10^{140} \times 0.21 \text{ atm}$$

$$= 1.9 \times 10^{-145} \text{ atm}$$

It is apparent that the concentration of methane in the atmosphere based on equilibrium considerations should be 10^{-145} atm. This is an incredibly small amount. When compared to the actual concentration of 1.85 ppmv ($\sim 2 \times 10^{-6}$ atm), it becomes clear that the amount of methane in air is not controlled by chemical equilibrium; instead, methane and many other gases in air are out of equilibrium and their concentrations are mainly controlled by kinetic factors. What are these factors? One important control on CH_4 abundance is the probability of collision or interaction of O_2 and CH_4 molecules in air (or on the surfaces of particulate substrates); another is the rate at which CH_4 and O_2 dissociate in the presence of each other (once molecules of the gases have collided) to form the products CO_2 and H_2O. These rate-limiting steps lend a strong kinetic control to the concentration of methane in Earth's atmosphere.

7.1.6 Air circulation

Air circulation plays an important role in transport of natural and anthropogenically derived compounds in the atmosphere, and much of it is driven by differences in density. For an air mass at sea level at constant temperature (e.g. 20 °C), the main control on the *density of air* is the amount of water vapor. This is because the other main gases in the atmosphere (N_2, O_2, Ar) are present at effectively fixed values; for example, dry air (0% H_2O) contains 78.08% N_2, 20.95% O_2, and 0.93% Ar. A very humid end-member will contain 4% water vapor and proportionally less N_2, O_2, and Ar than what is shown in Table 7.1. The density of these end-member air masses can be determined by carrying out a mass–balance calculation as follows:

For dry air with no H_2O:

$$[0.7808 \times 28.01 \text{ gmol}^{-1} \text{ (N}_2\text{)}]$$

$$+ [0.2095 \times 32.00 \text{ g mol}^{-1} \text{ (O}_2\text{)}]$$

$$+ [0.0093 \times 39.95 \text{ g mol}^{-1} \text{ (Ar)}]$$

$$= 28.95 \, g \, mol^{-1}$$

For humid air with 4% H_2O:

$$[0.7496 \times 28.01 \text{ g mol}^{-1} \text{ (N}_2\text{)}]$$

$$+ [0.2011 \times 32.00 \text{ g mol}^{-1} \text{ (O}_2\text{)}]$$

$$+ [0.0089 \times 39.95 \text{ g mol}^{-1} \text{ (Ar)}$$

$$+ [0.004 \times 18.01 \text{ g mol}^{-1} \text{ (H}_2\text{O)}]$$

$$= 27.86 \, g \, mol^{-1}$$

While perhaps counterintuitive, humid air is less dense than dry air, and this density difference drives circulation. Lower-density air (generally humid and warm) is transported convectively upward in the atmosphere (i.e. it rises and carries heat and water vapor upward), and when humid air masses collide with dry air masses, the dry air dives under the humid air and this forces the humid air to higher altitudes in the atmosphere. This results in condensation of water vapor that can lead to rain, snow, or other forms of precipitation.

Air temperature also drives circulation – hot air has greater kinetic energy and greater average distance between air molecules compared to cool (and often dry) air, so its density is lower and consequently, it rises; cool air sinks. Insolation is the main force responsible for heating the Earth's surface and atmosphere, and while radioactive decay within the Earth and volcanic and hydrothermal systems can also heat the Earth's surface and its atmosphere, the main control on atmospheric circulation is insolation and the absorption or reflectance of solar rays. Insolation is greatest near the equator (Figure 7.5), and with increasing latitude poleward, the sun strikes the Earth at progressively more oblique angles where more of it is scattered, absorbed, attenuated, or reflected. The result is that the equatorial belt receives the greatest amount of insolation (i.e. incoming heat) of any region at the Earth's surface. At latitudes greater than 35°N and 40°S, more energy is lost as IR radiation compared to what is gained as incoming solar radiation (Figure 7.5).

The occurrence of warm air (and soils, lakes, and oceans) in the tropics leads to high evaporation rates and warm, low-pressure, humid air within ~10° to 20° of the equator. (Highly productive tropical forest ecosystems also contribute to high water vapor content through transpiration.) As this warm, moist air rises through the atmosphere, it cools and eventually loses its moisture through condensation and precipitation as tropical rains. Cold, dry, high-pressure air from higher latitudes rushes toward the equator in response to the pressure

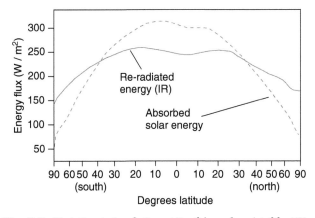

Fig. 7.5 Variation in insolation at Earth's surface (visible, UV, IR; dashed line) and reradiated infrared radiation from Earth's surface (solid line) plotted as a function of latitude. Note net influx of energy in tropics and subtropics and net loss of energy at latitudes greater than ~35°. Nonlinearity of x-axis reflects scaling by cosine of latitude. Source: Modified from Vonder Haar and Suomi (1971).

gradient – the cooler, higher-pressure air dives beneath the convectively uplifting tropical low-pressure air, setting into motion what, in the simplest sense, is known as a *Hadley cell* (named after eighteenth-century meteorologist George Hadley), a sketch of which is depicted in Figure 7.6.

Global air circulation is described by the *three-cell model* (Figure 7.6). As warm, humid, tropical air migrates upward into higher levels of the troposphere, it cools through radiative heat loss and water vapor is lost from the air mass as rain that falls in the tropics. Radiative cooling of tropic-derived, poleward-migrating air in the upper troposphere causes its density to increase, making it sink back to the Earth's surface at approximately 30°N and 30°S of the equator. This sinking air (now dry) heats adiabatically in the lower levels of the atmosphere, and the rate of heating is more rapid than the rate at which the rising tropical air cooled because the air mass is not thermally buffered by the high heat capacity of H_2O. Fully saturated tropical air cools and heats at a rate of ~5 °C km^{-1}; conversely, dry air cools and heats at a rate of approximately 10 °C km^{-1}.

The rapid heating of sinking, dry air at 30° north and south of the equator produces two belts of subtropical high-pressure regions that contain many of the great deserts of the world, including the Sahara Desert of northern Africa, the Mojave Desert of southwestern North America, the Arabian Desert of the Arabian Peninsula, and the Namib and Kalahari deserts of southwestern Africa (Figure 7.6). Some of this descending air returns toward the equator, setting up a convection cell still known as a *Hadley cell*. Air masses that return to the tropics from 30° north and south latitude are initially hot and dry, so they gather moisture as they migrate back toward the equator. The region where these two equator-bound air masses collide, known as the *intertropical convergence zone (ITCZ)*. The collision of equator-bound air masses combined with high moisture content and strong insolation causes air to rise convectively in the ITCZ – the latitudinal extent of the ITCZ is ~± 20° north/south of the equator, and this is where tropical rainforests occur.

Air that descends back to the Earth's surface at 30° splits, some of it returning to the equatorial belt, and some migrating poleward. This air is hot and dry, so it results in evaporation of water from the Earth's surface as this cell migrates poleward. Collision with cold, dry polar air produces precipitation in a belt known as the *Polar Front* at ~40° to 60° latitude. Some of the uplifted poleward-migrating air continues poleward where it ultimately sinks to the surface at the pole; from there it migrates back toward the equator. Convection cells

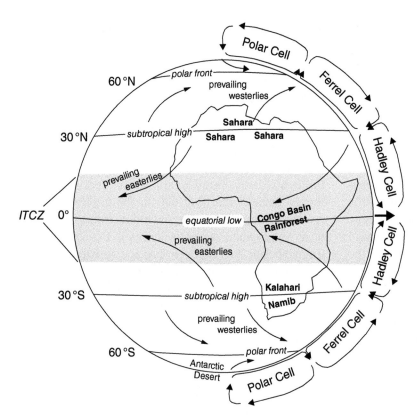

Fig. 7.6 Atmospheric circulation patterns on a global scale, with Africa presented here as an example of a continent with climate zones that are strongly controlled by global air circulation. Note the occurrence of tropical rainforests (e.g. Congo Basin) within the ITCZ, locations of deserts (e.g. Sahara, Kalahari, Namib) at 30° north and south latitude, as well as prevailing wind patterns that transport contaminants at many scales, from regional (acid rain, tropospheric ozone) to global (notably, transport of persistent organic pollutants from the northern continents to the Arctic in prevailing westerlies of the northern hemisphere Ferrel cell).

that deliver air from the mid-latitudes toward the poles contribute to the transport of *Pb*, dichlorodiphenyl-trichloroethane (*DDT*), polychlorinated biphenyls (*PCBs*), perfluorinated chemicals (*PFCs*), and other persistent pollutants from industrialized areas to Arctic and Antarctic regions.

This simplified explanation of the three-cell model might imply that air travels in north–south directions, but in reality air masses are deflected by rotation of the Earth; in the northern hemisphere, migrating air masses are deflected to the right (clockwise), while in the southern hemisphere, air masses are deflected to the left (counter-clockwise). This force, known as the *Coriolis Effect*, results in prevailing winds such as the tropical *easterlies* (also known as the northeastly and southeasterly trade winds) between the equator and 30° latitude; the prevailing *westerlies* between 30° and 60° latitude; and prevailing easterlies between 60° and the poles.

The Coriolis Effect exerts some important consequences on contaminant transport and climate. Sulfur dioxide (SO_2), nitrogen oxides (NO_x), mercury (Hg), and other contaminants from coal-fired power plants in the central United States and England are transported downwind in a northeasterly direction entrained in air carried in Ferrel cells toward vulnerable forest ecosystems in northeastern North America and Scandinavia, respectively, where they ultimately fall as components of acid rain and as various

species of Hg. The tropical easterlies that migrate across the Andes Mountains of northwestern South America produce orographic precipitation in the mountains and the very pronounced rainshadow of the Atacama Desert. In a similar manner, prevailing westerlies migrating across the Himalaya are responsible for the Gobi Desert rainshadow.

7.2 EVAPORATION, DISTILLATION, CO$_2$ DISSOLUTION, AND THE COMPOSITION OF NATURAL PRECIPITATION

The purest water in the hydrologic cycle (i.e. with the lowest amount of total dissolved solids) occurs in the atmosphere, where TDS is typically <5 mg l^{-1} (ppm). By comparison, river water typically contains 50–500 mg l^{-1} TDS, groundwater typically contains 100–1000 mg l^{-1} TDS, and ocean water contains 35 000 mg l^{-1} TDS (Table 4.1).

Most of the water in the atmosphere arrives by evaporation from oceans, and the fate of dissolved solids during evaporation helps to explain the occurrence of both very low TDS values of atmospheric H_2O and the high TDS values of ocean water (as well as evaporation-dominated water bodies such as the Dead Sea and the Great Salt Lake). Consider a surface water body (e.g. lake, ocean, or river)

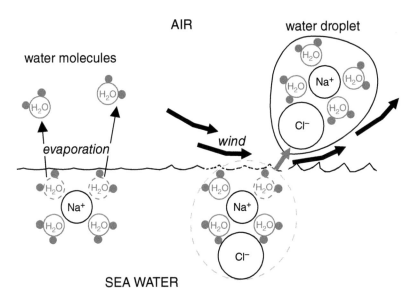

Fig. 7.7 Schematic sketch showing evaporation of isolated water molecules from sea water on the left, which depicts evaporation as distillation that leaves dissolved ions in the surface water body. On the right, wind is shown physically eroding a droplet of water containing dissolved salts into air. In reality, many molecules of NaCl may be contained in the water droplet.

with dissolved ions such as Na^+, Ca^{2+}, HCO_3^-, and Cl^-. In order for a water molecule to evaporate, it must achieve a kinetic energy that enables it to overcome the intermolecular bonds between the H_2O molecule in question and adjacent H_2O molecules, as well as bonds between the H_2O molecule and dissolved ions. Considering surface and atmospheric composition, mass of the water molecule compared to the dissolved ions is fundamental – water molecules (18 Da) are much lighter than the combined mass of a water molecule bonded to a calcium ion ($H_2O + Ca^{2+} = 58$ Da) or to a chloride ion ($H_2O + Cl^- = 53$ Da), so when H_2O molecules evaporate from surface water bodies, dissolved ions remain in the surface water body (Figure 7.7). This process is a form of *distillation* and it is the main reason for the low amounts of dissolved ions in atmospheric water (relative to surface water and subsurface water).

The white residues left behind in a stovetop pan or tea kettle are visual examples of dissolved solids left behind when water evaporates or boils.

Given that molecules of carbon dioxide in air dissolve in water droplets to form carbonic acid, rainfall is typically at least slightly acidic, and when in equilibrium with atmospheric CO_2, the pH of rainfall is 5.6. (Refer to Chapter 5 for calculations required to determine pH of rain in equilibrium with varying CO_2 concentrations.) Reactions of SO_2 and NO_x with water and other constituents of the atmosphere can produce rainfall that is much more acidic than what is produced by CO_2–H_2O equilibrium. In some cases, natural sources of S and N gases (e.g. volcanic eruptions) can produce rain with pH < 5, but in most cases, acid rain is caused by combustion of fossil fuels.

7.3 THE ELECTROMAGNETIC SPECTRUM, GREENHOUSE GASES, AND CLIMATE

The greenhouse effect is caused by molecular-scale interactions of IR radiation with trace gases that possess a central atom with symmetrical molecular structures (e.g. CO_2, CH_4, H_2O, O_3, and N_2O). In the absence of this phenomenon, Earth's average temperature would be approximately $-18\,°C$; in the presence of the greenhouse effect, Earth's current average temperature is approximately $15\,°C$ (and soon to be $16\,°C$).

In order to understand the greenhouse effect and trapping of heat by trace atmospheric gases, it is important to understand the electromagnetic radiation emitted by the sun (and reradiated by from the Earth's surface), for this is the main source of energy to heat Earth's atmosphere. Much of this radiation is visible – we can observe sunlight – but some also occurs at wavelengths (λ) slightly greater and less than wavelengths of visible solar radiation (Figure 7.8).

7.3.1 Electromagnetic spectrum

The visible part of the solar spectrum occurs at wavelengths in the range of 380–780 nm. The shortest wavelength visible light is violet ($\lambda = 380$–420 nm), while the longest wavelength light is red (620–780 nm). Adjacent to violet light at shorter wavelengths (and thus, higher frequencies) is UV light ($\lambda = 1$–400 nm). The high-frequency (= high-energy = short wavelength) UV radiation can burn skin and cause mutations in cells of living organisms. At the other end of the visible spectrum, radiation with wavelengths greater than those of red light is IR

Focus Box 7.4

Origin of Aerosols and Particulates in Air

In addition to evaporation and boiling, two additional processes that introduce dissolved salts and particulates to water in the atmosphere are:

1. winds that whip up water droplets from the ocean (Figure 7.7) – these droplets introduce Na^+, Cl^-, and other dissolved ions into the atmosphere; the residence time of these salts in air is two to three days, so they often "rain out" in coastal regions (note from Table 4.1 that coastal rain typically contains higher TDS than continental interior rain). In some cases these salts act as condensation nuclei for raindrops; and

2. dust storms in arid regions that introduce fine-grained particulates from desert soils into the atmosphere; some of these particles (termed aerosols) are soluble salts (e.g. halite or gypsum), but others are carbonates or clay minerals. The presence of Ca^{2+} and K^+ in aerosols is an important source of plant nutrients in many tropical and alpine soils where leaching over time would otherwise deplete these sensitive ecosystems of nutrients in the absence of aerosol deposition.

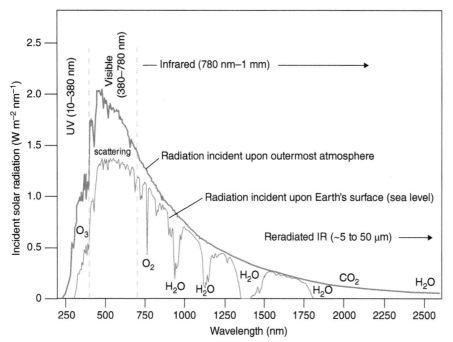

Fig. 7.8 Spectrum of incoming solar radiation showing (a) radiation incident upon outermost atmosphere, and (b) radiation incident upon Earth's surface. It is clear that the atmosphere absorbs and scatters appreciable amounts of incoming UV (10–380 nm) and visible radiation (380–780 nm). Sulfate aerosols and water droplets are common scatterers of incoming radiation. Absorption of incident radiation by specific gases in specific ranges (e.g. by water between ~1300 and 1500 nm and by ozone in the UV range of ~240 to 380 nm) is indicated on the figure. Source: The image source is the ASTM 1.5 solar output spectrum, a standard reference solar spectrum that was obtained online at: http://rredc.nrel.gov/solar/spectra/am1.5.

radiation (λ of IR ranges from 780 to 10^6 nm or 1 mm). The amounts of each type of radiation incident upon Earth's outermost atmosphere and upon Earth's surface at sea level are shown in Figure 7.8, and it apparent from these spectra that the greatest amount of radiation arrives as visible light followed in abundance by IR and UV radiation.

The effects of atmospheric gases (e.g. H_2O, CO_2, and O_3) as absorbers of incoming radiation are indicated by the differences between the incident radiation and the radiation which ultimately reaches Earth's surface. Notable examples are absorption of UV by ozone and absorption of IR by water and carbon dioxide. Visible light is not particularly affected by these gases; rather, visible radiation that

does not reach Earth's surface is most likely scattered and directed back toward outer space by aerosols (especially sulfates) and by water droplets.

7.3.2 Reradiation from earth's surface

The combined visible, IR, and UV radiation that reaches rocks, soil, and water at the Earth's surface heats these materials and they in turn reradiate the heat back toward outer space as IR. At the molecular level, the high frequency of incoming UV radiation increases the frequency of molecular vibrations in objects at the Earth's surface, which causes the objects to increase in temperature (you can feel this on a sunny day). Heat is then emitted back to the atmosphere from rocks, soils, and waters as IR radiation with a peak wavelength of 10 μm over the range of ~3 to 50 μm. These values are consistent with the expected black body radiation from a planet at ~15 °C based on the following Wien's law calculation:

$$\lambda_{peak} = b/T \qquad (7.7)$$

The term b is Wien's displacement constant, $2.897\,721 \times 10^{-3}$ K m, T is temperature in Kelvins, and λ_{peak} is the wavelength at which the emitted radiation is at its maximum intensity (i.e. the modal λ or the λ at the top of the peak). Given the units of b and T, the result is in units of meters. Using 288 K (15 °C) as the temperature of Earth, the maximum wavelength of radiation emitted is 1.01×10^{-5} m or 10.1 μm (10 100 nm). This value represents the highest intensity of the peak shown in Figure 7.9.

By comparison, doing the same calculation for the sun using a surface temperature of 6000 K, peak λ is 4.83×10^{-7} m or 483 nm, which corresponds to the actual solar peak visible in the spectrum shown in Figure 7.8.

The role of H_2O and CO_2 as the GHGs with the greatest ability to absorb outgoing IR radiation is evident in Figure 7.9. H_2O absorbs strongly in the ranges of ~3 to 4, 5 to 7, and > 22 μm whereas CO_2 absorbs strongly at ~13 to 15 μm. Details on absorption of IR radiation at the molecular level by H_2O, CO_2, and other gases are presented in Section 7.4.

The surface temperature of Earth (or other planets and moons) can be calculated if we can quantify insolation (i.e. the amount of energy Earth receives from the sun) and outgoing radiation (i.e. energy emitted by Earth as reradiated IR radiation). The basis of this calculation is the Stefan-Boltzmann Law, which states that energy radiated by a "*black body*" in space (e.g. a planet like Earth) is proportional to the fourth power of the absolute temperature of the body:

$$E = k \times T^4 \qquad (7.8)$$

where k is the Stefan-Boltzmann constant and T is absolute temperature in Kelvins.

In the absence of GHGs, the balance of incoming solar energy and reradiated heat is determined by the following equation:

$$(E_S/4) \times (1 - \alpha) = \sigma \times T_E^{\,4} \qquad (7.9)$$

Components of this equation include:
1 E_s is quantified in terms of $W\,m^{-2}$ (this is the solar constant, a value that represents the average amount of energy received by Earth from the sun; estimates vary, but it is safe to say that within the past 30 years it has not varied outside the range of 1360–$1370\,W\,m^{-2}$);
2 E_s multiplied by $^1/_4$, which is the ratio of the cross-sectional area of Earth as a disc in space to its total spherical surface area – this is necessary because solar radiation is intercepted by the area of the cross-sectional disc of the Earth (πr^2) but is radiated outward as a function of the area of the entire Earth sphere $(4\pi r^2)$;
3 $\alpha = 0.3$ (this is the estimated albedo of Earth, a unitless value – it quantifies the fraction of solar energy reflected back into outer space. What this value indicates is that the other 70% of incoming solar radiation is absorbed at the surface and reradiated as IR radiation);

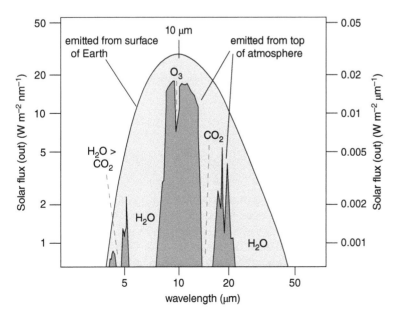

Fig. 7.9 Spectrum of infrared radiation that leaving Earth's atmosphere compared to the ideal spectrum predicted for emanation of black body radiation calculated using Wien's law. The difference is mainly due to absorption of IR radiation by greenhouse gases. Note that x-axis scale is logarithmic and that it the units shown here (μm) differ by a factor of 1000 relative to the x-axis in Figure 7.8.

4 $\sigma = 5.67 \times 10^{-8}$ W m^{-2} K^{-4} (this is the Stefan-Boltzmann constant, where K is temperature in Kelvins); and

5 T_E, which represents temperature (in K) of Earth's surface.

Rearranging to solve for T_E:

$$T_E = [(1365/4 \text{ W m}^{-2})$$
$$\times (1 - 0.3)/(5.67 \times 10^{-8} \text{ W m}^{-2} \text{ K}^{-4})]^{1/4}$$

$$T_E = 255 \text{ K} = -18°C$$

Yet average temperature of Earth is clearly not $-18°C$. The difference in this "black body" temperature calculation and Earth's actual surface temperature lies in the effect exerted by atmospheric gas molecules with symmetrical structures, mainly H_2O, CO_2, and CH_4 as well as O_3, N_2O, and chlorofluorocarbons (CFCs).

7.3.3 Greenhouse effect and heat trapping

The *greenhouse effect* can be incorporated into the Stefan-Boltzmann Law as *effective infrared transmission* (g_h), a term which expresses heat trapping by the greenhouse effect.

$$(E_S/4) \times (1.0 - \alpha) = g_h \times \sigma \times T_E^4 \qquad (7.10)$$

At an atmospheric CO_2 concentration of 280 ppmv, the value of g_h is 0.61. This means that 61% of reradiated IR radiation escapes the atmosphere headed for outer space; the other 39% interacts with CO_2, H_2O, and other GHGs and is reflected back toward Earth's surface. The mechanism of this interaction is presented in Section 7.4. At a g_h value of 0.61 (i.e. indicative of preindustrial CO_2 levels), global average surface temperature is determined as follows:

$$T_E = [(1367 \text{ W m}^{-2})/4$$
$$\times (1.0 - 0.30)/(0.61 \times 5.67 \times 10^{-8} \text{ W m}^{-2} \text{ K}^{-4})]^{1/4}$$

$$T_E = 288.28 \text{ K} = \sim 15°C$$

The greenhouse effect raises Earth's average temperature by about $33°C$; however, too much of a good thing can be problematic, and we have entered an era where concentrations of GHGs have reached levels not seen in at least hundreds of thousands of years (Section 7.4). One of the ways to forecast how increasing greenhouse gas concentrations will affect Earth's surface temperature is to perform calculations based on the first principles described in this section.

7.4 GREENHOUSE GASES: STRUCTURES, SOURCES, SINKS, AND EFFECTS ON CLIMATE

As mentioned above, what differentiates GHGs from non-GHGs like N_2 (78.1% of dry air), O_2 (20.9% of dry air), and Ar (0.93% of dry air) is the symmetrical character of their molecular structures.

Focus Box 7.5

Quantifying Climate Change and Air Temperature

Future climate scenarios can be predicted using the principles and equations in Section 7.3.3. For example, consider an atmospheric CO_2 concentration of 550 ppmv, a case in which the value of g_h becomes 0.60 (i.e. only 60% of outgoing IR radiation escapes to outer space). Then, the calculation of global average surface temperature is as follows:

$$T_E = [(1367 \text{ W m}^{-2})/4$$
$$\times (1.0 - 0.3)/(0.60 \times 5.67 \times 10^{-8} \text{ W m}^{-2} \text{ K}^{-4})]^{1/4}$$

$$T_E = 290.47 \text{ K} = 16.47°C = \sim 16°C$$

This simple calculation only takes into account radiative forcing from CO_2, but not feedback loops, other GHGs (including water), changes to albedo, and myriad other elements of the climate system, so it likely underestimates the warming associated with a ~doubling of CO_2 (from preindustrial 280 ppm).

A warmer atmosphere will cause melting of snow and lead to a decrease in albedo, so accounting for a decrease in albedo from 0.30 to 0.28, we obtain the following result:

$$T_E = [(1367 \text{ W m}^{-2})/4$$
$$\times (1.0 - 0.28)/(0.60 \times 5.67 \times 10^{-8} \text{ W m}^{-2} \text{ K}^{-4})]^{1/4}$$

$$T_E = 291.51 \text{ K} = 18.51°C = \sim 19°C$$

Thus, in a scenario involving a ~doubling of CO_2 and ~7% decrease in albedo (e.g. from 0.30 to 0.28 due to melting ice caps and other changes), average global temperature would increase by ~4°C. While these calculations are clearly not as comprehensive as global climate models, they are the types of considerations and calculations that are used in modeling climate.

7.4.1 Molecular structures and vibrations of greenhouse gases

Figure 7.10 presents examples of many greenhouse gas molecules. Each gas possesses a central atom that serves as a center of symmetry about which the molecules can stretch and bend when struck by IR rays. Ozone is symmetrical because bonds in both molecules shown on the left in Figure 7.10 are resonance hybrids. In other words, O_3 contains two end-member structures with a single bond on one side and double bond on the other, but the bonds resonate between the two sides of the central O atom, producing an overall bond order of 1.5 for each side (depicted on the right in parentheses). The scenario is similar with N_2O, which is not truly symmetrical, but one of its resonant modes contains double bonds on each side of the central N atoms, making its structure similar to that of CO_2, where the presence of a central atom permits stretching and bending of bonds on either side in the presence of IR radiation.

Figure 7.11 illustrates the ways in which CO_2 molecules vibrate in the presence of IR radiation of different

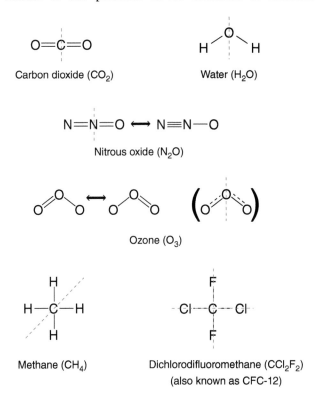

Carbon dioxide (CO_2)

Water (H_2O)

Nitrous oxide (N_2O)

Ozone (O_3)

Methane (CH_4)

Dichlorodifluoromethane (CCl_2F_2) (also known as CFC-12)

Fig. 7.10 The molecular structure of important greenhouse gases. In each case, dashed lines are drawn to emphasize the center(s) of symmetry in the molecules which allows them to stretch, flex, and bend in the presence of infrared radiation. (Note: for CH_4 and CFC-12, four centers of symmetry can be drawn from the 2-dimensional representations shown – for clarity, not all of these centers of symmetry are shown).

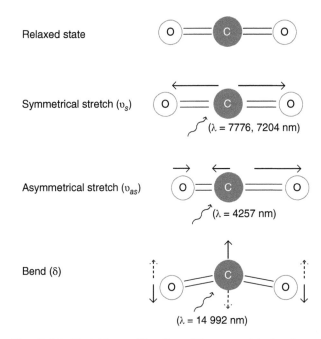

Relaxed state

Symmetrical stretch (υ_s)

(λ = 7776, 7204 nm)

Asymmetrical stretch (υ_{as})

(λ = 4257 nm)

Bend (δ)

(λ = 14 992 nm)

Fig. 7.11 Stretching and bending of the symmetrical carbon dioxide molecule in the presence of infrared radiation of varying wavelength.

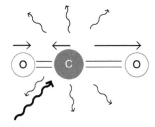

Fig. 7.12 Reradiation of infrared energy by a CO_2 molecule undergoing asymmetrical stretching. The incident IR ray is the bold wavy line and the thin squiggly lines are IR radiation re-emitted by molecular vibrations. Note that the net effect is redirection of outgoing IR radiation back toward Earth's surface.

wavelengths, keeping in mind that all GHGs behave in similar ways in the presence of IR radiation of a range of wavelengths. The wavelengths of IR radiation absorbed by CO_2 molecules are shown for each of the vibrational modes.

The vibrations shown in Figure 7.11 cause IR radiation to be reradiated in all directions. Some of the IR energy will be directed toward the outer limits of the atmosphere, but some will also be sent back toward the Earth surface, and this is how GHGs "trap" heat (Figure 7.12).

In addition to the vibrations shown, there are also rotations that broaden the range of wavelengths over which CO_2 (and other GHGs) can absorb IR. In contrast, the lack of a central atom in atmospheric gases like N_2, O_2, Ar, and H_2 prevents them from vibrating (via stretching and bending) in the presence of IR radiation, and thus they are not capable of "trapping" outgoing IR radiation.

7.4.2 Greenhouse gases, radiative forcing, GWPs

Knowledge of the capacity of CO_2 and other GHGs to trap heat in the atmosphere is nothing new – it was recognized over one hundred years ago by the Swedish chemist and Nobel prize winner Svante Arrhenius, who in the late 1800s became interested in the causes of ice ages. In 1896, he predicted that if CO_2 levels were to drop by 50%, Earth's surface temperature would fall by 4–5 °C; conversely, a doubling of CO_2 levels from 280 to 560 ppmv would produce an increase in surface temperature of about 5–6 °C. Although current models predict that a doubling of CO_2 will yield a temperature increase more in the range of 3 ± 1 °C, Arrhenius' first-principle calculations were on the right track.

The net influence of all GHGs and other climate-forcing agents (e.g. aerosol particles or net insolation) on the average net IR radiation leaving the atmosphere is *radiative forcing*. The combined radiative forcing of all GHGs in the atmosphere produces the greenhouse effect; however, not all GHGs contribute equally to this effect. Relatively abundant gases such as H_2O and CO_2 exert strong forcings based on their sheer abundance relative to trace gases. Some of the trace gases, particularly CH_4, N_2O, and CFCs, exert strong forcing due to their relatively high per-molecule *global warming potentials* (GWPs). GWP is controlled in part by molecular structure (i.e. ability to vibrate in the presence of reradiated IR), in part by residence times in the atmosphere, and also by where in the IR spectrum these gases are capable of absorbing IR radiation.

Figure 7.13 and Table 7.2 present data for some representative GHGs (Forster et al. 2007), where all values are normalized to $CO_2 = 1$. Recall that data on carbon cycling, including fluxes and reservoirs, are also presented in Chapter 5.

Many of the trace gases have GWPs tens to thousands of times greater than CO_2, indicating that small increases can have disproportionately large effects. In addition to per-molecule differences in absorption potential, a very important factor that contributes to the high GWP of CFCs like CFC-12 is that they are capable of absorbing IR radiation at frequencies outside of the ranges of absorption of the main GHGs H_2O and CO_2. Figure 7.14 displays the IR spectrum from 1000 to 15 000 nm, a range over which H_2O and CO_2 both display strong absorption bands (note: absorption bands are defined as regions within the IR spectrum where these gases absorb high amounts of reradiated IR radiation). Two points are worth noting here:

1 In some regions of the spectrum, all IR radiation is absorbed, so no increases in concentrations of H_2O or

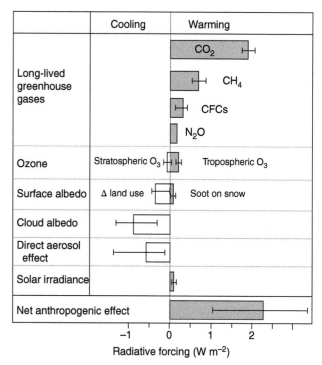

Fig. 7.13 Changes in radiative forcings from 1750 to 2005 including individual gases, changes to surface and cloud albedo, the direct effects of aerosols (e.g. sulfate particles), slight increase in energy emitted from the sun, and the net impact of human activities on radiative forcing. The net effect of changing atmospheric composition is an increase of 1.55 W m⁻² in radiative forcing. Source: Modified from Forster et al. (2007), IPCC (2013), and NOAA (https://www.esrl.noaa.gov/gmd/aggi/aggi.html).

CO_2 will have an impact on heat trapping, at least not in these IR ranges. Ranges in which the atmosphere is already saturated include ~4250 nm, where the asymmetrical stretching vibration of CO_2 (Figure 7.11) absorbs 100% of reradiated IR; another is between ~5000 and 7000 nm, where H_2O vibrations absorb 100% of reradiated IR. However, there are many regions where the atmosphere in undersaturated, e.g. with respect to CO_2 at 10 200–10 500 nm and with respect to CH_4 and N_2O at 7500–8000 nm.

2 There are regions of the spectrum (e.g. between ~8000 and 11 000 nm) where neither H_2O nor CO_2 absorb IR radiation. It is these regions where additions of trace gases, especially ozone, nitrous oxide, and CFCs, have great potential to absorb IR where little or none is currently being absorbed (or wasn't being absorbed prior to anthropogenic emissions). You will note that absorption lines for CFC-12 occur in the regions where neither H_2O nor CO_2 absorb. The same is true for N_2O and CH_4, although these are less obvious than CFC-12 because the absorption bands for these gases partially overlap or are close to bands for H_2O and CO_2.

Table 7.2 Selected properties of greenhouse gases.

	CO_2 ppmv	CH_4 ppmv	N_2O ppbv	CFC-12 pptv
Atmospheric concentration				
Pre-1750	280	0.8	288	0
1990	355	1.7	310	484
Current (2018)	410	1.9	333	500
Atmospheric lifetime	Years	Years	Years	Years
	100 ± 50^a	12	114	100
Global warming potential (GWP)	100-yr	100-yr	100-yr	100-yr
(per-molecule basis)	1	25	298	10 900
Increased radiative forcing^a	W m−2	W m−2	W m−2	W m−2
(Integrated for whole atmosphere)	2.01	0.51	0.20	0.16

^{a}The residence time of CO_2 in the atmosphere is difficult to precisely quantify due to myriad removal mechanisms. A 100-year time horizon cannot be applied to ozone (O_3) given its short residence time. Increased radiative forcing is expressed as the change in radiative forcing within the troposphere for each given gas.

Source: Data are from the US National Oceanic and Atmospheric Administration (https://www.esrl.noaa.gov/gmd/aggi/aggi.html) and Forster et al. (2007).

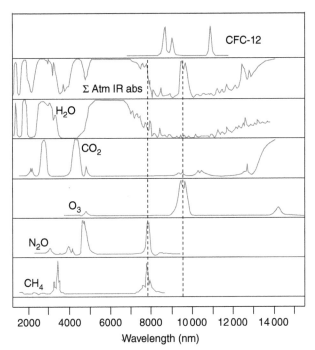

Fig. 7.14 Infrared spectrum of selected greenhouse gases showing the wavelengths in which each gas is infrared active, i.e. the regions in the spectrum where a given gas absorbs IR radiation. "S Atm IR Abs" is the spectrum produced by summing spectra of all gases with natural sources. Note that CFC-12 absorbs IR in two regions where naturally produced and anthropogenically enhanced gases (H_2O, CO_2, O_3, N_2O, CH_4) do not absorb (i.e. 8000–9000 and ~11 000 nm). Thus, any small additions of CFC-12 will have a large impact on IR absorption. Also note that IR absorption capacity is effectively saturated between ~5000 to 7000 nm, so any additional water (or other molecule that absorbs in this region) will not lead to increased absorption in this region.

As human population has grown, very short-term anthropogenic activities (e.g. transformation of hydrocarbons stored deep within Earth's crust into energy and atmospheric CO_2) have accelerated the pace of CO_2 flux to the atmosphere. Whereas atmospheric CO_2 appears to have been present at an atmospheric concentration of 260–280 ppmv during the Holocene (12 000 years ago to ~250 years ago; Figure 7.4), the rate of addition of CO_2 to the atmosphere now greatly exceeds the rate at which it is removed from the atmosphere by geological, hydrological, and biological processes; as a result, CO_2 has progressively increased in concentration since the onset of the industrial revolution. CO_2 concentration as of 2018 is ~410 ppmv. Figure 7.15 presents millennial-scale trends in greenhouse gas concentrations.

7.4.3 Global warming

Chapter 5 presents fluxes of anthropogenic carbon to the atmosphere, values which are relatively easy to quantify. Natural processes annually fix approximately one half of these emissions into marine and terrestrial reservoirs, but there is no doubt that human activities are adding GHGs to the atmosphere. *Net* flux of carbon to the atmosphere is currently on the order of 6 PgC yr^{-1} (6×10^3 TgC yr^{-1}), and the build-up of CO_2, CH_4, N_2O, and other GHGs leaves us with the inescapable conclusion that Earth's surface temperature is warming – to date, warming in the *troposphere* has reached a global average of ~1 °C (Figure 7.15). In response, other systems in contact with the atmosphere such as soils, shallow ocean waters, and lakes also, on average, are warming.

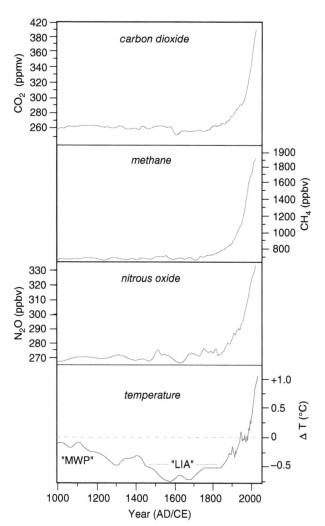

Fig. 7.15 Greenhouse gas trends and temperature anomalies (relative to average temperature for the period 1951–1980) for the time period 1000 CE–2018 CE. CH_4, CO_2, and N_2O data for the period 1000 CE–2004 CE are from MacFarling Meure et al. (2006), with data from 2005 through July 2018 obtained from the NOAA Annual Greenhouse Gas Index (AGGI) web page accessible at http://www.esrl.noaa.gov/gmd/aggi. Temperature data for the period 1000 CE–1880 CE are derived from ice core and firn data published by Moberg et al. (2006), and 1880 through 2018 data are from the Goddard Institute for Space Studies Surface Temperature Analysis web page (http://data .giss.nasa.gov/gistemp/graphs), accessed 14 December 2018. Data are global mean land–ocean temperature index referenced to the base period 1951–1980. The temperature curve is a five-year running average. Two climate anomalies often referred to in the literature, the Medieval Warm Period (MWP) and the Little Ice Age (LIA), are shown here.

The impacts and scope of this warming, including rising sea level, changing weather and climate (cooling in some places, warming in others, desertification in some areas, increased rainfall in others), changes to fluxes in the hydrologic cycle (including increased flood intensity and melting glaciers), increased storm intensity, loss of CH_4 from deep-sea sediments, peat bogs, and permafrost regions, and ecosystem shifts and losses, are too vast to cover within the scope of this chapter. In order to learn more of the science of climate change the reader is referred to Hansen et al. (2010) and in-depth books by Ruddiman (2013), Fletcher (2012), and many others.

QUESTIONS

7.1 At current 400 ppm CO_2 in the atmosphere, the term for effective infrared transmission $(g_h) = 0.61$, and current albedo is ~0.30. Using the Stefan-Boltzmann Law, determine mean global temperature for the current atmosphere, and also for a potential future world with atmospheric CO_2 concentration of 550 ppmv $(g_h = 0.60)$ and an albedo diminished by 10% (i.e. from 0.30 to 0.27).

7.2 Describe the relationship between $CO_{2(atm)}$ and mean global temperature over the past 400 000 years. Is it a direct correlation? Explain why or why not. Also examine the relationship between $CO_{2(atm)}$ and mean global temperature over the past 400 million years. Is it a direct correlation? Explain why or why not.

7.3 Calculate the density of relatively dry air (0.5% H_2O vapor) relative to humid air with 2.8% H_2O vapor. Express the answer in $g\,mol^{-1}$ and in terms of $g\,m^{-3}$. State any assumptions you might make to simplify the calculation.

7.4 Diagram 5 GHGs and explain what molecular properties cause them to "trap" IR radiation. Describe how these compounds interact with IR at the molecular level. For comparison, also diagram two non-GHGs.

7.5 Critically assess this comment: adding more CO_2 to the atmosphere will not cause warming because the IR-absorbing capacity of CO_2 in the atmosphere is already saturated in the 4250–4260 nm range. Is this statement true? False? Partially true? Justify.

7.6 Using 288 K as the temperature of Earth, determine the maximum wavelength of radiation emitted from Earth's surface.

REFERENCES

Augustin, L. et al. (2004). Eight glacial cycles from an Antarctic ice core. *Nature* 429: 623–628. https://doi.org/10.1038/nature02599 (full reference is presented in Chapter 7).

Berner, R.A. (1997). The rise of plants and their effect on weathering and atmospheric CO_2. *Science* 276: 544–546.

Berner, R.A. (2006). GEOCARBSULF: a combined model for Phanerozoic atmospheric O_2 and CO_2. *Geochimica et Cosmochimica Acta* 70: 5653–5664. https://doi.org/10.1016/j.gca.2005.11.032.

Berner, E.K. and Berner, R.A. (1996). *Global Environment: Water, Air and Geochemical Cycles*. Prentice-Hall.

Fletcher, C. (2012). *Climate Change: What the Science Tells Us*. Wiley.

Flückiger, J., Monnin, E., Stauffer, B. et al. (2002). High resolution Holocene N_2O ice core record and its relationship with CH_4 and CO_2. *Global Biogeochemical Cycles* 16 (1): https://doi.org/10.1029/2001GB001417.

Forster, P., Ramaswamy, V., Artaxo, P. et al. (2007). Changes in atmospheric constituents and in radiative forcing. In: *Climate Change 2007: The Physical Science Basis. Contribution of Working Group I to the Fourth Assessment Report of the Intergovernmental Panel on Climate Change* (ed. S. Solomon, D. Qin, M. Manning, et al.). Cambridge, United Kingdom and New York, NY, USA: Cambridge University Press http://www.ipcc.ch/pdf/assessment-report/ar4/wg1/ar4-wg1-chapter2.pdf.

Hansen, J., Ruedy, R., Sato, M., and Lo, K. (2010). Global surface temperature change. *Reviews of Geophysics* 48 (4): https://doi.org/10.1029/2010RG000345.

IPCC (Intergovernmental Panel on Climate Change) (2013). *Climate change 2013: The physical science basis*. Working Group I Contribution to the IPCC Fifth Assessment Report. Cambridge, United Kingdom: Cambridge University Press.

Lovelock, J.E. and Margulis, L. (1974). Atmospheric homeostasis by and for the biosphere: the Gaia Hypothesis. *Tellus* 14: 1–10.

MacFarling Meure, C., Etheridge, D., Trudinger, C. et al. (2006). Law dome CO_2, CH_4 and N_2O ice core records extended to 2000 years BP. *Geophysical Research Letters* 33: L14810. https://doi.org/10.1029/2006GL026152, ftp://ftp.ncdc.noaa.gov/pub/data/paleo/icecore/antarctica/law/law2006.txt.

Moberg, A., Sonechkin, D.M., Holmgren, K. et al. (2006). Highly variable Northern Hemisphere temperatures reconstructed from low- and high-resolution proxy data. *Nature* 433: 613–617.

Petit, J.R., Jouzel, J. et al. (19 authors) (1999). Climate and atmospheric history of the past 420,000 years from the Vostok Ice Core, Antarctica. *Nature* 399: 429–436.

Ruddiman, W.F. (2003). The anthropogenic greenhouse era began thousands of years ago. *Climatic Change* 61: 261–293.

Ruddiman, W.F. (2013). *Earth's Climate: Past and Future*, 3e. New York: W.H Freeman and Co.

Spiro, T.G. and Stigliani, W.M. (2003). *Chemistry of the Environment*, 2e. Upper Saddle River, NJ, USA: Prentice Hall.

Steffensen, J.P., Andersen, K.K., Bigler, M. et al. (2008). High-resolution Greenland ice core data show abrupt climate change happens in few years. *Science* 321: 680–684.

Vonder Haar, T.H. and Suomi, V.E. (1971). Measurements of the Earth's radiation budget from satellites during a five-year period. Part 1: extended time and space means. *Journal of Atmospheric Science* 28: 305–314.

Zachos, J.C., Rohl, U. et al. (13 authors) (2005). Rapid acidification of the ocean during the Paleocene-Eocene Thermal Maximum. *Science* 308: 1611–1615.

Air Quality: Urban and Regional Pollutants

This chapter focuses primarily on atmospheric contaminants that are harmful to human and ecological health on urban to regional scales. First comes the important role of oxygen and water in atmospheric reactions; decomposition of molecules of O_2 and H_2O in the presence of UV radiation create highly reactive free radicals, examples of which include oxygen, hydroxyl, and peroxy radicals. After principles of atmospheric chemistry are covered, Chapter 8 presents information related to sources of air pollutants, recent trends in atmospheric abundance of pollutants, and the chemical transformations responsible for decomposition of pollutants (e.g. of volatile organic compounds, NO_x, O_3, CO, SO_2) or mechanisms by which less chemically reactive constituents (e.g. particulates, Pb, and other metals) are removed from air. Additional topics in Chapter 8 include NO_x and SO_2 as agents of acid deposition (~acid rain) formation, the roles of NO_x and O_3 in photochemical smog formation, and the differences between impacts of ground-level ozone and stratospheric ozone.

8.1 AIR POLLUTION: DEFINITIONS AND SCOPE

The focus of regulatory policies – and Chapter 8 – has been on atmospheric constituents that are directly harmful to humans (or contribute to the formation of smog) and often are emitted from point sources (e.g. coal plants, sewage treatment facilities) that are feasible to regulate. Well-known examples include lead (Pb), carbon monoxide (CO), fine-grained particulate matter (PM-2.5), nitrogen oxides (NO_x), ground-level ozone (O_3), sulfur dioxide (SO_2), benzene, and other volatile organic compounds (VOCs), as well as in some cases trace elements like arsenic (As) and cadmium (Cd). The first six of these are the criteria pollutants regulated by the US Environmental Protection Agency (EPA) in outdoor air. The European Commission regulates additional pollutants, including As, Cd and Ni, polycyclic aromatic hydrocarbons (PAHs), and benzene.

The health effects of air pollutants are wide-ranging, from respiratory illnesses, especially for asthmatics (NO_x, SO_2, PM-2.5, and O_3), to diminished delivery of oxygen to vital organs (CO), to impaired cognitive and neurological capabilities (Pb). Many VOCs (e.g. benzene and PAHs) are carcinogenic.

The atmospheric gases discussed in Chapter 7 mostly pertain to elements and compounds that are not directly toxic to humans. CO_2, CH_4, and N_2O are emitted by natural processes as well as anthropogenic ones and would only be toxic in extremely high concentrations that are not encountered in the atmosphere. N_2 is virtually inert in the presence of most organisms, and O_2 is essential for respiration. CO_2 in an interesting consideration because it is emitted from a wide range of natural as well as anthropogenic sources (e.g. from point-source coal emissions to non-point-source automobiles and household wood stoves) and is homogenous in concentration globally. These factors present barriers to regulation compared to commonly monitored and regulated "air pollutants" such as Hg, Pb, SO_2, CO, and so on. Thus, while CO_2 is increasing due to anthropogenic activity – and is causing negative consequences – it is not considered in this chapter, which focuses instead on urban and regional air contaminants.

8.2 OXYGEN AND ITS IMPACT ON ATMOSPHERIC CHEMISTRY

The abundance and reactivity of O_2 exerts a very strong control on the composition of the atmosphere. Its presence makes air a highly oxidizing environment, causing reduced or not-completely-oxidized gases (e.g. VOCs, N_2O, CH_4, CO, NO_x, SO_2) to be unstable, and reaction with oxygen is a common means of contaminant transformation in the atmosphere.

O_2 has a very high electronegativity (only that of fluorine is higher) and hence has a strong tendency to react with electron donors, a good example being sodium metal (Na^0), an atom which will lose a valence electron to O_2 gas. The reaction stoichiometry is:

$$4Na^0 + O_2 \rightarrow 4Na^+ + 2O^{2-} \qquad (8.1)$$

Sodium metal does not occur in the atmosphere because it is so unstable in the presence of free O_2; early in Earth's history, virtually all sodium would have oxidized to Na^+ and either formed compounds (e.g. albite, $NaAlSi_3O_8$) or dissolved in water.

Organic matter (benzene in this case) is unstable in the atmosphere and undergoes oxidation to carbon dioxide:

$$C_6H_6 + 15/2 O_2 \rightarrow 6CO_2 + 3H_2O \qquad (8.2)$$

In this case, carbon atoms are converted from an average charge of −1 in benzene to +4 in carbon dioxide. However, as is apparent for the reaction of methane to carbon dioxide in Eq. (7.4), mere thermodynamic instability does not necessarily guarantee that organic matter or other chemically reduced trace constituents of the atmosphere will be oxidized. A typical benzene molecule will spend hours to days in the atmosphere before undergoing oxidation. Methane residence time in the atmosphere is 12 years and its concentration is many orders of magnitude greater than would be predicted based solely on thermodynamic considerations, so kinetic factors play an important role in the abundance of many unstable gases in the atmosphere (Section 7.1.5).

Factors controlling reactions of O_2 and other gases in the atmosphere include:

1 The probability of a collision involving oxygen and the reduced gas – trace gases may persist for longer than thermodynamically predicted due to relatively low frequency of collisions with O_2.

2 The fleeting opportunity for O_2 and a gas like CH_4 to react once they collide – these are gases moving at high speeds so they commonly collide and immediately bounce off each other in a Brownian motion style. This creates very different reaction conditions than those in waters, soils, or the crust of the Earth – a good example of the opposite condition would be the drawn-out time frame involved with O_2 reacting with pyrite in soil pores.

3 Water is a critical medium for oxidation reactions. Owners of old airplanes store their planes in deserts. This is interesting because the H_2O-poor desert atmosphere contains more oxygen than humid air, yet planes will rust if stored in a rainforest. The obvious key here is the role of water, functioning as a medium for the chemical reaction; thus, thinking about atmospheric reactions, if O_2 and an oxidizible gas both dissolve in water (e.g. a rain drop), the probability of a chemical reaction increases.

4 The O_2 molecule must overcome an energy barrier before it can react. A good example of an oxidation reaction that is not thermodynamically favored in the atmosphere at ambient temperatures (as compared to those in combustion chambers) is the oxidation of nitrogen gas to nitrogen monoxide.

$$N_2 + O_2 = 2NO \qquad \Delta H = +180 \, kJ \, mol^{-1} \quad (8.3)$$

The higher electronegativity of O_2 would have led us to predict that N_2 would lose electrons to O_2, but the reaction is not spontaneous. The positive enthalpy of formation indicates that this is an endothermic reaction; with added heat (e.g. in internal combustion engines, coal power plant furnaces, and hot forest fires), the reaction is more likely to occur. The inability of O_2 to oxidize N_2 gas in the atmosphere lies in the valence electron structure of the O_2 molecule.

8.3 FREE RADICALS

Singlet oxygen is responsible for some of the oxidation reactions in the atmosphere, but most of the airborne chemical species controlling oxidation reactions are various types of *free radicals*, a class of atoms and molecules which possess unpaired electrons in valence orbitals. Most free radicals are electrically neutral, but some are ionic; in most cases, the unpaired electrons make free radicals extremely reactive. Free radicals typically are fragments of compounds that have been broken into molecular-scale fragments by UV radiation or by chemical reactions between compounds and other free radicals. The unpaired electrons that cause these atoms or molecules to be classified as free radicals are the remnants of paired electrons that once occupied a single (σ) chemical bond in a molecule.

Focus Box 8.1

Not All O₂ Molecules Are the Same: Triplet vs Singlet Oxygen

Oxygen atoms in O_2 molecules in the ground state have an electronic configuration of $1s^2\ 2s^2\ 2p^4$, indicating that O_2 contains two unpaired electrons in the valence shell. The four p-level electrons can fill six possible locations distributed among three paired p orbitals. There are two possibilities for filling these sites, and the one involving an electron pair in the $2p_x$ orbital and two unpaired parallel electrons in the $2p_y$ and $2p_z$ orbitals is the configuration known as *triplet oxygen* (Figure 8.1). This is the stable, or ground-state, form of O_2.

The other main configuration of O_2 is termed *singlet oxygen*, consisting of oxygen atoms with electron pairs in $2p_x$ and $2p_y$ orbitals but no electrons in $2p_z$ (Figure 8.1) – this hybridization of oxygen yields a molecule whose energy level is $94\,kJ\,mol^{-1}$ above the energy level of triplet oxygen. It is far less stable than triplet oxygen and much more likely to react with reduced compounds (e.g. electron donors like organic molecules).

Because of its valence electron structure, singlet oxygen can react with electron pair donors and thus is far more reactive than triplet O_2. One of the ways that singlet oxygen forms in the atmosphere is via reaction of triplet O_2 with organic compounds that contain unpaired electrons and act as electron donors. Given that UV radiation can transform molecules to elevated energy states (i.e. with respect to valence electrons), these organic compounds with unpaired electrons often have been altered by UV radiation, making them more likely to react with triplet O_2 to yield the highly reactive singlet O_2.

Figure 8.1 schematically shows the differing O_2 electron arrangements.

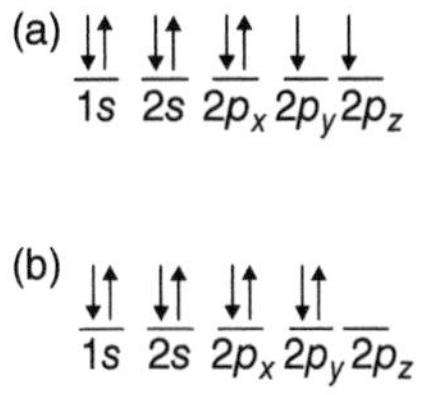

Fig. 8.1 (a) Electron orbitals and spins (spin-up, spin-down) in triplet oxygen. This is the electron configuration of stable, ground-state oxygen. (b) Electron orbitals and spins (spin-up, spin-down) in singlet oxygen. Note vacant $2p_z$ orbital and paired electrons of opposite spin in $2p_y$. Singlet hybridization of O_2 is far more reactive than triplet oxygen.

Free radicals are so unstable and reactive because of the high-energy state of the unpaired electron. This electron will seek to find another electron with which to pair up and create a stable chemical bond. Often, in order to pair up the high-energy electron, the free radical will react with a molecule containing paired electrons (these types of molecules are far more common than those with unpaired electrons); in doing so, the reaction will create a new free radical. In some cases, two free radicals will react and the result is molecules that are no longer free radicals, but usually the result is formation of a different type of free radical. This can set off a "free radical chain reaction" where reaction of a free radical with a molecule containing paired electrons produces a new free radical and new molecule containing paired valence electrons, and so on. An example of such a chain reaction is the catalytic destruction of stratospheric ozone presented in Section 8.10.

An illustrative example of free radical formation is the *photodissociation* of oxygen in the presence of UV radiation:

$$O_2 + h\upsilon \rightarrow {}^\cdot O^* + {}^\cdot O^* \tag{8.4}$$

The symbol "·" denotes that the atom (or molecule) is a free radical, the superscript "*" indicates the occurrence of a high-energy valence electron, and the resulting oxygen *atoms* are high-energy free radicals.

One of the most important free radicals in atmospheric chemistry is the hydroxyl radical, ${}^\cdot HO^*$, which forms in many ways, one of which is photodissociation of a water molecule:

$$H_2O + h\upsilon \rightarrow {}^\cdot HO^* + {}^\cdot H \tag{8.6}$$

Hydroxyl radicals can also form when a high-energy free radical oxygen atom reacts with a water molecule:

$$H_2O + {}^\cdot O^* \rightarrow {}^\cdot HO^* + {}^\cdot HO^* \tag{8.7}$$

Hydroxyl radicals react with many compounds in the atmosphere, including nitrogen oxides, sulfur oxides, and organic compounds. Hydroxyl radicals commonly react by either adding themselves to molecules or by abstracting a hydrogen atom from a molecule. An example of the latter (in this case, octane + hydroxyl radical) is:

$$C_8H_{18} + {}^\cdot HO^* \rightarrow {}^\cdot C_8H_{17}^* + H_2O \tag{8.8}$$

Table 8.1 Ambient air quality standards according to the US Environmental Protection Agency (EPA) and the EU European Commission (EC).

Pollutant	Level (EPA)	Avg Time	Level (EC)	Avg time
SO_2	30 ppb (78 µg m^{-3})	1 yr	—	—
SO_2	—	—	125 µg m^{-3}	24 h
SO_2	75 ppb (196 µg m^{-3})	1 h	350 µg m^{-3}	1 h
NO_2	53 ppb (100 µg m^{-3})	1 yr	40 µg m^{-3}	1 yr
NO_2	100 ppb (188 µg m^{-3})	1 h	200 µg m^{-3}	1 h
CO	9 ppm (10 mg m^{-3})	8 h	10 mg m^{-3}	8 h
CO	35 ppm (40 mg m^{-3})	1 h	—	—
$PM_{2.5}$	15 µg m^{-3}	1 yr	25 µg m^{-3}	1 yr
$PM_{2.5}$	35 µg m^{-3}	24 h	—	—
PM_{10}	150 µg m^{-3}	24 h	50 µg m^{-3}	24 h
O_3	0.070 ppm (140 µg m^{-3})	8 h	120 µg m^{-3}	8 h
Pb	0.15 µg m^{-3}	3 mo	0.5 µg m^{-3}	1 yr
As	—	—	6 ng m^{-3}	1 yr
Cd	—	—	5 ng m^{-3}	1 yr
Ni	—	—	20 ng m^{-3}	1 yr
Benzene	—	—	5 µg m^{-3}	1 yr
PAHs [a]	—	—	1 ng m^{-3}	1 yr

[a] Quantified by measuring Benzopyrene.

Focus Box 8.2

Reaction of Hydroxyl Radicals with Organic Molecules

Often, hydroxyl radical reactions with organic compounds (e.g. partially combusted hydrocarbons) are generically represented as:

$$R–H + \cdot HO^* \rightarrow \cdot R^* + H_2O \qquad (8.5)$$

In this case, "R" represents some constituent of an organic compound and "$\cdot R^*$" is the free radical formed when the hydroxyl radical stripped a hydrogen atom from the organic compound. The organic compound free radical is now very unstable and will likely react rapidly, and in this way, the hydroxyl radical contributes to decomposition of volatile organic compounds in the atmosphere.

Hydroxyl radicals also oxidize carbon monoxide to carbon dioxide, as follows:

$$CO + \cdot HO^* \rightarrow CO_2 + \cdot H^* \qquad (8.9)$$

Note that in addition to CO_2, a hydrogen radical is also produced. This neutral hydrogen atom, with its unpaired single electron, is very reactive and often combines with oxygen molecules to form the *hydroperoxyl radical*:

$$\cdot H^* + O_2 \rightarrow \cdot HO_2^* \qquad (8.10)$$

The hydroperoxyl radical, $\cdot HO_2^*$, also commonly forms as an intermediate product in the oxidation of most hydrocarbon combustion products and by-products.

This introduction to free radicals provides a basis for understanding sources, residence times, and reaction pathways of various air pollutants. The following sections discuss concepts related to pollutants in ambient air (i.e. outdoor air). Table 8.1 summarizes US and EU standards for ambient air and Plate 10 shows an urban air reporting system in Granada, Spain.

8.4 SULFUR DIOXIDE

Anthropogenic sulfur dioxide emissions globally are primarily derived from coal combustion, which contributes approximately double the amount of SO_2 annually ($\sim$30 Tg SO_2–S) to the global atmosphere as does petroleum combustion, and as much atmospheric SO_2 as is contributed by all other anthropogenic non-fossil-fuel SO_2 sources combined (e.g. metal refining, international shipping, biomass fuels, waste incineration, and more).

Global anthropogenic emissions of SO_2 (Figure 8.2) peaked in the early 1970s and generally declined until

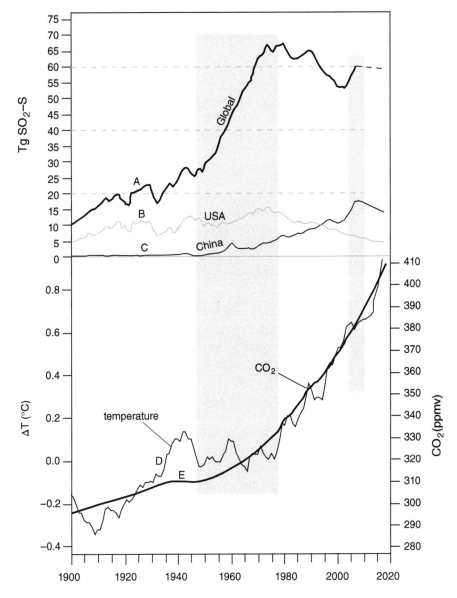

Fig. 8.2 SO_2 data from 1900 to 2018 showing (a) global emissions (estimated for 2010–2018), (b) United States emissions, and (c) China emissions, as well as global mean surface temperature (d) and atmospheric CO_2 concentration (e) Shaded bars indicate period where elevated SO_2 emissions (and resulting sulfate particle formation) may have mitigated greenhouse-related warming (e.g. Kaufmann et al. 2011). Sources for (a, b and c) are Smith et al. (2011), by data from Lu et al. (2010) and Li et al. (2017); (d) is from *US Global Climate Change Report* (http://www .globalchange.gov/publications/reports/ scientific-assessments/us-impacts/full-report) and (e) is from USA NASA Goddard Institute for Space Studies Surface Temperature Analysis web page accessible at http://data.giss.nasa.gov/ gistemp/graphs (note: data are accessible in tabular form as well as graphical form).

2000, but began to rise again in the early 2000s as China and other industrializing countries have increasingly turned to coal as an energy source. Anthropogenic SO_2 emissions in 2010 were $\sim$60 Tg SO_2 (Smith et al. 2011), an amount which is sixfold greater than natural SO_2 sources, which are on the order of $\sim$10 Tg SO_2 per year, predominantly derived from volcanic emissions (Stoiber et al. 1987). It is important to note that natural emissions of SO_2 constitute only $\sim$10% of total natural *sulfur* emissions; this is because the majority of natural S emissions to the atmosphere occur as low-residence-time sea salts (as sulfate) and dimethylsulfide (DMS, the gas derived from marine microorganisms).

Coal typically contains 1–3% sulfur contained mainly in pyrite (comprising 25–50% of the sulfur in coal) and

in thiophene (comprising 30–75% of S in coal) (Huffman et al. 1991). Pyrite (FeS_2) is a common mineral found in chemically reduced deposits such as coal, but it does not occur in hydrocarbons, helping to explain why coal typically contains double the sulfur content of liquid hydrocarbons.

Thiophene is an organic molecule with a pentagonal cyclic structure and the molecular formula C_4H_4S. The average oxidation state of sulfur in pyrite is S^-; in thiophene, it is S^0. Combustion of coal or any other fossil fuel causes oxidation of reduced sulfur (S^-, S^0) to S^{4+} and results in formation of SO_2, which, for thiophene, can be expressed as:

$$C_4H_4S + 6O_2 \rightarrow 4CO_2 + 2H_2O + SO_2 \qquad (8.11)$$

Focus Box 8.3

Sulfates and Climate: Scatterers that Contribute to Cooling

Sulfates in the atmosphere have the opposite effect of greenhouse gases because SO_4^{2-} particles function as scatterers of incoming solar radiation. When combustion adds SO_2 to air, one of the climate considerations is the potential cooling effect exerted by sulfate particles. Cooling from the 1950s to the 1970s and mitigated warming between 2000 and 2010 (Figure 8.2) may be due to sulfate emissions that are capable of counteracting radiative forcing of increased greenhouse gas concentrations (e.g. Kaufmann et al. 2011).

The main sources of SO_2 (and therefore, sulfates) in the 1950–1970s were Europe and the United States, but stricter emissions standards (e.g. The Clean Air Act and subsequent amendments in the USA) resulted in decreased SO_2 emissions in spite of increased fossil fuel use. Mitigated warming from 2000 to 2010 may be related to increased SO_2 emissions from southeast Asia, although emissions controls in China are already beginning to diminish SO_2 emissions from this region (Figure 8.2). If Kaufmann et al. (2011) are correct in their assessment of the role of sulfates as warming mitigators, the high rates of temperature increase seen from ~1980 to 2000 are likely to again appear.

The highest oxidation state of sulfur is S^{6+} and this is the form most stable in the oxidizing atmosphere, so once emitted from a smokestack or tailpipe, SO_2 is eventually oxidized from $S^{4+} \rightarrow S^{6+}$ according a wide array of reactions. Given the instability of the hydroxyl radical and the tendency of S^{4+} in SO_2 to lose two additional electrons, this is a common reaction:

$$SO_2 + {}^{\cdot}HO^* \rightarrow {}^{\cdot}HOSO_2^* \tag{8.12}$$

The compound $HOSO_2^*$ (which is also written as HSO_3^*) is an intermediate in the trajectory of sulfur dioxide to sulfuric acid; the oxidation state of sulfur is S^{5+}, so in order to reach the +6 oxidation state, $HOSO_2^*$ reacts with water and oxygen to produce sulfuric acid and the hydroperoxyl radical:

$${}^{\cdot}HOSO_2^* + O_2 + H_2O \rightarrow H_2SO_4 + {}^{\cdot}HO_2^* \tag{8.13}$$

The hydroperoxyl radical will go on to react rapidly with other gases and the sulfuric acid will dissolve in water droplets and rain out of the atmosphere.

The first dissociation constant of sulfuric acid is very high ($K_a = 2.4 \times 10^6$) and the second (K_a for $HSO_4^- = H^+ + SO_4^{2-}$ is 1×10^{-2}) is sufficiently high so that nearly all H_2SO_4 completely dissociates as follows:

$$H_2SO_4 \rightarrow 2H^+ + 2SO_4^{2-} \tag{8.14}$$

The products of this reaction dissolve in water droplets and then are rained out of the atmosphere. The kinetics of the stages of transformation from SO_2 to dissolved H^+ and SO_4^{2-} result in typical atmospheric residence time of SO_2-derived sulfur of two to eight days.

While these reactions demonstrate that SO_2 will eventually react to form dissolved H^+ and SO_4^{2-}, it is important to recognize two important consequences of SO_2 reaction kinetics:

1 An SO_2 molecule typically remains in air for one to two days (Lee et al. 2011), during which time it can exert negative effects on the human respiratory system; and

2 SO_2 and its derived reaction products (e.g. $\cdot HOSO_2^*$, H^+, SO_4^{2-}) are commonly carried hundreds of kilometers in prevailing winds, meaning that SO_2-derived sulfuric acid is often finally delivered in precipitation downwind of the original SO_2 emissions. More on this is presented in Section 8.11.

8.5 NITROGEN OXIDES

Although they are the two most abundant gases in air, N_2 and O_2 do not react with each other at ambient atmospheric temperatures due to the strength of covalent bonds in these molecules and the stable electron configuration of triplet O_2 (Figure 8.1). In the elevated temperature of internal combustion engines and coal furnaces (typically $\geq 2000\,°C$) however, N_2 and O_2 react to form NO_x.

Regarding terminology, NO_x consists of (NO and NO_2); when other oxidized forms of nitrogen are also included (e.g. HNO_3, NO_3 radical, N_2O_5), the blanket term is NO_y.

The nitrogen in NO is N^{2+} and in NO_2 it is N^{4+}. The highest oxidation state of nitrogen is +5, so in the atmosphere NO_x will oxidize to ultimately form N^{5+} (its form in the NO_3^- anion). The majority of NO_x produced by combustion of fossil fuels is NO, a stoichiometry dictated by the greater abundance of N_2 compared to O_2 in air pulled into the engine. NO emitted from stacks or tailpipes

---

Focus Box 8.4

Air Quality, NO_x, and Temperature of Combustion

Combustion is a problem no matter what the temperature. Low temperatures are undesirable because they create sooty emissions, so higher temperatures (~2000 to 2200 °C) are necessary for efficient combustion and cleaner emissions in furnaces and engines. One of the unavoidable results of combustion is the dissociation of N_2 and O_2 molecules into free radicals; some reform as N_2 and O_2, but others inevitably are reformulated into NO ("nitric oxide" or nitrogen monoxide) and NO_2 (nitrogen dioxide), especially given the kinetic factors involved (high abundance of N and O radicals interacting with each

other). Given the endothermic nature of NO_x formation (Eq. (8.3)), higher temperature fosters greater extent of NO_x formation – the heat of combustion provides the energy needed to facilitate this endothermic reaction. Diesel engine combustion occurs at a higher temperature than for gasoline engines, so diesel vehicles emit more NO_x. Some high-temperature natural processes also foster formation of NO_x, particularly lightning and forest fires (thus, an additional anthropogenic source of NO_x to the atmosphere is biomass burning, whether in the field or in power-generating plants).

---

typically reacts relatively rapidly with free radicals; the example here involves reaction with the hydroperoxyl radical to produce nitrogen dioxide and a hydroxyl radical:

$$NO^* + \cdot HO_2{}^* \rightarrow NO_2{}^* + \cdot HO^* \qquad (8.15)$$

NO can also react with O_2 to form NO_2 as follows:

$$2NO^* + O_2 \rightarrow 2NO_2{}^* \qquad (8.16)$$

The resulting NO_2 can follow different pathways. One is that it may decompose during the daytime in the presence of incoming UV radiation into nitrogen monoxide and a radical oxygen atom:

$$NO_2{}^* + h\upsilon \rightarrow NO^* + O^* \qquad (8.17)$$

The oxygen radical commonly will bond with O_2 to form ozone, and this is one of the key reactions in the photochemical smog system. More on that topic is presented in Section 8.9.

The NO^* radical may also react with O_2 (e.g. singlet O_2 or an electronically excited O_2 radical):

$$2NO^* + O_2{}^* \rightarrow 2NO_2 \qquad (8.18)$$

NO^* can also react with oxygen and water to form nitrous acid:

$$4NO^* + O_2 + 2H_2O \rightarrow 4HNO_2{}^* \qquad (8.19)$$

The nitrous acid that forms in reaction (8.19) is prone to photodissociation in the presence of UV to yield nitrogen monoxide and hydroxyl radicals, unstable and highly reactive products which then will immediately participate in many of the other reactions shown in this section, as well as in Sections 8.6, 8.9, and 8.10.

$$HNO_2{}^* + h\upsilon \rightarrow NO^* + HO^* \qquad (8.20)$$

The residence time of NO and NO_2 are dependent on the composition of the atmosphere (i.e. presence/absence of HO^*, O^*, etc.) and time of day (presence/absence of solar radiation). If ambient air contains ozone (and thus O^*), NO will react very rapidly (minutes to hours); if not, reaction with the hydroxyl radical typically takes place over one to two days. In the presence of the hydroperoxyl radical, NO_2 will react in time spans of ~1 to 3 hours; in the presence of intense solar radiation, NO_2 will decompose even more rapidly. In the absence of $HO_2{}^*$ or UV, NO_2 will react with HO^* on time spans of ~1 to 3 days.

NO_x is problematic for many reasons:

1 It is a biologically available form of nitrogen that contributes to eutrophication;

2 It ultimately results in formation of nitric acid, one of the two main components of acid deposition (along with sulfuric acid); and

3 NO_2 is problematic for the human respiratory system, particularly for lungs affected by asthma or other respiratory disorders.

Globally, NO_x derived from fossil-fuel combustion (Figure 8.3) shows a similar trend to global SO_2 emissions (Figure 8.2), although it is evident that emissions controls have been somewhat more effective regarding SO_2 compared to NO_2 (Figure 8.4).

To summarize NO_x atmospheric chemistry succinctly, the NO emitted during fossil fuel combustion is transformed into NO_2, which is then ultimately transformed to HNO_3 and rained out of the atmosphere as a component of acid deposition and as reactive nitrogen that contributes to eutrophication. However, as we have seen, the potential reactions of NO_x are abundant and include many species and various homogeneous and heterogenous reactions that proceed in the direction $NO \rightarrow NO_2$, but also $NO_2 \rightarrow NO$. These reactive gases are extremely

Focus Box 8.5

Homogeneous vs. Heterogeneous Reactions in the NO_x–HNO_3 System

Some reactions in the atmosphere are homogeneous and some are heterogeneous. The formation of the nitrous acid radical in Eq. (8.19) is a *homogeneous reaction* because both reactants, i.e. NO^* and O_2, are gases. Nitric acid can form by the *homogeneous gas phase reaction* of nitrogen dioxide and a hydroxyl radical:

$$NO_2^* + HO^* \rightarrow HNO_3 \qquad (8.21)$$

or by a pair of homogeneous gas phase reactions (that occur in the absence of solar radiation, i.e. at night) followed by the heterogeneous reaction of a gas (N_2O_5) and liquid (H_2O):

$$NO_2^* + O_3 \rightarrow NO_3^* + O_2 \qquad (8.22)$$

then

$$NO_2^* + NO_3^* \rightarrow N_2O_5 \qquad (8.23)$$

then

$$N_2O_5 + H_2O \rightarrow 2HNO_3 \qquad (8.24)$$

Nitric acid can also form by the *heterogeneous reaction* of $NO_2^*{}_{(g)}$ and $H_2O_{(l)}$:

$$2NO_2^* + H_2O \rightarrow HNO_2^* + HNO_3 \qquad (8.25)$$

The HNO_3 that forms by the various NO_x reactions is a soluble acid ($K_a = 2.4 \times 10^1$) that dissolves in water to yield H^+ and NO_3^-, soluble ions that are removed from the atmosphere in precipitation. Along with H_2SO_4, HNO_3 is a component of acid deposition (Section 8.11).

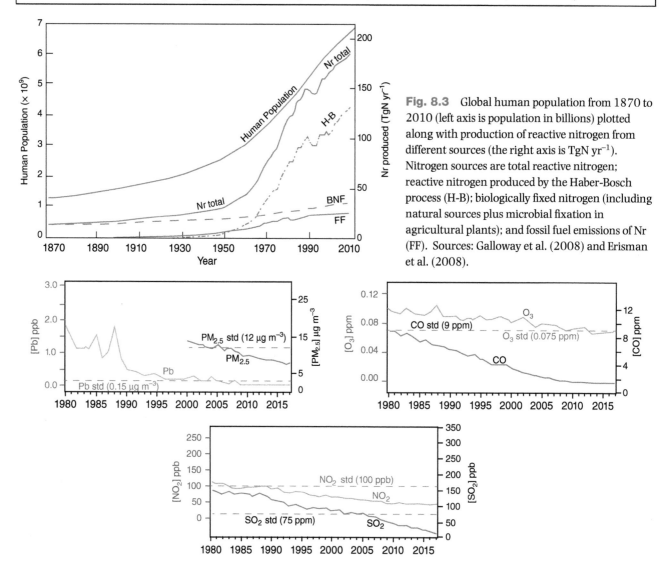

Fig. 8.3 Global human population from 1870 to 2010 (left axis is population in billions) plotted along with production of reactive nitrogen from different sources (the right axis is TgN yr^{-1}). Nitrogen sources are total reactive nitrogen; reactive nitrogen produced by the Haber-Bosch process (H-B); biologically fixed nitrogen (including natural sources plus microbial fixation in agricultural plants); and fossil fuel emissions of Nr (FF). Sources: Galloway et al. (2008) and Erisman et al. (2008).

Fig. 8.4 Trends in ambient air contaminants in the USA for the period 1980–2017, with the exception of $PM_{2.5}$, for which the standard arrived in the late 1990s. Note for NO_2, Pb, and O_3, the y-axis is elevated above the bottom line either to separate the plotted curves, or to overlay standard lines (for CO and O_3). Source: Data from https://www.epa.gov/air-trends.

important components of the troposphere and stratosphere, and given their importance in photochemical smog, acid deposition, and stratospheric ozone, NO_x appears again in subsequent sections in this chapter.

8.6 CARBON MONOXIDE

Carbon monoxide is an odorless, tasteless gas that is toxic to humans because it has a much stronger affinity for hemoglobin sites than does oxygen (the difference is ~300-fold), and for this reason, excess CO in air will diminish delivery of oxygen to cells in the body. At ambient concentrations (e.g. 0.1–0.2 ppmv in most air), CO will occupy 1% of hemoglobin-binding sites; however, in contaminated urban air with concentrations of 10–50 ppmv (the EPA standards are 9 and 35 ppmv; Table 8.1), up to $\sim10\%$ of hemoglobin sites will be occupied by CO, causing headaches and difficulty breathing. In tunnels, parking garages, and other enclosed areas where fossil-fuel emissions are not ventilated, concentrations can reach 100 ppmv, and in indoor air affected by improperly venting furnaces and stoves, humans can lose consciousness when values reach 250 ppmv. Death is nearly certain at concentrations ≥ 750 ppmv. Even at lower levels (e.g. ~5 to 10 ppmv), humans with heart problems can experience heart attacks when exposed for short periods. In most cases, the effects of CO poisoning are reversible and treatable by breathing uncontaminated air or by O_2 medical therapy.

Annual emissions of CO to the atmosphere are ~2800 Tg CO, and the three main sources are (i) oxidation of methane by the hydroxyl radical, which accounts for 800 Tg CO per year; (ii) biomass burning, at 700 Tg CO per year; and (iii) combustion of fossil fuels, which produces 650 Tg CO per year. These three sources comprise > 75% of the total CO flux to the atmosphere. Other sources include partial oxidation of organic compounds in air and emissions from vegetation and the oceans.

Incomplete combustion, whether during fossil-fuel combustion in a low air : fuel mixture, or during biomass burning, can be illustrated by this reaction of octane with oxygen:

$$C_8H_{18} + 8.5O_2 = 8CO + 9H_2O \qquad (8.26)$$

Increased air (i.e. increased O_2) will foster combustion that produces carbon dioxide:

$$C_8H_{18} + 12.5O_2 = 8CO_2 + 9H_2O \qquad (8.27)$$

In reality, the kinetics of combustion dictate that at least some CO will form during combustion, and the high concentrations of vehicle emissions and other sources of volatile hydrocarbons in urban and industrial areas can combine to produce CO-contaminated air.

Formation of CO from the oxidation of methane is a multistep process, beginning with reaction of methane with the hydroxyl radical:

$$CH_4 + HO^* \rightarrow CH_3^* + H_2O \qquad (8.28)$$

After additional steps, the reaction series culminates with photodissociation of formaldehyde (HCHO) and, in some cases, subsequent rapid reaction with O_2:

$$HCHO \rightarrow H_2 + CO \qquad (8.29)$$

or

$$HCHO\,(+O_2) \rightarrow 2HO_2 + CO \qquad (8.30)$$

The details of this reaction series are beyond the scope of this book but are covered in Seinfeld and Pandis (2006) and other similar sources. The mean residence time of CO in the atmosphere is approximately 30–100 days, and the main mechanism by which carbon monoxide is removed from the atmosphere is via reaction with the hydroxyl radical to produce carbon dioxide and a hydrogen radical:

$$CO + HO^* \rightarrow CO_2 + H^* \qquad (8.31)$$

Emissions controls and increased fuel efficiency have helped to reduce CO emissions in most countries since the early 1970s (see Figure 8.4 for data on CO in the US).

8.7 PARTICULATE MATTER

Particulates in the atmosphere are commonly termed aerosols ("air" + "soil"). The main natural processes responsible for the presence in air include:

1 wind erosion of soil, especially from arid regions, in which case the main aerosols are minerals such as calcite, gypsum, illite, iron hydroxides, and much more;

2 emissions of volcanic ash, which range from fragments of exploded volcanic rock to sulfate particles;

3 wind-derived sea salts (sea spray, mainly sulfates); and

4 soot from forest fires and emission of organic particulates from trees.

Anthropogenic activities that expose soil to erosion, and add sooty particles from combustion or industrial processes, add to the natural fluxes.

Aerosols range in size from ~1 nm ($\sim$ the thickness of a single unit cell of a clay mineral) to $\sim20\,\mu m$ (fine silt); in reality though, most particulates are in the range of

---

Focus Box 8.6

Particulates and Human Health

Countering their important beneficial roles in the atmosphere is the fact that particulates can be quite harmful to human health. In general, any particulates finer than 10 μm, particularly those finer than 2.5 μm ($PM_{2.5}$), are potentially harmful because they are difficult to expel from lungs, and thus they can accumulate in alveoli. So while quartz is not a notoriously toxic mineral, it can cause silicosis of the lungs when sufficient quantities of $PM_{2.5}$ quartz are inhaled over years to decades. Of particular concern are asbestiform mineral grains, especially from the amphibole group (e.g. crocidolite, see Chapter 2), and compounds known to be carcinogenic such as benzo(a)pyrene, a PAH whose structure is shown in Figure 3.9 and whose concentration in air is regulated by the European Commission (Table 8.1). The only $PM_{2.5}$ class that is likely not harmful in elevated amounts are soluble, nontoxic molecules like sea salts. Fortunately, as is the case with many other ambient air pollutants, concentrations have shown notable decreases due to enforceable air standards (Figure 8.4).

0.1 μm (10 000 nm) to 10 mm. Particles finer than 0.1 μm are simply less common, and at sizes greater than 10 μm, particles either are not easily entrained into air or, if they are entrained into air by wind, settle relatively rapidly.

Aerosols include the following groups:

1 relatively insoluble inorganic particles, i.e. minerals (e.g. illite, quartz, kaolinite);

2 relatively insoluble organic particles derived from incomplete combustion of biomass and hydrocarbons, as well as from natural plant-generated organic compounds. These compounds range from soot dominated by clusters of hexagonally arranged carbon atoms (like in graphite) to organic molecules such as benzo(a)pyrene and terpenes emitted by coniferous trees;

3 soluble ionic salts, mostly chlorides (e.g. NaCl) as well as sulfates such as H_2SO_4, NH_4SO_4, $NaSO_4$, and others. These compounds function as particulates when they act as condensation nuclei. Ultimately, the salt will likely dissolve in the droplet once water coalesces around the salt nucleus; and

4 toxic metals (e.g. Cd, Cr, Ni), metalloids (e.g. As, Se), and organic compounds (e.g. PAHs) are often sorbed onto particulates (mainly of classes 1 and 2 listed above), and thus can increase the toxicity of the particulate.

Aerosols can also be classified based on how they arrive in the atmosphere. Those that are introduced to the atmosphere in the solid state (e.g. sea salts, mineral particles, soot) are often referred to as *primary particulates*. Aerosols that form in the atmosphere from gases are sometimes referred to as *secondary particulates*, a good example of which is particulate SO_4^{2-} derived from SO_2 or DMS.

The amount of human-derived particulate matter in a given area varies widely, with populated areas containing the greatest amount of anthropogenic emissions; for example, Weijers et al. (2011) indicate that human activities account for roughly 75–85% of total particulates in the Netherlands. Globally, anthropogenic sources may account for ~20% of total particulates in the atmosphere (Wallace and Hobbs 2006), a lower value that reflects the dominance of natural sources in rural areas (e.g. windblown dust in deserts, sea salts in coastal areas, terpenes in pine forests, etc.). The main anthropogenic sources are (i) combustion of fossil fuels and biomass burning that results in soot and other organic aerosols, and (ii) enhanced erosion of mineral and organic particles, mainly from roads and cleared lands (in some cases this can lead to desertification).

The functions of particulates in the atmosphere are covered in various detail in other sections of the text, so only will be presented in summary form here. Their main roles are as:

1 condensation nuclei;

2 scatterers of incoming solar radiation, especially sulfates;

3 absorbers in incoming solar radiation, especially black soot; and

4 transporters of soil nutrients via windblown dust (e.g. Sahara to the tropics).

8.8 LEAD (PB)

Improvement in air quality with respect to lead since the 1970s is one of the great success stories of environmental regulation. Lead is a toxic metal that attacks the central nervous, cardiovascular, reproductive, and immune systems; it also impairs kidney function and can decrease the oxygen carrying capacity of blood. Some of the most notable effects of lead exposure are manifested in children, causing diminished developmental progress which results in lowered IQs. In adults, lead can result in high blood pressure and heart disease as well as the other effects listed above. It may persist for extended periods in the human body because it can accumulate in bones (Pb^{2+} behaves somewhat like Ca^{2+} in the human body), then decades later be released from bone back into blood during periods of bone turnover, e.g. menopause and osteoporosis (Silbergeld et al. 1988).

Prior to the successes of restrictions imposed by law, automobiles were the main source of lead in the atmosphere. Tetraethyl lead, $(C_2H_5)_4Pb$, was added to gasoline to prevent knocking (i.e. to improve combustion and engine output). Figure 8.4 shows that emissions have dropped drastically since the 1970s; the main sources of lead to the atmosphere in the twenty-first century are point sources such as smelter and metal refineries, waste incinerators, and lead-acid battery manufacturers.

8.9 HYDROCARBONS AND AIR QUALITY: TROPOSPHERIC OZONE AND PHOTOCHEMICAL SMOG

Unlike in the stratosphere, where ozone (O_3) is beneficial – it absorbs UV radiation (Sections 7.3 and 8.10) – ozone in the troposphere is problematic because it can impair lung function. Common symptoms include throat irritation, inflammation of lung tissue, chest pain, coughing, and congestion. Long-term exposure may permanently scar lung tissue, and it is particularly harmful to people with asthma, bronchitis, and emphysema. Furthermore, ozone is a stronger oxidizer than O_2, so when it is elevated above background in the lower troposphere, it is also harmful to leaves of plants (and thus reduces crop yields and affects forest health) and can oxidize anything else prone to oxidation, including plastics and rubber. These negative impacts can appear at concentrations as low as 100 ppbv (0.1 ppmv) (refer to Table 8.1 for current standards).

Photochemical smog and tropospheric ozone concentrations are closely related, so will be presented together in this section. There are two main types of smog (smoke + fog). One consists mainly of soot, smoke, and PAHs in air produced by partial combustion of fossil fuels in urban and industrial areas. This is the smoky, sooty air that plagued London for decades, most notably the 1952 smog event that resulted in thousands of premature deaths, where the contaminants were *primary pollutants* derived directly from tailpipes and smokestacks. This type of smog also resulted in 20 immediate deaths and thousands of cases of respiratory ailments in the small mill town of Donora, Pennsylvania, USA, in 1948.

The other main type of smog is *photochemical smog*, a type of air pollution that involves a series of reactions

resulting in a wide range of *secondary air pollutants* that are derived from primary pollutants, mainly NO_x and VOCs emitted by combustion of fossil fuels. The other key ingredient in photochemical smog is sunlight which leads to formation of radicals, hence the "photo" in photochemical. Photochemical smog is harmful to people because it consists of elevated concentrations of NO_2, ozone, and other substances that cause respiratory problems.

Photochemical smog forms in a series of reactions beginning with $NO \rightarrow NO_2$ (Eqs. (8.15), (8.16), and (8.18)). Nitrogen dioxide is a brown gas that contributes to the characteristic color of photochemical smog. In the presence of sunlight and high-energy UV ($h\upsilon$), NO_2 breaks down, releasing an isolated, reactive radical oxygen atom plus a NO radical:

$$NO_2 + h\upsilon \rightarrow NO^* + O^* \qquad (8.32)$$

The radical oxygen atom is then free to react with O_2 to form ozone:

$$O^* + O_2 \rightarrow O_3 \qquad (8.33)$$

Ozone is unstable so quickly reacts with nitric oxide (nitrogen monoxide):

$$O_3 + NO \rightarrow O_2 + NO_2 \qquad (8.34)$$

Continued production of NO by combustion and resulting formation of NO_2 followed by photodissociation into NO^* and O^* results in continued production of ozone, and NO_x and ozone would reach a steady-state concentration controlled by NO production (from combustion) if not for the added impact of VOCs emitted by incomplete combustion.

Combustion by-products (i.e. VOCs, e.g. fragments of alkanes) play an important role because they contribute to the formation of *hydroperoxyl radicals* which react with NO to produce NO_2, which dissociates in the presence of sunlight to yield NO and O which then produce ozone according to the reactions shown above. Thus, VOCs raise ozone concentrations above what would be produced by NO_x alone. Hydroperoxyl radical formation from VOCs is shown in the series of reactions that follow, understanding that these examples are representative of a very large family of reactions.

R is used in the examples as a variable that represents a constituent of an organic molecule, commonly a broken (or partially combusted) hydrocarbon chain. For example, R could represent C_8H_{17}, an octane molecule minus one H, in which case RH would be the complete octane molecule C_8H_{18}; alternatively, R could designate a fragment of an aromatic molecule (e.g. C_6H_5, benzene minus one H atom), or any other potential by-product of incomplete combustion.

Hydrocarbon fuels are rich in alkanes, so the "R" involved in photochemical smog is typically derived from an alkane. When emitted from a tailpipe or smokestack, the hydrocarbon can react with radical oxygen atoms and electronically excited oxygen gas to produce a hydrocarbon radical and a hydroxyl radical:

$$RH + O^* + O_2^* \rightarrow ROO^* + HO^* \qquad (8.35)$$

If R is an alkyl chain (e.g. the octane fragment suggested above), the term for ROO^* (or RO_2^*) is *alkylperoxyl radical*; it is highly reactive and commonly will react with NO to produce an *alkoxyl radical* (RO^*) plus nitrogen dioxide:

$$ROO^* + NO^* \rightarrow RO^* + NO_2^* \qquad (8.36)$$

This reaction series shows how VOCs (e.g. partially combusted alkanes) contribute to smog formation by producing NO_2, which yields O^* in the presence of sunlight, resulting in formation of ozone. Thus, both the VOCs and NO produced by combustion are sources of NO_2; therefore, both are triggers for photochemical smog reactions.

Some by-products of combustion can react to form additional harmful secondary pollutants. For example, the alkoxyl radical may react with O_2 to produce an aldehyde and a hydroperoxyl radical. In this case, an ethyl group (CH_3CH_2) is shown as the basis of the alkoxyl radical (although keep in mind that alkoxyl radicals typically are much larger molecules):

$$CH_3CH_2O^* + O_2 \rightarrow CH_3CHO + HO_2^* \qquad (8.37)$$

CH_3CHO is acetaldehyde (also known as ethanal in the International Union of Pure and Applied Chemistry [IUPAC] system or as ethyl aldehyde), a probable carcinogen. If the radical above had been CH_3O^* rather than $CH_3CH_2O^*$, the product would have been HCHO (formaldehyde), an irritant to eyes, nose, throat, and skin and also a probable carcinogen. Then, to make matters worse, the hydroperoxyl radical will contribute to the formation of NO_2 and, through the reaction series shown above, ozone as well.

Acetaldehyde will react with the hydroxyl radical to produce a new alkoxy radical plus water:

$$CH_3CHO + HO^* \rightarrow CH_3CO^* + H_2O \qquad (8.38)$$

This radical will rapidly react with oxygen to yield the acetylperoxy radical $CH_3COO_2^*$:

$$CH_3CO^* + H_2O \rightarrow CH_3COO_2^* \qquad (8.39)$$

The acetylperoxy radical rapidly reacts with NO_2 to form the secondary pollutant *peroxyacetyl nitrate (PAN)*, a relatively stable gas and eye irritant commonly associated with photochemical smog:

$$CH_3COO_2^* + NO_2^* \rightarrow CH_3COO_2NO_2 \qquad (8.40)$$

The stability of PAN in air is very temperature dependent, ranging from ~30 minutes at 25 °C (Bridier et al. 1991) to 5 years at −25 °C (Kleindienst 1994); it often persists long enough to serve as a means of transporting reactive N in the atmosphere.

Figure 8.5 depicts a typical photochemical smog sequence instigated by morning automobile traffic

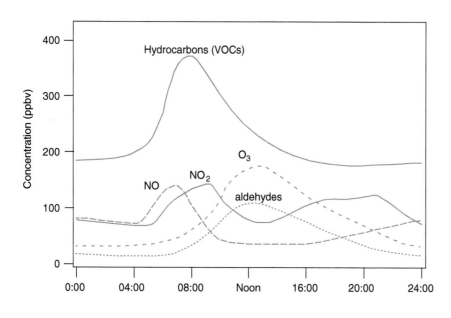

Fig. 8.5 Representative example showing concentrations of smog constituents instigated by morning vehicle traffic. Source: Adopted from Finlayson-Pitts and Pitts Jr (2000) and Spiro and Stigliani (2003).

Focus Box 8.7

Smog and Air Inversions: The Relationship of Air Quality to Topography

Topography and climate can combine to exacerbate air quality problems, especially in cities situated in valleys or topographic basins that limit air circulation. Air inversions occur when warm air rests atop colder air, constituting the opposite of the normal situation where air temperature decreases upwards in the atmosphere (Figure 7.1). Inversions like this may occur when cool marine air blows inland and becomes trapped below warmer air upwind of a mountain range (e.g. Los Angeles, USA), or when a warm front arrives and the warm air moves over the top of the existing colder, denser air mass. Pollutants released into the cooler, denser air do not rise above the inversion layer, instead becoming trapped in the cooler, often foggy air of the valley. Santiago, Chile, is an example of a large city in a topographic basin with large downwind mountains (Andes) that limit air circulation. Rappenglück et al. (2000) provide a detailed study of the atmospheric chemistry of Santiago smog. Plate 11 is an aerial view of southern Spain on a winter day and the inversion layer is quite evident, with brownish NO_2-tinged smog in the cooler air below the clear, warmer air above.

---

showing important aspects of photochemical smog chemistry. First, as commuters begin driving their cars on the way to work, concentrations of VOCs and NO rapidly increase. Shortly thereafter we see evidence for the $NO \rightarrow NO_2$ reactions that occur in the presence of gases such as HO_2^* and O_2. Then, dissociation of NO_2 in the presence of sunlight results in production of radical oxygen atoms (O^*), which react rapidly with O_2 to produce ozone, which commonly reaches its peak concentration in the afternoon. The drawn-out peak of ozone reflects production from NO_x as well as from VOCs and their by-products.

In summary, Figure 8.5 serves to illustrate (i) the effect of sunlight, particularly on dissociation of NO_2 and formation of O_3; (ii) the sequence of reactions; (iii) the approximate overall rates of reactions; and (iv) common concentrations of smog components, from background levels to peak levels. Also apparent is that NO_2 reaches its lowest concentration around noon, when maximum incoming solar radiation strongly drives $NO_2 + hv \rightarrow NO^* + O^*$. Concentrations of nitrogen dioxide then begin to rise in the early evening hours as photochemical dissociation of NO_2 slows down.

Most large cities are affected by photochemical smog to at least some extent. Some notable examples are Los Angeles, Mexico City, Beijing, and Rome.

8.10 STRATOSPHERIC OZONE CHEMISTRY

In the middle of the stratosphere, at an altitude of approximately 25–30 km, ozone concentrations are approximately 15 000 ppbv, or roughly 100 times greater than the amount of ozone present on a smoggy urban day. In the stratosphere, O_2 atoms are split apart into radical oxygen atoms by ultraviolet radiation (effectively, high-energy photons) with wavelengths <242 nm. (A graphical representation of the range of UV absorption by ozone is shown in Figure 7.8.) The pair of reactions is simple:

$$O_2 + hv \rightarrow O^* + O^* \qquad (8.41)$$

then (same as Eq. (8.33)):

$$O_2 + O^* \rightarrow O_3 \qquad (8.42)$$

The second of these reactions occurs in the presence of a third molecule (e.g. N_2) which absorbs some of the energy released by this reaction, explaining why there is no radical on the product (O_3) side.

Intense bombardment of UV with $\lambda < 325$ nm limits ozone concentrations, resulting in an O_2 molecule and another radical oxygen atom:

$$O_3 + hv \rightarrow O_2 + O^* \qquad (8.43)$$

Ozone concentrations are also constrained by the reaction of O_3 with radical oxygen atoms (produced by dissociation of O_2 or O_3 by UV):

$$O_3 + O^* \rightarrow O_2 + O_2 \qquad (8.44)$$

While both ozone and radical oxygen atoms are inherently unstable, the relatively low probability of a productive homogeneous gas-phase reaction limits reaction rate (i.e. by the low potential of interaction of both gas-phase reactants in the diffuse stratosphere).

The dissociation of O_2 and O_3 in the presence of intense UV radiation is what is often referred to as absorption of UV by the ozone layer. High-energy UV rays (with λ between

~230 and 320 nm) are consumed when bonds are broken, and this is crucial to life on Earth given the ability of UV to break bonds in organic molecules of living organisms, causing cancer and other problems.

The altitude of the ozone layer is controlled by the intensity of incoming UV radiation (it is much higher in the upper atmosphere compared to the lower atmosphere) coincident with the fact that there still are sufficient amounts of O_2 at that altitude to produce O_3. Above 30 km, oxygen is too dispersed to result in significant ozone formation. The "thickness" of the ozone layer, or the abundance of stratospheric ozone, varies with latitude, being greatest in the tropics where insolation is greatest, and lowest at the poles, where insolation is lowest. Ozone production is also greater during summer months compared to winter months.

Research by the Dutch chemist Paul Crutzen in the late 1960s had determined that emissions of nitrous oxide, a stable gas in the troposphere, could reach the stratosphere and catalytically destroy ozone as follows:

$$N_2O + O^* \rightarrow 2NO^* \qquad (8.45)$$

Once produced, NO^* reacts with ozone to produce a nitrogen dioxide radical and oxygen:

$$NO^* + O_3 \rightarrow NO_2{}^* + O_2 \qquad (8.46)$$

UV causes dissociation of O_3 to produce radical oxygen atoms:

$$O_3 + h\upsilon \rightarrow O_2 + O^* \qquad (8.47)$$

Radical oxygen atoms are then available to react with nitrogen dioxide:

$$O^* + NO_2 \rightarrow NO^* + O_2 \qquad (8.48)$$

The sum of reactions (8.46)–(8.48) is:

$$2O_3 \,(+NO^*) + h\upsilon \rightarrow 3O_2 \,(+NO^*) \qquad (8.49)$$

NO^* is ultimately neither produced nor consumed during the series of reactions, so it acts as a catalyst in the destruction of ozone. However, calculations indicated that nitrogen oxides were not present in sufficient quantities

---

Focus Box 8.8

Units of Measure of Stratospheric Ozone and the Ozone Hole

The long-term, natural steady-state concentration of ozone in the stratosphere is ~250 to 300 Dobson units, where one Dobson unit represents the hypothetical thickness of the ozone layer if it were pure ozone compressed under conditions of 0 °C and 1 atm. One Dobson unit (DU) = 10^{-2} mm or 10 μm; thus, if the thickness of the ozone layer above some part of the Earth is 300 DU, the thickness of pure ozone in the stratosphere under hypothetical conditions of 0 °C and 1 atm would be 3 mm. One DU also equals 4.46×10^{-4} mol of O_3 per m^2 cross section of stratosphere (i.e. the amount of ozone in a vertical column 1 × 1 m). The DU is named after G.M.B. Dobson of Oxford University, one of the early scientists to investigate atmospheric ozone and designer of the Dobson Spectrometer, the standard instrument used for ground-based measurements of stratospheric ozone. Figure 8.6 shows the precipitous drop in ozone concentrations over Antarctica from ~300 DU in the 1960s to ~100 DU in the 1980s, a trend that initially was recorded by NASA satellites but discarded because it was so extreme that it appeared as though the data were erroneous. Finally, in the early 1980s, scientists working for the British Antarctic Survey, after careful measurements at many locations over many years, published a paper demonstrating the presence of a "hole" in the ozone layer over Antarctica (Farman et al. 1985).

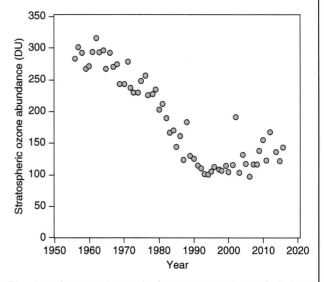

Fig. 8.6 Average stratospheric ozone abundance, in Dobson units (DU), for the month of October above Halley Station, Antarctica. Source: Data are from the British Antarctic Survey (www.antarctica.ac.uk) and NASA (https://ozonewatch.gsfc.nasa.gov/facts/history_SH.html).

to produce the observed decrease in ozone indicated in Figure 8.6. Around the same time, a paper published in 1974 by University of California at Irvine researchers M.J. Molina and F.S. Rowland had demonstrated that chlorofluorocarbons (CFCs) (Focus Box 3.2), which like N_2O are effectively stable in the troposphere, could migrate up to the stratosphere and catalytically destroy ozone much in the same way that nitrogen oxides can.

In the presence of intense UV radiation of the stratosphere, CFCs (using CFC-11, or trichlorofluoromethane, as the example) dissociate into a chlorine atom radical and a CFC radical:

$$CFCl_3 + h\upsilon \rightarrow Cl^* + CFCl_2{}^* \qquad (8.50)$$

The chlorine radical then proceeds to behave like NO^*:

$$Cl^* + O_3 \rightarrow ClO^* + O_2 \qquad (8.51)$$

Radical oxygen atoms produced by ozone dissociation can then regenerate radical Cl^* atoms by these two reactions:

$$O_3 + h\upsilon \rightarrow O_2 + O^* \qquad (8.52)$$

$$O^* + ClO^* \rightarrow O_2 + Cl^* \qquad (8.53)$$

The sum of reactions (8.51)–(8.53) is:

$$2O_3 (+Cl^*) + h\upsilon \rightarrow 3O_2 (+Cl^*) \qquad (8.54)$$

This series of reactions and the summation (8.54) show how CFCs act to catalytically destroy ozone. For their research, Crutzen, Molina, and Rowland were awarded the Nobel Prize in Chemistry in 1995.

The stability of CFCs in the troposphere is part of what made them appealing as refrigerator fluids and propellants (in aerosol cans). Since the Montreal Protocol of 1987 and subsequent amendments, CFCs have largely been phased out and replaced by hydrochlorofluorocarbons (HCFCs, e.g. $CHClF_2$), molecules which have shorter residence times in the troposphere ($\sim$1 to 10 year) and are less likely to reach the stratosphere. HCFCs, however, are greenhouse gases, so research and development continues to seek compounds that are not environmentally harmful. Also, the long residence times of CFCs ($\sim10^2$ years) indicate that they will continue to wreak havoc on the ozone layer well into the twenty-first century (Figure 8.7).

The unique Antarctic climate and halogen chemistry combine to play an important role in ozone destruction. Atmospheric chemist Susan Solomon of the US National Oceanic and Atmospheric Administration (NOAA) attributed the strong seasonal component of ozone destruction (Figure 8.8) to the presence of polar stratospheric clouds (PSCs), wispy high-elevation clouds that form only at very low temperatures (-80 to $-85\,°C$) during Antarctic winter. Observations had indicated a pronounced annual decrease in the stratospheric ozone above Antarctica during the early southern hemisphere spring (September and October), and Solomon determined that PSCs provide a template for chemical reactions that result in a large increase in concentrations of reactive chlorine in late winter (i.e. August).

The solid surfaces (ice crystals) in PSCs provide a template upon which less-reactive secondary chlorine gases (e.g. $ClONO_2$) can be converted to chlorine gas (Cl_2). This is a reaction that does not readily occur in the gas state. Then, as air begins to warm in late winter (August), the reservoir of Cl_2 accumulated by reactions on PSCs escapes from the ice crystal surfaces into the air, is converted to reactive Cl^* upon the return of sunlight and incoming UV radiation during early Antarctic spring, and then rapidly catalyzes ozone destruction – note the precipitous decline in ozone in September and October shown in Figure 8.8. Due to differences in topography (the presence of large mountains in Antarctica and none at the North Pole) that result in differences in atmospheric circulation, these conditions do not occur in the Arctic.

Methyl bromide (CH_3Br) is another source of an ozone-depleting halogen, Br in this case. It has long been used as a fumigant in agricultural applications and it also has natural sources, mainly biomass combustion and marine plankton metabolism. Like CFCs, CH_3Br (MeBr) dissociates in the presence of UV and bromine radicals participate in ozone reactions in a manner similar to chlorine radicals. Its use is now being phased out in many countries.

Given that the "ozone hole" is located above the South Pole, the greatest impacts of ozone depletion are felt in the southern hemisphere in spite of the fact that the majority of CFC emissions have been derived from industrialized countries of the northern hemisphere.

8.11 SULFUR AND NITROGEN AND ACID DEPOSITION

The pH of rainwater in equilibrium with carbon dioxide is $\sim$5.6, as can be determined from calculations as presented in Chapter 5 (Section 5.1.1). In some cases, unpolluted rainfall may have a pH below 5.6, especially in cases where nitrogen or sulfur compounds are emitted from forest fires, wetlands, oceans, or volcanoes. In fact, rains that fall downwind of some volcanoes (e.g. Volcan Poás in Costa Rica, Plate 9) can be very acidic, with pH values as low as

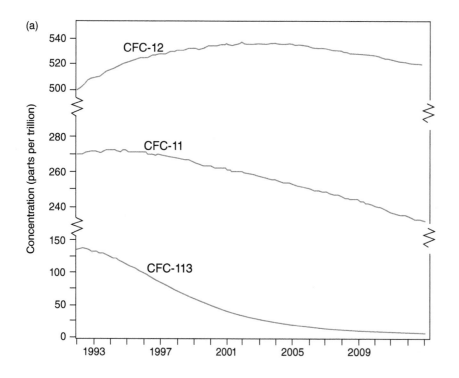

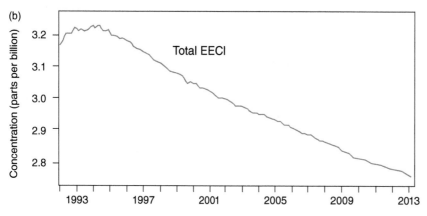

Fig. 8.7 CFC trends. (a) Trends in selected CFCs; (b) overall decrease in chlorine in the atmosphere expressed as total effective equivalent chlorine (EECl) for the time period 1992–2013. Source: Data are from the Global Monitoring Division of the US NOAA Earth System Research laboratory. Data accessed from: http://www.esrl.noaa.gov/gmd/hats. These trends continue, the result of effective international policy making.

2. However, such low values generally are constrained to within a few km of the volcanic source.

By convention, "acid rain" is defined as having a pH < 5. Because water is delivered to the surface of Earth in many forms, including rain, snow, fog, and dry deposition, the term acid precipitation is often used, as is the even more comprehensive term *acid deposition*, which is the term mainly used in this text. This latter term includes acids delivered to soils and water bodies as dry deposition, e.g. acids adsorbed onto particles, as well the acids dissolved in precipitation.

Fossil-fuel emissions are the primary sources of gases NO_x and SO_2 that transform in the atmosphere to nitric acid and sulfuric acid.

Formation of nitric acid can be represented as this homogeneous gas-phase reaction (this is the same reaction as Eq. (8.21) (note that the hydroxyl radical is written by some as OH^* and by others as HO^*):

$$NO_2{}^* + OH^* \rightarrow HNO_3 \qquad (8.55)$$

and the dissociation reaction:

$$HNO_3 \rightarrow H^+ + NO_3{}^- \qquad (8.56)$$

Formation of sulfuric acid can be visualized as a gas-phase reaction (also Eq. (8.13)):

$$SO_2 + OH^* + O_2 + H_2O \rightarrow H_2SO_4 + HO_2{}^* \qquad (8.57)$$

Or as an aqueous-phase reaction (i.e. in water droplets):

$$SO_2 + H_2O_2 \rightarrow H_2SO_4 \qquad (8.58)$$

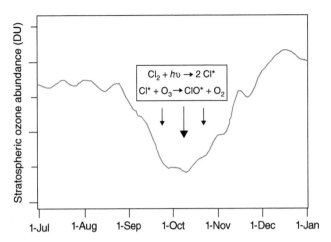

Fig. 8.8 Temporal evolution of ozone abundance above the South Pole showing average daily values from July 1 to January 1 for the years 1986–1995. Annual decrease in ozone abundance from September to November is due to the return of sunlight (and UV) and reactions of ozone with Cl^* in southern hemisphere springtime. See text for discussion of the role of polar stratospheric clouds (PSCs). Data source: Angell et al. (1995).

and the dissociation reaction:

$$H_2SO_4 \rightarrow 2H^+ + SO_4^{2-} \qquad (8.59)$$

The full series of chemical reactions involving N and S species in the atmosphere are more complex than this, and the details can be found earlier in this chapter.

Prevailing winds shown in Figure 7.6 illustrate the general directions in which acid deposition will be transported from source to deposition site. In the industrialized northern hemisphere between 30° and 60° north, prevailing transport of HNO_3 and H_2SO_4 is to the north and east. Notable examples are the eastern United States and southeastern Canada, where coal power plant and industrial and automobile emissions from the southeast, Midwest, and mid-Atlantic regions are carried downwind on prevailing westerlies to rural areas and forested ecosystems. Similarly, in northwestern Europe, emissions from the UK and northern Europe affect downwind forested ecosystems, notably those of Scandinavia. Industrialization has led to acid deposition problems in eastern China, and to a lesser extent, in northern India, and at numerous sites globally where base metal and oil refining and heavy automobile traffic produce acids in the atmosphere.

Ecosystem impacts of acid deposition are caused both by increased H^+ in precipitation water as well as increased concentrations of the nitrate and sulfate anions. Considering that most rain and melted snow are cycled through soils before being incorporated into plants or delivered to streams and lakes, soil geochemical and biochemical reactions largely dictate the effects of acid deposition on a given area. Changes to soil chemistry include acidification, loss of Ca^{2+} and other nutrients, increased solubility of Al (toxic to most plants) and other metals, buildup of NO_3^- and SO_4^{2-}, increased decomposition rates of natural soil organic matter, and leaching of H^+ and NO_3^- into surface water bodies. These changes produce further alterations to surface water chemistry, causing negative consequences such as acidification, eutrophication (due to increased NO_3^-), and decreased aquatic ecosystem health.

Given that acid deposition consequences are strongly controlled by soil mineralogy, chemistry, and biota, these topics are covered in Chapter 9. However, some of the

---

Focus Box 8.9

Kinetic Controls on Nitric Acid and Sulfuric Acid in Acid Rain

Kinetic factors exert a strong control on the reactions that ultimately result in formation of HNO_3 and H_2SO_4. The gas-phase reaction of NO_2 with OH^* occurs approximately 10 times faster than the reaction of SO_2 with OH^* (Seinfeld and Pandis 2006). Furthermore, reaction of NO_2 with OH^* directly produces HNO_3, whereas reaction of SO_2 with OH^* yields the intermediate phase HSO_3^*, which then must undergo at least one more reaction stage before sulfuric acid is formed (e.g. dissociation to SO_3^* then reaction with H_2O, among many possibilities). The important lesson here is that the reaction of NO_2 to HNO_3 occurs more rapidly than the reaction of SO_2 to H_2SO_4, and this can have implications for transport distance of N-derived acid deposition compared to S-derived acid deposition.

In urban air with abundant OH^* and NO_2, acid rain and acid fog may be deposited very close to the source of emissions, particularly in cities situated in enclosed basins like Los Angeles, where acid fogs have been observed to contain more than twice as much nitrate as sulfate (Waldman et al. 1982). Acid deposition produced mainly by coal combustion will produce more than twice as much H_2SO_4 as HNO_3 (not accounting for measures to reduce S or N emissions on site). Given that atmospheric residence time of SO_2 and its derivatives is approximately three to five days (before being deposited in rain, snow, or as dry deposition), and that air masses tend to travel 100 km per day, sulfuric acid deposition is commonly delivered to the land or ocean surface 500–1500 km from its source.

negative consequences take effect before acid deposition even reaches the soil. These include:

1 Damage to leaves and needles of trees that are directly exposed to acidic water droplets in forests of high mountain regions bathed in clouds and fog. In the presence of acids in air, essential nutrients can be stripped away from leaves and needles (e.g. exchange of H^+ for Ca^{2+}), making trees more susceptible to damage by other environmental factors such as disease, pests, and cold winter weather. Most of the negative impacts of acid deposition on forested ecosystems, however, are caused by changes to soil chemistry (Chapter 9).

2 Chemical disintegration of buildings, bridges, statues, memorials, and other objects made of limestone, marble, or certain metals (e.g. bronze, copper). An illustrative example of the effect of acid deposition is the way in which rain water of $pH = 3–5$ attacks the calcite in limestone and marble:

$$CaCO_3 + H^+ \rightarrow Ca^{2+} + HCO_3^- \qquad (8.60)$$

In this way, acid deposition effectively accelerates the natural chemical weathering of carbonate rocks, converting solid calcite to dissolved calcium and bicarbonate ions. Tombstones and sculptures are notable examples.

3 Because sulfates are light-scattering particles, acid deposition can make air hazy and impair views, which can negatively impact otherwise pristine regions.

Environmental science has demonstrated through decades of research the negative impacts of acid rain to ecosystems, humans, and infrastructure, and environmental policy has responded (albeit slowly at times) with legislation aimed at reducing SO_2 and NO_x emissions. As is the case with so many environmental laws, the results are demonstratively positive, resulting in decreases in acid deposition in many regions where impacts have been the most severe, notably in eastern North America and northwestern and eastern Europe. Indications that emissions of SO_2 in China are beginning to decrease (Figure 8.2) suggests that acid deposition may begin to decline there as well. Plates 14–16 present three series of maps showing increasing pH and decreasing nitrate and sulfate concentrations of deposition in the United States in 1994, 2002, and 2009. There are a few important considerations when looking at these maps.

1 The trends of H^+, NO_3^-, and SO_4^{2-} show the potential effectiveness of legislation in improving air quality.

2 US Clean Air Act amendments were initially successful at addressing sulfuric acid deposition, but not nitric acid deposition. Between 1994 and 2002, regulations led to notable improvements in sulfate emissions, but nitrate emissions increased during this period, and it was not until the 2002–2009 period that nitrate emissions began to improve. The reason for the difference between sulfate and nitrate is that the majority of sulfuric acid is derived from *point-source* coal-fired power plant emissions, which are easier to control at the point source than are the *non-point-source* automobile and agricultural emissions responsible for much of NO_x emissions.

3 Sulfate emissions in the 1990s were spread regionally, but note that nitrate in precipitation exhibits localized bullseye patterns, often surrounding cities such as Salt Lake City, Utah (SLC), and the Colorado Springs–Denver–Fort Collins corridor (the Front Range, FR) in Colorado, much of it related to automobile emissions. Decreases in nitrate from 2002 to 2009 are at least partly due to US Clean Air Act amendments requiring that, by 2004, sport utility vehicles (SUVs) and other light-duty trucks become subject to the same national pollution standards as cars.

4 Nitrogen deposition is broadly elevated across the agriculturally intensive High Plains (HP) states, and in spite of reductions between 2002 and 2009, this region continues to produce some of the highest N concentrations in the US. In much of the central HP states, ammonium comprises 25–75% of total nitrogen deposition (Schmeltz 2009). One certain source of NH_4^+ is animal feedlots (concentrated animal feeding operations or CAFOs; Todd et al. 2008), where ammonia derived from urea and other waste products is transformed to ammonium ($NH_3 + H^+ \rightarrow NH_4^+$), and the abundance of ammonium in the air in this region explains why the high N deposition in the Midwest is not correlated to acid deposition. Rather than contributing to acid deposition as does NO_x, ammonia emissions tend to neutralize acidic precipitation, yet are still problematic because they contribute to N pollution of surface and groundwaters, including the Gulf of Mexico "dead zone" at the mouth of the Mississippi River.

5 Atmospheric deposition of H^+, NO_3^-, and SO_4^{2-} are decreasing, but it will likely require decades for forested and aquatic ecosystems to return to pre-acid deposition conditions. Unfortunately, the compositions of sandy, inherently nutrient-poor soils typical of many forested alpine regions have already been significantly altered, as have the pH and dissolved ion chemistry of lakes in these areas. Accumulated acidity, sulfate, and nitrate in soils will continue to leach into surface water bodies for many years or decades, so ecosystem recovery, especially of high alpine forests and lakes, will take time.

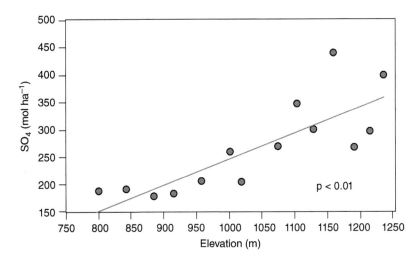

Fig. 8.9 Correlation between elevation and sulfate deposition in the Catskill Mountains of New York State. Source: Data and statistical analysis are from Lawrence et al. (1999).

Note that the maps in Plates 14–16 show *concentrations* of H^+, NO_3^-, and SO_4^{2-}, but not total *quantities* deposited on the landscape (where quantities are expressed in units such as kg ha^{-1} or mol ha^{-1}, rather than mg l^{-1} or mmol l^{-1} units of concentration). This becomes an important consideration when precipitation is not uniform; for example, a mountain range that receives twice as much annual precipitation as an adjacent valley will receive twice as much H^+, NO_3^-, and SO_4^{2-} if concentrations of these ions are equal in rainfall in the two areas. In the northern Appalachians of North America, total deposition of H^+, NO_3^-, and SO_4^{2-} in high-elevation areas is often three to seven times greater than deposition to nearby low-elevation sites (Lovett and Kinsman 1990). Figure 8.9 illustrates the orographic control on the quantity of sulfate deposited in a mountainous region of northeastern North America affected by acid deposition, demonstrating the added acid deposition load that falls on alpine regions compared to valleys and plains.

The *concentration* of sulfate in eastern New York State does not appreciably vary (~1.5 to 2.1 mg l^{-1} in 1994); if anything, the sulfate concentration in precipitation falling on the Catskill Mountains of southeastern New York State appears to have been slightly lower (1.8 mg l^{-1}) in 1994 than at lower elevation near New York City (2.1 mg l^{-1}) (Plate 15). However, the orographic precipitation effect results in significantly greater sulfuric acid *deposition rates* at higher elevation (~400 to 450 mol ha^{-1}) than at lower elevation (~175 to 200 mol ha^{-1}) (Figure 8.9). This combination of orographic effect and inherently nutrient-poor soils in alpine regions combine to explain why the greatest impacts of acid deposition occur at high elevations. More on this topic is presented in the framework of soil biogeochemical and mineralogical reactions in Chapter 9.

8.12 ORGANOCHLORINE PESTICIDES, MERCURY, AND OTHER TRACE CONSTITUENTS IN THE ATMOSPHERE

Toxic trace elements and compounds, both organic and inorganic, are emitted to the atmosphere from a wide range of sources, including energy generation, transportation, agriculture, residential activities, and industry. These include the six pollutants already presented (SO_2, NO_x, particulates, Pb, CO, O_3), but many others as well, including a wide range of organic compounds, fuel derivatives, industrial by-products, solvents, and pesticides, as well as trace metals and metalloids.

8.12.1 Pesticides in air

Organic contaminants in air include VOCs derived from incomplete combustion (Section 8.9), pesticides applied to farmlands, golf courses, residential lawns, parks, etc., and dioxins and furans released from paper-pulp bleaching, waste incineration, and manufacture of organochlorine chemicals used as pesticides and solvents (Section 3.1.6). Usually, harmful effects from organic compounds in air are chronic and express themselves in the form of cancer, disruption of endocrine and nervous system function, or other diseases after years of low-level exposure, but there have also been cases of acute poisoning associated with accidents, perhaps the most tragic being the explosion at a pesticide manufacture plant in Bhopal, India, in 1984 that released toxic methyl isocyanate into the air of Bhopal.

Pesticides enter air in two main ways. First, given that pesticides are often applied by aerial spraying, some enter air immediately as aerosols, droplets, or volatiles. Second,

---

Focus Box 8.10

Methyl Isocyanate, Pesticide Manufacture, and the 1984 Bhopal, India, Disaster

Accidental release of the toxic gas methyl isocyanate (MIC) from a Union Carbide Corp. pesticide manufacturing plant in Bhopal India in December 1984 led to the immediate deaths of ≥3800 people and premature death for thousands more in the years to follow. It is considered the worst industrial disaster in history and was caused at least in part by negligence marked by smaller incidents and warning signs in the months and years prior (Broughton 2005). The release occurred while most were sleeping, and the NY Times reported (R.D. McFadden, 10 Dec, 1984) that residents "stumbled into the streets, choking, vomiting, sobbing burning tears, joining human stampedes fleeing the torment of mist that seemed to float everywhere."

MIC has a very high vapor pressure (58 kPa), and even at very low concentrations, e.g. 0.1 ppm (100 ppb) in air, it causes coughing, chest pain, shortness of breath, asthma, eye irritation, and skin damage. Exposure at levels >20 ppm in air causes lung edema, emphysema, hemorrhaging, and bronchial pneumonia, potentially resulting in death. MIC is a reactant in the manufacture of carbaryl along with 1-naphtalene (Figure 3.18), a carbamate insecticide (sold under the name Sevin). Carbaryl is a carcinogen and structural mimic of the neurohormone melatonin, one consequence of which is that carbaryl bonds to melatonin receptors and worsens risk for diabetes and metabolic disorders (Popovska-Gorevski et al. 2017). Like many pesticides it is an acetylcholinesterase inhibitor (also referred to as an anti-cholinesterase chemical); compounds in this class inhibit the breakdown of acetylcholine by the cholinesterase enzyme, and thus impair central nervous system function.

---

the high vapor pressure of most synthetic pesticides means that at least some portion will vaporize into air from the soil or leaf surfaces, and volatilization can continue for days after application. In some cases, 80–90% of applied pesticide can volatilize within two to three weeks of application (Unsworth et al. 1999), upon which prevailing winds will transport the vapors to downwind areas.

Decomposition of pesticides in air occurs by hydrolysis, reaction with OH* radicals, and by deposition, commonly in rain. Pesticides which resist hydrolysis and reaction with OH* can be transported hundreds to thousands of kilometers, and many organochlorine insecticides have been detected in polar air, waters, and terrestrial ecosystems. While typical concentrations of pesticides in air are in the parts per trillion and parts per billion range, their propensity to bioaccumulate in organisms (from worms to humans) and biomagnify in ecosystems means that trace amounts in air can have noticeable effects on ecosystems as well as humans.

8.12.2 Mercury in air

The heavy metal mercury (Hg) causes impaired neurological development in children and the developing unborn, resulting in impaired brain and nervous system development. Diminished cognitive ability, memory deficits, and incomplete development of skills in the areas of language, fine motor movement, and visual-spatial relationships are common consequences of exposure. Effects are worst when exposure occurs during fetal development. Adults typically do not exhibit symptoms unless exposed to unusually high concentrations associated with point-source contamination or occupational exposure; consequences include tremors, mood swings, irritability, neuromuscular changes (e.g. muscle atrophy and twitching), headaches, and diminished cognitive abilities.

The main pathways causing release of Hg to air in the twenty-first century are coal combustion and waste incineration. According to the UN Environment Programme, coal combustion is the largest single source of Hg to air, and the US EPA estimates that 87% of anthropogenic Hg introduced to air comes from point-source combustion, mainly from coal-fired power plants, waste incineration (e.g. mercury thermometers, fluorescent light bulbs, and batteries), and industrial energy generation. Incineration of coal and municipal wastes volatilizes mercury and it escapes to the atmosphere mainly as elemental mercury (Hg^0) which is then oxidized to Hg^+ and Hg^{2+}. These inorganic forms (Hg^+, Hg^{2+}) are less toxic than Hg^0 and organomercury (e.g. methylmercury), but when deposited on soils or in lake and marine sediments, Hg^+ and Hg^{2+} can be methylated by microbially mediated reactions to produce methylmercury (CH_3Hg) or dimethyl mercury ($[CH_3]_2Hg$). (CH_3)$_2$Hg can volatilize from soils and sediments, but when in slightly acidic to acidic forest soils, the dimethyl mercury can react with H^+ to form a water-soluble methylmercury cation:

$$(CH_3)_2Hg + H^+ \rightarrow CH_3Hg^+ + CH_4 \qquad (8.61)$$

Focus Box 8.11

Mercury Exposure at Minamata Bay, Japan, and the Minamata Convention

An extreme case of mercury poisoning occurred in Minamata, Japan, where a polyvinylchloride (PVC) manufacturing plant was using Hg^{2+} as a catalyst. Discharges to Minamata Bay beginning in 1932 led to elevated mercury levels in organic-rich sediments, an environment where the inorganic Hg^{2+} species undergoes transformation to the far more toxic dimethyl mercury, $(CH_3)_2Hg$. Mercury bioaccumulates in tissues of organisms, and in the marine ecosystem of Minamata Bay, ecosystem decline was observed as fish populations decreased and seaweed along shores disappeared. Humans began to exhibit symptoms in 1956, including numbness of hands and feet, vision problems, and eventually convulsions followed by comas and death. Another devastating consequence

was developmental disabilities in infants exposed during fetal development. By 1959, high Hg concentrations had been detected in shellfish and high trophic-level fish that comprised a significant part of the diet of small fishing communities lining the bay, where the symptoms were most severe. The Minamata case was one of relatively acute exposure at high levels, but health issues caused by chronic exposure to Hg at low levels include hypertension, decreased intelligence, kidney disease, and spontaneous abortion. The acute and chronic toxicity of mercury led to the adoption in 2013 of the international Minamata Convention on Mercury, which seeks to minimize Hg use, emissions, and exposure.

CH_3Hg^+ is soluble, so it tends to be leached into surface waters where it can be incorporated into organisms, and once ingested, methylmercury crosses the intestinal wall. Then, because it binds to S atoms in proteins, it accumulates in the tissues of organisms and tends to biomagnify, causing upper trophic-level organisms (e.g. predatory fish and birds) to contain the highest concentrations of methylmercury in the food chain. When humans eat trout, salmon, and other high trophic-level fish, they too can accumulate Hg, and this is most problematic for young children and developing fetuses. Emissions controls are beginning to have positive impacts in some areas, notably Canada, where emissions have decreased ~80% since the early 1970s, and in the EU, where emissions are down ~60% since the late 1970s; in the US, the EPA is in the process (2011) of implementing new rules that should appreciably decrease Hg emissions.

Many other toxic trace metals and metalloids are emitted to the atmosphere each year, but the focus of the final segment of this chapter is three other trace elements regulated by the European Commission: arsenic, cadmium, and nickel.

8.12.3 As, Cd, and Ni

The metalloid *arsenic* and the trace metals *cadmium* and *nickel* all share common sources; for example, they all occur in sulfide metal ores and thus can be released from ore smelters. They also occur in trace amounts in coal and are released upon combustion in the absence of strong emissions controls. All three largely occur in air in the

particulate phase, often sorbed onto soot or other aerosols (e.g. oxides and hydroxides). Because nickel is more abundant in the crust of the earth, natural and anthropogenic processes release approximately 10 times as much Ni to air compared to As and Cd, so its lower toxicity is counteracted by its greater abundance. Cd^{2+} and Ni^{2+} both mainly occur as divalent cations (whether in compounds or sorbed) and neither has a gas phase; arsenic, however, has multiple oxidation states and can occur in a gaseous state. Given that the sources and atmospheric fates of Cd and Ni are dictated by the behavior of particulates (Section 8.7), their speciation and behavior is simpler than As. For example, arsenic in air can occur as oxides, hydroxides, or weak acids. As compounds comprised of As^{3+} are generally termed arsenites and those bearing As^{5+} are arsenates. The difference is significant because arsenite compounds are more toxic than arsenate. Given the oxidizing character of the atmosphere, compounds consisting of arsenic in lower oxidation states (As^0, As^{3-}) are much less common in air than the more-oxidized arsenite and arsenate forms.

In addition to particulate forms (mainly sorbed As) and dissolved forms (e.g. $H_2AsO_4^-$), arsenic also can occur as As^{3-} in gases, mainly arsine (AsH_3) and organo-As gases such as methylarsine (CH_3AsH_2), dimethyl arsine [$(CH_3)_2AsH$], and trimethyl arsine [$(CH_3)_3As$]. These gases are produced in chemically reducing environments, especially waterlogged soils. In rice paddy soils of the Bengal Fan of Bangladesh and southeast India, groundwater used for irrigation contains elevated arsenic from the natural aquifer materials, and formation of trimethyl arsine is fostered by the presence of soil organic matter (reactive methyl groups), warm temperatures, and microbial activity in the reducing environment of the rice paddy. Volatilization of arsines from soils may account for ~1% to 5% of total arsenic in the atmosphere globally, although this amount varies depending on local factors (e.g. mainly redox conditions and temperature).

Residence times of As in the gas forms of arsines and organoarsines in air is ~ hours to days, so they have potential for downwind transport. Their decomposition is driven by reaction with hydroxyl radicals (HO^*), a process which ultimately results in formation of arsenite or arsenate that may sorb to particulates or dissolve in water droplets and be removed from the atmosphere (Mestrot et al. 2011).

Arsenic was also formerly an ingredient in insecticides such as lead arsenate ($PbHAsO_4$), which was widely applied to fruit trees and other crops to control moths and other pests. By the 1950s, most arsenical insecticide applications had given way to the new generation of organochlorine pesticides, particularly dichlorodiphenyltrichloroethane (DDT) (Sections 3.1.6 and 3.3.3); however, decades of application of arsenical insecticides led to buildup in soils (particularly the organic-rich topsoil [O and A horizons]) in many agricultural regions, and As can enter the atmosphere via wind erosion of particulates. Once applied as lead arsenate, As has a strong affinity for soil organic matter, so often is transported in air sorbed to organic matter aerosols.

QUESTIONS

8.1 What is a free radical? Provide two examples and explain their origin in the atmosphere. Explain why free radicals are so important in atmospheric chemistry, providing two chemical equations to support your response.

8.2 The reaction of N_2 and O_2 to form NO has a positive enthalpy change: $\Delta H = 180\,kJ\,mol^{-1}$. Based on ΔH, is this a spontaneous reaction? If not, why does NO exist in the atmosphere?

8.3 What is the difference between singlet oxygen and triplet oxygen, and why is this significant for atmospheric chemistry?

8.4 Discuss the origin of methylmercury in aquatic ecosystems. In particular, what is the initial source of Hg to the atmosphere, and what transformations take place in with organic matter in soils and sediments to enhance Hg mobility and toxicity? Show with a schematic annotated sketch and chemical equations.

8.5 Compare reaction pathways and kinetics of formation of H_2SO_4 vs. HNO_3 in air. Which reaction occurs faster, and why? What implications does this have for length of travel in air before H_2SO_4 vs. HNO_3 are removed from air via precipitation or dry deposition?

8.6 List the major sources of N and S pollution in air, and be sure to indicate what species of these elements are typically emitted to air as primary pollutants (in contrast to the secondary or tertiary pollutants that form in air).

8.7 By what molecular-scale mechanism does the ozone layer protect the earth's surface from UV radiation? Use appropriate chemical reactions. Would you describe the preindustrial stable concentrations of ozone in the stratosphere as a steady state or dynamic equilibrium? Explain.

8.8 Show the processes and compounds responsible for the catalytic destruction of ozone. Discuss origins of ozone-destroying compounds, their stabilities in air, and show with chemical equations their catalytic effects.

8.9 Describe, with appropriate chemical reactions, the role of VOCs in photochemical smog. Consider also the role of air temperature in the process.

REFERENCES

Angell, J.K., Gelman, M.E., Hofmann, D., et al. (1995). Southern Hemisphere Winter Survey. National Oceanic and Atmospheric Administration (United States). Accessed online: http://www.cpc.ncep.noaa.gov/products/stratosphere/winter_bulletins/sh_95.

Bridier, I., Caralp, F., Loirat, H. et al. (1991). Kinetic and theoretical-studies of the reactions $CH_3C(O)O_2 + NO_2 + M-\leftarrow \rightarrow CH_2C(O)O_2NO_2 + M$ between 248-K and 393-K and between 30-Torr and 760-Torr. *Journal of Physical Chemistry* 95: 3594–3600.

Broughton, E. (2005). *The Bhopal disaster and its aftermath: A review*. Environmental Health: *A Global Access Science Source* https://doi.org/10.1186/1476-069X-4-6.

Erisman, J.W., Sutton, M.A., Galloway, J. et al. (2008). How a century of ammonia synthesis changed the world. *Nature Geoscience* 1: 636–639.

Farman, J.C., Gardiner, B.G., and Shanklin, J.D. (1985). Large losses of total ozone in Antarctica reveal seasonal ClO_x/NO_x interaction. *Nature* 315: 207–210.

Finlayson-Pitts, B.J. and Pitts, J.N. Jr. (2000). *Chemistry of the Upper and Lower Atmosphere – Theory, Experiments, and Applications*, vol. 969. San Diego: Academic.

Galloway, J.N. (2003). The global nitrogen cycle. In: *Biogeochemistry, Treatise on Geochemistry*, vol. Volume 8 (ed. W.H. Schlesinger), 557–583. Amsterdam, the Netherlands: Elsevier.

Galloway, J.N., Townsend, A.R., Erisman, J.W. et al. (2008). Transformation of the nitrogen cycle: recent trends, questions, and potential solutions. *Science* 320: 889–892.

Huffman, G.P., Mitra, S., Huggins, F.E. et al. (1991). Quantitative analysis of all major forms of sulfur in coal by X-ray absorption fine structure spectroscopy. *Energy and Fuels* 5: 574–581.

Kaufmann, R.K., Kauppi, H., Mann, M.L., and Stock, J.H. (2011). Reconciling anthropogenic climate change with observed temperature 1998–2008. *Proceedings of the National Academy of Sciences of the United States of America* https://doi.org/10.1073/pnas.1102467108.

Kleindienst, T.E. (1994). Recent developments in the chemistry and biology of peroxyacetyl nitrate. *Research on Chemical Intermediates* 20: 335–384.

Lawrence, G.B., David, M.B., Lovett, G.M. et al. (1999). Soil calcium status and the response of stream chemistry to changing acidic deposition rates. *Ecological Applications* 9: 1059–1072.

Lee, C., Martin, R.V., van Donkelaar, A. et al. (2011). SO_2 emissions and lifetimes: estimates from inverse modeling using in situ and global, space-based (SCIAMACHY and OMI) observations. *Journal of Geophysical Research* 116: D06304. https://doi.org/10.1029/2010JD014758.

Li, C., McLinden, C., Fioletov, V. et al. (2017). India is overtaking China as the world's largest emitter of anthropogenic sulfur dioxide. *Scientific Reports* 7: 14304. https://doi.org/10.1038/s41598-017-14639-8.

Lovett, G.M. and Kinsman, J.D. (1990). Atmospheric pollutant deposition to high-elevation ecosystems. *Atmospheric Environment* 24A (11): 2767–2786.

Lu, Z., Streets, D.G., Zhang, Q. et al. (2010). Sulfur dioxide emissions in China and sulfur trends in East Asia since 2000. *Atmospheric Chemistry and Physics* 10: 6311–6331. https://doi.org/10.5194/acp-10-6311-2010.

McFadden, R.D. (1984). India disaster: chronicle of a nightmare. *New York Times* (10 December), p. 1.

Mestrot, A., Merle, J.K., Broglia, A. et al. (2011). Atmospheric stability of arsine and methylarsines. *Environmental Science and Technology* 45 (9): 4010–4015.

National Atmospheric Deposition Program (NRSP-3) (2019). NADP Program Office, Wisconsin State Laboratory of Hygiene, 465 Henry Mall, Madison, WI 53706.

Popovska-Gorevski, M., Dubocovich, M.L., and Rajnarayanan, R.V. (2017). Carbamate insecticides target human melatonin receptors. *Chemical Research in Toxicology* 30: 574–582. https://doi.org/10.1021/acs.chemrestox.6b00301.

Rappenglück, B., Oyola, P., Olaeta, I., and Fabian, P. (2000). The evolution of photochemical smog in the metropolitan area of Santiago de Chile. *Journal of Applied Meteorology* 39: 275–290.

Schmeltz, D. (2009). Spatial and Temporal Trends in Sulfur and Nitrogen Deposition for the United States. *Regional Scientific Workshop on Acid Deposition in East Asia 2009*, 12–13 October 2009, Tsukuba, Japan. Accessed online: http://www.eanet.cc/meeting/RWS/presentation/5.pdf.

Seinfeld, J.H. and Pandis, S.N. (2006). *Atmospheric Chemistry and Physics: From Air Pollution to Climate Change*, 2e. Hoboken, NJ, USA: Wiley 1203 pp.

Silbergeld, E.K., Schwartz, J., and Mahaffey, K. (1988). Lead and osteoporosis: mobilization of lead from bone in postmenopausal women. *Environmental Research* 47: 79–94.

Smith, S.J., van Aardenne, J., Klimont, Z. et al. (2011). Anthropogenic sulfur dioxide emissions: 1850–2005. *Atmospheric Chemistry and Physics* 11: 1101–1116.

Spiro, T.G. and Stigliani, W.M. (2003). *Chemistry of the Environment*. Upper Saddle River, NJ, USA: Prentice-Hall.

Stoiber, R.E., Williams, S.N., and Heubert, B. (1987). Annual contribution of sulfur dioxide to the atmosphere by volcanoes. *Journal of Volcanology and Geothermal Research* 33: 1–8.

Todd, R.W., Cole, N.A., Clark, R.N. et al. (2008). Ammonia emissions from a beef cattle feedyard on the southern high plains. *Atmospheric Environment* 42: 6797–6805.

Unsworth, J.B., Wauchope, R.D., Klein, A.W. et al. (1999). Significance of the long range transport of pesticides in the atmosphere. International Union of Pure and Applied Chemistry (IUPAC) pesticide report 41. *Pure and Applied Chemistry* 71: 1359–1383.

Waldman, J.M., Munger, J.W., Jacob, D.J. et al. (1982). The chemistry of acid fog. *Science* 218: 677–680.

Wallace, J.M. and Hobbs, P. (2006). *Atmospheric Science: An Introductory Survey*. Elsevier 504 pp.

Weijers, E.P., Schaap, M., Nguyen, L. et al. (2011). Anthropogenic and natural constituents in particulate matter in the Netherlands. *Atmospheric Chemistry and Physics* 11: 2281–2294.

9

Chemical Weathering, Soils, and Hydrology

As the interface between the atmosphere and freshwater systems, soils play a critical role in the composition of the natural environment and in contaminant fate and transport. Plant nutrients are derived from mineral weathering and organic matter reactions in soil, and most fresh water on Earth has passed through soil at some point (rainfall or snowmelt usually infiltrates soils before recharging lakes, streams, or aquifers); for this reason, soil reactions tend to impart a strong signature to surface waters and groundwater. They influence the cycling of atmospheric CO_2 and other gases, and acid rain and other air pollutants often infiltrate soil, so biogeochemical reactions in soil influence the fate of these and many other air pollutants (e.g. CO_2).

The formation, evolution, and composition of soils is controlled by geological parent materials (rock and sediment), plants, microbes and biochemical reactions, climate, topography, tectonically driven uplift and erosion, and time, and this chapter explores this important medium through the lens of geochemistry.

9.1 CHEMICAL WEATHERING OF PRIMARY MINERALS IN SOILS

9.1.1 Thermodynamic vs. kinetic considerations

Thermodynamic principles are useful to a certain point when considering soil mineralogy and chemistry; for example, almost all high-temperature igneous and metamorphic minerals are unstable in the low-temperature,

oxidizing environment of most soils; conversely, hydrous low-temperature minerals (clays, hydroxides) are more likely to be stable in soils depending on biogeochemical factors. The complicating factor in applying thermodynamic principles is that soil mineral assemblages are nearly always in a state of disequilibrium, commonly consisting of minerals that, given more time, would undergo reactions to decrease the energy of the system (Sparks 2003). So rates of chemical weathering, i.e. kinetic factors, are equally as important – or more so – relative to thermodynamic factors.

The difference in thermodynamic approaches common to the study of higher-temperature systems compared to the relatively low temperatures of soils can be assessed using the (simplified) example of quartz and forsterite (olivine) in glacial till. The onset of soil formation by weathering of quartz–olivine till would have a mineral assemblage that is originally dictated by glacial erosion of parent rock (e.g. a quartzite and a peridotite) and subsequent deposition, and thus at the onset of weathering the composition of the soil is mostly controlled by physical processes rather than by chemical equilibrium. By means of comparison, consider a geological system at high temperature, e.g. amphibolite-grade metamorphic conditions of ~600 °C at 20 km depth in the crust. Under these conditions, reaction (9.1) will occur in the presence of metamorphic fluids in the direction of increasing stability:

$$\underset{\substack{\text{forsterite}\\\text{(olivine)}}}{Mg_2SiO_4} + \underset{\text{quartz}}{SiO_2} \rightarrow \underset{\substack{\text{enstatite}\\\text{(pyroxene)}}}{2MgSiO_3} \qquad (9.1)$$

Environmental and Low-Temperature Geochemistry, Second Edition. Peter Ryan.
© 2020 John Wiley & Sons Ltd. Published 2020 by John Wiley & Sons Ltd.

However, the low temperatures (typically 0–25 °C) of soils do not facilitate this reaction, – enstatite is unstable in soils – and instead what will happen eventually (perhaps over thousands or tens of thousands of years) is that forsterite will decompose via chemical weathering in the presence of weak acids, typically carbonic acid delivered by precipitation. The products of this weathering reaction are dissolved magnesium, silica, and bicarbonate:

$$Mg_2SiO_{4(s)} + 4H_2CO_{3(aq)} \rightarrow 2Mg^{2+}_{(aq)} + H_4SiO_{4(aq)}$$
$$+ 4HCO_3^{-}_{(aq)} \tag{9.2}$$

Quartz is far more resistant to chemical weathering than is forsterite and tends to be stable in soils, so we would predict that, over time, the quartz–forsterite soil would trend toward one dominated by quartz. The soil solution (the water in the pores of the soil) will contain dissolved magnesium, silica, and bicarbonate, at least in this simplified weathering scenario with no inputs of Na^+, Ca^{2+}, Cl^-, SO_4^{2-}, or other constituents. In this example, thermodynamic factors of the soil system dictate that forsterite is unstable, but the low-energy (low-temperature) reaction conditions of soils relative to higher-temperature systems mean that kinetic factors are at least equally important for understanding and quantifying soil composition.

9.1.2 Predicting weathering rates: Goldich stability sequence

The early–mid-twentieth-century geochemist Samuel Goldich examined the changing abundances of primary minerals in crystalline igneous and metamorphic rocks undergoing chemical weathering and realized that he could model chemical weathering as the inverse of Bowen's reaction series (Goldich 1938). Figure 9.1 depicts Bowen's reaction series, which shows the order of crystallization of silicate minerals from igneous melts as they

cool; when examined through the lens of chemical weathering, it is referred to as the *Goldich stability sequence* or *Goldich dissolution series*.

On the discontinuous, ferromagnesian side of Bowen's reaction series, the first mineral to crystalline is olivine (forsterite) at ~1200–1400 °C, followed by pyroxene, hornblende, and biotite. Eventually, non-ferromagnesian minerals will crystallize, starting with K–feldspar followed by muscovite and then, finally, at ~600–800 °C, quartz (stable zircon crystallizes at a similar temperature to quartz). On the plagioclase feldspar side of the reaction series, anorthite (Ca–plagioclase) crystallizes first from magma, and then through a continuous reaction series, evolves to Na–Ca–plagioclase and then eventually albite (Na–plagioclase) with decreasing temperature.

The relationship of Bowen's reaction series to mineral stability in soils and chemical weathering stems from the fact that this igneous crystallization sequence also represents, from top to bottom, increasing polymerization of SiO_4 tetrahedrons. The olivine mineral structure consists of isolated SiO_4 tetrahedrons with no polymerization (Chapter 2, Figure 2.10), and the dominantly ionic bonds that represent Mg–O bonding in olivine are susceptible to breakdown in soil solutions. Polymerization increases downward in the diagram, to progressively increased polymerization of SiO_4 tetrahedrons in pyroxene through the micas and K–feldspar to completely polymerized SiO_4 tetrahedrons in quartz. Quartz is so resistant to weathering because all O atoms are covalently bonded to Si atoms of adjacent silica tetrahedrons (details of SiO_4 polymerization are presented in Chapter 2, Section 2.3.4). The increasing polymerization corresponds to an increasing abundance of covalent bonds that are far less susceptible to decomposing in the presence of water (especially water that contains H^+ ions that promote chemical weathering) than are ionic bonds. The end-member case of ionic bonds are the halide salts, such as NaCl, which appear as the most susceptible to weathering of any mineral in Figure 9.1.

Focus Box 9.1

Chemical Weathering of Minerals Containing Ionic and Covalent Bonds

Many minerals consist of a mix of covalent and ionic bonds – a good example is forsterite (Mg olivine) used in the examples in Section 9.1.1, a silicate mineral that contains ionic bonds between Mg^{2+} and SiO_4^{4-} tetrahedrons, and covalent bonds between Si and O atoms within SiO_4^{4-} tetrahedrons. Thus, dissolution of olivine will result in dissolved Mg^{2+} and silica (as H_4SiO_4) because the

ionic bonds are susceptible to breaking in the presence of water, but the covalent bonds in silica tetrahedrons are not (note: "silica" commonly refers to Si–O species such as SiO_2, H_4SiO_4, or SiO_4^{4-}). Another example of chemical weathering of a mineral with ionic and covalent bonds, gypsum, is presented in Eq. (9.4).

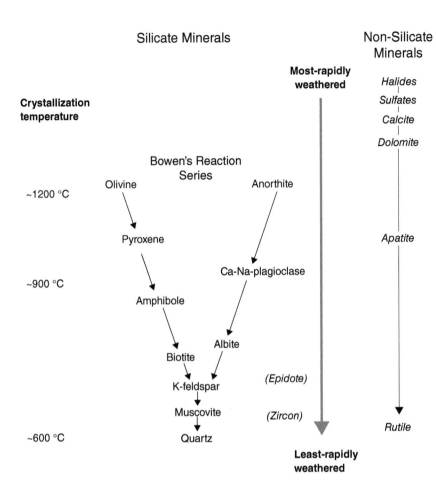

Fig. 9.1 Bowen's reaction series/Goldich dissolution series showing susceptibility to chemical weathering, from least stable at the top to most stable at the bottom (quartz); after Goldich (1938), Lasaga et al. (1994), and Drever (1997). Minerals which are not part of the original Bowen and Goldich series are in italics. Igneous crystallization temperatures are approximate and do not apply to the non-silicate minerals on the right.

9.1.3 Laboratory determinations of primary mineral weathering rates

Differences in weathering rates between minerals depicted in Figure 9.1 span many orders of magnitude (Table 9.1).

One way to understand relative weathering rates is to quantify the amount of time required to dissolve a 1 mm diameter sphere of a given mineral in, for example, a dilute solution at a pH of 5. Quantified in this way, a 1 mm sphere of calcite will dissolve in approximately one month, whereas an equal-size crystal of forsterite olivine will require 2000 years, and a 1 mm sphere of quartz will persist for approximately 34 million years before completely dissolving (Lasaga et al. 1994). Comparisons of field-determined weathering rates and laboratory-determined rates often differ by as much as an order of magnitude, but given that the differences in rates among minerals differ by many orders of magnitude, the values given in Table 9.1 are useful indicators of relative weathering rates of common minerals.

Ultimately, understanding weathering rates is necessary for many reasons, including: assessing the potential of an ecosystem to neutralize effects of acid rain or acid mine drainage (AMD); predicting rates of release of plant nutrients (e.g. Ca, K, P) from minerals; understanding rates of CO_2 uptake over human to geological time scales; determining potential for sorption or release of trace metal contaminants or organic contaminants; interpreting paleoclimate information contained in paleosols; and much more.

9.1.4 Chemical weathering reactions

Changes to structures and compositions of minerals exposed at the Earth surface (usually in soils) that result in release of ions to solution and/or formation of new minerals is chemical weathering. *Primary minerals* are inherited from igneous, sedimentary, or metamorphic rocks, or sediments derived from these rocks, and they are often unstable at the Earth surface because they form under conditions different from Earth surface conditions. They therefore are prone to decomposition by a variety of mechanisms, including dissolution, hydration, hydrolysis, and oxidation. The minerals that form by chemical weathering are known as *secondary minerals* and commonly consist of clay minerals (e.g. smectite, vermiculite, kaolinite),

Table 9.1 Weathering rates of common rock-forming minerals assuming hypothetical 1 mm sphere of each mineral in a dilute solution of pH = 5.

Mineral	Composition	Time (yr)	Dissol. rate (mol m^{-2} s^{-1})	Log dissol. rate (mol m^{-2} s^{-1})
Quartz	SiO_2	34×10^6	4.1×10^{-14}	-13.4
Kaolinite	$Al_2Si_2O_5(OH)_4$	6×10^6	1×10^{-13}	-13
K–feldspar	$KAlSi_3O_8$	740×10^3	5.0×10^{-13}	-12.4
Muscovite	$KAl_2Si_3AlO_{10}(OH)_2$	720×10^3	3.2×10^{-14}	-12.5
Phlogopite	$KMg_3Si_3AlO_{10}(OH)_2$	670×10^3	3.2×10^{-14}	-12.5
Albite	$NaAlSi_3O_8$	500×10^3	6.3×10^{-13}	-12.2
Hornblende	see caption	500×10^3	6.3×10^{-13}	-12.2
Diopside	$CaMgSi_2O_6$	140×10^3	3.6×10^{-12}	-11.4
Anorthite	$CaAl_2Si_2O_8$	80×10^3	4.0×10^{-12}	-11.4
Enstatite	$MgSiO_3$	16×10^3	3.2×10^{-11}	-10.5
Tremolite	$Ca_2Mg_5Si_8O_{22}(OH)_2$	10×10^3	1.1×10^{-11}	-11
Forsterite	Mg_2SiO_4	2×10^3	3.6×10^{-10}	-9.4
Dolomite	$CaMg(CO_3)_2$	1.6	3×10^{-7}	-6.5
Calcite	$CaCO_3$	0.1	1×10^{-6}	-6

Variation in rates between time (yr) versus dissolution rate and log dissolution rate reflects different approaches to measuring rates. See Brantley (2005) for details. The composition of hornblende is $(K,Na)_{0-1}(Ca,Na,Fe,Mg)_2(Mg,Fe,Al)_5(Si,Al)_8O_{22}(OH)_2$.
Source: Data compiled from Chou et al. (1989), Lasaga et al. (1994), Drever (1997), Brantley (2005), Buss et al. (2008) and numerous sources cited therein.

hydroxides (e.g. gibbsite, goethite), oxides (e.g. hematite), carbonates (e.g. calcite), and sulfates (e.g. gypsum). Secondary mineral assemblages are strongly controlled by climate, parent material, soil age, and, perhaps to a lesser extent, the two other soil-forming factors (biotic effects and topography).

9.1.4.1 Hydrolysis

One of the main driving forces behind chemical weathering is the action of carbonic acid dissolved in precipitation (rain and snow) on minerals. (Refer to Chapter 5 for details on formation of carbonic acid). The dissolution of potassium feldspar (microcline, orthoclase, or sanidine) in the presence of carbonic acid and water (reaction (9.3)), is a *hydrolysis reaction* that produces a solid clay mineral (kaolinite) as well as dissolved potassium, silica, and bicarbonate.

$$2KAlSi_3O_{8(s)} + 2H_2CO_{3(aq)} + 9H_2O_{(l)}$$
$$\text{K-feldspar} \quad\quad \text{carbonic acid}$$

$$\rightarrow Al_2Si_2O_5(OH)_{4(s)} + 2K^+_{(aq)} + 4H_4SiO_{4(aq)}$$
$$\text{kaolinite} \quad\quad\quad \text{dissolved silica}$$

$$+ 2HCO_3^-_{(aq)} \quad\quad\quad\quad\quad (9.3)$$
$$\text{bicarbonate}$$

During hydrolysis, H^+ ions (in this case, derived from H_2CO_3) replace cations (in this case, K^+) in the mineral structure, resulting either in complete dissolution of the primary mineral, or in formation of a secondary mineral.

Frequently, secondary silicate minerals (e.g. kaolinite) that form by the hydrolysis of feldspars inherit SiO_4 tetrahedrons from the primary feldspar and incorporate Al from the feldspar and H^+ and H_2O from soil water into the clay mineral octahedral sheet. This reaction also indicates that some silica ends up dissolved in solution as H_4SiO_4, and depending on rainfall and soil permeability, may wind up being leached out of the soil and into surface water or groundwater. The chemical weathering of olivine presented earlier in this section is also a hydrolysis reaction; in that case, four H^+ ions replace two Mg^{2+} ions, resulting in the transformation of $Mg_2SiO_{4(s)}$ to H_4SiO_4 and Mg^{2+} in solution. Hydrolysis of potassium feldspar is presented schematically in Figure 9.2.

9.1.4.2 Dissolution of ionic solids

Minerals dominated by ionic bonds, especially halide and sulfate salts, are prone to simple *dissolution* in the presence of water. The attraction of hydrogen atoms in water molecules to anions in the mineral, or of oxygen atoms in water molecules to cations in the mineral, is greater than the bonds between cations and anions in the salt, so the mineral dissolves, generically into X^+ and Y^- (e.g. Na^+ and Cl^-). Dissolution of the hydrated sulfate gypsum is a good example:

$$CaSO_4 \cdot 2H_2O \rightarrow Ca^{2+}_{(aq)} + SO_4^{2-}_{(aq)} + 2H_2O_{(l)} \quad (9.4)$$

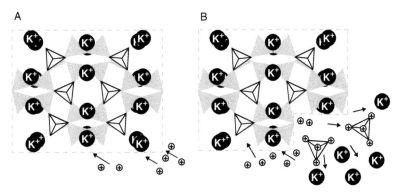

Fig. 9.2 Schematic sketch of K–feldspar hydrolysis. Diagram A is unweathered feldspar in a soil solution with hydrogen ions (small circles with "+" sign) attacking an edge of the crystal. Blue triangles represent silica tetrahedrons whose apices face into the page and the dashed line represents the edge of the crystal. Diagram B depicts the same crystal after hydrolysis has attacked an edge, showing two silica tetrahedrons which are no longer bonded to K^+, but rather are bonded to H^+, resulting in dissolved silica (H_4SiO_4) in solution. Potassium ions are also shown in solution, but in order to maintain clarity in the diagrams, not all K^+ can be shown. Note also that anions in solution (e.g. HCO_3^-) are not shown.

Focus Box 9.2

The Role of Surface Area: Volume in Weathering Rates

Chemical weathering is augmented by *physical weathering*, the mechanical disintegration of rocks that increases exposed surface area per volume of mineral or rock, allowing greater contact of water and air with rock and mineral surfaces. (A good analogy for those who make fires is the greater reaction rate of kindling – i.e. finely split wood with high surface area to volume ratio – compared to the thick, unsplit log from which the kindling was prepared). Common physical weathering processes include wind and water erosion, mass wasting (e.g. landslides), expansion of fractures by growing plant roots, freeze–thaw cycles of ice and liquid water that increase lengths and widths of fractures, and thermal expansion and contraction (especially in deserts) that leads to exfoliation. Any factor that increases the ratio of surface area : volume will increase reaction rate; this makes physical weathering and grain size important parameters to consider when assessing kinetic controls on e.g. mineral dissolution or sorption–desorption reactions.

The covalent bonds in sulfate prevent decomposition of S–O bonds in water, so the resulting products are dissolved ions of Ca^{2+} and SO_4^{2-} plus molecules of water released from the structure upon dissolution.

9.1.4.3 Reduction–oxidation

Many primary minerals that form in reducing environments of the crust contain chemically reduced atoms that make them prone to oxidation when in contact with or in close proximity to the atmosphere. Soil water typically contains enough dissolved oxygen to facilitate *oxidation reactions*, and of the ions in minerals prone to oxidation in soils, Fe^{2+} and S^{2-} are the most important. Other redox-sensitive elements are far less abundant (e.g. As, Mn) or, as is the case with carbon, generally occur in their most-oxidized state in minerals; carbonate minerals, for example, contain carbon in the C^{4+} state and so will not decompose by oxidation. Organic matter, on the other

hand, with carbon ranging from C^{4-} to C^0, is prone to oxidation in soils.

The oxidation of pyrite is quite common in soils, and it occurs via the oxidation of sulfur from S^{6-} to S^{2+} (also presented in Chapters 1, 2 and 4):

$$2FeS_{2(l)} + 2H_2O_{(l)} + 7O_{2(g)} \rightarrow 2Fe^{2+}_{(aq)} + 4H^+_{(aq)} + 4SO_4^{2-}_{(aq)} \qquad (9.5)$$

The ferrous iron released by sulfur oxidation is then oxidized to ferric iron and incorporated into iron hydroxide:

$$Fe^{2+}_{(aq)} + 2H_2O_{(l)} \rightarrow FeOOH_{(s)} + 3H^+_{(aq)} + e^- \qquad (9.6)$$

The iron hydroxide produced (goethite in this case) imparts an orange color typical of iron-bearing soils, and the electron shown as a product will be consumed by an electron receptor such as a soil microbe or perhaps soil nitrate undergoing reduction to N_2 or N_2O. In fact,

Focus Box 9.3

Acid Mine Drainage (AMD) and Sulfide Weathering

Weathering of sulfide minerals is distinct in that it produces acid rather than consuming acid (most minerals, when they undergo weathering, consume H^+). In rocks and soils with abundant pyrite (likely 5% or more), weathering may result in net acidification. This natural occurrence is sometimes referred to as acid rock drainage. When sulfide-rich rock is mined, the high surface area : volume ratio of the crushed ore increases reaction rate, and when mined rock is exposed to atmosphere or oxidized water, it will generate more sulfuric acid than the soil or rock can buffer (the main buffer would be silicate or carbonate weathering). The reaction is effectively catalyzed by bacteria such as *Thiobacillus ferrooxidans* who take advantage of sulfide anions as an electron source. The series of reactions associated with generation of AMD are as follows (Figure 9.3):

1. $2FeS_2 + 7O_2 + 2H_2O = 2Fe^{2+} + 4H^+ + 4SO_4^{2-}$
2. $2Fe^{2+} + 0.5O_2 + 2H^+ = 2Fe^{3+} + H_2O$
3. $Fe^{3+} + 3H_2O = Fe(OH)_3 + 3H^+$
   or
   $FeS_2 + 8H_2O + 14Fe^{3+} = 15Fe^{2+} + 2SO_4^{2-} + 16H^+$

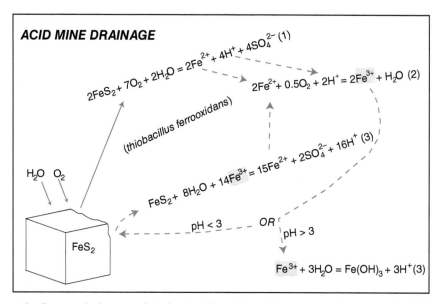

Fig. 9.3 This schematic diagram depicts reactions involved in acid drainage, including the capacity of ferric iron (shaded boxes) to oxidize pyrite.

microbial activity often catalyzes this reaction because it is an energy source. Ferrous iron originally present in silicate minerals also typically undergoes oxidation in a similar manner after hydrolysis liberates it from primary minerals.

Both of these oxidation reactions (9.5) and (9.6) are anomalous as far as chemical weathering is concerned because they produce acid, rather than consuming acid. In rocks and soils rich in pyrite or other sulfide minerals, oxidative weathering can result in *acid rock drainage*, the natural analog to anthropogenic AMD. That said, it is important to realize that nearly all chemical weathering results in *buffering* of acidity inherent to rainfall, mainly

through hydrolysis reactions which consume H^+ and sequester it in mineral structures (e.g. in the octahedral sheet of clays) or in dissolved products (e.g. H_4SiO_4).

9.1.4.4 Hydration

Incorporation of water molecule(s) into the structure of an existing mineral is *hydration*, a process that results in a new structure where the water molecule is rearranged into H^+ and OH^- constituents of the new hydrated mineral structure. The hydration of hematite is a common example of a hydration reaction:

$$Fe_2O_{3(s)} + H_2O_{(l)} \rightarrow 2FeOOH_{(s)} \qquad (9.7)$$

The product of this reaction, the mineral goethite, is sometimes written as $Fe_2O_3 \cdot H_2O$, which helps to illustrate the hydration of the primary mineral hematite (Fe_2O_3), but in reality this is an oversimplification that does not reflect dissociation of H_2O into H^+ and OH^- upon incorporation into the goethite structure. Hydration as shown in reaction (9.7) differs from hydrolysis (reaction (9.3)) in that hydration involves full incorporation of H_2O into the mineral structure, whereas hydrolysis involves replacement of cations in minerals with H^+.

9.2 PRODUCTS AND CONSEQUENCES OF CHEMICAL WEATHERING

9.2.1 Congruous vs. incongruous weathering: dissolved vs. solid products

Products of chemical weathering range from secondary minerals (i.e. solid particles) to dissolved ions. Reactions that result in complete dissolution of a primary mineral are *congruent reactions*, an example of which is hydrolysis of albite under acidic conditions, causing aluminum to remain in solution (rather than forming a secondary mineral):

$$NaAlSi_3O_{8(s)} + 4H^+_{(aq)} + 4H_2O_{(l)} \rightarrow Na^+_{(aq)} + Al^{3+}_{(aq)}$$

$$+ 3H_4SiO_{4(aq)} \qquad (9.8)$$

Under slightly acidic conditions in well-leached soils (e.g. sandy forest soils or some tropical soils), pH may not be sufficiently acidic for aluminum to remain in solution (Chapter 4, Section 4.5) and the solid aluminum hydroxide mineral gibbsite [$Al(OH)_3$] will form:

$$NaAlSi_3O_{8(s)} + H^+_{(aq)} + 7H_2O_{(l)} \rightarrow Na^+_{(aq)} + Al(OH)_{3(s)}$$

$$+ 3H_4SiO_{4(aq)} \qquad (9.9)$$

Weathering reactions that produce dissolved ions and solid minerals, like Eq. (9.9), are *incongruent reactions*.

Reactions (9.8) and (9.9) demonstrate that, for a given primary mineral, chemical disintegration of minerals could result in either congruent or incongruent weathering. In fact, the congruent reaction that results in production of gibbsite is not the only possible congruent reaction of albite; with different weathering stoichiometry, kaolinite can form:

$$2NaAlSi_3O_{8(s)} + 2H_2CO_{3(aq)} + 9H_2O_{(l)}$$

$$\rightarrow Al_2Si_2O_5(OH)_{4(s)} + 2Na^+_{(aq)} + 4H_4SiO_{4(aq)}$$

$$+ 2HCO_3^-_{(aq)} \qquad (9.10)$$

Whether or not weathering will proceed congruently or incongruently, and the ultimate fate (solubility, mobility) of atoms and molecules released from primary minerals during chemical weathering, depends on many factors, from soil pH and microbial activity to the composition of the soil solution, as well as the mineral undergoing weathering (e.g. ionic solids almost always decompose congruently).

9.2.2 Geochemical quantification of elemental mobility in soil

Understanding which elements tend to be mobile and prone to leaching, and which tend to be immobile and remain in soil, may have applications to predicting nutrient retention, contaminant mobility, or paleoclimate interpretation. Table 9.2 presents geochemical data from a soil sequence in Costa Rica showing the compositional changes that take place during chemical weathering. The parent materials for the soils in this particular region are river sediments, and the two soils, one 10 000 years old, the other 125 000 years old, are evolving on river terraces originally comprised of the parent material indicated in the table. The main process responsible for the changes seen in this soil sequence is the effect of time, and the data facilitates analysis of which elements are relatively immobile over time, and which tend to be soluble and leached out of the soil over time.

Calculations in the right-hand columns of Table 9.2 are presented in two ways. The first quantifies the change in concentration of each element or mineral, as follows:

$$\%change = 100 \times (wt\% \; soil - wt\% \; PM)/wt\% \; PM \quad (9.11)$$

Using this measure, it is evident that concentrations of some elements have decreased markedly, especially Mg, Ca, and Na, whose concentrations in 125 ka soil are 86–96% less than concentrations in the parent material. Others have increased markedly – Ti, Fe, and Zr are present in concentrations 69–99% greater than what they are in the parent material. K decreases by 64%, Si exhibits a small decrease (12%), and Al increases by ~29%. Using % change as a measure of compositional change, it is possible that concentrations of soluble elements may decrease up to 100% (i.e. be completely removed by leaching). Concentrations of immobile elements will increase as soluble elements are removed, potentially to values exceeding 100% (this can occur when base cation concentrations in parent material are high).

Table 9.2 Geochemical changes in a soil sequence containing 10 and 125 ka terrace soils derived from parent material (PM) comprised of river sediments dominated by basaltic and andesitic volcanic detritus intermixed with soil minerals derived from erosion further upstream.

	PM	10 ka	125 ka	% change (total)	Mass Transfer Coeff (τ_j)
Si	26.5	25.3	23.2	−12.4	−0.48
Ti	0.88	1.37	1.77	+99.7	+0.18
Al	5.24	5.99	6.75	+28.9	−0.24
Fe	4.26	6.42	7.20	+68.9	0
Mg	2.35	0.97	0.32	−86.3	−0.92
Ca	2.37	0.18	0.09	−96.2	−0.98
Na	0.52	0.12	0.05	−90.7	−0.95
K	0.47	0.18	0.17	−64.3	−0.79
Zr	129	164	219	+69.8	0
Quartz	13.1	11.5	18.8	+43.5	NA
Plagioclase	12.8	2.8	< 0.5	−100	NA
Augite	4.1	< 0.5	< 0.5	−100	NA
Smectite	42.2	40.6	< 2	−100	NA
Kaolin	15.9	27.6	57.8	+263	NA
Goethite	6.2	6.9	3.2	−48.4	NA
Hematite	< 0.5	< 0.5	7.8	+inf	NA

All values for PM, 10 and 125 ka (except for Zr [ppm]) are presented as weight % element or mineral. The two columns on the right are explained in the text. The mass transfer coefficient does not apply to minerals.

Source: Data are obtained from B-horizons of soils along the central Pacific coast of Costa Rica (Fisher and Ryan 2006; Ryan and Huertas 2009).

Using % change in concentration, element mobility in this soil sequence ranked from most mobile to least mobile is:

Ca > Na > Mg > K > Si > Al > Fe > Zr > Ti

Note that high-charge cations Al^{3+}, Fe^{3+}, Zr^{4+}, and Ti^{4+} are the least mobile, consistent with data presented in Figure 4.9. Recall that Si^{4+} forms moderately soluble H_4SiO_4.

Element mobility can also be assessed by calculating the mass transfer coefficient, τ_j, for each element (Brimhall and Dietrich 1987). This is done by determining two ratios, one of the concentration of a given element (component j) in soil to its concentration in parent material, and the other of the concentration of a presumed immobile element in parent material versus its concentration in soil (commonly Zr or Ti is chosen as the immobile component i). The equation is:

$$\tau_j = [(C_{j,w} \times C_{i,p}) \div (C_{j,p} \times C_{i,w})] - 1 \qquad (9.12)$$

where w indicates weathered material and p indicates parent material and C indicates "concentration of"; for example, $C_{j,w}$ represents the concentration of element j in weathered material and $C_{i,p}$ represents the concentration of the immobile element in parent material.

In this approach, a value of −1 indicates complete depletion, a value of zero indicates no change (this is the case for Fe in Table 9.2 and also, by definition, Zr), and

any value > zero implies addition from an outside source (e.g. eolian deposition). We can see that both rankings give the same relative mobilities, with the low-charge base cations (Ca^{2+}, Na^+, Mg^{2+}, K^+) showing greatest mobilities (i.e. greatest potential to be leached out of soil), and the high-charge Al^{3+}, Fe^{3+}, Zr^{4+}, and Ti^{4+} showing the least mobility. Thus, base cations released by chemical weathering are the most likely to remain in solution following weathering of a mineral; given sufficient water leaching downward through soil, the base cations will be depleted with time, leaving a soil enriched in the relatively insoluble Al^{3+}, Fe^{3+}, Zr^{4+}, and Ti^{4+}.

9.2.3 Quantifying chemical weathering: CIA

The relatively systematic behavior of elements in soil described in Section 9.2.2 can be applied to quantifying the extent of chemical weathering of a given soil or paleosol. The *chemical index of alteration* (CIA; Nesbitt and Young 1982) is a widely applied weathering index that includes relatively immobile Al in the numerator and Al as well as mobile Ca, Na, and K in the denominator. Values used in this equation are bulk soil concentrations (rather than, for example, the composition of the exchangeable fraction) as oxide molar concentrations, and the CaO used in the calculation must only apply to CaO derived from silicate minerals (i.e. calcium in carbonates is to be excluded).

$$CIA = 100 \times Al_2O_3 \div (Al_2O_3 + CaO + Na_2O + K_2O) \quad (9.13)$$

Given that Al accumulates in soils as they are leached and that Ca, Na, and K are depleted by leaching, CIA will increase with increased chemical weathering and leaching. Applying this formula to the 10 and 125 ka soils shown in Table 9.2 (and converting wt% element to molar oxide values):

$$CIA\,(PM) = 100 \times 0.19 \div (0.19 + 0.083 + 0.023 + 0.012) = 62.3$$

$$CIA\,(10\,ka) = 100 \times 0.22 \div (0.22 + 0.0063 + 0.0052 + 0.0046) = 93.2$$

$$CIA\,(125\,ka) = 100 \times 0.25 \div (0.25 + 0.0031 + 0.0022 + 0.0043) = 96.3$$

Predictably, the CIA value increases with increasing leaching associated with increased soil age. The upper limit of CIA, i.e. the value corresponding to complete leaching of soluble base cations, is 100. The lower limit is dictated by concentrations of Al, Ca, Na, and K in parent material. For example, fresh, unweathered basalt has a CIA of ~30–40, whereas granite has CIA values of ~45–50. River sediments, like the parent material shown in Table 9.2, typically are in the CIA range of 60–70 because they have experienced some chemical alteration since being eroded from parent rock.

Many other indices using molar concentrations of soil chemical components are available to quantify the extent of chemical weathering. Included among these are: the sum of base cations divided by the sum of immobile cations $(\Sigma\ base\ cations \div \Sigma\ [Al_2O_3 + Fe_2O_3 + TiO_2])$; the ratio of Fe_2O_3 to FeO (which quantifies oxidation of ferrous to ferric iron); the sum of base cations divided by Al_2O_3; and others (Birkeland 1999). All take advantage of the fact that certain elements (mainly high-charge cations) are inherently immobile in soil and tend to accumulate with time, whereas others tend to decrease in concentration with time, e.g. soluble low-charge base cations or oxidizable constituents (e.g. tendency of Fe^{2+} to oxidize to Fe^{3+} during weathering).

9.2.4 Secondary minerals: controls on formation, mineral stability diagrams[1]

The abundance of primary minerals in soil is controlled by a combination of geological processes (e.g. intrusion and exposure of granite, or deposition of glacial till) and the resistance of chemical bonds in primary minerals to chemical weathering. The abundance of secondary minerals, on the other hand, is controlled by the soil-forming factors (Section 9.3.2): climate, biotic influences, topography, parent material, and time (and sometimes anthropogenic influences).

9.2.4.1 Factors controlling soil mineralogy

The composition of the soil solution from which minerals crystallize is one of the main factors controlling pedogenic (soil-formed) mineralogy, and there are a few basic principles of soil formation that help to understand and predict soil mineralogy:

1 Effective precipitation is correlated to leaching of base cations, so highly leached soils are likely to evolve to a mineralogy dominated by kaolinite, hematite, and/or goethite and gibbsite;

[1] Note: this section does not present details of mineral structures or compositions – that information is in Chapter 2. The material in this section builds upon the foundations presented in Chapter 2.

2 Arid and semi-arid region soils characterized by low effective precipitation and limited leaching of base cations will foster formation of pedogenic minerals that contain base cations, including smectite, illite, calcite, and gypsum. Abundance of parent minerals is likely to be greater in arid region soils as compared to humid tropical soils of the same age;

3 Soil physical properties are important; for example, permeable sandy soils will foster more extensive leaching of base cations or other substances than will impermeable soils;

4 Poorly drained areas (e.g. wetlands, toeslopes) tend to accumulate base cations and have higher Si : Al ratios than well-leached soils. This favors formation of smectite over kaolinite;

5 Compared to primary minerals, secondary (pedogenic) minerals tend to be fine-grained and typically are concentrated in the clay size fraction (<2 μm); thus, they are more reactive and have higher cation exchange capacities than do primary minerals.

Variation in soil pedogenic mineralogy as a function of climate is illustrated in Figures 9.4 and 9.5. Figure 9.4 depicts some of the main aspects of soil mineralogy as controlled by precipitation, notably the occurrence (and stability) of the Ca-minerals gypsum and calcite in arid region soils (MAP < 500 mm), the dominance of smectite in arid and semi-arid regions with MAP ~200–1000 mm,

the prevalence of kaolinite (or halloysite) in humid temperate to tropical climates with MAP ~1000–3000 mm, and ultimately the increasing abundance of hydroxides and oxides of Al and, to a lesser extent, Fe, in extremely leached, highly evolved tropical soils (see Focus Box 9.4).

Latitudinal variation is correlated to soil mineralogy and thickness of the soil profile (Figure 9.5), from tropics poleward through the desert belt, temperate and boreal regions, and finally to tundra and polar regions.

Soil mineralogy in temperate regions is difficult to generalize due to stronger effects of parent material and topography (e.g. elevation) compared to the tropics and arid regions. Rainfall and leaching are intermediate to the humid tropics and arid regions, so temperate region soil mineralogy (and soil depths) fall somewhere between end-member deep kaolinite–gibbsite soils (tropics) and shallow smectite–calcite soils (arid). Temperate soil minerals include vermiculite, interstratified biotite-vermiculite (also known as hydrobiotite), illite, interstratified mica-smectite (which can form via hydrolysis and leaching of K^+ in mica interlayers), as well as intermixed kaolinite and smectite. Polar regions generally experience the least-intense chemical weathering of global climate zones, so are likely to remain rich in primary minerals.

9.2.4.2 Mineral stability diagrams

Soils typically do not achieve thermodynamic equilibrium, but *mineral stability diagrams* based in thermodynamic principles are useful in examining likely trajectories of secondary mineral growth and transitions in soil. The main rationale is that, for a soil with a given chemistry of soil water (soil solution), certain mineral(s) will be stable, and the system will trend in that direction. To view one extreme, strongly leached soils rich in dissolved aluminum and depleted of Si and base cations will foster crystallization of gibbsite ($Al(OH)_3$), but not calcite or smectite. Mineral stability diagrams can be constructed from congruent reactions between primary minerals and secondary minerals, and congruent reactions between two different secondary minerals.

Returning to the weathering of albite (Na–plagioclase feldspar), we can begin to construct a mineral stability diagram:

$$2NaAlSi_3O_{8(s)} + 2H^+ + 9H_2O_{(l)} \rightarrow Al_2Si_2O_5(OH)_{4(s)}$$
$$+ 2Na^+_{(aq)} + 4H_4SiO_{4(aq)} \quad (9.14)$$

The equilibrium constant for this reaction is:

$$K_{eq} = [Na^+]^2 \times [H_4SiO_4]^4/[H^+]^2 \quad (9.15)$$

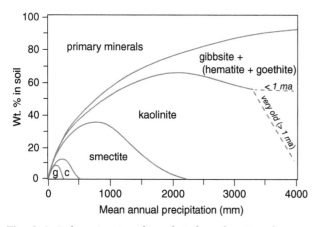

Fig. 9.4 Pedogenic mineralogy plotted as a function of mean annual precipitation. Given the complex interactions of the five soil-forming factors, this sketch is to be viewed as a simplified model intended to show dominant trends. In addition to pedogenic mineral trends, also note progressive decrease in % primary minerals (from ~100% to ~10%) with increasing precipitation. Source: Modified from Sherman (1952), data in Birkeland (1999), and many other sources. Symbols: "g", gypsum; "c", calcite. Dashed lines indicate variation in soil mineralogy of differing ages in the humid tropics.

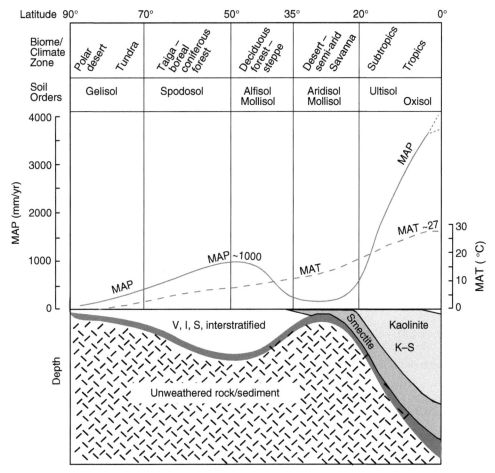

Fig. 9.5 Schematic sketch of soil minerals and relative depths of typical soils across a latitudinal transect, from pole to equator. Soil depth in the tropics may be tens of meters whereas in other regions is typically < 2 m. Common soil orders for each latitudinal zone are provided. MAP, mean annual precipitation; MAT, mean annual temperature; V, vermiculite; I, illite; S, smectite; K–S, interstratified kaolin–smectite. Source: Modified from Strakhov (1967) and Birkeland (1999).

For the purposes of formulating the equation such that variables are arranged in a useful way, it can be arranged into:

$$K_{eq} = [H_4SiO_4]^4 \times ([Na^+]^2/[H^+]^2) \qquad (9.16)$$

Rules of logarithms allow rearrangement of this equation into:

$$Log\,(K_{eq}) = 4 \times log\,[H_4SiO_4] + 2 \times log\,([Na^+]/[H^+]) \qquad (9.17)$$

then

$$\tfrac{1}{2} log\,(K_{eq}) = 2 \times log\,[H_4SiO_4] + log\,([Na^+]/[H^+]) \qquad (9.18)$$

Rearranging these terms into an equation of a line results in:

$$log\,([Na^+]/[H^+]) = -2 \times log\,[H_4SiO_4] + \tfrac{1}{2} log\,(K_{eq}) \qquad (9.19)$$

In this case, $log\,([Na^+]/[H^+]) = y$; $-2 = m$ (slope); $log\,[H_4SiO_4] = x$; and $\tfrac{1}{2} log\,(K_{eq}) = b$. We know that, plotted on log x and y axes, the slope will be -2 and the y intercept will be one half of the log of the equilibrium constant for the reaction.

All that remains now to plot this equation is the equilibrium constant, which can be calculated from standard free energies of formation of the species and phases involved in the reaction (values used in these calculations are given in Table 9.3). The principles of this topic are covered in Chapter 1, and tell us that, in a general case, the Gibbs free energy of formation (ΔG^o_R) for the reaction is computed

Table 9.3 Thermodynamic data used in development of the mineral stability diagram in Figure 9.8.

Compound or species	Name	$G°_f$ (kJ mol^{-1})
$NaAlSi_3O_8$	Albite (low albite)	−3712
$Al_2Si_2O_5(OH)_4$	Kaolinite	−3778
$Na_{0.33}Al_2(Si_{3.67}Al_{0.33})$ $O_{10}(OH)_2$	Na–beidellite	−5354
$Al(OH)_3$	Gibbsite	−1144
H^+	Hydrogen ion	0
$H_2O_{(l)}$	Water	−237.2
$Na^+_{(aq)}$	Sodium ion	−261.7
H_4SiO_4	Dissolved silica	−1316

Given uncertainty in Gibbs free energy values, precision is shown to only one decimal place. See Appendix C for a more extensive table of thermodynamic data.

Source: Data are from Dean (1979) and Robie et al. (1979).

according to this Eq. (1.52):

$$\Delta G°_R = \Sigma n_x \times G°f_x \text{ (products)} - \Sigma n_x \times G°f_x \text{ (reactants)}$$

Pertaining to the congruent weathering of albite to kaolinite, the equation is:

$$\Delta G°_R = [(G°_f \text{ (kaol))} + 2(G°_f (Na^+)) + 4(G°_f (H_4SiO_4))]$$
$$- [2(G°_f \text{ (albite))} + 2(G°_f (H^+)) + 9(G°_f(H_2O_{(l)}))] \quad (9.20)$$

and

$$\Delta G°_R = [(-3778) + 2(-261.7) + 4(-1316)]$$
$$- [2(-3712) + 2(0.0) + 9(-237.2)] = -6.6 \text{ kJ}$$

Chapter 1 (Eq. 1.61) indicates how the Gibbs free energy of formation and the equilibrium constant for a reaction are related (for units of kJ/mol) as follows:

$$\Delta G°_R = -5.708 \times \log(K_{eq})$$

Applying this to the weathering of albite to kaolinite:

$$-6.6 \text{ kJ} = -5.708 \times \log(K_{eq})$$

then

$$\text{Log}(K_{eq}) = -6.6 / -5.708 = 1.2$$

and

$$K_{eq} = 10^{(-6.6/-5.708)} = 10^{1.2}$$

We can now go back to the equation of the line and add in the log of this K_{eq} value (log $[10^{1.2}] = 1.2$) for "b," the y-intercept:

$$\log ([Na^+]/[H^+]) = -2 \times \log [H_4SiO_4] + \tfrac{1}{2}(1.2) \quad (9.21)$$

$$\log ([Na^+]/[H^+]) = -2 \times \log [H_4SiO_4] + 0.6 \quad (9.22)$$

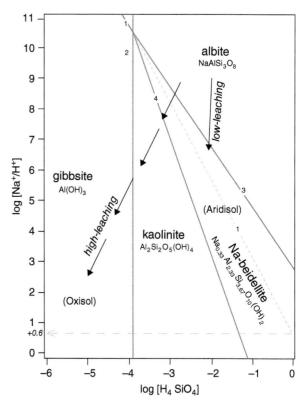

Fig. 9.6 Mineral stability diagram showing stability fields for gibbsite, kaolinite, Na–beidellite (a smectite), and albite, at 25 °C and 1 atm. Note that both axes are logarithmic. The x-axis represents dissolved Si (as H_4SiO_4) in soil water; the y-axis represents increasing $[Na^+]$ toward the top of the diagram, and increasing $[H^+]$ (decreasing pH) toward the bottom. Arrows depict representative trends of soil water under a leaching dominated scenario (high-leach) such as might be found in well-leached soils such as Oxisols, Ultisols, and Spodsols, and a limited leaching scenario (low-leach) like would be expected to occur in an Aridisol, Mollisol, or Vertisol. Approximate compositions of soil waters in Oxisols and Aridisols are shown.

This gives the equation for a straight line with a slope of −2 and y-intercept of +0.6 (line labeled "1" on Figure 9.6). The variable on the x-axis is log $[H_4SiO_4]$ and the variable on the y-axis is log $([Na^+]/[H^+])$.

With this same approach, the line between gibbsite and kaolinite can be plotted using this reaction as the basis:

$$Al_2Si_2O_5(OH)_{4(s)} + 5 H_2O_{(l)} \rightarrow 2 Al(OH)_3 + 2 H_4SiO_{4(aq)} \quad (9.23)$$

The equilibrium constant is:

$$K_{eq} = [H_4SiO_4]^2 \quad (9.24)$$

Using rules of logarithms, this equation can be rearranged into:

$$\text{Log} (K_{eq}) = 2 \times \log [H_4SiO_4] \quad (9.25)$$

then

$$\tfrac{1}{2}\log(K_{eq}) = \log[H_4SiO_4] \qquad (9.26)$$

We can now rearrange these terms into an equation of a line, as above (Eqs. (9.21) and (9.22)) but understanding that $\log([Na^+]/[H^+]) = 0$ for this reaction:

$$0 = -\log[H_4SiO_4] + \tfrac{1}{2}\log(K_{eq}) \qquad (9.27)$$

The equilibrium constant for this reaction, derived using the same approach used for the example of albite → kaolinite, is $10^{-7.7}$, resulting in this equation:

$$0 = -\log[H_4SiO_4] - 3.9 \qquad (9.28)$$

or ...

$$\log[H_4SiO_4] = -3.9 \qquad (9.29)$$

Thus, the line separating kaolinite and gibbsite is vertical (labeled "2" on Figure 9.6 and crosses the x-axis at −3.9) because, by definition, $\log[H_4SiO_4]$ is the variable on the x-axis. The lack of involvement of Na^+ and H^+ in this reaction intuitively indicate that changes to $[H_4SiO_4]$ in the soil solution are all that is required to drive this reaction.

Another common mineral involved in weathering of albite is sodium smectite, represented here as sodium beidellite, a common clay mineral in soils that are less leached than those that produce abundant kaolinite:

$$2.33NaAlSi_3O_{8(s)} + 2H^+ + 6.66H_2O_{(l)}$$
$$\rightarrow Na_{0.33}Al_2Si_{3.67}Al_{0.33}O_{10}(OH)_{2(s)} + 2Na^+{}_{(aq)}$$
$$+ 3.33H_4SiO_{4(aq)} \qquad (9.30)$$

The equilibrium constant for this reaction is:

$$K_{eq} = [Na^+]^2 \times [H_4SiO_4]^{3.33}/[H^+]^2 \qquad (9.31)$$

Then rearranging and applying rules of logarithms, this equation can be rearranged into:

$$\text{Log}(K_{eq}) = 3.33 \times \log[H_4SiO_4] + 2 \times \log([Na^+]/[H^+]) \qquad (9.32)$$

then

$$\tfrac{1}{2}\log(K_{eq}) = 1.67 \times \log[H_4SiO_4] + \log([Na^+]/[H^+]) \qquad (9.33)$$

Now, rearranging these terms into an equation of a line:

$$\log([Na^+]/[H^+]) = -1.67 \times \log[H_4SiO_4] + \tfrac{1}{2}\log(K_{eq}) \qquad (9.34)$$

Pertaining to the congruent weathering of albite to Na–beidellite, the equation is:

$$\Delta G^o{}_R = [(-5354) + 2(-261.7) + 3.33(-1316)]$$
$$- [2.33(-3778) + 2(0.00) + 6.66(-237.2)]$$
$$= -30.9\ kJ \qquad (9.35)$$

The slope of the albite–Na–beidellite line is −1.67 (Eq. (9.34)) and dividing −30.9 by −5.708 and putting the result into log form gives a K_{eq} of $10^{5.4}$, and because the term on the right side of the equation is $\tfrac{1}{2}\log(K_{eq})$, the y-intercept is 2.7.

To complete this four-component mineral stability diagram, it is necessary to calculate the equation of the line for the reaction of Na–beidellite to kaolinite, a transition that occurs in soils over time as Na^+ and H_4SiO_4 are leached away.

$$Na_{0.33}Al_2Si_{3.67}Al_{0.33}O_{10}(OH)_{2(s)} + 0.33H^+{}_{(aq)}$$
$$+ 3.76H_2O \rightarrow 1.16Al_2Si_2O_5(OH)_{4(s)} + 0.33Na^+{}_{(aq)}$$
$$+ 1.33H_4SiO_{4(aq)} \qquad (9.36)$$

For this reaction, the equilibrium constant is:

$$K_{eq} = [Na^+]^{0.33} \times [H_4SiO_4]^{1.33}/[H^+]^{0.33} \qquad (9.37)$$

Skipping a couple of steps, we arrive at:

$$\text{Log}(K_{eq}) = 1.33 \times \log[H_4SiO_4]$$
$$+ 0.33 \times \log([Na^+]/[H^+]) \qquad (9.38)$$

Then (see above for examples including all steps):

$$\log([Na^+]/[H^+]) = -4 \times \log[H_4SiO_4] + 3 \times \log(K_{eq}) \qquad (9.39)$$

The slope of this line (line "4" on Figure 9.6) is −4 and the y-intercept (from Gibbs free energy calculation, as in the examples given), is −4.7.

This diagram shows that higher concentrations of H_4SiO_4 favor albite, as does increasing the ratio of $[Na^+]/[H^+]$. These observations are reasonable in a qualitative sense – albite has a higher molar concentration of Si than does kaolinite (e.g. Si to Al in albite is $3:1$, whereas in kaolinite it is $1:1$), and 1 mol of albite contains 1 mol of Na; kaolinite contains no Na.

Other considerations related to Figure 9.6 include:

1 Where leaching is intense enough to decrease dissolved Na and Si simultaneously in the soil solution, albite may directly transform to kaolinite. Numerous studies have

documented direct transformation of feldspars to kaolinite. In cases where some Na is leached but the less-mobile Si is not, Na–beidellite will form. Numerous studies have also documented transformation of feldspars to smectites;

2 Starting at $\log [H_4SiO_4] = -2$ and $\log [Na^+/H^+] = 2$, shifting the system composition vertically upward (i.e. increasing dissolved sodium or raising pH) will (or could) cause kaolinite to transform to Na–beidellite. From this same point (−2, 2), shifting the system composition horizontally to the right (i.e. increasing dissolved Si) will (or could) also cause reaction of kaolinite to Na–beidellite. So too will combinations of increasing Na and Si (however, in reality kaolinite may persist in the Na–beidellite field for thousands or millions of years due to kinetic factors);

3 In the framework of chemical weathering, leaching of Na^+ and dissolved silica drives the reaction of albite toward kaolinite (and ultimately to gibbsite), but in cases where either net leaching is limited or nonexistent (e.g. arid regions), or where kinetic factors have not yet permitted sufficient leaching for kaolinite to form (i.e. young soils), Na–beidellite and similar smectite clays occur in soils (even in some cases in young humid tropical soils). At the other end of the weathering spectrum, intense leaching and sufficient time leach Na^+ (and other base cations) and silica to such an extreme that gibbsite becomes a dominant soil mineral.

Mineral stability diagrams are useful, but it is important to consider uncertainty in measurements of thermodynamic data used in mineral stability diagrams, and the applicability of these diagrams and thermodynamic

principles to the analysis of chemical weathering. One concern is that under the low-temperature conditions in soils, there can be no true thermodynamic equilibrium between primary minerals like albite and secondary minerals like kaolinite (or Na–beidellite, etc.). While the weathering reactions that result in formation of kaolinite and Na–beidellite clearly do occur, these reactions are not reversible under soil pressure–temperature conditions, so even as concentrations of Na and Si increase into the albite stability field, kaolinite (or Na–beidellite) will not react to form albite at 25 °C.

Uncertainties also arise from the fact that the Na, Ca, and K smectites shown in Table 9.3 are idealized end-member compositions that lack Fe and Mg, yet soil smectites typically contain Fe and Mg, so the commonly used Gibbs free energy values do not account for compositional complexity. This will affect the slopes and y-intercepts of lines in mineral stability diagrams (note the difference in Figures 9.6 and 9.7, diagrams involving similar minerals constructed from different thermodynamic data). Similarly, variation in grain sizes and abundance of cleavage planes and other crystallographic discontinuities can result in different mineral solubilities, which in turn will affect Gibbs free energies – clearly coarse-grained, hexagonal kaolinite will have a different Gibbs free energy than will a poorly crystalline, fine grained, Fe-bearing kaolinite.

In spite of limitations, the types of diagrams shown in Figures 9.6 and 9.7 are applicable to research in environmental geochemistry. As long as we recognize that slopes and locations of phase boundaries (lines) are

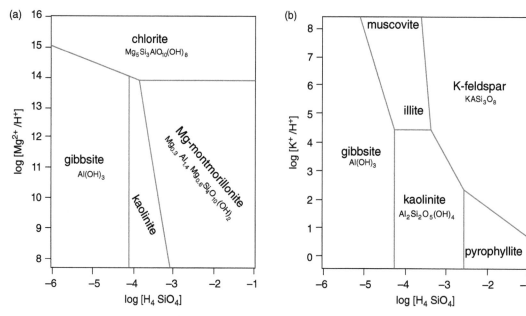

Fig. 9.7 Mineral stability diagram for systems Al–Mg–Si–O–H (left, Source: modified from Lee et al. 2003) and K–Al–Si–O–H 25 °C, 1 atm.

limited by some degree of uncertainty, the diagrams do show approximate stability fields of primary minerals and secondary minerals, allowing us to understand relative stabilities of minerals in the presence of solutions and likely directions of chemical reactions in the presence of fluids of known composition.

9.3 SOIL PROFILES, NOMENCLATURE, SOIL-FORMING FACTORS

9.3.1 The concept of the soil profile

A *soil profile* that has formed by the weathering of basalt on the Nicoya Peninsula of Costa Rica is shown in Plate 13. (This soil is a tropical Alfisol – refer to Section 9.3.3 for terms.) Examining this profile provides insight into processes that lead to layering and composition of soils. Beginning from the top of the profile, the combined processes of addition of leaves, decaying tree trunks, decaying roots, and other organic matter result in relatively organic-rich upper horizons or "topsoil", and chemical weathering of rock or sediment produces mineral-rich lower horizons. The upper horizons, termed O-horizons and A-horizons, contain decomposing organic matter ranging from identifiable plant remains to black, fine-grained humus. The presence of organic matter in topsoil imparts a dark brown to black color to these horizons, which typically contain 0.5–5% organic matter.

Underlying the O- and A-horizons are horizons whose colors are dominated by either parent minerals or secondary minerals (e.g. yellows, oranges, and reds of iron oxides and hydroxides, or whites and grays of carbonates). Because it is closer to the surface, the B-horizon is more evolved than the underlying C-horizon. B-horizons are often dominated by secondary clays and hydroxides (or carbonates in arid climates), whereas C-horizons more closely represent parent material. In cases where soils have formed via chemical weathering of bedrock, C-horizons are underlain by saprolite, which, while altered by weathering, will still contain evidence of sedimentary layering, metamorphic foliation, igneous textures, or other properties inherent to the parent rock. E-horizons (also known as albic horizons) occur in highly-leached forest soils where the combination of high precipitation and added effect of organic acids results in leaching of even iron, producing layers that are white to gray and often dominated by quartz.

In addition to the inherent mobilities of elements and weatherability/stability of minerals in soil, the main

soil-forming factors, as first described by the Swiss soil scientist Hans Jenny in the 1930s and 1940s, ultimately exert a very strong control on soil geochemistry. These soil-forming factors are climate (cl), organisms/biotic effects (o), relief/topography (r), parent material (p), and time (t); given that anthropogenic (a) activities also influence soils, they are often included as well, and the factors are often referred to by the acronyms *clorpt* or *clorpt(a)*. Section 9.3.2 briefly presents these factors as they relate to soil geochemistry.

9.3.2 Soil-forming factors

9.3.2.1 Climate

Climate is very influential in determining soil type because it strongly controls the intensity of mineral decomposition; it also influences whether or not products of primary mineral dissolution remain in soil or are leached out of soil toward surface water or groundwater. The two main factors associated with climate are precipitation and temperature, which are commonly quantified as mean (or median) annual precipitation (MAP) and mean (or median) annual temperature (MAT). Increased MAP provides the medium in which chemical weathering reactions occur (i.e. water) as well as protons (H^+) that

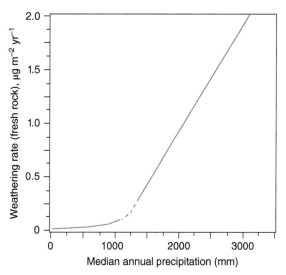

Fig. 9.8 Weathering rate as a function of precipitation in Hawaii where parent material is freshly exposed basalt. Source: Diagram redrawn from data presented by Stewart et al. (2001). The sharp transition in rate between ~1100 and 1400 mm MAP may represent a threshold across which increased leaching results in decreased concentrations of base cations to the point that reactions are driven strongly to the right (i.e. weathering products).

Focus Box 9.5

Lichens as Initiators of Chemical Weathering and Soil Formation

Lichens are an illustrative example of how organisms affect chemical weathering and soil formation. They consist of a fungus and a photosynthesizer (e.g. green algae) in a symbiotic relationship that enhances weathering of bare rock surfaces. Their nature and effect can be summarized as a mutualism between:

A fungus that exudes organic acids (e.g. oxalic acid) which accelerate rate of mineral dissolution (both by H^+ attack and by chelation), thus providing mineral nutrients (e.g. Ca, K, P) to the photosynthesizer. Penetration of fungal hyphae into the rock also increases exposed surface area, providing pathways for water to enter the rock, also enhancing chemical weathering.

A photosynthesizer, typically algae, which produces H_2CO_3 (via respiration, enhancing mineral dissolution) and sugars needed as an energy source for itself and the fungus.

Both members of a lichen community produce nutrient-rich organic matter as they decay, and these organic remains combined with the first stages of rock breakdown constitute the beginning of soil formation on exposed rock surfaces (Figure 9.9).

Fig. 9.9 Lichens on the surface of vesiculated basalt, Courtesy of US Geological Survey. (https://geomaps.wr.usgs .gov/parks/sunset).

drive hydrolysis reactions. A particularly illustrative example from Hawaii, where other soil-forming factors (e.g. basaltic parent material, temperature, soil age) are relatively constant, demonstrates the effect of precipitation on chemical weathering rates (Figure 9.8).

But while MAP certainly exerts a strong control on chemical weathering, it is often useful to determine effective precipitation (precipitation minus evapotranspiration) because soils exposed to the greatest effective precipitation will experience the greatest amount of *leaching* (downward movement of water through soil toward the water table), and hence, the greatest loss of soluble ions (e.g. Na^+, Mg^{2+}, K^+, Ca^{2+}, SO_4^{2-}, HCO_3^-). Conversely, in areas with low effective precipitation, and especially where evaporation is greater than precipitation (i.e. deserts), carbonates (e.g. calcite) or sulfates (e.g. gypsum) may crystallize in soil because soluble ions are less likely to be leached out, and thus reach saturation.

Temperature also strongly controls weathering. An old adage in biogeochemistry says that for every increase of $10\,°C$, rates of biogeochemical reactions double, and at ambient temperatures of soils, rates may even more than double for each $10\,°C$ increase (Davidson and Janssens

2006). This concept is embedded in the Arrhenius equation (Eq. 1.76), which indicates that increasing temperature will increase k (the rate constant), increasing the rate of chemical reaction:

$$k = A \times e^{-E_a/RT}$$

The combined effect of high MAP and MAT make the humid tropics the region with the highest weathering rates, and conversely, the cold deserts of Antarctica are characterized by the lowest weathering rates on Earth. Hot desert soils may experience rapid chemical weathering following rainfall events, but evaporation tends to inhibit leaching of soluble base cations out of soil, leading to formation of soil layers rich in calcite or gypsum and fostering formation of sodium smectites.

9.3.2.2 Organisms

From microbial activity that enhances reaction rate (especially of redox reactions) to uptake of nutrients by roots that decreases concentrations of base cations and other nutrients in soil solution, the activity of organisms plays an important role in chemical weathering and soil formation. The rhizosphere, the zone that extends ~2 mm out from roots, is compositionally different from the rest of the soil; roots commonly release organic acids to enhance solubility of nutrients, and concentration gradients caused by nutrient uptake cause nutrients to migrate toward the root, enhancing reactions even outside of the rhizosphere. Decomposition of leaves, woody material, and other forms of organic matter produces humic and fulvic acids that contribute to hydrolysis; burrowing worms enhance aeration and permeability; and decay of organic matter produces solid humus that retains nutrients, contributes to retention of soil moisture, and binds soil particles together, improving soil structure and permeability.

9.3.2.3 Relief (topography)

Landscape position – hilltop? valley bottom? – of a soil can either enhance or impede chemical weathering and soil formation. Chemical weathering is enhanced by leaching, and leaching occurs more effectively where water tables are deeper (farther below the surface) in summits or higher elevation areas. Hilltops and shoulders of hills tend to be well-drained and see greater chemical weathering rates than do soils in footslope or toeslope areas (the bases of hills), where shallow water tables inhibit downward leaching (Figure 9.10). Often, base cations and soluble anions leached out of higher elevations are carried in solution along groundwater flow paths toward lower elevations, then deposited in toeslope regions, a process which can inhibit chemical weathering in toeslopes by increasing concentrations of reaction products (e.g. Na^+, Ca^{2+}, H_4SiO_4, HCO_3^-) in the soil solution. The low (deep) water table in summit positions also tends to facilitate oxidation reactions, whereas the high (shallow) toeslope water table

creates more-reducing conditions that not only inhibit oxidation reactions but can foster reduction reactions.

Soils in toeslopes with shallow water tables often exhibit mottling or gley texture consisting of iron hydroxide splotches or spheres (typically 1 mm to 1 cm in diameter) in a matrix of gray, chemically reduced soil – in these soils redox reactions often exert strong control on element mobility. Orientation with respect to the sun is important too – in arid and semi-arid regions, north-facing slopes retain greater moisture than do south-facing slopes; therefore, leaching tends to be greater in soils on north-facing slopes, whereas on south-facing slopes, soils are more evaporation-dominated and soil solutions will be richer in base cations and other soluble soil constituents.

9.3.2.4 Parent material

Information in Figure 9.1 and Table 9.1 illustrates the effect of mineralogical content of rock or sediment on the rate of chemical weathering, soil formation, and the composition of the soil solution. Basaltic parent material dominated by olivine, pyroxene, and Ca-plagioclase will weather far more rapidly than granitic parent material dominated by K-feldspar, muscovite, and quartz. Moreover, the cationic composition of soil water with basaltic parent material will contain a strong Ca–Mg signature whereas granitic parent materials will produce waters enriched in K and Na. Limestones, dolostones, and marbles are prone to rapid chemical weathering in humid climates that far outpaces weathering rates of basalts, although in arid climates carbonate rocks are relatively resistant to weathering. Physical attributes are important too – greater amounts of exposed surface area in fractured

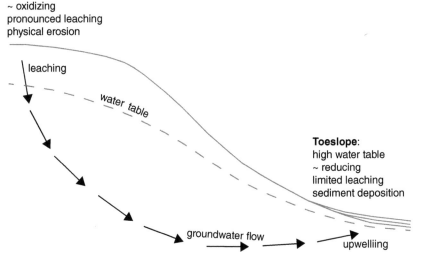

Summit:
low water table
~ oxidizing
pronounced leaching
physical erosion

leaching

water table

Toeslope:
high water table
~ reducing
limited leaching
sediment deposition

groundwater flow

upwelliing

Fig. 9.10 Sketch of a topographic transect indicating factors that tend to enhance chemical weathering rates in summit positions, and those that tend to inhibit chemical weathering in toeslope positions.

rocks compared to non-fractured rocks will facilitate influx of water and enhance weathering and soil formation; similarly, permeable sediments or rock will likely weather more rapidly than impermeable sediment or rock of the same mineralogical composition.

9.3.2.5 Time

Most rock-forming minerals (calcite, olivine, pyroxene, etc.) are unstable in soil (to varying degrees), but their decomposition in weathering regimes has a kinetic component, much like the ice cube that does not immediately melt on a hot summer day in spite of the obvious instability of the situation. Figure 9.1 and Table 9.1 both contain information pertaining to the time factor; so too does Table 9.2, which presents data on the evolution of a tropical soil sequence with time. Plagioclase, which comprises ~13% of parent material (river sediment), decreases to ~3% in 10 000 year-old soils, and is undetectable in 125 000 year-old soils. Kaolinite appears to be at least metastable in many tropical soils, increasing from 16% in parent material to 28% in 10 000 year-old soil, and ultimately (in this sequence), to 58% in 125 000 year-old soil. In fact, kaolinite appears to be stable in many tropical soils where rainfall levels do not force the system to the next stage, gibbsite (Ryan et al. 2016). The trend toward increasing kaolinite over time is consistent with the observation that kaolinite is the dominant mineral in Oxisols (Section 9.3.3), so the progression toward a kaolinite-dominated mineral assemblage in tropical soils is progression toward a long-term steady state.

Trends of chemical weathering rate, soil formation rate, leaching of base cations, buildup of immobile elements, and other pedogenic processes tend to follow parabolic (logarithmic) or exponential curves; in other words, rates are rapid at the onset and then taper with time. Figure 9.11

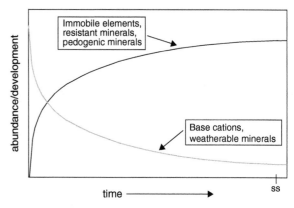

Fig. 9.11 Curves depicting typical evolution of soil characteristics with time. The black curve represents constituents that increase in concentration or degree of development with time (e.g. $[Al^{3+}]$, carbonate horizon thickness, Pb^{2+} sorption), whereas the blue curve depicts the trajectory of constituents that tend to decrease over time (e.g. $[Na^+]$, [parent materials]). See text for full explanation. The location on the x-axis labeled as "ss" indicates the point in time at which steady state conditions have developed.

depicts typical parabolic (logarithmic) curves where the black curve would represent increasing concentrations of pedogenic clays (e.g. kaolinite), pedogenic hydroxides and oxides (e.g. goethite, hematite, gibbsite), cations which are relatively immobile and tend to accumulate in soil with time (e.g. Fe^{3+}, Ti^{4+}, Zr^{4+}), and minerals which are resistant to chemical weathering (e.g. quartz, zircon, rutile). The black curve could also represent thickness of weathering rinds, depth of the soil profile or, in arid climates, thickness of calcite or gypsum horizons. The blue curve is representative of soil components that decrease over time; examples include leaching of base cations and weathering of primary minerals, both of which initially occur rapidly, but then decrease in rate with time. Examples are in Fisher and Ryan (2006).

Can Soils Reach a State of Thermodynamic Stability? Or Steady State?

Chemical weathering reactions are not reversible in soil, so a true state of dynamic equilibrium cannot exist in soil (where e.g. dynamic equilibrium cannot occur for the reaction of Na-feldspar to kaolinite because it is not reversible at ambient soil temperatures). However, many soils do appear to achieve a steady state, or quasi-steady state with respect to soil development, where curves like those shown in Figure 9.11 eventually reach a slope of zero, or nearly so. One scenario in which a steady state might develop involves initial rapid chemical weathering of plagioclase and other Ca-bearing minerals accompanied by rapid leaching of calcium; then, eventually, the low, long-term rate of leaching may be balanced by eolian additions of Ca. In this case, the flux relationship would be $Ca^{2+}_{in} = Ca^{2+}_{out}$. (Eolian addition is balanced by leaching losses.) Birkeland (1999) presents arguments pertaining to the existence of steady-state conditions in soil, and while there is disagreement over whether or not steady states exist in soils, from a practical point of view there are cases where so little change occurs in soils greater than a certain age that they effectively achieve a steady state. For this reason, it may be practical to model landscape-scale soil properties as if the older soils have reached steady state. In the humid tropics, for example, soils older than ~100 000 years old are nearly uniformly characterized by a kaolinite-rich, nutrient-poor status regardless of whether the soil is 100, 500, or 1000 ka (1 Ma).

This is typical behavior for systems which are out of equilibrium; at first, rates of change are rapid in the direction of equilibrium, but eventually decrease as the system approaches equilibrium. In some cases, soils appear to reach a state of equilibrium, typically after tens of thousands to millions of years, where the time frame depends on the other soil-forming factors (e.g. tropical soils will evolve toward the low-slope part of the parabolic curve more rapidly than will semi-arid soils).

9.3.2.6 Anthropogenic factors

The natural processes described above may dominate the overall chemical and mineralogical character of most soils, but human activities have certainly altered soil compositions in many areas. Acid rain, for example, accelerates leaching of base cations and decomposition of natural organic matter, leading to changes in the composition of the soil solution and of soil particles. In other areas, emissions of trace metals from smelters and coal-fired power plants directly alter the composition of soil via deposition. In agricultural areas, additions of nutrients (e.g. Ca, K, N, P) and pesticides to enhance crop growth alter soil chemistry, as does improper disposal or storage of mine wastes, fuels, and solvents. These topics are explored in Sections 9.5 through 9.9 toward the end of this chapter.

9.3.3 Soil classification – soil orders and geochemical controls

The US Department of Agriculture has developed a systematic nomenclature for soils known as Soil Taxonomy that has gained broad acceptance globally. It is presented here in part because of its widespread use, but also because of its systematic approach in which many of the criteria used to classify soils are based on soil mineralogy and/or geochemistry. This section also includes terms that are not part of the USA Soil Taxonomy system (e.g. laterite, chernozem, etc.) but also are common in studies of soil geochemistry.

Each soil order has the suffix "sol" preceded by a root that carries some relationship to the soil itself. Be aware that soil orders are further classified into suborders, of which there are 55, great groups, of which there are at least 200, and other criteria which ultimately result in complex, yet informative terms such as *halloysitic, isohyperthermic Tropeptic Haplorthox*. Although useful information, the scope of this text does not permit this level of detail, so please refer to Birkeland (1999), USDA online information on soil taxonomy, and numerous other sources.

9.4 SOILS AND THE GEOCHEMISTRY OF PALEOCLIMATE ANALYSIS

Numerous studies have extrapolated knowledge of climate and pedogenesis to the interpretation of paleoclimate signatures preserved in paleosols, ancient soils typically preserved in sediments or sedimentary rock. This is particularly useful in terrestrial sedimentary sequences where paleosols are preserved in fluvial floodplain deposits, alluvial fans, and other aggrading sequences. Paleosols from humid tropical paleoenvironments should be highly leached, Al–Fe-rich, and dominated by kaolinite, whereas paleosols formed in arid environments should be rich in

Focus Box 9.8

Geochemistry of USDA Soil Orders

The properties of the first seven orders are strongly controlled by MAP and thus are arranged more or less in order of decreasing leaching (i.e. from humid tropical to temperate to semi-arid to arid); the properties of the last five (Andisols, Histosols, Gelisols, Inceptisols, and Entisols) are less-easily organized by precipitation and so are presented separately.

Oxisols: highly leached, intensely weathered tropical soils that form in tropical and subtropical environments. Little or no remaining primary minerals; rich in oxides and hydroxides of Al and Fe as well as kaolinite. Nutrient-poor (low Ca, K, and Mg), acidic (pH ~4–5), limited ability to retain fertilizer-added nutrients, low cation exchange capacity (CEC) and low base saturation.

Ultisols: strongly leached, clay-rich soils in forested ecosystems of temperate subtropics, e.g. subtropical forests of India and the southeastern United States. Low base saturation (<35%), usually low inherent fertility, rich in kaolinite, less leached than Oxisols. The terms *laterite* and *latosol* apply to Ultisols and Oxisols.

Spodosols (also known as *podsols* or *podzols*): cool, humid temperate coniferous forest soils. Leached E-and B-horizons, relatively nutrient poor. Organic acids produced by coniferous needle decomposition result in acidic (pH ≤ 4) E/A/B horizons. Very distinctive layering of dark brown/black A-horizon over white/gray E-horizon over bright red/orange B-horizon over tan C-horizon

Alfisols: moderately leached soils in temperate, deciduous forest ecosystems; % BS > 35%. Clay-rich B-horizons (vermiculite, illite, or smectite with higher CEC than the kaolinites of oxisols or ultisols). Relatively high nutrient level, good agricultural soils. Abundant in the agriculturally productive US Midwest, and in temperate deciduous forests of Europe, central Asia, Africa, and elsewhere (Plate 13).

Vertisols: rich in smectite, form in areas with pronounced dry and rainy seasons (e.g. monsoonal climates of western India, eastern Sudan, and southeastern Texas, USA). The shrink–swell nature of smectites lends distinctive characteristics to Vertisols; during the dry season, smectites dehydrate (lose H_2O from interlayers) and deep, wide desiccation cracks form. During rainy periods, smectites rehydrate and the cracks disappear. Shrink–swell creates engineering problems (e.g. cracked foundations, heaved roads), yet the soil is nutrient-rich.

Mollisols (also known as *Chernozems*): grassland ecosystem soils between ~35 and 55° latitude (e.g. the Pampa of Argentina, the Russian Steppe, North American Great Plains, parts of Mongolia). The combination of relatively high primary productivity (of grasses) and low decomposition and leaching rates produces thick, soft, humus-rich A-horizons (mollic epipedons) that are excellent for agriculture.

Aridisols (also known as *Sierozems*): Desert soils with calcite- or gypsum-bearing B-horizons. Limited leaching and pronounced evaporation of soil water result in crystallization of otherwise soluble minerals at depths typically ~10–100 cm. In well-developed Aridisols, layers 15–50 cm thick of well-cemented calcite or gypsum may form. High nutrient concentrations, so with irrigation, have potential to be productive agricultural soils. Major concern is sodification (Section 9.7).

Andisols: young soils in volcanic ash deposits. Low density (< 0.90 g cm^{-3}), low organic matter content, abundance of noncrystalline volcanic glass and metastable weathering products such as allophone and imogolite. High water-retention capacity, high cation and anion exchange capacities, nutrient-rich, tend to retain fertilizer nutrients well. Occur mainly in moist forested ecosystems on the flanks or bases of volcanoes (e.g. Japan, Philippines, Indonesia).

Histosols: organic rich peat-bog/wetland soils. Typically contain ≥ 20 wt% organic carbon, total soil thickness is at least 40 cm. High organic matter content results in very low bulk density, at times < 0.3 g cm^3. Typically form where anaerobic conditions and restricted drainage inhibits the decomposition of plant matter; thus, Histosols are a reservoir of carbon that can be lost to the atmosphere when wetlands are drained. Histosols are also referred to as peats and mucks.

Gelisols: soils where permafrost occurs within 2 m of the surface. Geographically limited to polar regions and some high elevation regions. Low temperature inhibits decomposition of organic matter, so like Histosols, Gelisols store large quantities of organic carbon. Warming of polar and alpine regions will likely result in release of some of this stored organic carbon to the atmosphere. *Tundra soil* is a term sometimes used to refer to Gelisols.

Entisols: soils that exhibit almost no pedogenic development. Typically are recently deposited sand (e.g. river floodplains, beaches, dunes) or virtually unweathered sediments exposed by mass wasting. No distinctive horizon (with the possible exception of a thin A-horizon), properties are determined by the properties of the unconsolidated parent material. With time, Entisols evolve into one of the other soil orders; e.g. recently deposited glacial till in cool temperate forest will likely evolve into a Spodosol given thousands or tens of thousands of years of pedogenesis.

Inceptisols: intermediate to Entisols and other soil orders (mainly Oxisol, Ultisol, Spodosol, Alfisol, Mollisol, Aridisol). Given their transitory state, Inceptisols are widely distributed across climate zones, parent materials, and the other soil-forming factors. In the humid tropics, Inceptisols (e.g. tropepts) will eventually evolve into Oxisols; in grassland ecosystems, Inceptisols will evolve with time into Mollisols; and so on.

Focus Box 9.9

Precambrian Climate Analysis by Paleosol Chemistry and Mineralogy

550 million year old (550 Ma) paleosols preserve a record of latest Precambrian climate of the East European craton (now NW Ukraine). The paleosols have not been altered appreciably by diagenesis or hydrothermal activity, so they retain the signature imparted by the weathering environment of that time. The parent material is basalt, and Liivamägi et al. (2018) studied mineral composition, major and trace element geochemistry (including e.g. mass transfer coefficient, τ_j, Eq. (9.12)), and stable isotope geochemistry, among other methods, to determine under what climate conditions these paleosols formed. They represent a critical time in Earth history, more or less coincident with the explosion of complex life forms in the transition from the Ediacaran (end of the Precambrian) to Cambrian period. Results indicate that primary minerals (e.g. plagioclase, pyroxene, and trioctahedral clay) disappeared completely in that weathering environment,

and secondary minerals crystallized from the products of chemical weathering. Hematite is ubiquitous, dioctahedral smectite occurs deeper in soil profiles, kaolinite–smectite (K–S) is abundant at intermediate depths, and kaolinite is the dominant clay in the shallowest, most-evolved part of profiles. This sequence suggests that smectite formed first, then was sequentially altered to K–S, then kaolinite with increasing development, akin to the sequence observed in modern soils in lowland humid tropics (Ryan and Huertas 2009), and geochemical indices are consistent, classifying the paleosols as paleo-Ultisols (Section 9.3.3). Liivamägi et al. (2018) suggest that the hot humid Ediacaran climate on the East European Craton at ~550 Ma may have resulted in significant release of phosphorus (liberated from basalt by weathering) into the Ediacaran ocean, a potential factor in the explosion of life and rise in atmospheric O_2 at that time (Chapter 7, Section 7.1.2).

---

smectite, gypsum, or calcite. Of course, time is an important factor in soil formation, and given that some young tropical soils are rich in pedogenic smectite, caution must be exercised and other approaches incorporated (e.g. paleobotany of seeds or root casts, stable isotope analysis).

9.5 EFFECTS OF ACID DEPOSITION ON SOILS AND AQUATIC ECOSYSTEMS

When CO_2 in air dissolves into water, it produces carbonic acid (H_2CO_3), and this is the origin of naturally acidic rainfall (or snow) with a pH = 5.6 (Chapter 5, Section 5.1.1). In areas where natural sources of S or N cause the formation of H_2SO_4 or HNO_3 in air (Chapter 8, Section 8.11), the pH of precipitation may drop below pH of 5.6, becoming extreme in cases where volcanic gases result in pH < 3, but these occurrences tend to be insignificant globally. Anthropogenic acid rain with pH < 5 is the focus of the following section, particularly the response of soils to anthropogenic acid deposition from sulfuric and nitric acids derived from combustion of fossil fuels or, less commonly, from industrial emissions. Chapter 8 (Section 8.11) presents the origin and atmospheric chemistry as well as definitions of acid deposition relative to acid precipitation and acid rain. The following section delves into one of the major impacts of fossil fuel emissions – acid deposition – on soils and the aquatic ecosystems to which they are linked.

The main constituents added to soil in acid precipitation (in excess of their abundance in natural precipitation)

are H^+, $SO_4{}^{2-}$, and $NO_3{}^-$, so understanding the way in which these ions interact with soil minerals and organic matter helps to explain most of the negative impacts of acid deposition. It is also important to understand that some soil orders (e.g. Spodosols, Oxisols, Ultisols) are naturally acidic, in large part due to decomposition of organic matter that produces humic and fulvic acids. In these soils, however, microbially facilitated decomposition of organic acids (to CO_2, H_2O, and other compounds) tends to minimize their long-term negative impacts compared to the more stable, stronger minerals acids H_2SO_4 and HNO_3. Due to their inherent natural acidity, Spodosols and related coniferous forest soils are more prone to negative effects of acid deposition than would be naturally alkaline soils such as Aridisols and Mollisols.

9.5.1 Increased solubility of Al in acidic soil solution

Aluminum is nearly insoluble at neutral pH, but as pH drops below ~6, Al is released to solution as the polyatomic cations $Al(OH)_2{}^+$ and $Al(OH)^{2+}$ or the ion Al^{3+} (Eqs. (4.8)–(4.10)), depending on pH (Figure 4.10). The main cause of increased Al solubility in acidic soils is the ability of H^+ to dissolve Al-bearing minerals, including the Al-hydroxide gibbsite ($Al[OH]_3$) (Figure 9.12) as well as aluminosilicates. The sum of Eqs. (4.10)–(4.12) helps to visualize the net effect of H^+ of Al solubility:

$$Al(OH)_3 + 3\,H^+ = Al^{3+}{}_{(aq)} + 3\,H_2O \qquad (9.40)$$

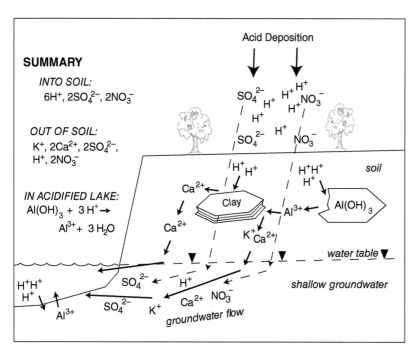

Fig. 9.12 Schematic diagram of effects of acid deposition on soil and adjacent lake. With the exception of the reaction in the lake sediments, the diagram attempts to balance ionic charges; for example, −6 mol of charge from sulfate and nitrate and +6 mol of charge from H⁺ enter the soil in acid deposition, and −6 mol of charge from sulfate and nitrate leach from the soil and enter the lake, balanced by +6 mol of charge (2 Ca²⁺, 1 K⁺, and 1 H⁺).

This is effectively an acid–base reaction, but unfortunately it is one that buffers solutions at a pH ∼3 to 4 (compare that to buffering by carbonate species at pH ∼5.5 to 8.0). Al is toxic to most plants, and one example of how elevated Al may inhibit plant growth in acidic soil environments is via incorporation of Al into cell walls. Ca^{2+} enhances cell wall stability, but in acidic soils where plant roots incorporate excess Al in place of Ca, cell wall stability is impaired and plants become less resistant to cold, disease, and other stresses.

Aluminum leached from acidic soils into lakes or released to acidic lake water from sediments in acidic lakes is also potentially harmful to fish. When Al solubility in soils increases as described above, soluble Al can be leached out of soils or sediments and into surface water bodies; also, Al in sediments may be leached into lake water if lake pH drops below pH ∼5.5. Dissolved Al in lakes or streams is problematic because fish breathe by passing water through their gills and extracting dissolved oxygen from the water. The pH inside a fish gill typically is between six and seven, so if the water brought into the gill contains elevated amounts of dissolved Al, here is an example of what can happen:

$$Al^{3+}_{(aq)} + 3H_2O = Al(OH)_{3(s)} + 3H^+_{(aq)} \qquad (9.41)$$

Al is insoluble in circum-neutral fish gills so solid $Al(OH)_3$ precipitates in or on the gills. This causes the gill to excrete mucous in an attempt to deal with the irritation, and the ultimate result is a light gray mucous-plus-$Al(OH)_3$ coating that causes fish to suffer from respiratory distress and cell necrosis that can lead to death.

9.5.2 Displacement of adsorbed nutrient cations

When excess H⁺ enters soil it alters the composition of exchangeable cations adsorbed onto surfaces of clays and humus (Figure 9.12). Consider H⁺ (or H_3O^+) as a cation and you can begin to understand impacts to the adsorbed plant-available cation composition of particle surfaces. Two primary factors enhance sorption of H⁺ onto particle surfaces and result in desorption of cations such as Ca^{2+} and K⁺: (i) the high activity (or concentration) of H⁺ in acidic soils, and (ii) the relatively low ratio of hydration radius to charge (Table 4.6), where the relatively small radius of H_3O^+ makes it strongly attracted to exchange sites and allows it to outcompete large radius Ca^{2+} and K⁺ ions, especially in H⁺-rich (acidic) solutions. Thus, the "H⁺ cation" alone may result in loss of nutrient cations from particle surfaces, and then once Ca^{2+} and other cation nutrients are free to migrate in the soil solution, they are prone to leaching and loss from the soil.

Increased Al solubility can also have a profound effect on the exchangeable cation pool. The high charge and relatively small hydration radius of the Al^{3+} cation mean that it is more strongly attracted to particle surfaces than are larger, lower-charge nutrient cations K⁺ and Ca^{2+} (Table 4.6). The exchange reaction below (similar to

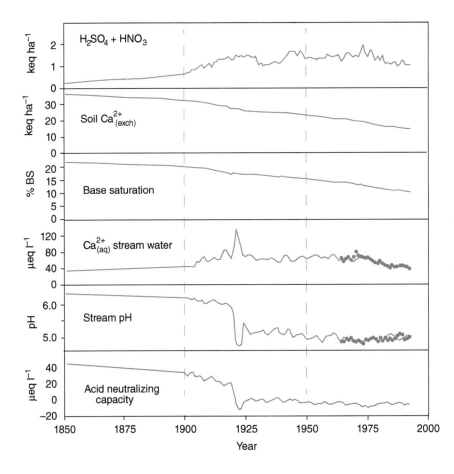

Fig. 9.13 Study of the Hubbard Brook Experimental Forest (HBEF), New Hampshire, northeast USA, showing changes in atmospheric deposition of mineral acids, soil exchangeable Ca^{2+}, base saturation, dissolved Ca^{2+} in stream water, stream pH, and ANC. Data consist of measured and modeled data, and data points are shown for late twentieth-century stream measurements to show comparison of measured and modeled. Early 1920s spikes in Ca, pH, and ANC are attributed to short-term forest disturbance (tree harvesting) and release of Ca^{2+} and H^+ from forest floor. Source: Adopted from Gbondo-Tugbawa and Driscoll (2003).

Eq. (2.6)) depicts adsorption of Al and desorption of Ca with increasing activity of Al in solution.

$$Al^{3+}_{(aq)} + Ca^{2+} \cdot CLAY = Ca^{2+}_{(aq)} + Al^{3+} \cdot CLAY$$

Al for Ca exchange is depicted schematically in Figure 9.12, and with time, soils become Ca-depleted and Al-rich, an imbalance that is harmful to native ecosystems and to agriculture. Initially, dissolved Ca often increases in lake and stream water as it is flushed out of soil, an effect visible in Figure 9.13 (especially early 1920s). Later, as soils become saturated with Al, it too begins to leach into surface water bodies (Figure 9.14).

9.5.3 Leaching of base cations enhanced by increased NO_3 and SO_4

Soil solutions must maintain electrical neutrality, and the high flux of soluble anions (NO_3^-, SO_4^{2-}) through and out of soils requires that cations also be leached to balance the change of the leaching nitrate and sulfate. Given that Al^{3+} and H^+ tend to displace Ca^{2+} (and K^+ to a lesser extent) from exchange sites, the pronounced leaching of anions

through soil and into groundwater results in depletion of Ca^{2+}, a critical nutrient cation in many forest ecosystems.

Two other considerations related to loss of base cations from soils, which are usually less important than the mechanisms described above, are (i) the physical removal of base cations when trees are harvested, and (ii) emissions controls that have reduced the amounts of particulates from smokestacks, resulting in a return to preindustrial (i.e. ~low) deposition of base cations onto soils.

9.5.4 Decrease of soil buffering capacity and base saturation

Soils are capable of buffering acid deposition in four main ways:

1 by release of Ca^{2+} and other base cations adsorbed onto particle surfaces in exchange for H^+ (shown schematically in Figure 9.12);

2 via chemical weathering, especially of carbonates and other reactive minerals that consume H^+ as they decompose (e.g. $CaCO_3 + H^+ = Ca^{2+} + HCO_3^-$);

3 reaction of dissolved bicarbonate with H^+ ($HCO_3^- + H^+ \rightarrow H_2CO_3$); and

---

Focus Box 9.10

Base Saturation (% BS) and Soil Quality

Base saturation (BS) quantifies the abundance of exchangeable base cations (in $cmol_c$ kg^{-1}; refer to Chapter 2, Section 2.5.1.5) relative to the total CEC, also expressed in units of $cmol_c$ kg^{-1}. It is quantified as follows:

$$\%BS = [(Na^+ + Mg^{2+} + K^+ + Ca^{2+}) \div (total\ CEC)]$$
$$\times 100 \qquad (9.42)$$

When exchange sites are dominated by the base cations (i.e. Na^+, Mg^{2+}, Ca^{2+}, and K^+), % BS will be high; for example, Mollisols (semi-arid grassland soils) or Aridisols typically have % BS values > 50% and often >

80%. Conversely, when exchange sites are dominated by Al^{3+}, Fe^{3+}, and H_3O^+, as they often are in acidic soils, base saturation values are typically < 20%. Note that in Figure 9.13, % BS values of the Spodosols in the watershed of the Hubbard Brook Experimental Forest (HBEF) were ~20% prior to anthropogenic acid deposition; after a century of acid deposition, % BS is approximately 10%. In other words, 50% of the base cations formerly occupying exchange sites on soil particles have been leached out of soils and replaced by Al^{3+}, H_3O^+, and perhaps Fe^{3+}. Soils with low % BS tend to be nutrient-poor and incapable of buffering acid deposition.

4 reaction of organic matter (e.g. humus) with H^+. In some cases, humus may sorb H^+ onto exchange sites but in other cases, H^+ may result in breakdown of organic matter.

The combined effect of mechanisms 1 through 4 is the *acid neutralizing capacity* (ANC) of the soil, i.e. ANC is the capacity of particulates plus dissolved species in soil (or a water body) to neutralize acid. It is effectively synonymous with alkalinity but also includes soil solids. Soils rich in calcite or with abundant clays or humus that are capable of exchanging base cations for H^+ will withstand the effects of acid deposition better than will soils rich in quartz, feldspars, and other minerals that are largely nonreactive and have very low CEC. Soils or water bodies with ANC < 50 $\mu eq\ l^{-1}$ are typically considered sensitive to acidic deposition (they have a low capacity to buffer acids), whereas soils or waters with ANC values > 500 $\mu eq\ l^{-1}$ are relatively resistant to acidification. For definition of alkalinity and how it is determined, refer to Chapter 5 (Section 5.1, Figures 5.1 and 5.2, and Eqs. (5.15) and (5.16)). ANC can become negative if the concentration of acids exceeds bases – an example is an acidic stream or lake lacking bicarbonate and with minerals acids H_2SO_4 and HNO_3 and a pH of 4.

9.5.5 Acid deposition and heavy metals

A useful generalization in environmental geochemistry is that, as is the case with Al and Fe, the mobility of heavy metals increases with decreasing pH. The Eh–pH diagrams of Cu and Pb shown in Figure 4.13 indicate that these metals are most soluble under acidic conditions, and the adsorption edge curves in Figure 4.17 reveal that

heavy metals tend to desorb from particle surfaces as pH decreases. The order of heavy metal desorption with decreasing pH (below pH = 7) is:

$$Ni^{2+} > Cd^{2+} > Zn^{2+} > Cu^{2+} > Pb^{2+} > Cr^{3+}$$

What this indicates is that as soil pH decreases, Ni^{2+} will become mobile more readily than will the other metals listed, and Pb^{2+} and Cr^{3+} are least likely to be mobilized. (although in reality Cr^{6+} in the polyatomic anion CrO_4^{2-} is more apt for oxidized soil environments than the reduced Cr^{3+} form; refer to Figure 4.18 for anion data).

Field studies and laboratory experimentation summarized by Wilson and Bell (1996) indicate that Ni^{2+} and Zn^{2+} tend to be most readily mobilized and leached from soil as pH decreases due to acid deposition. Cd^{2+} forms strong complexes with particulate organic matter and this enhances its retention in soil even as it is released from mineral (e.g. Fe-hydroxide) surfaces. Pb^{2+} forms similarly strong complexes with organic matter, a major factor in its low mobility in soils. The cause of the > fivefold increase in Pb concentration in sediments of Big Moose Lake (Figure 9.14) is not leaching from soils, but rather atmospheric deposition of anthropogenic Pb into the lake and onto the surrounding landscape, from where it is physically eroded into the lake.

9.6 SOILS AND PLANT NUTRIENTS

The chemical composition of plants provides insight into nutrients that plants must obtain from soil. The carbon and oxygen in plants is derived from atmospheric CO_2 via photosynthesis, and the hydrogen used by plants to

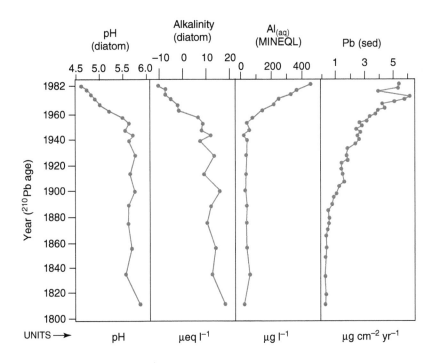

Fig. 9.14 Data from a sediment core from Big Moose Lake in the southwestern Adirondack region, New York state, USA. Age of sediment layers was determined by ^{210}Pb and ^{137}Cs dating (Chapter 11). Alkalinity and pH were determined from diatoms in the lake sediment and $Al_{(aq)}$ by modeling using MINEQL. Source: Original data and analytical details are in Charles et al. (1987).

form carbohydrates and other organic molecules can be derived from air or soil water. Nearly all other components of plants, i.e. the mineral nutrients (Table 9.4), must be obtained from the soil, and the majority of the mineral nutrients are, in fact, derived from chemical weathering of minerals, at least initially – the "initially" caveat refers to the fact that decomposing organic matter provides nutrients (originally derived from minerals) to plants.

Plants nutrients are divided into the macronutrients and micronutrients, where *macronutrients* occur in concentrations greater than 0.1% (by wt) and *micronutrients* typically comprise < 200 ppm of bulk plant tissue. Because C, H, and O are not derived from the soil solution, they typically are not classified as nutrients, *strictu* sensu. The macronutrients are N, P, K, Ca, Mg, and S and the micronutrients include boron, chloride (Cl^-, not Cl_2 gas), copper, iron, manganese, molybdenum, nickel, and zinc, and the only one of these which is generally not derived from rock weathering is nitrogen, which is fixed into soils from the atmosphere by symbiotic relationships of plants and nitrogen-fixing bacteria. (Anthropogenic NO_x emissions and fertilizer additions are now responsible for the majority of N fixation in many soils.) The remainder of the macronutrients and micronutrients are all mineral-derived and are liberated to the soil by chemical weathering (Table 9.5).

Once a given nutrient (e.g. Ca^{2+}) has been liberated from a mineral (e.g. $CaCO_3$) by chemical weathering and incorporated into a plant, it may then be retained and cycled in the ecosystem among plants (or other organisms), fine-grained organic matter (humus), exchange sites on surfaces of minerals or humus, in the soil solution, or in minerals that crystallize in the soil.

The roles of nutrients in plants include nitrogen as a component of proteins; phosphorus as a component of adenosine di- and triphosphate and, thus, a critical element in photosynthesis; calcium as a constituent of cell walls and magnesium as a component of chlorophyll; boron in production of carbohydrates; and copper and chloride in regulating plant metabolism. Note that of the main elements in the continental crust and soils, neither Al nor Na are plant nutrients and when they are abundant in the exchangeable/plant available fraction, they are likely to impair plant growth.

9.7 SALINE AND SODIC SOILS

Where high amounts of rain or snowmelt lead to strong leaching, e.g. humid tropics or moist temperate forests, weathering of primary minerals and leaching of base cations tends to result in naturally acidic soils (e.g. Spodosols, Oxisols, Ultisols). On the contrary, in hot arid climates, chemical weathering is often intense during short periods following rainfall, but when evaporation exceeds infiltration, as it typically the case in arid region soils, base cations liberated from primary minerals tend to not be leached from soil. In some cases, strong evaporative forces draw soil water upward toward the surface, carrying dissolved base cations upward, resulting in accumulation of base cations in soil A- and B-horizons.

Table 9.4 Examples of nutrient concentrations in grasses (e.g. alfalfa, clover, ryegrass), apple leaves, pine needles, and tomato leaves.

	Grasses	Apple leaves	Pine needles	Tomato leaves
N (wt%)	1.0–5.0	1.5–2.5	1.0–1.5	3.0–5.0
K (wt%)	1.0–4.0	1.0–2.0	0.30–0.45	2.5–4.5
Ca (wt%)	0.20–2.0	1.0–1.5	0.10–0.20	1.5–4.0
Mg (wt%)	0.10–0.50	0.25–0.50	0.05–0.10	0.45–0.70
P (wt%)	0.20–0.50	0.15–0.30	0.10–0.20	0.30–0.50
S (wt%)	0.10–0.50	0.10–0.30	0.05–0.15	0.50–1.0
B (ppm)	10–50	20–50	3–10	30–60
Cu (ppm)	5–75	5–20	1–10	5–15
Fe (ppm)	25–250	50–250	20–100	100–200
Mn (ppm)	25–150	25–100	50–600	50–200
Mo (ppm)	0.1–0.5			
Zn (ppm)	20–80	20–50	20–50	30–50

Typical values for carbon, hydrogen, and oxygen in bulk plant tissue are: C = 40–50%, H = ~5%, and O = 40–50%. Chloride is not listed because, while an important nutrient, it is rarely present in limited amounts that would affect plant growth. It may be present in plants in the range ~0.1 to 1%.

Source: Data obtained from Brady and Weil (2002) and the University of Arkansas Cooperative Extension Service (http://www.uaex.edu/Other_Areas/publications/pdf/FSA-3043.pdf), and various other sources.

Table 9.5 Plant nutrients, their common forms, and sources. For each of the macronutrients, decomposing organic matter is also a source.

Nutrient	Source	Plant-available Form
C	Atmosphere (photosynthesis)	CO_2
O	Atmosphere (photosynthesis)	CO_2
H	Atmosphere, soil water	H_2O
Macronutrients		
N	Fixation from N_2	$NO_3^- > NH_4^+$
P	Apatite ($Ca_5[PO_4]_3OH$)	HPO_4^{2-}, $H_2PO_4^-$, etc.
K	Micas, K–feldspars	K^+
Ca	Calcite, plagioclase feldspars	Ca^{2+}
Mg	Dolomite, chlorite, and many silicates	Mg^{2+}
S	Sulfides, sulfates	SO_4^{2-}
Micronutrients		
B	Borates, micas	$H_2BO_3^-$
Cl	Salts, micas, precipitation	Cl^-
Cu	Sulfides (e.g. chalcopyrite)	Cu^{2+}
Fe	Pyrite, Fe-oxides/hydroxides, silicates	Fe^{2+} or Fe^{3+}
Mn	Iron oxides	Mn^{2+}, Mn^{4+}
Mo	Sulfides, oxides	MoO_4^{2-}, $HMoO_4^-$
Ni	Millerite, trace in some silicates	Ni^{2+}
Zn	Sphalerite, trace in some silicates	Zn^{2+}

Chemical weathering of primary minerals such as silicates and carbonates are often the main source of base cations in soil, although some coastal areas receive Na in rainfall (Table 4.1), and weathering of marine shales may produce soils with exchangeable Na^+ higher than normal. A very important anthropogenic source of Na is irrigation water. As Table 4.1 indicates, rainwater typically contains ~1 ppm Na^+, whereas river water or groundwater in arid regions often contains > 100 ppm Na^+, so when crops are irrigated, Na^+ is added to soil, often in high concentrations, and if evaporation exceeds infiltration rate, Na^+ will not be leached out of soil, but rather accumulate.

Saline soils have high concentrations of base cations balanced by anions Cl^-, SO_4^{2-}, HCO_3^-/CO_3^{2-}, and/or NO_3^-, and are defined by the UN FAO (Food and Agriculture Organization) as having electrical conductivity $> 4\,dS\,m^{-1}$ ($TDS > 2500\,mg\,l^{-1}$). They are typically alkaline, with pH in the range of 8–10. Excess soil salinity makes less water available to plants than is actually in the rooting zone because as the salt concentration increases, so too does the osmotic pressure of the soil solution. The high level of total dissolved solids (TDSs) in soil water and low TDS in root water is the cause of the pressure difference. In saline soils, the osmotic pressure gradient forces water out of roots and into the soil to dilute the salty soil solution. This not only makes it difficult for plants to draw water into roots, but roots may actually lose water to the salty solution according to the osmotic pressure gradient. Another problem with saline soil water is that, in addition to the osmotic effect, excessively high concentrations of Ca, Na, and Mg may limit the uptake of other essential plant nutrients.

Sodic soils are a special subgroup of saline soils that contain very high concentrations of exchangeable Na^+; according to the UN FAO, *sodic soils* possess exchangeable sodium percentage (ESP) $> 15\%$. The ESP formula is much like the one for % BS, with variables expressed in $cmol_c$ kg^{-1}, and in this case the numerator is solely exchangeable Na^+.

$$\%ESP = [Na^+ \div (total\ CEC)] \times 100 \qquad (9.43)$$

In saline or sodic soils, the exchange sites are overwhelmingly dominated by Na^+, Ca^{2+}, and Mg^{2+}, so in some cases the denominator of ESP is presented as the sum of those three exchangeable cations. The sodium absorption ratio (SAR) is also used to define sodic soils, where the threshold is 12 and variables are expressed in $mmol_c\ l^{-1}$:

$$SAR = \frac{[Na^+]}{\sqrt{([Na^+] + [Ca^{2+}])/2}} \qquad (9.44)$$

In addition to the fact that Na^+ is not a plant nutrient and may poison plants or impede uptake of nutrient cations, abundant exchangeable Na^+ also negatively impacts soil structure, especially in clay-rich soils. Because of its low ionic charge and relatively high hydration radius (Table 4.6), the charge on hydrated Na^+ is weak and Na^+ does not strongly attract 2:1 layers of smectite clays; in effect, Na^+ causes dispersion of clay particles, which results in destruction of soil aggregates that enhance permeability. This leads to poor drainage (low permeability)

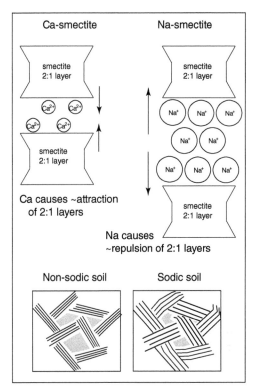

Fig. 9.15 Schematic sketches showing the effects of exchangeable Ca^{2+} vs Na^+ in soils. The top sketches diagram the capacity of Ca^{2+} to collapse smectite 2:1 layers around the interlayer vs. the way that Na^+ causes expansion and dispersion of smectite 2:1 layers. In soil (bottom), crystals of clay are shown. Ca^{2+} leads to formation of aggregates that enhance soil permeability whereas clay expansion and dispersion caused by Na^+ reduces pore size and decreases permeability (shown by blue polygons). Source: Modified from McBride (1994).

that impairs movement of water throughout the soil and limits migration of nutrients toward roots. In contrast, the relatively high charge and low hydration radius of Ca^{2+} (Table 4.6) means that soils with abundant exchangeable Ca^{2+} will contain smectite 2:1 layers that are collapsed around the hydrated Ca^{2+}, an attribute that fosters formation of soil aggregates and enhances permeability (Figure 9.15).

9.8 TOXIC METALS AND METALLOIDS

When contemplating the environmental availability or behavior of a given element or compound in soil (or surface water or groundwater), it is important to recognize that, often, only a fraction of the total (bulk) amount of an element or compound in soil or sediment is mobile. In other words, often only a small fraction is capable of being transported in solution or as extremely fine-grained colloidal

---

Focus Box 9.11

Saline Soils Globally and Throughout Human History

Soil sodification is an important environmental and social issue globally because 15–20% of the world's agricultural soils are saline soils under irrigation. Continued addition of Na^+ in irrigation water brings these soils closer to a sodic state, and once reached, sodic soils may end up abandoned or in need of remediation. This is a problem that humans have faced for thousands of years. Examples go far back, including (i) early agricultural societies built upon irrigated soils in Mesopotamia (modern day lowland Iraq, where poorly drained soils exacerbate salt buildup) and, (ii) the 450–1450 CE Hohokam civilization near modern-day Phoenix, Arizona (USA), where irrigation and poor drainage elevated soil salinity to high levels (14–27 dS m^{-1}; Woodson et al. 2015). Two approaches to remediation are (i) to try to flush Na out of the soil (often limited by lack of water in arid regions, or made difficult by poor soil structure), or (ii) to add Ca (often in the form of powdered gypsum) to increase the Ca : Na ratio and decrease activity of Na. Yet often the low-permeability of sodic soils inhibits flushing/leaching of Na, and additions of Ca to improve Ca : Na ratio eventually cannot overcome increasing salinity.

---

suspensions, or taken up by plants or other organisms. In many cases, the majority is immobile, and mainly bound in (or onto) minerals, organic matter, or other solids or organisms that prevent its release into solution. Determining the mobile (environmentally available) vs. immobile fractions is equally as important as determining bulk concentrations of an element or compound in soil.

Questions that can help to determine mobility of metals or metalloids (or most elements or compounds in environmental geochemistry) include:

1 Does the element or compound tend to form complexes with organic matter? If so, does this limit its mobility, as would be the case of complexation with particulate organic carbon (POC, e.g. humus), or does organic matter enhance mobility, as would be the case of complexation of certain metals with dissolved organic carbon (DOC, e.g. humic acid)?

2 Is the element or compound likely sorb to particle surfaces?

3 How do solution pH and redox state (i.e. Eh or pe) affect mobility of the element or compound?

4 Is it likely that the element or compound will coprecipitate in minerals (as is the case with many trace metals and arsenic in pyrite or Fe-hydroxides)? In this case, the mobility of the trace element will be strongly controlled by the stability of the host mineral.

5 Will the element or compound combine with other elements to form a mineral in which the element (e.g. Pb, Mn, Zn) is a dominant component (e.g. $PbCO_3$, MnO, ZnS)? If so, how soluble is the mineral in the soils being examined?

These questions seek to determine *speciation* (Chapter 4, Sections 4.4, 4.5, 4.10) and address issues such as: What % of substance X is plant-available or likely to be leached from soil into surface water? Will substance X be released to solution if the water table is lowered by excessive pumping of water wells and sediments become aerated (i.e. are the toxic elements coprecipitated in sulfides that might be decomposed by oxidation)? How much of substance X is complexed to particulate organic matter?

The various reservoirs (or "pools") of a given element in soil, as well as mechanisms responsible for transfer of an element from one reservoir to another, are shown in Figure 9.16. To some extent, the speciation of any element will be controlled by inherent characteristics such as ratio of ionic charge to radius ("ionic potential"; Figure 4.9), but the environment also exerts strong controls.

Two examples of environmental controls on potentially toxic trace elements are highlighted in the two following paragraphs to provide insight into trace element behavior, one pertaining to redox-sensitive manganese and one to anthropogenic lead arsenate insecticide.

In reducing waters (e.g. low-O_2 groundwater), manganese tends to be soluble as Mn^{2+} but when exposed to an oxidizing environment (e.g. the casing inside a well where air circulates), Mn tends to precipitate as hydrated Mn oxide, a black gooey (when wet) or powdery (when dry) precipitate. Unpublished analysis by the author of a viscous black coating on a water filter attached to a well (producing from a black shale aquifer) was identified by X-ray diffraction as partially hydrated birnessite, a manganese oxide mineral with the chemical formula $Na_{0.7}Ca_{0.3}Mn_7O_{14} \cdot nH_2O$. Manganese is a potential neurotoxin that, with chronic long-term exposure, may cause symptoms similar to those associated with Parkinson's disease; accordingly, the redox state of well water will influence the amount of dissolved Mn in drinking water supplies. Speciation of Mn in a shallow oxidized aquifer would likely be as birnessite or similar Mn oxide or hydroxide, whereas in anoxic groundwater Mn is more likely to be dissolved as Mn^{2+}.

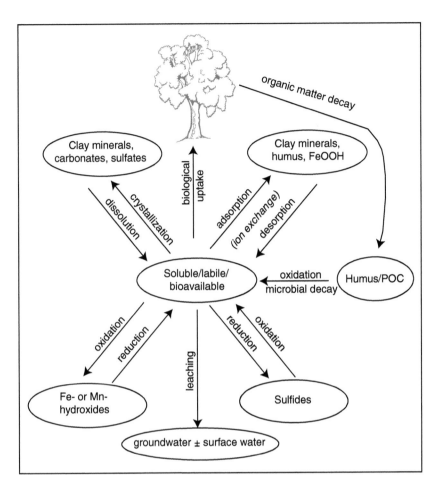

Fig. 9.16 Sketch of the main reservoirs, pathways, and mechanisms associated with speciation of trace metals and other substances in soil. Microbial activity, including oxidation of sulfides, organic matter decay, and mineral crystallization, strongly influences many of these processes. Source: Modified from McBride (1994).

Focus Box 9.12

Sequential Chemical Extraction

One common approach to determining speciation of elements or compounds in soil is sequential chemical extraction (SCE). A frequently used SCE approach was developed by Tessier et al. (1979) and has been slightly modified by subsequent researchers (e.g. Rauret et al. 1999; Young et al. 2006). A representative example of SCE is:

Target phase/pool	Reagent
1. Exchangeable ions, soluble in weak acids (e.g. carbonates)	0.11 M acetic acid (pH = 3)
2. Reducible (e.g. Fe-, Mn-hydroxides)	0.5 M hydroxylamine hydrochloride (pH = 1)
3. Oxidizable (sulfides, organic matter)	8.8 M hydrogen peroxide (pH = 2.5)
4. Residual (some silicates, well-crystallized oxides)	Aqua regia solution (e.g. 50% HNO_3 + 20% HCl)

In each of the four stages, a known mass of dried soil or sediment (e.g. 3.0 g) is reacted with a known volume (e.g. 25 or 100 ml) of the reagent. Often the same initial powder is exposed sequentially, so that first the weakly held (bioavailable, ~soluble) exchangeable and carbonate-bound ions are released to solution, then after washing to remove extracted ions and leftover previous reagent, the same powder can be reacted in the presence of hydroxylamine hydrochloride to extract ions bound to (or in) Fe- or Mn-hydroxides, and so on, up through stage 4. One important caveat is that the reagents are not always selective; for example, Ryan et al. (2008) found that certain Fe–Mg smectite and chlorite clays dissolve during stages 1–3. Given that trace metals substitute into the octahedral sheet, interpretation without mineralogical analysis might assign those metals to hydroxides or sulfides rather than clays. For this reason, many geochemists refrain from using specific phases, but rather indicate that a certain metal is e.g. associated with the oxidizable fraction.

Focus Box 9.13

Kinetic vs. Thermodynamic Controls on Pb Mobility in Soil

Understanding the thermodynamic and kinetic factors that control the sorption of toxins to soil solids – or desorption from solids into the soil solution – is an important part of predicting environmental availability of constituents ranging from heavy metals to toxic organic compounds. Data from adsorption isotherms (Chapter 4, Section 4.12.4) are useful in this task, but are limited in that the experiments used to obtain isotherms assume thermodynamic equilibrium, i.e. do not quantify kinetic effects. Sparks (2003) emphasizes that soil reactions rarely reach equilibrium, and that quantifying kinetic (i.e. time) controls is imperative. In a series of experiments using soils with differing amounts of soil organic matter (SOM), Strawn and Sparks (2000)

determined that Pb sorbs first by fast reaction followed by slow reaction rate, much like the logarithmic curve shown in Figure 9.11; rapid at first followed by lower rates as the system approaches equilibrium. Sorption of Pb is proportional to the quantity of SOM (more SOM = greater Pb sorption), but also important is that the greater the amount of SOM, the slower the rate of sorption; similarly, the fraction of Pb desorbed from soil decreases with increasing SOM, and the SOM fraction appears to control the slow phase of sorption and desorption. In a practical sense, greater amounts of SOM foster greater Pb sorption and slower rates of desorption (release of Pb to soil waters).

Lead arsenate ($PbHAsO_4$) was a widely used pesticide in fruit orchards from the late 1800s until adoption of dichlorodiphenyltrichloroethane (DDT) in the 1940s. The decades of lead arsenate application provide the opportunity to compare the behavior of Pb and As under effectively identical environmental factors. For a given soil, the mineralogy, organic matter composition and content, temperature, moisture content, biological community, permeability, pH, redox conditions, and any other potential factors available to affect one of the elements are also there to affect the others, and results reveal important differences in their inherent behaviors. Pb forms very strong complexes with POC (e.g. humus) and sorbs to iron hydroxides, and thus is nearly immobile (unless soil erosion occurs, in which case Pb may be physically eroded and transported along with particulates). Arsenic also tends to be largely immobile as it is similarly sorbed to POC and hydroxides. However, arsenic can become mobilized when soils are fertilized with phosphate, which exists as a mix of $H_2PO_4^-$ and HPO_4^{2-} at soil pH of 6–8. Similarly, in this pH range arsenic will occur as a mix of $H_2AsO_4^-$ and $HAsO_4^{2-}$. Given the smaller radius of P^{5+} relative to As^{5+}, the radius of phosphate anions will be smaller than that of arsenate anions: the radius of PO_4^{3-} is 1.75 Å compared to 1.86 Å for AsO_4^{3-}. The effect is predictable – phosphate anions are attracted to adsorption or exchange sites more strongly than arsenate, so as the amount of phosphate increases by fertilizer application, arsenate mobility will increase. Thus, the inherent difference between cationic Pb^{2+} and anionic $H_2AsO_4^-$ and $HAsO_4^{2-}$ leads to different behavior when environmental factors (in this case, addition of phosphate) change.

9.9 ORGANIC SOIL POLLUTANTS AND REMEDIATION (FUELS, INSECTICIDES, SOLVENTS)

Many of the important considerations related to behavior of organic pollutants in the environment are presented in Chapters 3 (Section 3.3) and 8 (Section 8.9), and Appendix A, which presents a case study involving remediation of groundwater contaminated with organic solvents. The following section, therefore, draws attention to pertinent issues at the intersection of organic compounds and soil. Reactions in soil control the potential vaporization of volatile organic compounds (VOCs) into buildings, potential uptake of organic pollutants into organisms (including crops), potential leaching into aquifers, and potential for transport of organic contaminants into surface waters.

In soil, organic compounds nearly always occur in one of five possible states: sorbed (e.g. to soil organic carbon or minerals), gaseous (especially in the case of VOCs), liquid (i.e. non-aqueous-phase liquid [NAPLs]), aqueous/dissolved (e.g. ~soluble benzene, trichloroethylene [TCE]), or incorporated into organisms (i.e. plants, animals). Quantifying the factors that control uptake into these reservoirs, therefore, is vital for predicting and understanding behavior of organic compounds in the environment. Figure 9.17 is a schematic sketch of the main reservoirs and reactions that occur involving organic contaminants in soils. It is worth emphasizing an important difference between the behavior of organic contaminants as shown in Figure 9.17, compared to potential pathways of inorganic metals and metalloids shown in Figure 9.16: the potential for organic compounds to

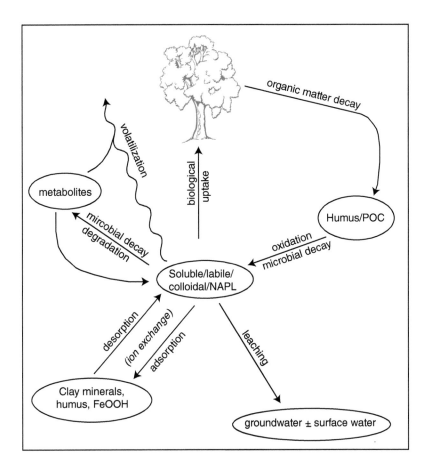

Fig. 9.17 Schematic diagram of potential reservoirs and reactions of organic contaminants once they have been introduced into soil. For most organic compounds, the soluble/labile/colloidal/ NAPL reservoir (i.e. likely to be mobile) is small given low solubilities (Table 3.4) and tendency to sorb to organic matter.

decompose into metabolites. A nonradioactive element such as Pb, As, or Hg will always be Pb, As, or Hg, and although its speciation may change, it will not decompose into nontoxic end-products such as CO_2 and H_2O (the ultimate step in decomposition of hydrocarbons, for example). Another important difference between organic compounds and inorganic elements or molecules is that organics may occur in soil as insoluble NAPLs, whereas inorganic elements and substances in the environment do not.

Two factors strongly control sorption of organic compounds – the abundance of soil organic carbon (POC, humus) and, usually to a much lower extent, the mineralogy of the soil. Nonpolar organic compounds such as benzene, toluene, ethylbenzene, and xylene (BTEX) and most VOCs have very high K_{OW} values (Eq. (3.4)), indicating they are strongly attracted to soil organic carbon, and numerous studies indicate that the extent of sorption of solvents, fuels, and herbicides in soil is proportional to the quantity of SOM. What this means is that soils with high organic matter content (e.g. Mollisols, Alfisols, Spodosols) tend to sorb and limit migration of organic compounds; conversely, for soils with low organic matter

content – often the case with Aridisols, Entisols, and Oxisols – clay minerals and other inorganic solids may be an important variable exerting control on sorption.

Soil minerals exhibit a wide range in sorption potential – smectites and vermiculites have high CEC so they may adsorb cationic species (e.g. the herbicides diquat and paraquat), but in cases of nonpolar organic compounds, the basal tetrahedral sheet of 2 : 1 layers may be capable of adsorption. Two examples illustrate the complexity of adsorption to soil clays: Na-smectite has a greater capacity to adsorb atrazine (a weakly polar herbicide) than does Ca-smectite (Prima et al. 2002), yet smectite was found to have no capacity to adsorb either of the nonpolar compounds toluene or benzene.

Understanding the behavior of organic pollutants in soils, surface water, and groundwater is clearly a challenging proposition. Some of the basic principles are provided here and in Chapter 3. Further analysis is also provided in Appendix A, a case study that explores the complex relationships among inorganic minerals and elements, organic compounds, microbial communities, redox, and pH, as well as options and consequences of remediation actions.

QUESTIONS

9.1 The data in Table 9.2 indicate that goethite (FeOOH) decreases with time in tropical soils, and at the same time, hematite (Fe_2O_3) increases. From these data, one might hypothesize that goethite transforms to hematite in soils at ~25 °C over time spans of 10^4 to 10^5 years. Assess this hypothesis using thermodynamic data; in other words, is the transformation of goethite to hematite spontaneous? Is goethite stable in these soils? Is it metastable? Is hematite stable relative to goethite? Or vice versa?

9.2 At what soil solution conditions do kaolinite and gibbsite exist in equilibrium?

9.3 Determine the ESP for a soil that is 40% smectite, 30% calcite, 20% quartz, and 10% illite with exchangeable Na^+ of $1210\,mg\,kg^{-1}$. Is this a sodic soil? Explain the negative effects that high amounts of exchangeable Na^+ have on soils (consider chemical and physical effects). In what type of climate does this problem occur, and why?

9.4 Explain why Goldich found Bowen's reaction series useful in predicting weathering rates of igneous minerals in soil. Compare relative rates derived from this approach to the rates presented in Table 9.1 and indicate a factor that might explain any differences.

9.5 Citing mineral formulas, explain why smectite and calcite are common in Aridisols whereas kaolinite and gibbsite are common in Oxisols. Comment on how this difference could affect plant nutrient retention in Aridisols vs. Oxisols.

9.6 Explain why kinetic factors are often more influential in determining soil mineralogy than thermodynamic factors, citing three examples.

9.7 Determine relative gain and loss of elements associated with weathering of basalt to a tropical Alfisol B-horizon (Plate 13). Use the data provided to rank elements in terms of % change as well as by calculating the mass transfer coefficient, τ_j, for each element. Major elements are presented as wt% oxides and trace elements as $mg\,kg^{-1}$ (ppm).

This dataset pertains to question 9.7 and is from the B-horizon and underlying basalt parent material shown in Plate 13.

	SiO₂	TiO₂	Al₂O₃	Fe₂O₃(t)	MnO	MgO	CaO
B-horizon	44.6	1.78	27.7	21.3	0.29	1.75	0.79
Basalt	52.3	1.32	19.5	16.1	0.16	5.53	1.81

Na₂O	K₂O	P₂O₅	Cr	Ni	Sr	Zr
0.04	0.06	0.64	436	326	69	104
0.26	2.18	0.24	442	281	240	71

For both methods, rank elements in terms of mobility, and compare results produced by the two methods.

9.8 Construct a mineral stability diagram like the one in Figure 9.8, replacing albite with K–feldspar and Na–beidellite with K–montmorillonite.

9.9 What are common fates of H^+ delivered to soils by acid deposition? What effect does added H^+ have on Ca^{2+} and Al^{3+}? Explain the problem associated with high amounts of NO_3^- and SO_4^{2-} fluxing through soils acid affected by acid deposition.

9.10 Provide two chemical equations that depict the oxidation of pyrite in acid mine (or rock) drainage systems (include one involving ferric iron as reactant). Also describe the role of bacteria in these reactions.

9.11 List and briefly describe three important ways by which the behavior of organic pollutants (e.g. VOCs) in soil differs from the behavior of inorganic elements (e.g. Cr, Pb).

REFERENCES

Birkeland, P.W. (1999). *Soils and Geomorphology*. Oxford University Press.

Brady, N.C. and Weil, R.R. (2002). *The Nature and Properties of Soils*, 14e. Upper Saddle River New Jersey: Prentice Hall.

Brantley, S.L. (2005). Reaction kinetics of primary rock-forming minerals under ambient conditions. In: *Surface and Ground Water, Weathering and Soils. Treatise on Geochemistry*, H.D. Holland and K.K. Turekian (executive editors) (ed. J.I. Drever), 73–118. Elsevier.

Brimhall, G.H. and Dietrich, W.E. (1987). Constitutive mass balance relations between chemical composition, volume, density, porosity, and strain in metasomatic hydrochemical systems: results on weathering and pedogenesis. *Geochimica et Cosmochimica Acta* 51: 567–587.

Buss, H.L., Sak, P.B., Webb, S.M., and Brantley, S.L. (2008). Weathering of the Rio Blanco quartz diorite, Luquillo Mountains, Puerto Rico: coupling oxidation, dissolution, and fracturing. *Geochimica et Cosmochimica Acta* 72: 4488–4507.

Charles, D.F., Whitehead, D.R., Engstrom, D.R. et al. (1987). Paleolimnological evidence for recent acidification of big moose Lake, Adirondack Mountains, NY (USA). *Biogeochemistry* 3: 267–296.

Chou, L., Garrels, R.M., and Wollast, R. (1989). Comparative study of the kinetics and mechanisms of dissolution of carbonate minerals. *Chemical Geology* 78: 269–282.

Davidson, E.A. and Janssens, I.A. (2006). Temperature sensitivity of soil carbon decomposition and feedbacks to climate change. *Nature* 440: 165–173. https://doi.org/10.1038/nature04514.

Dean, J.A. (1979). *Lange's Handbook of Chemistry*, 12e. New York: McGraw-Hill.

Drever, J.I. (1997). *The Geochemistry of Natural Waters: Surface and Groundwater Environments*, 3e. Upper Saddle River, New Jersey, USA: Simon and Schuster.

Fisher, G.B. and Ryan, P.C. (2006). The smectite to disordered kaolinite transition in a tropical soil chronosequence, Pacific Coast, Costa Rica. *Clays and Clay Minerals* 54: 571–586.

Gbondo-Tugbawa, S.S. and Driscoll, C.T. (2003). Factors controlling long-term changes in soil pools of exchangeable basic cations and stream acid neutralizing capacity in a northern hardwood forest ecosystem. *Biogeochemistry* 63: 161–185.

Goldich, S.S. (1938). A study in rock-weathering. *Journal of Geology* 46: 17–58.

Lasaga, A.C., Soler, J.M., Ganor, J. et al. (1994). Chemical weathering rate laws and global geochemical cycles. *Geochimica et Cosmochimica Acta* 58: 2361–2386.

Lee, B.D., Sears, S.K., Graham, R.C. et al. (2003). Secondary mineral genesis from chlorite and serpentine in an ultramafic soil Toposequence. *Soil Science Society of America Journal* 67: 1309–1317.

Liivamägi, S., Środoń, J., Bojanowski, M.J. et al. (2018). Paleosols on the Ediacaran basalts of the East European Craton: a unique record of paleoweathering with minimum diagenetic overprint. *Precambrian Research* 316: 66–82.

McBride, M.B. (1994). *Environmental Chemistry of Soils*. New York: Oxford University Press.

Nesbitt, H.W. and Young, G.M. (1982). Early Proterozoic climates and plate motions inferred from major element chemistry of lutites. *Nature* 299: 715–717.

Prima, S., Evangelou, V.P., and McDonald, L.M. (2002). Surface exchange phase composition and nonionic surfactant effects on the nonequilibrium transport of atrazine. *Soil Science* 167: 260–268.

Rauret, G., López-Sànchez, J.F., Sahuquillo, A. et al. (1999). Improvement of the BCR three step sequential extraction procedure prior to the certification of new sediment and soil reference materials. *Journal of Environmental Monitoring* 1: 57–61.

Robie, R.A., Hemingway, B.S., and Fisher, J.R. (1979). Thermodynamic properties of minerals and related substances at 298.15 K and 1 bar (10^5 Pascals) pressure and at higher temperatures. In: *United States Geological Survey Bulletin 1452*. Washington, D.C: United States Government Printing Office. Accessed online 27 November 2018: http://pubs.usgs.gov/bul/1452/report.pdf.

Ryan, P.C. and Huertas, F.J. (2009). The temporal evolution of pedogenic Fe-smectite to Fe-kaolin via interstratified kaolin-smectite in a moist tropical soil chronosequence. *Geoderma* 151: 1–15.

Ryan, P.C., Hillier, S., and Wall, A.J. (2008). Stepwise effects of the BCR sequential chemical extraction procedure on dissolution and metal release from common ferromagnesian clay minerals: a combined solution chemistry and X-ray powder diffraction study. *Science of the Total Environment* 407: 603–614.

Ryan, P.C., Huertas, F.J., Hobbs, F.C., and Pincus, L.N. (2016). Kaolinite and halloysite derived from sequential transformation of pedogenic smectite and kaolinite-smectite in a 120 ka tropical soil chronosequence. *Clays and Clay Minerals* 64: 488–516.

Sherman, G.D. (1952). The genesis and morphology of the alumina-rich laterite clays [Hawaii]: in problems of clay and laterite genes. In: *American Institute of Mining and Metallurgical Engineering Annual Meeting* February 1951, 154–161.

Sparks, D.L. (2003). *Environmental Soil Chemistry*. London: Academic Press.

Stallard, R.F. and Edmond, J.M. (1983). Geochemistry of the Amazon: 2. The influence of geology and weathering environment on the dissolved load. *Journal of Geophysical Research Oceans and Atmospheres* 88: 9671–9688.

Stewart, B.W., Capo, R.C., and Chadwick, O.A. (2001). Effects of rainfall on weathering rate, base cation provenance, and Sr isotope composition of Hawaiian soils. *Geochimica et Cosmochinica Acta* 65: 1087–1099.

Strakhov, N.M. (1967). *Principles of Lithogenesis*. Edinburgh: Oliver and Boyd.

Strawn, D.G. and Sparks, D.L. (2000). Effects of soil organic matter on the kinetics and mechanisms of Pb(II) sorption and desorption in soil. *Soil Science Society of America Journal* 64: 144–156.

Tessier, A., Campbell, P.G.C., and Bisson, M. (1979). Sequential extraction procedure for the speciation of particulate trace metals. *Analytical Chemistry* 51: 844–851.

Wilson, M.J. and Bell, N. (1996). Acid deposition and heavy metal mobilization. *Applied Geochemistry* 11: 133–137.

Woodson, M.K., Sandor, J.A., Strawhacker, C., and Miles, W.D. (2015). Hohokam canal irrigation and the formation of irragric anthrosols in the Middle Gila River Valley, Arizona, USA. *Geoarchaeology: An International Journal* 30: 271–290.

Young, S.D., Zhang, H., Tye, A.M. et al. (2006). Characterizing the availability of metals in contaminated soils I. The solid phase: sequential extraction and isotopic dilution. *Soil Use Management* 21: 450–458.

10

Stable Isotope Geochemistry

Variations in the ratios of stable isotopes of many major and trace elements are very useful in the field of environmental geochemistry, particularly as tracers of biogeochemical processes. Their applications range from determining the origin of atmospheric contaminants, sources of nitrates in groundwater, variable sources of groundwater recharge, global paleotemperature, the role of microbes in mineral reactions, and much more. Most elements have two or more naturally occurring isotopes, but not all do. Notable examples of environmentally significant elements that have atoms of only one isotope include arsenic (^{75}As), fluorine (^{19}F), manganese (^{55}Mn), and phosphorus (^{31}P), and this attribute provides one less method for studying their behavior in the environment.

10.1 STABLE ISOTOPES – MASS DIFFERENCES AND THE CONCEPT OF FRACTIONATION

Stable isotopes are non-radioactive isotopes of an element. Common examples include ^{1}H–^{2}H (^{1}H–D), ^{12}C–^{13}C, ^{16}O–^{17}O–^{18}O, ^{14}N–^{15}N, and ^{32}S–^{34}S. The *chart of the nuclides* (Figure 10.1) consists of horizontal rows that contain various isotopes of a given element. Using a familiar example, oxygen (Z = 8) contains three naturally occurring isotopes, ^{16}O, ^{17}O, and ^{18}O, the main difference being the number of neutrons (8, 9, and 10, respectively). Magnesium also has three stable isotopes (^{24}Mg, ^{25}Mg, ^{26}Mg) with 12, 13, and 14 neutrons, respectively, and the two stable isotopes of carbon (^{12}C and ^{13}C) and one radioactive isotope (^{14}C) are evident (so too is the very rare ^{11}C).

A fundamental principle of this field of study is that stable isotopes of the same element have the same Z and valence structures (i.e. they will adopt the same oxidation states regardless). This means that stable isotopes of a given element behave very similarly, but not identically, in biogeochemical systems. Each of the three isotopes of oxygen will bond with silicon to form SiO_2. Given that each of the oxygen atoms could be any of the three oxygen isotopes, possible SiO_2 molecules include $Si^{16}O_2$, $Si^{16}O^{17}O$, $Si^{16}O^{18}O$, and so on. Silicon also has three stable isotopes, ^{28}Si, ^{29}Si, and ^{30}Si, so the combinations of silicon and oxygen isotopes further expands the array of SiO_2 molecular mass configurations to include molecules such as $^{28}Si^{16}O_2$, $^{28}Si^{16}O^{17}O$, $^{29}Si^{16}O_2$, and so on.

In biochemical reactions, proteins incorporate both ^{14}N and ^{15}N, and both N isotopes can bond with oxygen to form nitrate ($NO_3{}^-$). The same is true for stable carbon isotopes (^{12}C and ^{13}C) and the radioactive carbon isotope, ^{14}C – all three isotopes can be incorporated into CO_2, organic matter, or carbonate minerals. The subtle variations in behavior caused by the difference in mass between isotopes of the same element (or the compounds that the different isotopes form, e.g. $H_2{}^{16}O$ vs. $H_2{}^{18}O$), is known as *mass-dependent fractionation (MDF)*, and this is the crux of stable isotope analysis.

Water helps to explain the differences in behavior that stable isotopes impart to molecules. Both of the naturally occurring hydrogen isotopes, ^{1}H and ^{2}H (the heavier hydrogen isotope, ^{2}H, is typically referred to as "D," for deuterium) will combine with oxygen to form H_2O – most water molecules are $^{1}H_2O$, but a small amount are ^{1}HDO or D_2O. Considering oxygen isotopes, the isotopic

Environmental and Low-Temperature Geochemistry, Second Edition. Peter Ryan.
© 2020 John Wiley & Sons Ltd. Published 2020 by John Wiley & Sons Ltd.

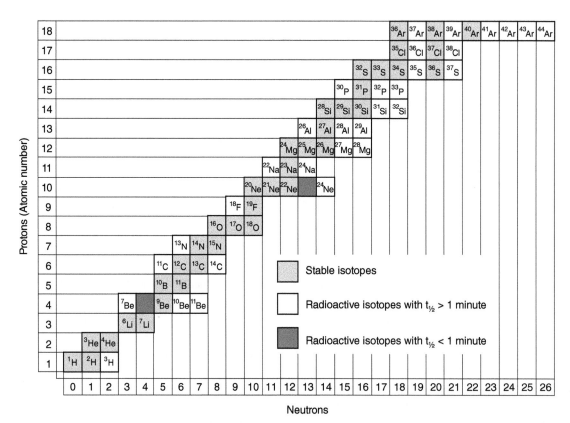

Fig. 10.1 Chart of the low atomic mass nuclides. Note that most elements have two or more stable isotopes (e.g. sulfur has four, ^{32}S, ^{33}S, ^{34}S, and ^{36}S), but some have only one stable isotope (e.g. Be, F, Na, Al, P, and As [not shown]). Radioactive isotopes included in this chart that occur in sufficient quantities to be useful in environmental geochemistry include ^{3}H (tritium), ^{10}Be, ^{14}C, ^{26}Al, and ^{36}Cl. This chart also includes many isotopes with very short half-lives that effectively do not occur in nature (e.g. ^{11}C, ^{37}S, ^{18}F), and it also does not show numerous others with very short half-lives.

Focus Box 10.1

Isotopes and Non-integer Atomic Masses of Elements

The periodic table of the elements shows that nearly all elements have *non-integer atomic masses*. For example, iron has an atomic mass of 58.55 g mol^{-1}, which can also be stated as 55.85 Daltons (Da), or 55.85 atomic mass units (where 1 amu = 1 Da). To be certain, there are no Fe atoms with 29.85 neutrons, the number required to produce an atomic mass of 55.85 Da when added to 26 (Z = 26 for Fe). Rather, the value 55.85 is produced by a mass balance of Fe isotopes with the following abundances (http://ie.lbl.gov/education/parent/Fe_iso.htm):

Fe^{54}	5.58%
Fe^{56}	91.95%
Fe^{57}	2.18%
Fe^{58}	0.30%

The mass balance calculation is expressed as

$$\text{Atomic mass (Da)} = 54 \times (0.0558) + 56 \times (0.9195)$$
$$+ 57 \times (0.0218) + 58 \times (0.0030)$$
$$= 55.92$$

The slight discrepancy between the value on the periodic table (55.85 Da) and the value produced by the mass balance computed above (55.92 Da) is caused by the fact that all atomic masses are normalized to ^{12}C, which is defined as having a mass of exactly 12 Da. Similarly, the same calculation for the Mg isotopes ^{24}Mg (78.99%), ^{25}Mg (10.00%), and ^{26}Mg (11.01%) produces an atomic mass of 24.320 Da, as compared to the ^{12}C scaled value of 24.305 Da.

Table 10.1 Examples of stable isotope combinations in water and carbon dioxide.

Water:

$^1H_2^{16}O$, $^1H_2^{17}O$, $^1H_2^{18}O$

$DH^{16}O$, $DH^{17}O$, $DH^{18}O$

$D_2^{16}O$, $D_2^{17}O$, $D_2^{18}O$

Carbon dioxide:

$^{12}C^{16}O_2$, $^{12}C^{16}O^{17}O$, $^{12}C^{16}O^{18}O$, $^{12}C^{17}O^{18}O$, $^{12}C^{17}O_2$,

$^{12}C^{18}O_2$,

$^{13}C^{16}O_2$, $^{13}C^{16}O^{17}O$, $^{13}C^{16}O^{18}O$, $^{13}C^{17}O^{18}O$, $^{13}C^{17}O_2$,

$^{13}C^{18}O_2$

possibilities for water molecules expand to $^1H_2^{16}O$, $^1H_2^{17}O$, $^1H_2^{18}O$, $D_2^{16}O$, and so on. In all, there are nine stable isotopic combinations of water when considering all combinations of H, D, and the three stable isotopes of oxygen. There are 10 stable isotope combinations in carbon dioxide produced by ^{12}C, ^{13}C, and the three stable isotopes of oxygen (Table 10.1).

Two properties that make stable isotopes particularly useful in geochemical studies are:

1 the mass of molecules affects their behavior, so e.g. isotopically heavy nitrate ($^{15}NO_3^-$) will behave slightly differently than ($^{14}NO_3^-$); and

2 covalent bonds are stronger when the heavier isotope is involved, so e.g. isotopically heavier nitrate (i.e. $^{15}NO_3^-$) will decompose more slowly than nitrate formed by the lighter isotope (i.e. $^{14}NO_3^-$).

The evaporation of water molecules from a body of water (e.g. ocean, wetland, lake) provides a good example of the fractionation of stable isotopes that makes them such a useful way to study processes in natural (and anthropogenically altered) systems. The mass of $H_2^{16}O$ is $18\,g\,mol^{-1}$ whereas $H_2^{18}O$ is $20\,g\,mol^{-1}$. Based on its lower molecular mass, the probability of an $H_2^{16}O$ molecule evaporating is greater than the probability of an $H_2^{18}O$ molecule evaporating. Fractionation caused by evaporation produces atmospheric water with a higher ratio of $^{16}O/^{18}O$ than the original water body, and as evaporation proceeds, the surface water body becomes progressively enriched in water with ^{18}O. This causes the $^{16}O/^{18}O$ ratio of the water body to decrease as the amount of ^{18}O increases with progressive evaporation. Similarly, when water vapor in the atmosphere condenses, the heavier $H_2^{18}O$ and ($^1HD^{18}O$, $D_2^{18}O$, $D_2^{16}O$, etc.) molecules will be more likely to condense due to their greater mass (and hence lower vibrational energy that stabilizes intermolecular bonds between water molecules) as compared to $H_2^{16}O$. To emphasize this point a little further, as an air mass migrates and cools, water molecules in the first

precipitation to fall will be enriched in D and ^{18}O (relative to their lighter counterparts) as compared to precipitation that falls farther down gradient.

Stable isotopes of the light elements such as hydrogen, carbon, nitrogen, oxygen, and sulfur have been applied for longer and with greater frequency than isotopes of heavier elements (e.g. Fe, Cu, Hg). Part of the reason for this is the percent difference in mass between the isotopes under consideration (and their compounds). The difference in mass between $H_2^{18}O$ and $H_2^{16}O$ is $(20–18)/18 = 11\%$. In other words, a molecule of $H_2^{18}O$ is 11% heavier than a molecule of $H_2^{16}O$. On an atomic level, ^{18}O is 12.5% heavier than ^{16}O, and the mass of D is twice that of ^1H. Conversely, the difference in mass between the two most abundant stable isotopes of iron, ^{54}Fe and ^{56}Fe, is only 3.7%.

The relatively small difference in the mass of stable isotopes of heavy elements means that fractionation will be much less pronounced and more difficult to detect by mass spectrometers, so stable isotopes of heavy elements have shorter histories in geochemical and biochemical studies than the lighter isotope stable systems. However, with the development of new analytical techniques (particularly multicollector inductively coupled plasma mass spectrometry, MC-ICP-MS) that offer higher precision and greater sensitivity, stable isotope geochemists are examining with increasing frequency stable isotope fractionation in heavier elements such as Fe, Zn, Cu, Mo, Ti, Hg, and more.

10.2 DELTA (δ) NOTATION

Considering the stable isotopes of a given element, there is always one whose abundance is far greater than the other stable isotope(s) of that element. For example, ^1H comprises 99.985% of all hydrogen; D (^2H) is only 0.015% of all hydrogen atoms. Similarly, ^{14}N is 99.63% of all nitrogen, and ^{15}N comprises only 0.37% of stable nitrogen atoms (Table 10.2).

To present isotope abundances as simple ratios of heaver to lighter isotope would produce very small values. In the case of hydrogen, the average ratio of D : ^1H is ~0.00015. Thus, in order to produce values that are more convenient for comparison and for analysis of fractionation, a simple equation is applied to stable isotope pairs. This is known as the *delta notation*, where the ratio of heavy isotope to light isotope in a sample (of water, mineral, gas, leaf tissue, etc.) is compared to the ratio of heavy to light isotope in a standard. Using hydrogen as an example (and remembering that D stands for the heavier isotope ^2H):

$$\delta D\ (\text{‰}) = \left(\frac{(D/H)_{smpl} - (D/H)_{std}}{(D/H)_{std}} \right) \times 1000 \qquad (10.1)$$

Table 10.2 Stable isotopes, abundances (relative %), differences in mass (relative %), delta notations, and ratios of standard values of a representative (but not complete) list of the stable isotopes applied in environmental geochemistry.

Element	Isotopes	Abundance	Mass diff. (%)	δ notation	Standard with ratio
Hydrogen	^1H ^2H (D)	99.985 0.015	99.8	δD	VSMOW[a]: $D/H = 1.557 \times 10^{-4}$
Boron	^{10}B ^{11}B	19.9 80.1	10.0	$\delta^{11}B$	SRM 951[b]: $^{11}B/^{10}B = 4.0456$
Carbon	^{12}C ^{13}C	98.90 1.10	8.3	$\delta^{13}C$	PDB[c]: $^{13}C/^{12}C = 1.122 \times 10^{-2}$
Nitrogen	^{14}N ^{15}N	99.63 0.37	7.1	$\delta^{15}N$	Atmospheric N_2: $^{15}N/^{14}N = 3.613 \times 10^{-3}$
Oxygen	^{16}O ^{17}O ^{18}O	99.76 0.038 0.20	12.5 (^{18}O vs. ^{16}O)	$\delta^{18}O$ (ratio of $^{18}O/^{16}O$)	VSMOW: $^{18}O/^{16}O = 2.0052 \times 10^{-3}$ PDB[c]: $^{18}O/^{16}O = 3.76 \times 10^{-4}$
Magnesium	^{24}Mg ^{25}Mg ^{26}Mg	78.99 10.00 11.01	8.3 (^{26}Mg vs. ^{24}Mg)	$\delta^{26}Mg$ (ratio of $^{26}Mg/^{24}Mg$) also $\delta^{26}Mg$	SRM 980[d]: $^{26}Mg/^{24}Mg = 1.392 \times 10^{-1}$ DSM3[d]:
Sulfur	^{32}S ^{33}S ^{34}S ^{35}S	95.02 0.750 4.21 0.02	6.2 (^{34}S vs. ^{32}S)	$\delta^{34}S$ (ratio of $^{34}S/^{32}S$)	CDM[e]: $^{34}S/^{32}S = 4.43 \times 10^{-2}$
Iron	^{54}Fe ^{56}Fe ^{57}Fe ^{58}Fe	5.58 91.950 2.18 0.30	3.7 (^{56}Fe vs. ^{54}Fe)	$\delta^{56/54}Fe$ (ratio of $^{56}Fe/^{54}Fe$)	IRMM-014[f]: $^{54}Fe/^{56}Fe = 6.37 \times 10^{-2}$
Copper	^{63}Cu ^{65}Cu	69.15 30.85	3.2 (^{65}Cu vs. ^{63}Cu)	$\delta^{65}Cu$ (ratio of $^{65}Cu/^{63}Cu$)	ERM-AE637[g]: $^{65}Cu/^{63}Cu = 4.456 \times 10^{-1}$
Mercury	^{196}Hg ^{198}Hg ^{199}Hg ^{200}Hg ^{201}Hg ^{202}Hg ^{204}Hg	0.15 9.97 16.87 23.10 13.18 29.86 6.87	1.98 (^{202}Hg vs. ^{198}Hg)	$\delta^{202/198}Hg$ (ratio of $^{202}Hg/^{198}Hg$)	NIST-3133[h]: $^{202}Hg/^{198}Hg = 2.9689$

[a]Vienna standard mean ocean water (VSMOW) is the standard for hydrogen isotopes. It is also commonly used as a reference for oxygen, although oxygen in PDB is sometimes used (especially in paleoclimate studies of oxygen in carbonates).

[b]The ratio of $^{11}B/^{10}B$ prepared from Searles Lake borax (SRM 951 from the National Bureau of Standards, Palmer and Slack 1989), an evaporite deposit, is commonly used as the boron isotope standard (Hoefs 2009).

[c]PDB is a fossil from the Pee Dee Fm in South Carolina, USA, used as the $^{13}C/^{12}C$ standard, and also sometimes for $^{18}O/^{16}O$.

[d]SRM 980 is heterogeneous (limited to precision of ±0.2‰) so DSM3 has also been adopted as a Mg isotope standard.

[e]Troilite (FeS) from the Canyon Diablo Meteorite (CDM) is commonly used as a reference value for the $^{34}S/^{32}S$ ratio.

[f]IRMM-014 is metallic iron provided by the EU Institute for Reference Materials and Measurements.

[g]Limited supply of SRM 976 led to development of new Cu isotope standards. Refer to Moeller et al. (2012) for information.

[h]Details of Hg isotopic method development and standards are presented by Blum (2011).

If the ratio of D/H in the sample is lower than the D/H ratio in the standard, the resulting "delta D" value will be negative. Conversely, samples with D/H ratios that are higher than the standard D/H value will produce positive values of δD. To produce values in the approximate range of −100 to +100, the ratio in large brackets is multiplied by 1000. (It is common to multiply ratios by 100 to get per cent values, e.g. for interest rates or winning percentages of sports teams.) In the case of the delta notation in Eq. (10.1), multiplying by 1000 results in values expressed in

The Language of Stable Isotopes: Depleted vs. Enriched, Light vs. Heavy

A certain language has developed in the field of stable isotope geochemistry to describe isotope variations and trends. Often, one isotope is referred as being enriched (or depleted) with reference to the other in the stable isotope pair. For example, rainwater is enriched in ^{16}O relative to the water body from which it evaporated (and if it enriched in ^{16}O it is depleted in ^{18}O). Rainwater is also isotopically lighter (^{16}O-enriched, ^{18}O-depleted) compared to the heavier (^{18}O-enriched) surface water from which it

is derived. When referring to delta values, stable isotope researchers frequently indicate that a particular sample is more (or less) positive as compared to a different sample. One might say that residual water left behind by evaporation is 10 per mil more positive than the vapor that evaporated from the water body. In this case, a value such as −15‰ (the remnant surface water body) is 10‰ more positive (or "less negative") than the −25‰ water vapor that has evaporated from it into the atmosphere.

"per mil," i.e. per thousand. Per cent is represented by the symbol %, per mil is represented by the symbol ‰.

The standards used in reference to stable isotope pairs vary depending on the composition of the standard and the isotopes in question. For hydrogen and oxygen, Vienna standard mean ocean water (VSMOW) is commonly used to produce a delta value. The delta notation for oxygen referenced to VSMOW is:

$$\delta^{18}O \quad (‰) = \left(\frac{(^{18}O/^{16}O)_{smpl} - (^{18}O/^{16}O)_{VSMOW}}{(^{18}O/^{16}O)_{VSMOW}} \right) \times 1000 \qquad (10.2)$$

In some cases (e.g. where oxygen isotopes are measured in carbonate minerals), the Pee Dee Belemnite (*PDB*, from a carbonate rock) is used as a reference standard (Table 10.2). Oxygen isotope values referenced to VSMOW can be converted to PDB values as follows (Friedman and O'Neil 1977):

$$\delta^{18}O_{VSMOW} = (\delta^{18}O_{PDB} \times 1.03086) + 30.86 \qquad (10.3)$$

10.3 FRACTIONATION: VIBRATIONAL FREQUENCIES, MASS, AND TEMPERATURE DEPENDENCE

Fractionation during chemical bonding is caused by differences in vibrational energies associated with light and heavy isotopes of a given element; the consequence is that the heavier isotope of a given element forms a stronger covalent bond than the lighter isotope in the same bonding site. However, this type of fractionation only applies to bonds that have a predominantly covalent character. Ionic and metallic bonds do not involve sharing of electrons, so there is little or no isotopic fractionation when e.g. ionically bonded salts crystallize from solution.

10.3.1 Stable isotopes and chemical bond strength

Quantum theory states that the energy of a diatomic oscillator (e.g. a gaseous molecule comprised of two atoms) can only occur at discrete energy levels, and the possible energy levels depend on the mass of the atoms that comprise the molecule. The atoms in the diatomic oscillator will vibrate at a ground frequency v_0 (for most gases, this applies to systems at room temperature), and energy of the molecule is:

$$E = \tfrac{1}{2}hv \qquad (10.4)$$

where h is Planck's constant and v is the vibrational frequency of an atom in a molecule.

Higher v (vibration frequency) equates to higher E values, which corresponds to weaker bonds. The vibrational energy is lower for a bond involving the heavier isotope of an element, meaning that *bonds involving heavier isotopes are stronger*. For systems at equilibrium, the heavier isotope (^{18}O, ^{15}N, ^{34}S, etc.) will be more likely to form a stable chemical bond within a molecule than will its lighter counterpart (^{16}O, ^{14}N, ^{32}S, etc.), and conversely, because bonds involving lighter isotopes are weaker, they have a greater probability of being broken than do their heavier counterparts. The oxidation state of elements involved in bonding also affects fractionation because the heavier isotope preferentially forms bonds with elements of higher oxidation state (Kendall and Doctor 2003).

10.3.2 Temperature-dependent stable isotope fractionation

In addition to the differences in mass among isotopes of a given element, temperature also controls stable isotope fractionation. Differences in vibrational energy are most pronounced at low temperatures, and as temperature

---

Focus Box 10.3

Crystallization of Quartz from Solution as Example of Mass-Dependent Fractionation

When quartz begins to crystallize from a solution that has reached saturation, the first molecules of quartz to form are enriched in ^{18}O relative to ^{16}O, and the solution from which the quartz is being crystallized becomes depleted in ^{18}O. Reaction (10.5) depicts the formation of quartz from dissolved silica (as H_4SiO_4). In one idealized H_4SiO_4 molecule, all oxygen atoms are the heavier ^{18}O isotope and in the other, all atoms of oxygen are ^{16}O:

$$H_4Si^{18}O_4 + H_4Si^{16}O_4 \rightarrow 2Si^{18}O_2 + 4H_2^{16}O \quad (10.5)$$

In this reaction, SiO_4^{4-} tetrahedrons enriched in ^{18}O preferentially form covalent bonds with other ^{18}O-rich tetrahedrons, and thus ^{18}O is preferentially fractionated into quartz. Silicon atoms must shed some oxygen atoms when quartz forms, and the ones more likely to be shed are the ^{16}O atoms with weaker Si–O bonds. The ^{16}O atoms are left to form bonds with H^+ and are fractionated into water. We can say that the quartz is enriched in oxygen-18 relative to the water, and the quartz will have a higher (more positive) $\delta^{18}O$ value than will the water.

---

increases, atoms of both light and heavy isotopes vibrate with higher frequency and the difference in their frequencies goes down the warmer the temperature. The result is that high temperatures produce smaller differences in vibrational energy between heavy and light isotopes of a given element compared to cold conditions, where the difference in behavior of light vs. heavy isotope (and hence, fractionation) is greatest. Based on this, we would predict that quartz crystallized at low temperatures (e.g. 200 °C) will exhibit greater fractionation of ^{16}O and ^{18}O than will quartz crystallized at higher temperature (e.g. 700 °C) (Clayton et al. 1972). At some point, if a system is hot enough there is no appreciable difference in bonding of a light vs. heavy element of a given atom.

Early theoretical studies by Urey (1947) and Bigeleisen and Mayer (1947) indicate that the temperature dependence of fractionation is proportional to $1/T^2$, where T is in Kelvins, and where fractionation is expressed as $\delta^{18}O_{phaseA} - \delta^{18}O_{phaseB}$, using oxygen as the example. According to Savin and Lee (1988), fractionation of oxygen isotopes between most minerals and water is nearly linear when plotted as $\delta^{18}O_{phaseA} - \delta^{18}O_{phaseB}$ versus $1/T^2$. Although far less common, fractionation can be proportional to $1/T$ in some lower temperature systems (e.g. with respect to hydrogen isotope fractionation at temperatures < 0 °C), yet regardless of the magnitude, higher temperatures produce a larger number in the denominator of $1/T^2$ (or $1/T$) and fractionation is inversely proportional to temperature.

Figure 10.2 indicates the strong temperature dependence of oxygen isotope fractionation between quartz and water over a temperature range of ~0 to 400 °C. As is the case with many graphs in stable isotope geochemistry, the units of temperature presented on the x-axis are $10^6/T^2$, where temperature is in degrees Kelvin. So for reference, 300 K (27 °C) is equal to 11.1 on the x-axis,

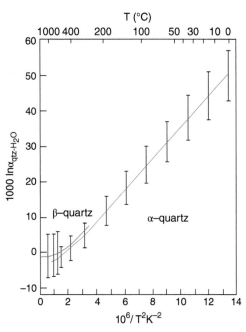

Fig. 10.2 Calculated values showing the relationship between temperature and fractionation of ^{16}O and ^{18}O in quartz (in equilibrium with water). Fractionation clearly decreases with increasing temperature. Source: Modified from Kawabe (1978). The term $1000 \ln\alpha_{qtz\text{-}H2O}$ is effectively equal to $\delta^{18}O_{qtz} - \delta^{18}O_{H2O}$, where $\alpha_{qtz\text{-}H2O} = (^{18}O/^{16}O)_{qtz}/(^{18}O/^{16}O)_{H2O}$.

and 700 K (423 °C) corresponds to an x-axis value of 2.0. Figure 10.2 indicates that when quartz crystallizes from solution at 5 to 25 °C – temperatures that are typical of many earth surface systems (e.g. soils, surface waters, and groundwaters) – the difference in $\delta^{18}O$ between quartz and water will be approximately 40–50‰. At 400 K (127 °C), the difference in $\delta^{18}O$ between quartz and water will be approximately 15‰, and at temperatures greater than 700 K (427 °C), there is essentially no fractionation between quartz and water.

Focus Box 10.4

Temperature Effect on Fractionation During Biomineralization of $CaCO_3$

Horibe and Oba (1972) quantified the temperature effect on isotopic fractionation between carbonates and water (Figure 10.3) using data from shells of two pelecypods from Mutsu Bay, Japan. Results indicated that for calcite precipitated during formation of *Patinopecten yessoensis* (the Ezo giant scallop or Yesso scallop), the empirical equation that describes the temperature dependence of oxygen isotope fractionation (Horibe and Oba 1972) is:

$$T \, (°C) = 17.04 - 4.34 \times (\delta_c - \delta_w)$$
$$+ \, 0.16 \times (\delta_c - \delta_w)^2 \qquad (10.6)$$

In this equation, δ_c represents the $\delta^{18}O$ of CO_2 derived from calcite by reaction with strong phosphoric acid at 25 °C, and δ_w is the $\delta^{18}O$ of CO_2 in equilibrium with water at 25 °C. Because the oxygen occurs in carbonates, the values are given relative to PDB.

The equation for aragonite (a polymorph of $CaCO_3$) found in shells of *Anadara broughtonii* is slightly different than the calcite equation (using PDB as the standard):

$$T \, (°C) = 13.85 - 4.54 \times (\delta_c - \delta_w)$$
$$+ \, 0.04 \times (\delta_c - \delta_w)^2 \qquad (10.7)$$

These equations demonstrate two important principles: (i) temperature affects isotopic fractionation; and (ii) mineral structure affects isotopic fractionation, where calcite and aragonite are both $CaCO_3$ but possess different crystal structures. These equations also serve to illustrate the applicability of oxygen isotopes to paleoclimate analysis.

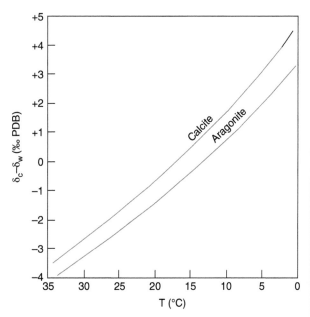

Fig. 10.3 Influence of temperature on oxygen isotope fractionation between calcium carbonates and water. Note that fractionation is greatest at lowest temperature, and that the two polymorphs of $CaCO_3$ exert slightly different controls on fractionation. Source: Based on data for marine carbonates from Horibe and Oba (1972).

Similar temperature-dependent fractionation effects occur when calcium carbonate (calcite or its polymorph aragonite) crystallize from solution. As is the case with quartz, the carbonate minerals (with carbon effectively occurring in a high oxidation state, C^{4+}) preferentially form bonds with ^{18}O, so the carbonates are enriched in ^{18}O relative to the water from which they crystallize, and as is the case with quartz–H_2O, this effect is greatest at low temperatures.

The examples in Figures 10.2 and 10.3 apply to quartz-water and carbonate-water *systems at equilibrium*. Although it does not always occur, isotopic equilibrium *can* be achieved in closed, well-mixed systems, where the equilibrium represents the condition where the forward and backward rates of isotope exchange are equal among minerals or other phases (e.g. water, organic matter, or gases). Considering that oxygen isotopes in marine water

are generally well-mixed, stable isotope geochemists consider that crystallization of carbonates in the presence of marine water is an equilibrium process.

10.3.3 Equilibrium and non-equilibrium isotope fractionation

Examining the isotopic reaction involving water and calcium carbonate as an exchange reaction

$$3H_2^{18}O + CaC^{16}O_3 = 3H_2^{16}O + CaC^{18}O_3 \qquad (10.8)$$

shows that ^{16}O and ^{18}O are exchanged between the molecules of water and calcium carbonate, and this idealized reaction represents complete exchange of three atoms of ^{16}O for three atoms of ^{18}O between the two phases. The heavier isotope is attracted to the phase where it bonds with an element of high oxidation number (C^{4+} in $CaCO_3$) and the light isotope is attracted to the phase where it bonds with an element of low oxidation number (H^+ in H_2O). If given sufficient time, and assuming the system is closed such that water molecules are not diffusing away too fast, equilibrium in this reaction may be reached; if not, kinetic isotope effects apply. The following two sections present considerations related to equilibrium and kinetic factors controlling isotope fractionation.

10.3.3.1 Equilibrium isotope fractionation

The equilibrium constant for the chemical reaction in Eq. (10.8) is:

$$K_{eq} = \frac{\alpha_{CaC^{18}O_3}}{\alpha_{CaC^{16}O_3}} \times \frac{\alpha^3_{H_2^{16}O}}{\alpha^3_{H_2^{18}O}} = \alpha \qquad (10.9)$$

The activities represent molar concentrations of each phase. If there is no fractionation of oxygen isotopes between water and calcium carbonate, the value of the K_{eq} will be exactly 1.0000 (α is the fractionation factor, which is a term used for the K_{eq} related to stable isotope fractionation). However, for water and calcite *at equilibrium* at 25 °C, the K_{eq} is equal to 1.0286 (O'Neil and Epstein 1966), indicating that the heavier isotope of oxygen is preferentially partitioned into calcite relative to water (and water is enriched in ^{16}O).

Fractionation of ^{18}O into calcite over water is predictable given the high effective oxidation state of carbon (C^{4+}) relative to hydrogen (H^+) and the attraction of high oxidation state ions for the heavier isotope of an element, and this is what has been observed in various studies, including those of Horibe and Oba (1972) and Friedman and O'Neil (1977). The term "effective" is used here and above because the bonds are dominantly covalent rather than ionic. The K_{eq} of 1.0286 produces a $\delta^{18}O$ value of 28.6‰ (assuming a $\delta^{18}O$ value of zero, a reasonable assumption given that the $\delta^{18}O$ value of VSMOW is equal to zero). Given what we know about temperature effects on isotope fractionation (i.e. less fractionation with increasing temperature), the K_{eq} or α will approach 1.0000 as temperature increases.

These examples (including Figures 10.2 and 10.3) illustrate equilibrium fractionation of stable isotopes, a condition most commonly achieved for inorganic reactions in well-mixed closed systems where the stable equilibrium condition occurs; metamorphic rocks that remain at a constant temperature for millions of years likely will reach equilibrium fractionation. In biological systems, in open systems, and in systems where time factors are important (e.g. rapid crystallization that does not permit equilibrium to be achieved), isotopic fractionation is dominated by *non-equilibrium effects*. In these cases, *kinetic factors* typically control fractionation.

10.3.3.2 Non-equilibrium isotope fractionation

The Arrhenius equation, first encountered during the discussion of kinetics of chemical reactions in Chapter 1 (Eq. (1.76)), helps to explain the kinetic isotope effect.

$$k = A \times e^{(-E_a/RT)}$$

The activation energy (E_a) of ^{12}C–O bonds is lower than the E_a for ^{13}C–O bonds (corresponding to a weaker ^{12}C–O bond than ^{13}C–O bond), and given the negative value of E_a in the exponent of the Arrhenius equation, decreasing activation energy will increase reaction rate (k). Furthermore, the lower diffusion rate of the heavier isotope will be represented in this equation as a lower value of A (the pre-exponential factor), which will also result in lower reaction rates for the heavier isotope.

The *Rayleigh fractionation* model is used to describe and quantify non-equilibrium isotope fractionation. First described at the end of the nineteenth century (Rayleigh 1896), Rayleigh fractionation is the kinetically controlled isotope fractionation that takes place during unidirectional reactions between a source reservoir and a sink. During the unidirectional reaction, the source reservoir decreases in magnitude. A good example is evaporation of water from a pond, decreasing the size of the pond (in this case, the pond is considered the "reservoir"). Because these reactions are effectively distillation reactions, the term *Rayleigh distillation* also describes this process.

Some of the most well-known examples of Rayleigh fractionation involve the physical changes of state between liquid water and water vapor. The isotope fractionation that takes place during $H_2O_g \rightarrow H_2O_l$ or $H_2O_l \rightarrow H_2O_g$ is often kinetically controlled. This is because *rates* of evaporation or condensation exert a greater control on the fractionation than do equilibrium controls. In addition, the product (e.g vapor produced by evaporation) is continually removed from contact with the source reservoir, so fractionation is influenced open system behavior.

Considering the condensation of water from an air mass as it migrates through the atmosphere, the first molecules of water to condense from the vapor phase will be enriched in the heavy isotopes ^{18}O and D. As these droplets grow in size and eventually fall in the form of precipitation, they are removed from contact with the water vapor reservoir.

Focus Box 10.5

Examples of Non-equilibrium, Kinetic Isotope Effects

1. Molecules with the lighter isotope will react more rapidly than molecules with the heavier isotope – this is because the chemical bond involving the lighter isotope is weaker and hence easier to break. It requires less energy to decompose molecules with the lighter isotope. This effect is especially pronounced in biological systems where organisms expend less energy when they carry out reactions involving molecules with the lighter isotope. A good example is photosynthesis, where plants preferentially react with $^{12}CO_2$ molecules that are more easily decomposed compared to the more strongly bonded $^{13}CO_2$ molecules. The plant needs to break CO_2 into one C^{4+} and two O^{2-} atoms,

and it is easier to do with $^{12}CO_2$. As a consequence, photosynthetically produced organic matter is enriched in ^{12}C relative to ^{13}C. One result of this effect is that, in marine environments, phytoplankton have $\delta^{13}C$ values that are 20‰ lower (lighter) than the dissolved CO_2 in ocean water (Drever 1997).

2. The lower vibrational frequency of the heavier isotope means that it tends to diffuse more slowly than its lighter counterpart and thus is less likely to come into contact with potential reactants. Furthermore, because of differences in diffusion rates, the lighter isotope can be separated from the heavier isotope (particularly in open systems) and thus equilibrium may not be established.

Focus Box 10.6

Evaporation, Condensation, and Kinetic Factors in the Hydrologic Cycle

Kinetic effects related to evaporation are enhanced by many factors, including surface temperatures (greater temperature produces greater evaporation rates), salinity of the reservoir (greater salinity can impede evaporation), wind speeds (winds produce shear at the water surface and enhance evaporation), and finally and most importantly, humidity. Low humidity increases evaporation rates and enhances non-equilibrium conditions. Conditions of high

humidity allow exchange of vapor with the liquid phase, and as humidity approaches 100%, the vapor–liquid system approaches equilibrium. Considering condensation, H_2O_g in the atmosphere will typically condense to H_2O_l at rates controlled by kinetic factors/non-equilibrium conditions that include influence of humidity, temperature, and presence or absence of condensation nuclei.

As this process continues, the vapor reservoir will become (i) progressively depleted in ^{18}O and D (and enriched in ^{16}O and ^1H), and (ii) smaller in size by virtue of rainout. This represents a classic case of Rayleigh distillation, which is quantified by the following equation (Broeker and Oversby 1971):

$$R = R_0 \times f^{(\alpha-1)} \qquad (10.10)$$

where R_0 is the initial ratio of $^{18}O/^{16}O$ (or D/H) in the reservoir (water vapor in this case), R is the ratio of $^{18}O/^{16}O$ (or D/H) in the vapor remaining in the reservoir when the fraction of vapor remaining is given by the term f (where f ranges from 0.0 to 1.0). The α term is the fractionation factor that, for a system of water vapor and liquid water, can be represented as:

$$\alpha = R_{liquid}/R_{vapor} = (^{18}O/^{16}O)_l \div (^{18}O/^{16}O)_v \quad (10.11)$$

In this equation R is the ratio of $^{18}O/^{16}O$ in the liquid to vapor states (it could also apply to D/H or other

pairs). For a system of liquid water and water vapor under equilibrium conditions at 25 °C, the value of α is 1.0092. This means that, under these conditions, the ratio of $^{18}O/^{16}O$ in the liquid phase will be 1.0092 times greater than the $^{18}O/^{16}O$ ratio in the water vapor phase. This intuitively makes sense given that the heavier isotope will be fractionated into the liquid when H_2O condenses. This effect becomes more pronounced as system temperature drops, where the cooler the system, the greater the fractionation: at 20 °C, $\alpha = 1.0098$ and at 0 °C, $\alpha = 1.0117$ (Kendall and Caldwell 1998).

10.4 δ^{18}O AND δD

Stable isotopes of oxygen (and hydrogen) in water are frequently used to examine processes within the hydrologic cycle, and this section provides examples, beginning with Rayleigh distillation of an air mass, followed by other applications of hydrogen and oxygen isotope analysis to water in the hydrologic cycle and paleoclimate analysis.

10.4.1 Rayleigh distillation in the hydrologic cycle

Rearranging Eq. (10.2) as follows will provide a useful equation that facilitates prediction or quantification of oxygen isotopic compositions of liquid and vapor (gas) phases affected by Rayleigh distillation.

$$\delta^{18}O_{smpl}/10^3 = [(^{18}O/^{16}O_{smpl}) - (^{18}O/^{16}O_{VSMOW})]/$$
$$(^{18}O/^{16}O_{VSMOW}) \qquad (10.12)$$

Simplified using R in place of the two oxygen isotope ratios above:

$$\delta^{18}O_{smpl}/10^3 = (R_{smpl} - R_{VSMOW})/R_{VSMOW} \qquad (10.13)$$

Rearranging and skipping a couple of steps gives:

$$R_{smpl} = [R_{VSMOW} \times (\delta^{18}O_{smpl} + 10^3)]/10^3 \qquad (10.14)$$

From Eq. (10.11) and substituting Eq. (10.14) for R_{liquid} in the numerator and for R_{vapor} in the denominator:

$$\alpha_{liquid}^{vapor} = R_{liquid}/R_{vapor}$$
$$= [R_{vsmow} \times (\delta^{18}O_{vapor} + 10^3)]/$$
$$[R_{vsmow} \times (\delta^{18}O_{liquid} + 10^3)] \qquad (10.15)$$

Further simplification results in:

$$\alpha_{liquid}^{vapor} = [(\delta^{18}O_{vapor} + 10^3)]/[(\delta^{18}O_{liquid} + 10^3)] \qquad (10.16)$$

Combining now with Eq. (10.10) for Rayleigh distillation to solve for the isotopic composition of the vapor phase that remains after a certain amount of condensation (this is the term f):

$$\alpha_{Vo}^{V} = Rv/R_{Vo} = f^{(\alpha-1)}$$
$$= [(\delta^{18}O_V + 10^3)]/[(\delta^{18o}{}_{Vo} + 10^3)] \qquad (10.17)$$

Here, the subscript "v" refers to the vapor that remains at any given point in the process of condensation from an initial vapor ("v_o").

To determine the isotopic composition of the vapor that remains after a certain amount (f) of the original vapor has been affected by condensation:

$$\delta^{18}O_V = [f^{(\alpha-1)} \times (\delta^{18}O_{Vo} + 10^3)] - 10^3 \qquad (10.18)$$

The main variables here are the fractionation factor α (controlled by temperature and mass difference of the isotopes undergoing fractionation) and the extent to which the air mass has already undergone distillation (f).

To determine the composition of the liquid water that condenses, begin with this α relationship:

$$\alpha_{vapor}^{liquid} = (\delta^{18}O_l + 10^3)/(\delta^{18}O_v + 10^3) \qquad (10.19)$$

and rearrange to arrive at:

$$\delta^{18}O_l = \alpha_{vapor}^{liquid} \times (\delta^{18}O_v + 10^3) - 10^3 \qquad (10.20)$$

At this point it is possible to answer questions like "what is the $\delta^{18}O$ value of liquid water condensing from vapor with a $\delta^{18}O$ value of $-10‰$ at a temperature of 25 °C (where $\alpha = 1.0092$)," and "for an air mass containing water vapor with an initial $\delta^{18}O_{Vo}$ of $-14‰$, determine the isotopic composition of vapor remaining ($\delta^{18}O_V$) after 50% of the original vapor has already condensed to water droplets."

Entering $\alpha = 1.0092$ and an initial vapor with $\delta^{18}O_v = -10‰$ into Eq. (10.20) indicates that the $\delta^{18}O_l$ (of liquid water that condenses from the initial vapor) will have a $\delta^{18}O_l = -0.89‰$. These two delta values are plotted in the upper left of Figure 10.4.

From the initial vapor of $-10‰$, we can use Rayleigh distillation to determine the composition of the remaining vapor at any given time (Eq. (10.18)), and from those vapor values, determine the condensed liquid composition from Eq. (10.20). These values are plotted in Figure 10.4, which shows typical fractionation and changes in isotopic composition as Rayleigh distillation progressively affects an air mass with water vapor and condensed liquid water.

10.4.2 The meteoric water line

The fractionation of hydrogen isotopes (H and D) is very similar to oxygen isotopes, so the first water to condense from a parcel of air will be enriched in D relative to H and the shrinking water vapor reservoir will become progressively enriched in H. A bivariate plot of $\delta^{18}O$ against δD in meteoric water results in a straight line known as the *meteoric water line* (Figure 10.5), the equation of which was quantified by the pioneering isotope geochemist Harmon Craig in the early 1960s.

$$\delta D = 8 \times \delta^{18}O + 10 \qquad (10.21)$$

Precipitation in the tropics has $\delta^{18}O$ and δD values that are very similar to standard mean ocean water; with increasing latitude, $\delta^{18}O$ and δD values of precipitation

Rayleigh Distillation and the Hydrologic Cycle

A few important observations to reinforce regarding Rayleigh distillation and water in the hydrologic cycle in general are:

1. Liquid that condenses from vapor is enriched in ^{18}O (or D) relative to the initial vapor, and vapor that evaporates from surface water is lighter than the surface water.
2. In a migrating air mass, the initial liquid that condenses from vapor falls as precipitation, thus removing it from contact with the vapor, and preventing the system from reaching equilibrium. This is a good example of a kinetic isotope effect.
3. As water vapor in an air mass migrates away from its source, it becomes progressively lighter, so the H$_2$O that

condenses and falls as precipitation becomes progressively lighter (Figure 10.4).

4. Precipitation is always at least slightly depleted in ^{18}O relative to sea water; i.e. δ^{18}O of precipitation is always lower than δ^{18}O of sea water whose value is 0‰. This applies to hydrogen and δD too.
5. Fresh water (e.g. in lakes, rivers, and groundwater) is usually depleted in ^{18}O and D relative to sea water, i.e. fresh water has δ^{18}O and δD values less than zero.
6. The Rayleigh equations presented in this section for oxygen isotopes can also be applied to hydrogen isotopes, making sure to use α values that apply to hydrogen isotopes at the temperature of the system.

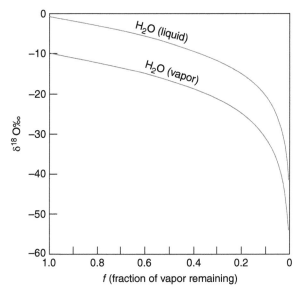

Fig. 10.4 Fractionation of oxygen isotopes during condensation at 25 °C as predicted by the Rayleigh distillation model. The initial δ^{18}O value of the vapor is −10‰ and the value of the first liquid water to form by condensation is −0.9‰. The condensed liquid water is then removed from contact with the vapor (i.e. it falls as rain), so equilibrium cannot be established. The vapor becomes progressively lighter (more negative) with the removal of isotopically heavier liquid, and subsequent condensation of new liquid from continuously lighter vapor forms the trends illustrated above. These trends occur in nature when air masses migrate from the tropics toward higher latitudes (see Figures 7.5 and 7.6).

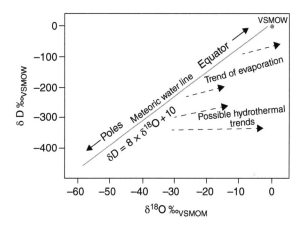

Fig. 10.5 Linear correlation of oxygen and hydrogen isotopes in meteoric water. The lighter (more negative) values toward the poles (high latitudes) is explained by Rayleigh distillation. Note the position of ocean water (VSMOW), and also that evaporation has a more pronounced effect on oxygen isotope compositions than it does on hydrogen isotopes. This is true for hydrothermal waters also, with some waters following flat lines to the right off the meteoric water line. The slope of the meteoric water line is 8; the slope of the evaporated water line is ~5. Source: Modified from Craig (1961).

become progressively more negative. The only waters that do not plot on or close to the line in Figure 10.5 are surface waters that have been appreciably affected by evaporation – they veer off the meteoric water line

with lower slope. Predictably, evaporation of lakes in arid regions causes progressive fractionation of ^{18}O and D into the surface water, but this alone does not explain deviation from the meteoric water line. The decreased slope of the evaporation-dominated water indicates that the surface water becomes enriched in ^{18}O at a greater rate that it does for D, a phenomenon that is due to the greater difference in mass between H$_2$^{18}O and H$_2$^{16}O (11%) as compared to HDO and H$_2$O (5.6%). This proportional mass difference

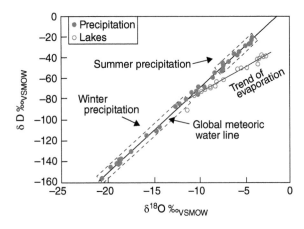

Fig. 10.6 Oxygen and hydrogen isotopic values of precipitation and lake water from northern Wisconsin, USA. Note that winter precipitation is isotopically lighter than summer precipitation, and that precipitation from this region falls on the global meteoric water line. Evaporation of water from lakes causes their isotopic compositions to change in a manner similar to the trend shown in Figure 10.4. Source: Data are from Kendall et al. (1995).

10.4.3 Regional-scale to global-scale variations in precipitation

The following sections explore some of the ways that oxygen and hydrogen isotopes are applied to understanding processes within the hydrologic cycle, particularly regarding the origins and behaviors of water at various stages of the hydrologic cycle. One way to begin to visualize isotopic variation in the hydrologic cycle is to examine the fractionation of meteoric water as a function of distance from the vapor source. Figure 10.7 depicts a schematic example of orographic precipitation, where an air mass cools as it rises over hills or mountains. The uplifting of the air causes *adiabatic cooling* (resulting from expansion of air rising into the lower air pressure of higher altitude). The cooling of air vapor results in condensation that leads to precipitation that is spatially associated with the foothills and mountains.

The first water to condense, as predicted by thermodynamic considerations and by the kinetic controls described by Rayleigh distillation, is enriched in ^{18}O and D relative to precipitation that falls further downgradient (Figure 10.7). Accordingly, the precipitation that falls on the lee (downwind) side of the mountain range is depleted in ^{18}O and D compared to the precipitation that falls on the upwind side of the range. Spatial variability in δ^{18}O on a broader scale is depicted in Figure 10.8, which illustrates patterns as a function of latitude (and longitude to a lesser extent) in the western hemisphere. (Also, although not presented here, the same pattern occurs with hydrogen isotope values, which likewise become progressively lighter downgradient.)

The decrease in δ^{18}O and δD values in meteoric water from the equator toward the poles has been termed the *latitude effect* (Faure 1998), a phenomenon that is illustrated spatially in Figure 10.8 and is primarily controlled by the following factors:

means that H$_2$^{18}O is more preferentially fractionated into the surface water reservoir than HDO, producing trajectories with lower slopes like those illustrated in Figure 10.5.

An example of local-scale variability in the isotopic composition of meteoric water is provided for northern Wisconsin, USA (Figure 10.6). While the variation in δ^{18}O and δD values are much smaller than they are on for global meteoric water (Figure 10.5), precipitation water still plots on the meteoric water line with a slope of 8, where winter precipitation contains isotopically lighter H$_2$O as compared to summer precipitation (the temperature control on δ^{18}O and δD is presented below). Also notable is that meteoric water derived from evaporation of surface water bodies veers off the meteoric water line with a slope of 5.2 (Kendall et al. 1995).

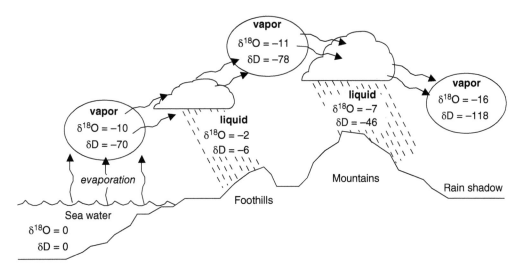

Fig. 10.7 Schematic diagram representing the fractionation of oxygen and hydrogen isotopes in water vapor and precipitation as an air mass first passes over ocean water (and gains water vapor by evaporation), then continues downwind over hills and mountains that cause adiabatic cooling, condensation, and precipitation. Note that as rain falls from the air mass, the remaining water vapor becomes progressively lighter. Subsequently, water that condenses and falls as precipitation also becomes progressively lighter. The $\delta^{18}O$ and δD values are per mil relative to VSMOW.

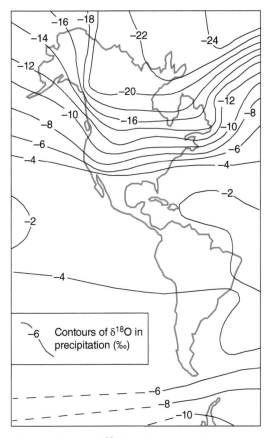

Fig. 10.8 Variation in $\delta^{18}O$ (‰) in precipitation in the western hemisphere. The general shift toward lighter (lower $\delta^{18}O$) values poleward is a result of Rayleigh distillation. Source: Data are from Yurtsever and Gat (1981) and values are relative to VSMOW.

1 removal of ^{18}O and D from air mass by condensation of liquid water as the air mass migrates poleward, a process driven by Rayleigh distillation and increasing fractionation factors as the air mass cools;

2 evaporation of water at higher latitudes occurs at lower temperature, so the fractionation factor will be greater and the vapor produced at higher latitudes will be relatively depleted in ^{18}O and D; and

3 transpiration of water from plants preferentially fractionates ^{18}O and D into compounds in plant tissue (the heavier isotopes will form more stable bonds), thus transmitting water vapor depleted in ^{18}O and D to the atmosphere.

In addition to the progressive decrease in $\delta^{18}O$ downgradient (in this case, poleward) of the tropical moisture source, there also is a pronounced seasonality in the $\delta^{18}O$ pattern, which brings us to a discussion of the temperature dependence of $\delta^{18}O$ in precipitation.

10.4.4 Temperature dependence of $\delta^{18}O$ in precipitation

Seasonal variations of $\delta^{18}O$ (or δD) values are caused by differences in temperature that affect the fractionation of oxygen (or hydrogen) isotopes when surface water evaporates. Warmer summer air results in precipitation that is enriched in the heavier isotopes ^{18}O and D relative to winter precipitation. As is the case with isotopic fractionation in minerals, higher temperatures (of sea water

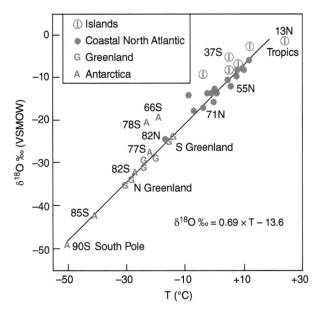

Fig. 10.9 Linear correlation between mean annual air temperature and $\delta^{18}O$ of precipitation as a function of latitude. Source: Modified from Dansgaard (1964).

in the case of Figure 10.9) impart smaller differences in vibrational energy between ^{18}O and ^{16}O, or D and H, and this produces lower fractionation factors for isotopes of oxygen and hydrogen. The converse is true, and lower temperature leads to greater fractionation between surface water and evaporated vapor, and the result is a linear correlation between temperature and $\delta^{18}O$ (Figure 10.9).

The linear correlation discovered by Dansgaard (1964) yields the equation:

$$\delta^{18}O_{MAP} = 0.695 \times T_{MAT} - 13.6 \qquad (10.22)$$

where $\delta^{18}O_{MAP}$ is $\delta^{18}O$ of mean annual precipitation and temperature (T_{MAT}) is mean annual surface temperature in °C. An increase of 10 °C correlates to a $\delta^{18}O$ increase of approximately 7‰. More recent work by Rozanski et al. (1993) indicates that there are regional trends embedded within the global trend suggested by Dansgaard's data, including values for the Antarctic Peninsula and interior Canada (not shown) that deviate from Dansgaard's best-fit line. Some of the local effects that contribute to nonlinearity show up in Figure 10.8.

Similar effects are apparent with increasing elevation and increased distance from the equator (both of which correspond to lower temperature), where $\delta^{18}O$ and δD of evaporated water become lighter as temperature drops. Furthermore, because precipitation becomes progressively enriched in light oxygen poleward, polar ice caps are a reservoir of ^{16}O-enriched water (compared to sea water, which is relatively enriched in $\delta^{18}O$ compared to the ice

caps). Marine sediments deposited between 13 000 and 8000 years ago record a series of 0.5–2.0‰ decreases in the $\delta^{18}O$ value of oxygen mineralized into planktonic foraminifera, evidence for periodic pulsing of meltwaters into the global ocean (Broecker et al. 1989).

10.4.5 Paleotemperature analysis using oxygen and hydrogen isotopes

The temperature dependence of $\delta^{18}O$ and δD in meteoric water also varies as a function of temperature over longer time scales and is one of the most important means of analyzing *paleotemperature*. Given what we know about the relationship of temperature and $\delta^{18}O$ (or δD), we would predict that colder periods in the geological past should record precipitation with lower $\delta^{18}O$ (and δD) values, and vice versa. One of the best records of precipitation that extends well back into the Pleistocene are the thick ice caps on Antarctica and Greenland, where meteoric water is deposited as snow that eventually transforms to ice. Glacial ice from the polar regions contains periodic layers of volcanic ash which can be radiometrically dated and geochemically fingerprinted, permitting determination of depositional age.

Two of the longest glacier paleotemperature records are from ice cores retrieved from thick ice caps in Antarctica (Petit et al. 1999; Augustin et al. 2004). Ice core data presented in Chapter 7 (Figure 7.3) indicate that $\delta^{18}O_{atm}$ and δD of H_2O that fell as precipitation (now preserved in layers of glacial ice) vary cyclically with global temperature changes that correspond with variations in earth's orbit. Measurement of $\delta^{18}O$ and δD in ice cores has proven to be a very important tool in examining natural variations in temperature (globally and regionally) as well as the relationship between temperature and CO_2 prior to anthropogenic climate change.

10.4.6 Oxygen and hydrogen isotopes as tracers in soils and groundwater

In addition to their utility in assessing precipitation trends and global paleotemperature, $\delta^{18}O$ and δD are also useful tracers of many other processes within the hydrological cycle. A groundwater sample with $\delta^{18}O$ and δD values that are similar to modern precipitation – and plots on the meteoric water line – is likely young water with relatively low residence times. The rationale is that recently infiltrated waters will have a signature whose greatest influence is the composition of recent rain or snowmelt,

Oxygen Isotopes and Hydrothermal Systems

Hydrothermal waters, which follow a nearly flat trajectory toward higher $\delta^{18}O$ values (Figure 10.5), are an excellent example of waters that experience appreciable isotopic exchange with ^{18}O-enriched minerals in the subsurface. They typically evolve toward greater $\delta^{18}O$ values yet show no noticeable change in δD from initial compositions that are nearly identical to local meteoric water (Craig 1966). The nearly constant hydrogen isotope values paired with increasing $\delta^{18}O$ indicate that locally infiltrating meteoric water is heated by magma or a high geothermal gradient, and the resulting hot waters accelerate dissolution of – and isotopic exchange with – ^{18}O-enriched minerals in the subsurface (Craig 1966). These types of data are useful in examining hydrothermal processes and origins of hydrothermal waters; for example, if the isotopic data indicate modification of recently recharged meteoric water, it would make less likely the possibility that the hydrothermal waters are of deep crustal or volcanic origin. D'Amore et al. (1987) used $\delta^{18}O$ and δD as tracers to determine that reinjection of geothermal waters at a site in Tuscany (Italy) does not negatively impact geothermal electricity generation. The reinjected waters (identifiable by $\delta^{18}O$ and δD), once heated up by the high geothermal gradient and then pumped back to the surface, contain less dissolved gases than do the natural geothermal waters and are more efficient in energy generation.

i.e. the modern meteoric signature. How would you interpret a sample of groundwater whose $\delta^{18}O$ and δD values are appreciably lighter than modern precipitation? Figures 10.5–10.9 indicate that cooler air masses or greater extent of Rayleigh distillation lead to isotopically lighter precipitation, so the isotopically light groundwater may represent older water that infiltrated the aquifer during the Little Ice Age (~1300–1850) or Pleistocene (> 12 ka) when meteoric water had lower $\delta^{18}O$ and δD values than modern water (Figure 7.3). It is also important to consider whether the water plots on the meteoric water line or not. Groundwater that plots to the right of the meteoric water line (Figures 10.6 and 10.9) likely indicates (i) evaporation-modified $\delta^{18}O$ and δD values, or (ii) oxygen exchange with isotopically heavier carbonates, clays, and other minerals in the subsurface (e.g. Savin and Lee 1988). Most minerals are rich in oxygen, so when they dissolve they change the composition of pore water, but hydrogen isotope values of waters are only minimally affected by reaction with minerals. Oxygen is far more abundant than hydrogen in minerals. Even hydrous clay minerals typically contain many times more moles of oxygen than hydrogen, so when minerals dissolve in water they will have a much greater impact on $\delta^{18}O$ than δD. In fact, although perhaps obvious, the dissolution of minerals that contain no hydrogen such as carbonates and anhydrous sulfates (e.g. $CaCO_3$, $BaSO_4$) will have no impact on δD.

10.4.7 Application of oxygen and hydrogen isotopes to paleosol climate records

Clay minerals and hydroxides that form in soils record oxygen and hydrogen isotope compositions that are dominated by local meteoric water that infiltrates soils, and for this reason, $\delta^{18}O$ and δD values of pedogenic (soil-formed) minerals contain useful information about paleoclimate. One important assumption is that the $\delta^{18}O$ and δD values relate to local climate at the time of soil formation. The source of O and H must be from soil water that has not been significantly affected by O and H derived from parent mineral dissolution (e.g. amphiboles, feldspars, micas, and carbonates in sedimentary, igneous, and metamorphic rocks). As long as the isotopic composition of the pedogenic minerals is dominated by the composition of meteoric water, paleoclimate researchers have a useful tool for studying changing temperature or precipitation of a region.

10.5 $\delta^{15}N$

Stable isotopes of nitrogen (^{14}N, ^{15}N) are a useful tool in assessing various problems in low-temperature geochemistry, including natural nitrogen cycling and the determination of anthropogenic nitrogen species. Values of $\delta^{15}N$ are useful in tracing reactive nitrogen compounds from the following processes and sources:

1 Combustion of fossil fuels or biomass produces the gaseous nitrogen oxides commonly referred to as NO_x. The N in NO_x is from N_2 in air participating in the combustion, so NO_x compounds formed by reaction with air should have an atmosphere $\delta^{15}N$ signal.

2 Human and domesticated animal wastes contain organic N-bearing compounds that transform by microbially mediated processes to ammonium (NH_4^+), nitrate (NO_2^-), and nitrate (NO_3^-). This N pathway causes

enrichment in ^{15}N (a signal influenced by metabolism and excretion), so it is possible to distinguish septic nitrate (high $\delta^{15}N$) from fuel combustion-derived nitrate.

3 Chemical fertilizers (e.g. ammonium nitrate, potassium nitrate) are made by the Haber-Bosch process, so have $\delta^{15}N$ values that are very similar to air; manure tends to be isotopically heavier (higher $\delta^{15}N$) than chemical fertilizers, thus provides a way to study sources of N contamination.

Given that air is well mixed with respect to N_2 gas and that the ratio of $^{15}N/^{14}N$ is stable, the standard for reporting $\delta^{15}N$ values is the isotopic composition of N_2 in the atmosphere. Hence, the $\delta^{15}N_{atm}$ value of atmospheric N_2 is 0‰, and any nitrogen compounds derived with no appreciable fractionation from atmospheric nitrogen will also have $\delta^{15}N_{atm}$ values that are close to 0‰. This is the case for the ammonium and nitrate in synthetic fertilizers, with $\delta^{15}N_{atm}$ values usually from $-5‰$ to $+5‰$ (Figure 10.10). Similarly, nitrogen oxides (NO_x) produced by combustion of fuels also exhibit $\delta^{15}N_{atm}$ values that are close to zero – in the case of combustion, the source of the NO_x is chemical reaction of atmospheric N_2 and O_2 in the combustion chamber.

Plant residues also possess $\delta^{15}N_{atm}$ values that are close to zero; animal wastes, however, tend to be enriched in ^{15}N relative to plant matter as well as atmospheric nitrogen and its direct derivatives. This enrichment is caused by two factors,: first, consumers (in terms of trophic level) tend to preferentially retain ^{15}N over ^{14}N; while this can produce urine that is slightly depleted in ^{15}N relative to the $\delta^{15}N$ of the food source, animal tissue and solid waste is typically enriched by 2–3‰ compared to the food source.

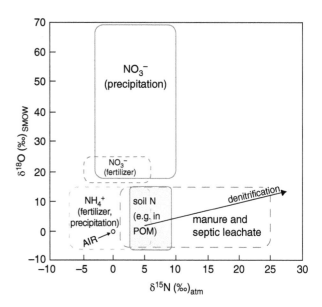

Fig. 10.10 Differentiation of nitrogen sources based on $\delta^{15}N$ and $\delta^{18}O$. Denitrification is the reaction of nitrate or other N species to N_2 gas – a topic covered in Chapter 6 – and the liberation of isotopically light ^{15}N (and ^{16}O) results in a trend toward isotopically heavier remaining organic solids. Source: Modified from Kendall and McConnell (1998) and other sources cited therein.

The second factor contributing to ^{15}N-enriched animal waste is volatilization of isotopically light (^{15}N-depleted) ammonia from manure or other wastes at or near the land surface. The result of these ^{15}N-concentrating processes is that organic wastes in sewage and septic effluent and animal manure have $\delta^{15}N_{atm}$ values that typically range from +10‰ to +25‰ (Kendall and Doctor 2003). Because solid and liquid organic fertilizers are derived from animal wastes (e.g. organic manure), they also have positive $\delta^{15}N_{atm}$ values that make them similar to their manure source.

Isotopic analysis of nitrate or other NO_x phases is enhanced by the oxygen that provides a $\delta^{18}O$ value to augment the $\delta^{15}N$ value. This is useful when trying to distinguish between e.g. synthetic fertilizer vs. manure as the source of nitrate pollution to an aquifer. Those two potential sources overlap in the $\delta^{15}N$ range of +2‰ to +6‰ (animal waste $\delta^{15}N$ values range from +2‰ to +25‰, and for synthetic fertilizer is −5‰ to +6‰). The $\delta^{18}O$ of nitrate in synthetic fertilizers is generally +15‰ to +25‰, a value that reflects the atmospheric O_2 from which it was synthesized (for O_2 in air, $\delta^{18}O_{VSMOW} = +23.5$‰; Kroopnick and Craig 1972). The $\delta^{18}O$ value of animal wastes is lower, typically in the range of −8‰ to +12‰, reflecting incorporation of surface water with lower $\delta^{18}O$ values than those of O_2 in air. Figure 10.10 illustrates fields for different potential contaminants.

Denitrification is the biochemical reduction of oxidized N species such as nitrate or nitrogen dioxide to nitrogen gas (or the intermediate nitrous oxide N_2O). The transformation is facilitated by various species of *Pseudomonas* bacteria, often beginning with nitrate dissolved in water and progressing through intermediate stages (not all shown here) to the final product, usually N_2. Each reaction in the sequence is driven by N species serving as electron receptors – in Eq. (10.23), nitrate (NO_3^-) is reduced to nitrite (NO_2^-), and nitrogen is reduced from N^{5+} to N^{3+}, accommodating the metabolical needs of bacteria who require an electron acceptor.

$$NO_3^- + 2\,H^+ + 2e^- = NO_2^- + H_2O \qquad (10.23)$$

then

$$NO_2^- + 2H^+ + e^- = NO + H_2O \qquad (10.24)$$

then

$$2NO + 2H^+ + 2e^- = N_2O + H_2O \qquad (10.25)$$

Then, either N_2O volatilizes to air, or

$$N_2O + 2H^+ + 2e^- = N_2 + H_2O \qquad (10.26)$$

Considering fractionation of ^{15}N and ^{14}N during this series of reactions, we can ask the question: if you were a microbe, which would you target, $^{14}NO_3^-$ or $^{15}NO_3^-$? The bonds in the lighter species will have higher vibrational frequency (Eq. (10.4)), making $^{14}NO_3^-$ more susceptible to decomposition, and the nitrate that remains will be ^{15}N-enriched. If it were necessary to determine whether denitrification or dilution was the main cause of decreasing nitrate down a groundwater flow path, analysis that indicated increasing $\delta^{15}N$ downgradient would indicate likelihood of denitrification.

The complex nitrogen cycle is prone to kinetic controls imparted by biologically mediated effects, yet in many cases, analysis of $\delta^{15}N$ (often in combination with $\delta^{18}O$) has proven useful in examining source of nitrogen species at or near the Earth's surface. Thorough reviews of nitrogen isotopes in soils, waters, air, and biological systems are given by Kendall and Doctor (2003).

10.6 δ^{13}C

The ratio of the two stable isotopes of carbon, ^{13}C and ^{12}C, helps to understand sources and cycling of carbon in the Earth's atmosphere, surface environments, and crust. As a general rule, carbon derived from organic matter is isotopically light (^{12}C-enriched), whereas carbon derived from inorganic sources like marine water (mainly as dissolved HCO_3^-) and carbonate, and igneous and metamorphic rocks (as carbonate minerals, dissolved HCO_3^- or volatilized CO_2) is isotopically heavy. Hydrocarbons and coal, while derived from rocks in the crust of the Earth, are isotopically light because the carbon in these reservoirs was originally fixed by photosynthesis (into marine algae that have transformed to hydrocarbons or terrestrial plants that have transformed to coal).

Figure 10.11 illustrates representative $\delta^{13}C$ values of carbon in environmental reservoirs, including land plants (C_3 vs C_4) as well as hydrocarbons, marine organic matter, and marine carbonates (which shows the ~0‰ value of limestone). In some cases, diagenesis causes deviation of ±5‰ around the initial 0‰ value, but as a general rule, carbon from carbonate rocks is isotopically heavy relative to other potential sources.

Because the standard for carbon isotope values is a marine carbonate (the PDB), sea water and the fossils and minerals that form from dissolved inorganic carbon (DIC) in sea water have $\delta^{13}C_{PDB}$ values that are close to 0‰ – sea water is approximately −2 to +2‰ and minerals (e.g. $CaCO_3$ in limestone) that crystallize from sea water are generally in the range of −5 to +5‰. The $\delta^{13}C$ of atmospheric CO_2 is ~−8.5‰ and approaching −9‰ as isotopically light carbon ($\delta^{13}C_{PDB} = -20$ to

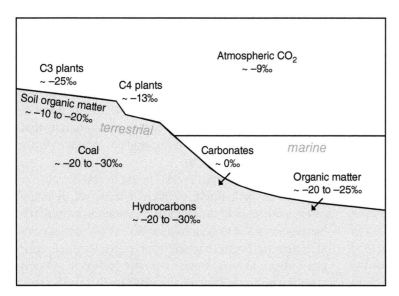

Fig. 10.11 $\delta^{13}C_{PDB}$ values for common reservoirs of carbon at and near the earth surface. Note that all reservoirs contain $\delta^{13}C$ values $\leq 0‰$, and that C3 plants, hydrocarbons, and marine organic matter have the lowest (isotopically lightest) values; carbonates are the heaviest with values of $\sim 0‰$.

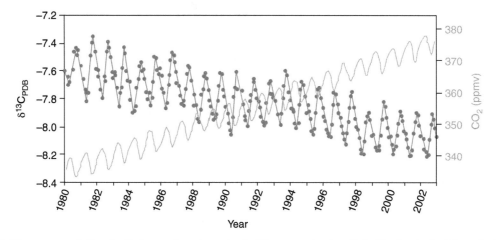

Fig. 10.12 A 23-year record of cyclically increasing atmospheric CO_2 concentration (light blue) paired with cyclically decreasing $\delta^{13}C$ of atmospheric CO_2 (dark blue). Both trends result from combustion of isotopically light hydrocarbons. Data are from Oak Ridge National Lab accessed at: http://www.nicholas.duke.edu/thegreengrok/climatedisruption_solomon.

$-30‰$) from fossil fuels is added to the atmosphere ($\delta^{13}C$ of CO_2 in air in 1800 was $\sim -6.5‰$; Francey et al. 1999). Figure 10.12 illustrates this trend, which also depicts increasing atmospheric CO_2 concentrations and annual cycles in carbon isotope ratios and CO_2 concentrations related to photosynthesis.

Atmospheric CO_2 concentrations reach annual low values in August and September, reflecting uptake of CO_2 during the growing season in the northern hemisphere – the far greater abundance of land (and hence terrestrial vegetation) in the northern hemisphere as compared to the southern hemisphere means that the annual pulse of plant growth in Asia, Europe, North America, and Africa north of the equator produces a drop in atmospheric annual CO_2 values, globally. The inverse relationship on an annual basis between CO_2 and

$\delta^{13}C$ also can be traced to photosynthesis, a process that selects ^{12}C over ^{13}C; hence, $\delta^{13}C$ increases (becomes less negative) when photosynthesis peaks.

10.6.1 Carbon isotope analysis of paleoenvironment

Stable carbon isotopes are fractionated differently by the two main photosynthetic processes employed by plants. C_3 plants are the dominant forms of vegetation on the Earth's surface and tend to thrive in most climate zones (notable exceptions are arid regions and salt marshes). They are called C_3 plants because the initial organic compound synthesized by photosynthesis in these plants, 3-phosphoglyceric acid, has three carbon atoms. $\delta^{13}C$

Focus Box 10.11

Application of Carbon Isotopic Shifts to Paleoclimate Analysis.

1. Fractionation from plant to consumer: There is a +14‰ shift in δ^{13}C from plant to tooth enamel of large mammals (e.g. bovids, equids, rhinocerids) which reflects the preferential fractionation of ^{13}C into mammalian tooth and bone (recall that the heavier isotope – ^{13}C vs. ^{12}C – forms stronger bonds (Section 10.3.1)). The result is that C_3 grasses with δ^{13}C = –25‰ would produce large mammal tooth enamel with δ^{13}C = –11‰. This shift occurs in modern animals and it is applied to paleomammal–vegetation systems to determine carbon isotope compositions of paleoecosystems.

2. Global ecosystem shift at 6 Ma: Prior to 6 Ma, the δ^{13}C of large grazing mammals from sites across the globe

cluster around –10‰; in the 6 million years since then, δ^{13}C values from the same sites produce an average value of approximately +1‰. When corrected for fractionation, these values indicate a shift 6 million years ago from grasses with δ^{13}C = –24‰ to –9‰, values that line up very well with δ^{13}C of C_3 and C_4 plants, respectively. Cerling et al. (1993, 1997) interpreted these data as evidence for a sharp transition to a more arid climate at 6 Ma that caused a decline in the abundance in C_3 plants and a corresponding increase in C_4 plants.

values of C_3 plants are generally in the range of –23‰ to –34‰ (relative to PDB) with average values of –25‰ to –26‰. C_3 plants, however, are not well-adapted to hot, dry conditions. Their stomata remain open all day during photosynthesis and thus they suffer from water loss and dehydration.

C_4 plants, on the other hand, have adapted to hot, dry environments by alternately opening and closing stomata during photosynthesis, which allows CO_2 to enter the stomata but limits water loss. In C_4 photosynthesis, a 3-carbon compound (pyruvate or phospho-enol pyruvate), is carboxylated to give a 4-carbon dicarboxylic acid (Hatch and Slack 1966). This process causes a lower degree of carbon isotope fractionation, and typical C_4 plant $\delta^{13}C_{PDB}$ values are –9‰ to –16‰ with an average of –12‰ to –13‰. Many terrestrial C_4 plants are tropical grasses and desert plants as well as salt marsh grasses. A third photosynthetic pathway similar to the C_4 process is known as CAM (Crassulacean acid metabolism) – many desert succulents are CAM plants, and their typical δ^{13}C values are –12‰ to –23‰; hence, they generally overlap with C_4 or are intermediate to C_3 and C_4.

If organic matter from decayed plants or bone or eggshells from animals that once consumed plants is sampled from a series of paleosols or buried sediments and is found to undergo a shift from e.g. –30‰ to –10‰ over time, it likely indicates a change from C_3 to C_4 plants, evidence for a transition from a temperate climate to a more arid climate. This approach has been applied in many regions to measure the timing and magnitudes of climate shifts. Examples include Cerling et al. (1993), Stern et al. (1994), Cerling et al. (1997), and Bechtel et al. (2008).

10.6.2 Carbon isotopes in hydrology and chemical weathering

The isotopic values of DIC are useful for determining DIC sources in watersheds, including those affected by acid precipitation. (DIC is usually HCO_3^- but sometimes CO_3^{2-} or H_2CO_3; Section 5.1). Given the crucial role that carbonate weathering plays in acid neutralization, quantifying the contribution of carbonate weathering to the total DIC budget of a watershed is essential in assessing origin and cycling of DIC and buffering potential. Carbon isotope values for DIC derived from rock weathering (δ^{13}C ~0‰ to –5‰) are typically greater than the DIC derived from soil CO_2 (usually –10‰ to –25‰). The low soil CO_2 values are caused by: (i) photosynthesis and microbial respiration in the soil, which produces δ^{13}C values that reflect the isotopic composition of the plants undergoing decomposition (e.g. δ^{13}C ~–25‰ in soils dominated by C_3 plants); and (ii) presence in soil of atmospheric CO_2 with δ^{13}C values of ~–9‰.

The weathering of carbonate minerals – particularly calcite that occurs in limestones, calcite veins, carbonate grains in soils, and various other forms – is a vital process that aids in buffering the effects of acidification, so identifying the contribution of carbonate weathering to a watershed is an important aspect of assessing buffering potential. In watersheds with soils that contain low carbonate mineral content, understanding the flow paths of subsurface waters that eventually recharge lakes and rivers is helpful in determining whether the surface water bodies will become acidic or not.

Waters that follow shallow flow paths to surface waters through calcite-poor soils will impart δ^{13}C values that

---

Focus Box 10.12

Clumped Isotopes and Ancient Seawater Temperature

Usually, stable isotope biogeochemistry is concerned with the ratio of two isotopes of a given element, e.g. ^{12}C and ^{13}C, ^{16}O and ^{18}O, ^{34}S and ^{32}S, etc. Clumped isotope geochemistry targets bonding of rare isotopes of two elements in a compound, e.g. ^{13}C–^{18}O in calcite, relative to the extent that they bond with the far more abundant ^{12}C and ^{16}O. Obtaining precise measurements of two rare isotopes has become possible with twenty-first-century advances in gas source isotope ratio mass spectrometry. The "mass 47 anomaly parameter (Δ_{47})" is referenced to $^{13}C^{18}O^{16}O$, specifically to the extent that the measured sample differs from a sample with background amount of $^{13}C^{18}O^{16}O$.

At high temperatures, there is no clumping of ^{13}C and ^{18}O and $\Delta_{47} = 0$; as temperature decreases, clumping increases, i.e. cooler temperatures favor clustering of ^{13}C–^{18}O bonds, and $\Delta_{47} > 0$ (Affek 2013). The ^{13}C–^{18}O bond in carbonate minerals is sensitive to the temperature of the solution from which calcite precipitates, independent of bulk isotopic composition. The temperature sensitivity makes clumped isotope analysis an excellent tool for determining sea-surface temperatures deep in geologic time and terrestrial soil temperatures by analysis of Δ_{47} in calcite. Both are very good proxies for past global atmospheric temperatures, especially when considering anything older than the Pleistocene ice core proxy. Eiler (2007) reviews the development of the field, one that is in a growth and development process. As with any mineralogical or geochemical paleoclimate proxy, it is important to consider potential modifications of the original signal by diagenesis.

---

are isotopically light and indicative of soil organic matter and atmospheric CO_2 (e.g. $-20‰$ to $-30‰$); in these cases, the $\delta^{13}C$ values are dominated by soil respiration, reflecting limited or no contribution from isotopically heavier carbon from calcite. In contrast, waters that circulate to deeper levels and experience greater interaction with calcite in bedrock or glacial till before recharging surface waters are characterized by relatively higher $\delta^{13}C$ values (e.g. $0‰$ to $-5‰$) that reflect the contribution of carbon from isotopically heavier calcite (Kendall and Doctor 2003). In this way, analysis of $\delta^{13}C$ values provides quantitative tracer information on soil and ground water flow paths that is applicable to forecasting response of a watershed to acidification, particularly in regions where soils (i.e. upper 1 m) are carbonate-poor.

10.7 $\delta^{34}S$

Of the four stable isotopes of sulfur, ^{32}S, ^{33}S, ^{34}S, and ^{35}S (Figure 10.1, Table 10.2), the two most abundant and most commonly measured in Earth's surface and shallow crustal environments are ^{34}S and ^{32}S. The ratio of these two isotopes is reported in standard delta notation as $\delta^{34}S_{CDM}$, where the $\delta^{34}S$ reference is the Canyon Diablo Meteorite (CDM) from Meteor Crater Arizona (Sharp 2007). Values of $\delta^{34}S$ generally fall within the range of $-30‰$ to $+30‰$, although rare examples are as low as $-65‰$ (in sulfides) and as high as $120‰$ (in sulfates; Hoefs 2009). Figure 10.13 summarizes typical sulfur isotopic compositions in various reservoirs.

In general, the lowest $\delta^{34}S$ values occur in sedimentary sulfides (e.g. pyrite), whereas values on the higher end of the scale occur in sulfate (SO_4^{2-}), including marine sulfate ($+21‰$; Kendall and Doctor 2003) and evaporite sulfate minerals ($+10‰$ to $+30‰$; Hoefs 2009). The partitioning makes sense intuitively: the lighter ^{32}S is partitioned into the gas phase (as H_2S) and the heavier ^{34}S remains in aqueous or solid phase in SO_4^{2-} (Figure 10.14a). In very strongly reducing environments (e.g. deep anoxic marine sediments), however, complete reduction of sulfate can result in isotopically heavy iron sulfide with values as high as $+50‰$. In this case, isotopically light ^{32}S is preferentially lost to the atmosphere as H_2S and ^{34}S is partitioned into the solid phase (FeS_2) that remains in the water or sediment (Figure 10.14b).

In addition to sulfate and sulfide minerals, sulfur also occurs in organic matter and atmospheric gases (e.g. SO_2), and hence sulfur isotope analysis is often used to understand the sources and processes responsible for the cycling of sulfur among the atmosphere, oceans, freshwater bodies, and the crust.

10.7.1 Fractionation of sulfur isotopes

Two main processes control the fractionation of sulfur isotopes in natural systems (Drever 1997; Hoefs 2009):

1 Equilibrium isotopic exchange that occurs among transformations between sulfide (S^{2-}) and sulfate (SO_4^{2-}); and

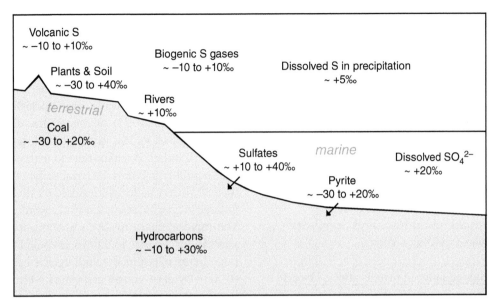

Fig. 10.13 $\delta^{34}S_{CDT}$ values for common reservoirs of sulfur at and near the earth surface. Note that many of these reservoirs have wide ranges of $\delta^{34}S$, which in part reflects the extent of reduction–oxidation reactions in the sulfur cycle.

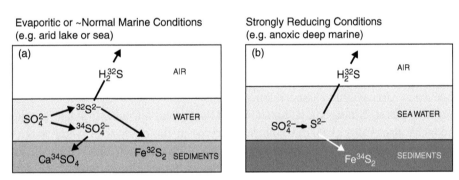

Fig. 10.14 Fractionation of sulfur isotopes in two different scenarios. In normal marine to evaporation-dominated seas or lakes (a), sulfate reduction is only partial; in this case, ^{32}S is partitioned into the sulfide anion, some of which goes on to form isotopically light H_2S, and some of which forms isotopically light iron sulfide. In terms of the solid phases in this case, ^{34}S is partitioned into the sulfate, so the sulfate is isotopically heavy and the sulfide is isotopically light. In strongly reducing conditions (b), all sulfate is reduced to sulfide and ^{34}S is partitioned in iron sulfide in sediments and ^{32}S into hydrogen sulfide gas.

2 Kinetic isotopic exchange affected by microbially mediated redox reactions. For example, the greatest magnitude of fractionation within the sulfur isotope system occurs during dissimilatory sulfate reduction, a process that results in formation of biogenic sulfides with isotopically light (negative) values. In this process, bacteria (e.g. *desulfovibrio* or *desulfatomaculum*) take advantage of the energy associated with the transfer of eight electrons produced by sulfate reduction:

$$2CH_2O + SO_4^{2-} = H_2S + 2HCO_3^- \qquad (10.27)$$

In this example, carbon is oxidized while sulfur is reduced. Examining only the change in oxidation states of the individual atoms:

$$S^{6+} + 8e^- = S^{2-}$$
$$2C^0 = 2C^{4+} + 8e^-$$

The S^{6+} cation occurs in sulfate and the sulfide (S^{2-}) anion occurs in H_2S (and also FeS_2 and other sulfide minerals); C^0 represents reduced carbon in organic matter (here given as CH_2O) while C^{4+} represents oxidized carbon in bicarbonate.

The formation of isotopically light sulfide minerals during sulfate reduction is partly due to the liberation of H_2S gas with $\delta^{34}S$ values that are ~25‰ lower than the sulfate that is being reduced. This leaves the residual sulfate enriched in ^{34}S, and when sulfide anions in H_2S participate in chemical reactions to form pyrite, the pyrite will be isotopically light (Figure 10.14a). When sulfate is biochemically reduced to sulfide, the high (heavy) δ^{34}S signal of the sulfate is transferred to the sulfide mineral, usually pyrite.

Selected examples of sulfur isotopes applied to problems in environmental geochemistry are discussed in Section 10.7.2.

10.7.2 Sulfur isotopes as tracers of acid deposition, sulfate reduction, and seawater composition over geologic time

Sources and cycling of sulfur in regions affected by acid precipitation. Given that δ^{34}S of precipitation is often different than δ^{34}S of soil sulfur and bedrock sulfur, relative contributions from these sources in surface waters affected by acidification can be assessed. For example, from studies in Canada, Japan, and Czech Republic, Nriagu and Coker (1978), Ohizumi et al. (1997), and Novák et al. (2005) all observed that anthropogenic sulfur (e.g. from coal combustion and ore refining) has higher δ^{34}S values than does natural soil sulfur. This demonstrates how δ^{34}S values of surface water can be used to assess relative contributions of natural versus anthropogenic sulfur from the atmosphere (and hence also acidity given that sulfuric acid is a major component of acid rain). However, higher δ^{34}S values are not always associated with anthropogenic sulfur from the atmosphere; Wadleigh and Blake (1999) observed isotopically lighter anthropogenic sulfate (relative to natural sulfate) as recorded in lichens from eastern Canada. Clearly, natural variability in anthropogenic as well as natural sources influence the δ^{34}S. It is also worth noting that, as is the case with hydrogen and oxygen in water and nitrogen and oxygen in nitrate, paired analysis of δ^{34}S and δ^{18}O in sulfate can enhance assessment of sources and cycling.

Tracking Microbially Mediated Reduction of Sulfate in Groundwater. If hydrocarbons in soils or groundwater are to decompose by oxidation, another component must undergo reduction. The microorganisms that facilitate hydrocarbon decomposition gain energy by transferring electrons from hydrocarbon (reduced organic C) to sulfate. The S^{6+} in sulfate is reduced to S^{2-} anions and carbon is oxidized to C^{4+} in CO_2. The pronounced increase in δ^{34}S of residual sulfate (i.e. the sulfate not yet reduced to sulfide) that accompanies sulfate reduction has been used to quantify anaerobic microbial activity to assess and predict effectiveness of hydrocarbon degradation.

Determining Variations in Seawater Chemistry over Geologic Time. There is effectively no fractionation between dissolved sulfate in seawater and the evaporite minerals that form from evaporating sea water, so evaporate minerals in marine sediments should preserve the δ^{34}S of marine water over geologic time. Kramm and Wedepohl (1991) attributed the decrease in δ^{34}S from a high of +31‰ in early Cambrian time to a low of +10‰ in late Permian time to extremely low rates of continental runoff and bacterial sulfate reduction. If isotopically heavy sulfur were stored on the continents (e.g. in large evaporitic seas), sea water would become lighter. The rapid increase to 25‰ as Pangea began to break apart in the Triassic and Jurassic could be explained by a rapid influx of sediments and dissolved sulfate with high δ^{34}S, or by deep anoxia associated with mass extinction that fractionated ^{34}S into sediments (Figure 10.14). One complicating factor is diagenesis, making it necessary to ascertain that burial, lithification, and alteration of the sediments by temperature and time have not altered the original sedimentary δ^{34}S value.

10.8 NONTRADITIONAL STABLE ISOTOPES

Many recent advances in stable isotope geochemistry have come in the analysis of metal stable isotopes, including heavy metals such as Cr, Fe, Ni, Cu, Zn, Mo, Cd, and Hg, and the lighter elements Li, Mg, Ca, and Cl. The development of the newer isotope methods is largely due to improved precision of instrumental analysis via development of protocols using MC-ICP-MS and thermal ionization mass spectrometry (TIMS). To some extent, the relatively late development of some of these methods may also be attributed to the old assumption that elements which form dominantly ionic bonds (e.g. Mg, Ca, Cl) would not be affected by isotopic fractionation in the environment. They are, but not nearly to the extent of fractionation with covalent bonding, and the relatively low-magnitude fractionations of many of the nontraditional systems highlighted below require high-precision techniques and analytical tools.

The relatively small proportional differences in masses of the isotopes of the heavier transition metals – as compared to more traditional stable isotopes (e.g. ^{18}O vs. ^{16}O) – make it difficult to precisely measure fractionation in natural systems; for example, the mass difference between ^{65}Cu and ^{63}Cu is only 3.2% whereas for ^{18}O and ^{16}O it is 12.5% (Table 10.2). If MDF is the dominant control on the behavior of isotope pairs of two different elements, the element with greater mass (e.g. Cu) will be fractionated less than the lighter one (e.g. O). The relatively low extent of fractionation of transition metals and heavier elements was not precisely quantifiable with pre-2000 methods and instrumentation; now, with advances in MC-ICP-MS and TIMS, geochemists have the capability to precisely measure fractionations as small as 1‰ or 2‰ and often as low as 0.1‰ or 0.2‰.

The principles and concepts of three metal stable isotopes (Cu, Fe, Hg) and two other nontraditional stable isotopes (Mg and $\delta^{37}Cl$) are presented in the following sections. Although not exhaustive, these examples will provide some background about how these systems operate in nature and how they can be used to answer questions regarding natural and anthropogenic processes.

10.8.1 $\delta^{65/63}$Cu

The standard notation for the two stable isotopes of copper is:

$$\delta^{65}Cu \ (‰)$$
$$= \left(\frac{(^{65}Cu/^{63}Cu_{smpl}) - (^{65}Cu/^{63}Cu_{NIST976})}{(^{65}Cu/^{63}Cu_{NIST976})} \right)$$
$$\times 1000 \qquad (10.28)$$

Ranges of $\delta^{65}Cu$ in natural materials presented in Borrok et al. (2008), Fujii et al. (2013), and Thompson et al. (2013) include sulfide minerals (+1‰ to −5‰), ocean water (0.7–0.9‰), oxidized carbonate minerals such as cuprite or malachite (−2‰ to +3‰), plants (0‰ to −5‰), and stream water (−1‰ to +3‰). When Cu is oxidized, ^{65}Cu tends to be fractionated into the oxidized phase (water or mineral, hence relatively high values in surface water and minerals stable in oxidizing environments) while plants and sulfide minerals appear to become enriched in ^{63}Cu preferentially over ^{65}Cu, leading to relatively low or negative values in these reservoirs.

Given that copper is influenced by redox reactions ($Cu^+ = Cu^{2+} + e^-$), stable Cu isotopes have been measured to study origins of ore deposits as well as processes controlling weathering and transport of copper in acid mine drainage areas. A study by Kimball et al. (2009) of a drainage basin in Red Mountain, Colorado, USA, impacted by sulfide oxidation and associated acid mine drainage serves to illustrate the potential application of Cu isotopes to environmental geochemistry. Data from this study demonstrate that stream water (~ +1.5‰) is 1.4–1.6‰ enriched in ^{65}Cu relative to the parent sulfide

minerals enargite and chalcopyrite, whose $\delta^{65}Cu$ values are ~0.01 ± 0.10‰ and 0.16 ± 0.10‰, respectively (terrestrial Cu sulfides cluster around 0.0 ± 1.0‰). The oxidation of Cu^+ to Cu^{2+} has resulted in enrichment of ^{65}Cu in the oxidized Cu phases in solution, and the solid weathering products remaining in the tailings have been enriched in ^{63}Cu. This behavior is typical of metal stable isotopes (e.g. Cr, Cu, Mo, Fe), where heavier isotopes tend to be partitioned into oxidized aqueous phases.

In order to further explore the controls on Cu isotope fractionation during sulfide weathering, Kimball et al. (2009) carried out controlled leaching experiments in the presence of, and absence of, the bacterium *Acidithiobacillus ferrooxidans* (known to accelerate weathering of many sulfides). Abiotic weathering, i.e. experiments carried out in the absence of *A. ferrooxidans*, reproduces very similar values to those obtained from the field study; however, in the presence of *A. ferrooxidans*, fractionation is virtually nonexistent for enargite and actually occurs in the opposite direction for chalcopyrite (waters are depleted in ^{65}Cu by ~0.6‰ relative to the parent sulfide mineral). In the absence of microbes such as *A. ferrooxidans*, copper sulfide weathering is driven by preferential oxidation of ^{65}Cu (relative to ^{63}Cu) which releases ^{65}Cu into solution. TEM (transmission electron microscope) analysis shows that when copper sulfide weathering occurs in the presence of microbes such as *A. ferrooxidans*, ^{65}Cu is preferentially incorporated into bacterial cell walls or other parts of the bacterial community (Figure 10.15), but is not released into solution.

Given that the natural weathering of enargite and chalcopyrite led to a ~ +1.5‰ enrichment of ^{65}Cu in stream water (relative to the parent sulfide minerals), it appears that microbes such as *A. ferrooxidans* do not play a significant role in copper sulfide weathering – at least not at Red Mountain – and this has implications for remediation. If bacteria do not play an important role in weathering of Cu sulfides, bactericides would not be expected to limit metal release and acid generation if the main source is Cu sulfides (this is in contrast to weathering of pyrite, where *A. ferrooxidans* frequently enhances reaction rate, and where bactericides may be applied to limit acid generation).

10.8.2 $\delta^{56/54}$Fe

Of the four stable isotopes of iron (Table 10.2), the two most abundant are ^{56}Fe and ^{54}Fe. Like Cu and other metal isotopes, iron isotopes are fractionated during redox reactions, resulting in enrichment of ^{56}Fe in ferric

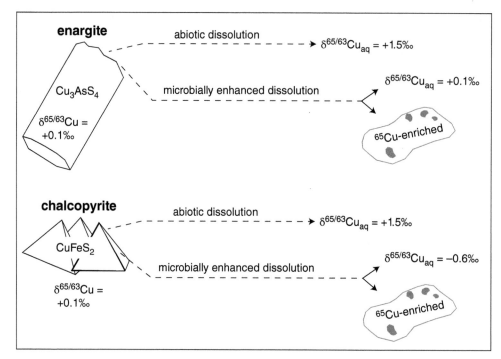

Fig. 10.15 Schematic diagram of fractionation of Cu isotopes during weathering of two Cu sulfide minerals, enargite, and chalcopyrite. The bacterium *Acidithiobacillus ferrooxidans* (shown schematically) sequesters ^{65}Cu, meaning that the extent of microbially mediated dissolution of Cu sulfides can be assessed by $\delta^{65/63}$Cu.

aqueous phases (e.g. Fe(III)(H$_2$O)$_6^{3+}$) and ^{54}Fe in ferrous phases (e.g. Fe(II)(H$_2$O)$_6^{2+}$). The consequence of this initial fractionation and the low solubility of ferric iron is that ^{56}Fe tends to preferentially partition into iron oxides and hydroxides (Bullen 2011), which results in ^{54}Fe^{2+} enrichment in solution. An example is this representation of the oxidation of pyrite:

$$^{56}\text{Fe}^{54}\text{FeS}_4 + 7\,^{1}\!/_{4}\text{O}_2 + 9\,^{1}\!/_{2}\text{H}_2\text{O}$$
$$\rightarrow\ ^{56}\text{Fe}^{3+}\text{OOH}_{(s)} + {}^{54}\text{Fe}^{2+}\!\cdot\!(\text{H}_2\text{O})_{6(aq)}$$
$$+ 6\text{H}^+_{(aq)} + 4\text{SO}_4^{2-}{}_{(aq)} + 3e^- \qquad (10.29)$$

As with Cu isotopes, Fe isotopes also have the potential to assess microbial mediation; for example, when goethite dissolution is abiotic and proton assisted (i.e. in presence of H$^+$), there is no fractionation of ^{56}Fe and ^{54}Fe – both are released to solution at the same rate; however, ligand-assisted dissolution of goethite leads to fractionation of ^{54}Fe into solution over ^{56}Fe. (Examples of ligands include the biomolecules oxalate and acetate and inorganic species like phosphate.) Reductive dissolution also preferentially releases ^{54}Fe to solution, for example in the reduction of Fe(III) in clays and hydroxides. Covalent bonds are stronger when ^{56}Fe is the iron isotope, rather than ^{54}Fe, and mineral breakdown will tend to cause relatively low δ^{56}Fe in soil water or groundwater. If iron

hydroxides are precipitating out of solution along a flow path, the solution should become progressively lighter as ^{56}Fe is fractionated into the Fe^{3+} mineral phase.

10.8.3 $\delta^{202/198}$Hg

Mercury has seven stable isotopes (^{196}Hg, ^{198}Hg, ^{199}Hg, ^{200}Hg, ^{201}Hg, ^{202}Hg, and ^{204}Hg), the most abundant being ^{202}Hg at 29.86% (Table 10.2). The main oxidation states of Hg at the Earth's surface are the cation Hg^{2+} and the gas Hg^0. The divalent Hg^{2+} often forms strong complexes with organic matter, commonly as methylmercury, e.g. (CH$_3$)$_2$Hg (Section 8.12). It is a toxic element in any form although when methylated it is mobile, prone to ingestion by organisms, and more toxic than other Hg forms. Hg often bioaccumulates and biomagnifies in aquatic ecosystems and mercury advisories are in place in many regions globally due to high Hg content in fish. Its average crustal abundance is very low (0.1 mg kg^{-1}) but Hg is concentrated in some sulfide-rich deposits at values from 0.1% to 2%, mainly as cinnabar (HgS), and mines and Hg-refining sites are often sources of Hg contamination. Two other major sources of Hg are emissions to the atmosphere derived from coal combustion and waste incineration – the scale of impact of these emissions are typically on scales of hundreds of kilometers.

Hg isotope analysis is a twenty-first century method and is still in development (Blum and Johnson 2017). The delta notation is $\delta^{202/198}$Hg (Table 10.2), but values of the other Hg isotopes are also expressed. One is as delta notation where the denominator of the isotope pair is ^{198}Hg.

$$\delta^{xxx}\text{Hg } \text{‰} = \left(\frac{(\delta^{xxx}\text{Hg}/\delta^{198}\text{Hg})_{\text{sample}}}{(\delta^{xxx}\text{Hg}/\delta^{198}\text{Hg})_{\text{SRM3133}}} - 1 \right) \times 1000$$

(10.30)

(Note that while expressed differently than other delta (δ) notations [e.g. Eq. (10.28)], this way is mathematically identical.)

Another approach is α values (Eq. (10.31)), which is the ratio of an isotope pair in one phase (A) relative to that same ratio in a second phase, i.e. phase B (note: "phase" here includes minerals, bacteria, groundwater, stream water, plant leaf, etc.).

$$\alpha_{\text{A-B}} = \frac{^{201}\text{Hg}/^{198}\text{Hg}_A}{^{201}\text{Hg}/^{198}\text{Hg}_B}$$

(10.31)

Focus Box 10.13

Mercury Isotopes Applied to Assessing Air Pollution

The isotopic composition of Hg in coal deposits (e.g. plotted as δ^{202}Hg vs. Δ^{201}Hg, Figure 10.16) can indicate sources of Hg pollution in air or water. In order to test whether Hg emitted from a coal-fired power plant (in west central Florida, USA) is deposited close to the point source, or whether it is carried far downwind with no local effects, Sherman et al. (2012) examined Hg isotope compositions of the coal burned at the plant. They also determined Hg isotope compositions of precipitation within 10 km downwind of the plant, and of precipitation unaffected by the plant (i.e. not downwind of the smokestack). They traced the distinct δ^{202}Hg signature of coal burned at the

power plant (δ^{202}Hg = −0.72‰) to isotopically light δ^{202}Hg (δ^{202}Hg = −2.6‰) in rainfall deposited with 5–10 km of the point source. The locally deposited rainfall δ^{202}Hg is distinct from precipitation that falls upwind of the coal plant (δ^{202}Hg = +0.07 + 0.17‰), demonstrating that Hg from the plant is deposited close to the source. Fractionation of light ^{198}Hg during combustion into smokestack emissions (and ^{202}Hg into ash solids) explains the lighter (−2.6‰) δ^{202}Hg in precipitation relative to the value of the coal itself (−0.72‰). The partitioning of the lighter isotope into the vapor is a typical consequence of MDF.

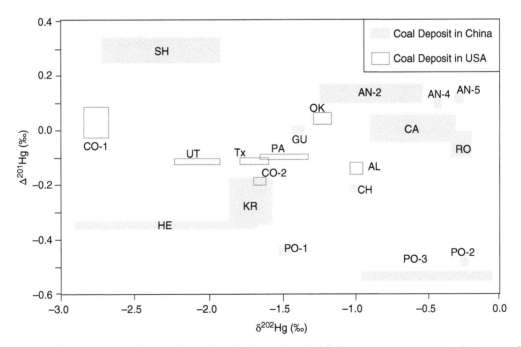

Fig. 10.16 Mercury isotope compositions of coals from China and the USA. Boxes encompass analytical uncertainty for a given deposit, and letter symbols refer to distinct coal deposits, details of which are in Biswas et al. (2008), the source of the data shown here.

When α is > 1.000, the heavy isotope is fractionated into phase A; when α is < 1.000, the heavy isotope is fractionated into phase B.

Another delta notation, this one with capital Δ, is shown below for the odd-numbered Hg isotopes:

$$\Delta^{201}Hg = \delta^{201}Hg - (\delta^{202}Hg \times 0.752) \qquad (10.32)$$

and

$$\Delta^{199}Hg = \delta^{199}Hg - (\delta^{202}Hg \times 0.252) \qquad (10.33)$$

The odd-numbered Hg isotopes ^{201}Hg and ^{199}Hg behave differently than the even-numbered Hg isotopes due to the magnetic isotope effect, a difference in orbital structure that affects reaction/bonding rate. This is an example of *mass-independent fractionation* (MIF), a process that provides an additional angle for investigation of biotic and abiotic reactions, notably photoreduction and its effect on Hg cycling in aquatic ecosystems (Bergquist and Blum 2007).

10.8.4 $\delta^{26}Mg$ and $\delta^{44/42}Ca$

Magnesium and calcium are both major constituents of silicate and carbonate rocks as well as rivers, lakes, sea water, groundwater, and (to varying extents) plants and other organisms. Stable isotopes of Mg and Ca have proven useful in assessing chemical weathering, changes over geological time to the composition of sea water, and the uptake of Ca by plants. The abundances of Mg isotopes are given in Figure 10.2; Ca has three stable isotopes, ^{44}Ca (0.65%), ^{42}Ca (2.01%), and ^{40}Ca (96.9%), as well as some short-lived radioactive ones. Figure 10.17 depicts known ranges of the $^{26/24}Mg$ and $^{44/42}Ca$ isotopes – note the distinct $\delta^{26}Mg$ signal in carbonate rocks vs silicate rocks and medium to large rivers, and partitioning of light Mg and Ca isotopes into plants.

Ratios of $^{44/42}Ca$ and $^{44/40}Ca$ are useful in assessing Ca cycling. Plants preferentially absorb ^{40}Ca from soil waters (Figures 10.17 and 10.18), leaving the soil solution enriched in ^{44}Ca and ^{42}Ca.

10.8.5 $\delta^{37/35}Cl$

Chlorine has two naturally occurring stable isotopes, ^{37}Cl (24.23%) and ^{35}Cl (75.77%). Radioactive ^{36}Cl occurs in very trace amounts as a cosmogenic radionuclide and is covered in Chapter 11. *Standard mean ocean chloride* (SMOC) is the reference standard and the delta notation for $^{37}Cl/^{35}Cl$ is $\delta^{37}Cl_{SMOC}$. A good example of the applicability of Cl isotopes is tracking the decomposition of trichloroethylene (TCE) to its degradation products (e.g.

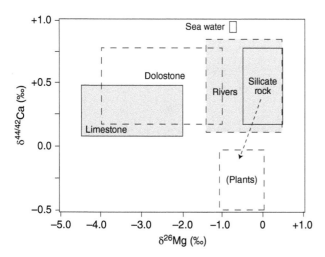

Fig. 10.17 Ranges of $\delta^{26}Mg$ and $\delta^{44/42}Ca$ values for inorganic reservoirs and plants (Source: data from Young and Galy 2004; Tipper et al. 2006). Note that carbonate rocks are depleted in ^{26}Mg relative to silicate rocks. Uptake of Mg and Ca by plants fractionates the lighter isotopes into the plant, as shown by the arrow. Relative contributions of carbonate vs. clastic inputs to river dissolved or particulate loads can be assessed using Ca and Mg stable isotopes (e.g. Tipper et al. 2008).

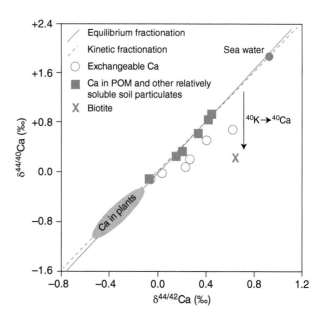

Fig. 10.18 Fractionation of Ca isotopes in a forested ecosystem. Note ^{40}Ca-enriched signal in plant tissue (red oak), and also the effect of radioactive decay on the $^{44/40}Ca$ ratio. Where ^{40}K is present (e.g. in biotite), the ratio of $^{44/40}Ca$ will decrease with time due to $^{40}K \rightarrow {}^{40}Ca$ decay.

Ca Stable Isotopes and Sources of Plant-Available Ca^{2+}

To examine sources and long-term availability of exchangeable Ca^{2+} in an acidic forest soil, Farkaš et al. (2011) analyzed $\delta^{44/42}$Ca and $\delta^{44/40}$Ca in many reservoirs, including plant tissue (red oak), cation exchange sites, particulate organic matter (POM), and Ca^{2+} derived from biotite weathering in the soil C horizon. Their results showed that isotope values of exchangeable Ca^{2+} are on a mixing line between organic soil values and biotite-derived

Ca isotope values (visible in Figure 10.18), indicating that exchangeable Ca^{2+} (up to 25% of total Ca$^{2+}_{exch}$ in this study) is derived from weathering of biotite (Ca^{2+} derived from biotite interlayers). The biotite isotopic signature is not present in older plant matter, suggesting that the contribution from this deeper Ca source is increasing with time as acidic deposition depletes Ca in the organic soil horizons (upper 15 cm).

dichloroethylene, DCE, Figure 3.26). Given the higher vibrational energy that occurs when ^{35}Cl is bonded to a carbon atom compared to C–^{37}Cl bonds, ^{35}Cl should be preferentially released as TCE begins to decay to DCE, and this is what has been observed in an experimental study (Numata et al. 2002) as well as a field-based study (Sturchio et al. 1998). When TCE breaks down, initial chlorine released to solution is enriched in ^{35}Cl, and the remaining DCE becomes enriched in ^{37}Cl (Figure 10.19).

Zones of enhanced degradation, notably organic-rich clay-silt lenses, were characterized by increased concentration of the chloride anion (Cl$^-$) with an anomalously low δ^{37}Cl value, reflecting release of ^{35}Cl-enriched chlorine from TCE decomposition.

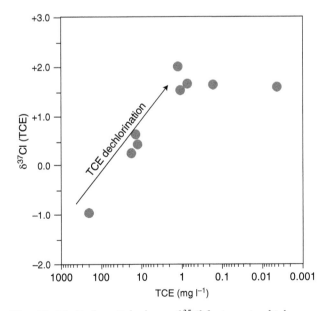

Fig. 10.19 Preferential release of ^{35}Cl during microbial degradation of TCE to DCE causes δ^{37}Cl of the Cl remaining in the metabolite DCE to increase. Source: Modified from Sturchio et al. (1998).

10.9 SUMMARY

The traditional or conventional stable isotopes of C, H, O, N, and S have been studied for decades, providing valuable information on topics ranging from paleotemperature to nutrient cycling and contaminant sources. The development of methods and paradigms in the field of "nontraditional" stable isotope analysis, including lighter elements such as Li, B, Mg, Ca, and Cl, as well as metal and clumped stable isotopes, is still in early stages and more remains to be done before their full significance to environmental geochemistry can be appreciated. At the same time, these elements are critical to environmental geochemistry in many ways: Fe controls the cycling of many other trace elements and organic compounds; Cr, Hg, and Mo are potential contaminants of soil and groundwater; Cl is a component of many toxic organic compounds; and Cu and Mn are tracers of geochemical processes and vital plant trace nutrients. Further research into the significance of metal stable isotopes is an exciting area of development in environmental geochemistry and one that is providing new approaches to understanding the environment into the twenty-first century.

QUESTIONS

10.1 What is the ^{18}O/^{16}O ratio of a sample that has a δ^{18}_{VSMOW} value of +23‰?

10.2 For a mass of water vapor at 25 °C with an initial δ^{18}O of −12‰, determine the δ^{18}O of the precipitation that will condense, and the vapor remaining, after 50% of the initial vapor has condensed.

10.3 Equilibrium fractionation factors for evaporation of water are $\alpha_{0C} = 1.0117$ and $\alpha_{25C} = 1.0092$. Calculate δ^{18}O of vapor in equilibrium with sea water at (A) 0 °C and (B) 25 °C.

10.4 Explain the difference in the equilibrium fractionation factors in question 3.

10.5 $\delta^{18}O$ values for oxygen in $H_2O(s)$ sampled from an ice core are shown in Figure 10.20 (depth is in meters). Which depth (or depth range) corresponds to the warmest temperatures, and which is indicative of the coolest temperatures? Explain in ~2 to 3 sentences the fundamental process that controls the variation in the oxygen isotope compositions of ice cores.

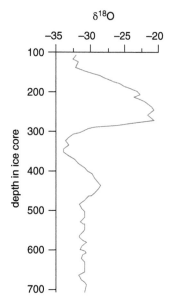

Fig. 10.20 Oxygen isotope composition of H_2O in an ice core, example for question 5.

10.6 Why does isotopic fractionation decrease with increasing temperature? Cite pertinent equation(s).

10.7 What is the likely origin of elevated nitrate in an aquifer where the nitrate has a $\delta^{15}N$ value of +2.3‰ and a $\delta^{18}O$ value of +17.8‰? (both relative to VSMOW). Explain how this source of nitrate gets its signature.

10.8 Provide three examples of how metal stable isotopes are applied to environmental analysis. Why has the development of this field lagged behind the more traditional analysis of δD, $\delta^{13}C$, $\delta^{15}N$, $\delta^{18}O$, and $\delta^{18}S$?

10.9 Oxygen and hydrogen isotopic compositions are plotted in Figure 10.21, an example of data from a temperate continental environment. Address the following questions:

A. What are the most likely δD and $\delta^{18}O$ values for modern precipitation in this region? Explain.

B. Suggest an explanation for the trend in the warm, fault zone groundwater.

C. Suggest an explanation for the trend in surface water below the text for GMWL (global meteoric water line).

D. Speculate on why the shallow groundwater samples plot closer to surface waters than do the deep groundwater samples.

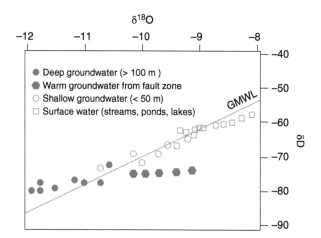

Fig. 10.21 Plot of oxygen (‰, $\delta^{18}O_{VSMOW}$) vs. hydrogen (‰, δD_{VSMOW}) isotope values for samples of different reservoirs in the hydrologic cycle. GMWL is global meteoric water line.

10.10 A bedrock aquifer in pyrite-bearing, organic-rich black slate has numerous wells with elevated arsenic (30% of wells contain $As > 10~\mu g\,l^{-1}$). Dissolved Fe, SO_4, and As are positively correlated, and the $\delta^{34}S$ value of dissolved sulfate ranges from +31‰ to +38‰. The range of values of $\delta^{34}S$ in pyrite is +32‰ to +53‰. Suggest an explanation for the high $\delta^{34}S$ pyrite values; also, provide possible explanation(s) for the origin/cause of the elevated As. Do the geochemical correlations and isotopic data provide insight?

REFERENCES

Affek, H.P. (2013). Clumped isotopic equilibrium and the rate of isotope exchange between CO_2 and water. *American Journal of Science* 313: 309–325.

Augustin, L. et al. (2004). Eight glacial cycles from an Antarctic ice core. *Nature* 429 (6992): 623–628. https://doi.org/10.1038/nature02599.

Bechtel, A., Gratzer, R., Sachsenhofer, R.F. et al. (2008). Biomarker and carbon isotope variation in coal and fossil wood of Central Europe through the Cenozoic. *Palaeogeography, Palaeoclimatology, Palaeoecology* 262: 166–175.

Bergquist, B.A. and Blum, J.D. (2007). Mass-dependent and –independent fractionation of hg isotopes by photoreduction in aquatic systems. *Science* 318: 417–420.

Bigeleisen, J. and Mayer, M.G. (1947). Calculation of equilibrium constants for isotopic exchange reactions. *The Journal of Chemical Physics* 15: 261–267.

Biswas, A., Blum, J.D., Bergquist, B.A. et al. (2008). Natural mercury isotope variation in coal deposits and organic soils. *Environmental Science and Technology* 42: 8303–8309.

Blum, J.D. (2011). Applications of stable mercury isotopes to biogeochemistry. In: *Handbook of Environmental Isotope Geochemistry, Advances in Isotope Geochemistry* (ed. M. Baskaran). Berlin: Springer-Verlag.

Blum, J.D. and Johnson, M.W. (2017). Recent developments in mercury stable isotope analysis. *Reviews in Mineralogy and Geochemistry* 82 (1): 733–757.

Borrok, D.M., Nimick, D.A., Wanty, R.B., and Ridley, W.I. (2008). Isotopic variations of dissolved copper and zinc in stream waters affected by historical mining. *Geochimica et Cosmochimica Acta* 72: 329–344.

Broecker, W.S., Kennett, J.P., Flower, B.P. et al. (1989). Routing of meltwater from the Laurentide Ice Sheet during the Younger Dryas cold episode. *Nature* 341: 318–321.

Broeker, W.S. and Oversby, V.M. (1971). *Chemical Equilibria in the Earth*. New York and London: McGraw-Hill.

Bullen, T.D. (2011). Stable isotopes of transition and post-transition metals as tracers in environmental studies. In: *Handbook of Environmental Isotope Geochemistry, Advances in Isotope Geochemistry* (ed. M. Baskaran). Berlin: Springer-Verlag. 177–203.

Cerling, T.E., Wang, Y., and Quade, J. (1993). Expansion of C_4 ecosystems as indicator of global ecological change in the Late Miocene. *Nature* 361: 344–345.

Cerling, T.E., Harris, J.M., MacFadden, B.J. et al. (1997). Global vegetation change through the Miocene/Pliocene boundary. *Nature* 389: 153–158.

Chamberlain, C.P., Poage, M.A., Craw, D., and Reynolds, R.C. (1999). Topographic development of the Southern Alps recorded by the isotopic composition of authigenic clay minerals, South Island, New Zealand. *Chemical Geology* 155: 279–294.

Clayton, R.N., O'Neil, J.R., and Mayeda, T.K. (1972). Oxygen isotope exchange between quartz and water. *Journal of Geophysical Research* 77 (17): 3057–3067.

Craig, H. (1961). Isotopic variations in meteoric waters. *Science* 133: 1702–1703.

Craig, H. (1966). Isotopic composition and origin of the Red Sea and Salton Sea geothermal brines. *Science* 154: 1544–1548.

D'Amore, F., Fancelli, R., and Panichi, C. (1987). Stable isotope study of reinjection processes in the Larderello geothermal field. *Geochimica et Cosmochimica Acta* 51: 857–867.

Dansgaard, W. (1964). Stable isotopes in precipitation. *Tellus* 16: 436–468.

Drever, J.I. (1997). *The Geochemistry of Natural Waters: Surface and Groundwater Environments*, 3e. Upper Saddle River, NJ: Simon and Schuster.

Eiler, J.M. (2007). "Clumped-isotope" geochemistry – the study of naturally-occurring, multiply-substituted isotopologues. *Earth and Planetary Science Letters* 262: 309–327.

Farkaš, J., Déjeant, A., Novák, M., and Jacobsen, S.B. (2011). Calcium isotope constraints on the uptake and sources of Ca^{2+} in a base-poor forest: A new concept of combining stable ($\delta^{44}/^{42}Ca$) and radiogenic (εCa) signals. *Geochimica et Cosmochimica Acta* 75: 7031–7046.

Faure, G. (1998). *Principles and Applications of Geochemistry*, 2e. Upper Saddle River, NJ: Prentice-Hall.

Francey, R.J., Allison, C.E., Etheridge, D.M. et al. (1999). A 1000-year high precision record of $\delta^{13}C$ in atmospheric CO_2. *Tellus* 51B: 170–193.

Friedman, I. and O'Neil, J.B. (1977). Compilation of stable isotope fractionation factors of geochemical interest. In: *Data of Geochemistry*, 6e (ed. M. Fleischer), Chapter KK. US Geol. Surv. Prof. Paper 440-KK.

Fujii, T., Moynier, F., Abe, M. et al. (2013). Copper isotope fractionation between aqueous compounds relevant to low temperature geochemistry and biology. *Geochimica et Cosmochimica Acta* 110: 29–44.

Hatch, M.D. and Slack, C.R. (1966). Photosynthesis in sugarcane leaves: a new carboxylation reaction and the pathway of sugar formation. *Biochemical Journal* 101: 103–111.

Hoefs, J. (2009). *Stable Isotope Geochemistry*, 6e. Berlin: Springer.

Horibe, Y. and Oba, T. (1972). Temperature scales of aragonite-water and calcite-water systems. *Fossils* 23 (24): 69–79.

Kawabe, I. (1978). Calculation of oxygen isotope fractio ation in quartz-water system with special reference to the low temperature fractionation. *Geochimica et Cosmochimica Acta* 42: 613–621.

Kendall, C. and Caldwell, E.A. (1998). Fundamentals of isotope geochemistry. In: *Isotope Tracers in Catchment Hydrology* (ed. C. Kendall and J.J. McDonnell), 51–86. Amsterdam: Elsevier Science B.V.

Kendall, C. and Doctor, D.H. (2003). Stable isotope applications in hydrologic studies. In Treatise on Geochemistry, Volume 5. Ed: James I. Drever. Executive Editors: Heinrich D. Holland and Karl K. Turekian, pp. 319–364. Amsterdam, Elsevier.

Kendall, C. and McDonnell, J.J. (eds.) (1998). *Isotope Tracers in Catchment Hydrology*. Amsterdam: Elsevier Science B.V.

Kendall, C., Sklash, M.G., and Bullen, T.D. (1995). Isotope tracers of water and solute sources in catchments. In: *Solute Modelling in Catchment Systems*, 261–303. New York: Wiley.

Kimball, B.E., Mathur, R., Dohnalkova, A.C. et al. (2009). Copper isotope fractionation in acid mine drainage. *Geochimica et Cosmochimica Acta* 73: 1247–1263.

Kramm, U. and Wedepohl, K.H. (1991). The isotopic composition of strontium and sulfur in seawater of Late Permian (Zechstein) age. *Chemical Geology* 90: 253–262.

Kroopnick, P. and Craig, H. (1972). Atmospheric oxygen: isotopic composition and solubility fractionation. *Science* 175 (4017): 54–55. https://doi.org/10.1126/science.175.4017.54.

Moeller, K., Schoenberg, R., Pedersen, R.B. et al. (2012). Calibration of the new certified reference materials ERM-AE633 and ERM-AE647 for copper and IRMM-3702 for zinc isotope amount ratio determinations. *Geostandards and Geoanalytical Research* 36: 177–199.

Novák, M., Kirchner, J.W., Fottová, D. et al. (2005). Isotopic evidence for processes of sulfur retention/release in 13 forested catchments spanning a strong pollution gradient (Czech Republic, central Europe). *Global Biogeochemical Cycles* 19: GB4012. https://doi.org/10.1029/2004GB002396.

Nriagu, J.O. and Coker, R.D. (1978). Isotopic composition of sulphur in precipitation within the Great Lakes Basin. *Tellus* 30: 365–375.

Numata, M., Nakamura, N., Koshikawa, H.J., and Terashima, Y. (2002). Chlorine isotope fractionation during reductive dechlorination of chlorinated ethenes by anaerobic bacteria. *Environmental Science and Technology* 36: 4389–4394.

Ohizumi, T., Fukuzaki, N., and Kusakabe, M. (1997). Sulfur isotopic view on the sources of sulfur in atmospheric fallout along the coast of the Sea of Japan. *Atmospheric Environment* 31: 1339–1348.

O'Neil, J.R. and Epstein, S. (1966). Oxygen isotope fractionation in the system dolomite-calcite-carbon dioxide. *Science* 152: 198–201.

Palmer, M.R. and Slack, J.F (1989). Boron isotopic composition of tourmaline from massive sulfide deposits and tourmalinites. *Contributions to Mineralogy and Petrology* 103: 434–451.

Petit, J.R., Jouzel, J., Raynaud, D. et al. (1999). Climate and atmospheric history of the past 420,000 years from the Vostok Ice Core, Antarctica. *Nature* 399: 429–436.

Rayleigh, J.W. (1896). Theoretical considerations respecting the separation of gases by diffusion and similar processes. *The London, Edinburgh, and Dublin Philosophical Magazine and Journal of Science* 42 (259): 493–498. https://doi.org/10.1080/14786449608620944.

Rozanski, K., Araguas-Araguas, L., and Giofiantini, R. (1993). Isotopic patterns in modern global precipitation. *Geophysical Monograph* 78: 1–36.

Savin, S.M. and Lee, M. (1988). Isotopic studies of phyllosilicates. *Reviews in Mineralogy* 19: 189–223.

Sharp, Z. (2007). *Principles of Stable Isotope Geochemistry*. Pearson Prentice Hall.

Sherman, L.S., Blum, J.D., Keeler, G.J. et al. (2012). Investigation of local mercury deposition from a coal-fired power plant using mercury isotopes. *Environmental Science and Technology* 46: 382–390.

Stern, L.A., Johnson, G.D., and Chamberlain, C.P. (1994). Carbon isotope signature of environmental change found in fossil ratite eggshells from a South Asian Neogene sequence. *Geology* 22: 419–422.

Sturchio, N.C., Clausen, J.L., Heraty, L.J. et al. (1998). Chlorine isotope investigation of natural attenuation of trichloroethene in an aerobic aquifer. *Environmental Science and Technology* 32: 3037–3042.

Thompson, C.M., Elwood, M.J., and Wille, M. (2013). A solvent extraction technique for the isotopic measurement of dissolved copper in seawater. *Analytica Chimica Acta* 775: 106–113.

Tipper, E.T., Gaillardet, J., Galy, A., and Louvat, P. (2008). Ca isotope ratios in the largest rivers in the world: Implications for the global Ca budget and weathering processes. *Geochimica et Cosmochimica Acta* 72: A947–A947.

Tipper, E.T., Galy, A., Gaillardet, J. et al. (2006). The magnesium isotope budget of the modem ocean: Constraints from riverine magnesium isotope ratios. *Earth and Planetary Science Letters* 250: 241–253.

Urey, H. (1947). The thermodynamic properties of isotopic substances. *Journal of the Chemical Society (London)* 1947: 562–581.

Wadleigh, M.A. and Blake, D.M. (1999). Tracing sources of atmospheric sulphur using epiphytic lichens. *Environmental Pollution* 106: 265–271.

Young, E. and Galy, A. (2004). The isotope geochemistry and cosmochemistry of Mg. *Reviews in Mineralogy and Geochemistry* 55: 197–230.

Yurtsever, Y. and Gat, J.R. (1981). Atmospheric waters. In: *Stable Isotope Hydrology: Deuterium and Oxygen-18 in the Water Cycle*. Technical Report Series 210:103-142. Vienna: IAEA.

11

Radioactive and Radiogenic Isotope Geochemistry

Radioactive isotopes, whether naturally occurring (e.g. ^{14}C, ^{235}U, and ^{238}U) or synthetic (e.g. ^{99}Tc, ^{239}Pu) are characterized by atoms with unstable nuclei that undergo radioactive decay to *daughter* isotopes, which, because they form by radioactive decay, are termed *radiogenic*. These daughter products may also be radioactive, or they may be stable. Radioactive and radiogenic isotopes are useful in environmental analysis much in the same way that stable isotopes are, i.e. as tracers of geochemical and biochemical reactions. A very common approach to quantitative dating of rocks, sediments, soils, groundwater, and organic matter is radioactive decay methods. Some radioactive elements, whether naturally occurring or anthropogenic, are toxic, so geochemical research has also focused on the behavior of various radionuclides in environmental systems.

This chapter begins with some fundamentals of radioactive decay and radioactive-radiogenic elements, then moves on to examples of radionuclides applied to environmental and low-temperature geochemistry.

11.1 RADIOACTIVE DECAY

The nuclei of *radioactive isotopes* undergo spontaneous decay to *daughter atoms* (which may be stable or radioactive) via various mechanisms, all of which release energy and particles. Regardless of the mechanism, radioactive decay produces a change in both Z (# of protons) and N (# of neutrons) from parent to daughter isotope (note: atomic mass may remain unchanged depending on decay

mechanism). Radioactive decay occurs by emission of three types of radiation known as alpha (α), beta (β), and gamma (γ) rays, the presence of which were discovered by Ernest Rutherford in 1899 and 1900 from groundwork laid by Henri Becquerel, Marie Curie, and Pierre Curie in the late 1800s.

11.1.1 Mechanisms and products of radioactive decay

Alpha radiation takes the form of a particle that is essentially a helium nucleus, with two protons and two neutrons and a charge of +2, and is mainly produced when atoms of atomic number ≥ 58 undergo radioactive decay. A good example of alpha decay is the transformation of ^{238}U, the most abundant isotope of uranium on the earth, to ^{234}Th (thorium):

$$^{238}_{92}U^{6+} \rightarrow\ ^{234}_{90}Th^{4+} + {}^{4}_{2}He^{2+} (\alpha) + \gamma + E \qquad (11.1)$$

Note that alpha decay causes a reduction of two protons, resulting in transformation from a uranium nucleus with Z = 92 to thorium with Z = 90, and the loss of two neutrons in concert with the two protons means that atomic mass is reduced by 4 from 238 to 234 Da. In addition to the emission of an alpha particle, gamma radiation (γ) is also released, as is kinetic energy (E) associated with the fast-moving α particle plus the recoil energy imparted to the nucleus by the ejection of the α particle.

As a product of a radioactive decay reaction, ^{234}Th is termed *radiogenic*. It is also radioactive and will undergo

Environmental and Low-Temperature Geochemistry, Second Edition. Peter Ryan.
© 2020 John Wiley & Sons Ltd. Published 2020 by John Wiley & Sons Ltd.

further spontaneous transformation via *beta decay* to ^{234}Pa:

$$^{234}_{90}\text{Th} \rightarrow \, ^{234}_{91}\text{Pa} + \, ^{0}_{-1}e^{-} \, (\beta^{-}) + \gamma + \underline{\nu} \qquad (11.2)$$

The decay reaction here results in a +1 increase in Z and no change in atomic mass, so the increase in Z must be balanced by the loss of a neutron. Viewed in this light, beta decay is the transformation (in the nucleus) of a neutron to a proton, and conceptually occurs by the spontaneous emission of an electron from the nucleus (represented in Eq. (11.3) by $_{-1}{}^{0}e^{-}$ *or* β^{-}) plus a gamma ray (γ) and antineutrino ($\underline{\nu}$), both of which are subatomic particles also emitted during beta decay:

$$^{1}_{0}\text{N} \rightarrow \, ^{1}_{1}\text{P} + \, ^{0}_{-1}e^{-} \, (or \, \beta^{-}) + \gamma + \underline{\nu} \qquad (11.3)$$

Radioactive decay also occurs by emission of a positively charged electron (a positron) from the nucleus, a process that converts a proton to a neutron plus a positron (and a gamma ray and neutrino, ν). This type of decay is *positron decay*, positive beta decay, or beta-plus (β^{+}) decay

$$^{26}_{13}\text{Al} \rightarrow \, ^{26}_{12}\text{Mg} + \beta^{+} + \gamma + \nu \qquad (11.4)$$

where the β^{+} particle has a +1 charge and no mass and represents the transformation of a proton to a neutron.

As is apparent in Eqs. (11.1) through (11.4), *gamma radiation* occurs in conjunction with alpha or beta decay. While the emission of gamma rays does not alter the number of protons or neutrons in the nucleus, it does help to stabilize the nucleus in its spontaneous progression from a higher, less stable state to a lower, more stable energy state.

Spontaneous fission describes radioactive decay in which the nucleus splits into two or more new nuclei of similar mass. Unlike alpha decay, which produces an α particle that is very small compared to the remaining nucleus, spontaneous fission results in (usually) two new atoms, each of similar size. It mainly applies to synthetic processes like those in nuclear reactors, but a few heavy naturally occurring isotopes such as ^{238}U, ^{235}U, and ^{232}Th also can undergo spontaneous decay via fission. However, the incredibly slow decay rate (half-lives are $\geq 10^{15}$ years) relative to decay rates of alpha and beta decay means that spontaneous fission is an insignificant natural process relative to the others. A notable exception comes from the geological past, where natural fission chain reactions took place in U-rich rocks in Gabon during Proterozoic time ($\sim$2 Ga) (e.g. Gauthier-Lafaye et al. 1996).

11.1.2 Half-lives, decay rates, and decay constants

Decay rate is often expressed in terms of the *half-life* ($t_{1/2}$) of the isotope, where one half-life is the amount of time required to transform one half of the mass of a radioactive isotope to its daughter product – after one half-life, 50% of the original radioactive isotope will remain, and the other 50% of it will have been transformed to its daughter product; 75% of the parent material will have decayed to daughter product after two half-lives (Table 11.1). Note that values range from very short ($t_{1/2}$ for ^{218}Po = 3.05 min) and relatively short ($t_{1/2}$ for ^{222}Rn = 3.8 days) to intermediate ($t_{1/2}$ for ^{14}C = 5730 years) and incredibly long ($t_{1/2}$ for ^{238}U = 4.5 × 10^{9} years).

The exponential nature of radioactive decay is an example of a first-order reaction (Chapter 1, Section 1.11.2.2). The decay of a parent isotope (P) to a daughter isotope (D) is described as:

$$-dP/dt = \lambda_{P} \times P \qquad (11.5)$$

P is the amount of parent isotope remaining at time t, and λ_{P} is the decay constant for the parent isotope, which represents the probability that a radioactive atom will undergo spontaneous decay. The units of λ are reciprocal time (e.g. yr^{-1}).

Focus Box 11.1

Comparing Alpha, Beta, and Gamma Radiation

Gamma rays possess a very short wavelength (< 0.1 nm), so they are the most energetic form of electromagnetic radiation and can penetrate skin, bone, and other living tissues as well as thin sheets of metal, glass, and wood. By contrast, alpha particles can be deflected by a sheet of paper and will not penetrate skin, and only tend to be problematic to living organisms when they are inhaled or ingested. The high-energy beta particles (or rays) are more penetrative than alpha particles (they can penetrate skin, for example), but far less so than gamma rays. Like alpha particles, risk to living organisms from beta radiation is greatest when inhaled or ingested.

Table 11.1 Decay mechanisms, half-lives, and energies for numerous radionuclides that are the focus of research in some sphere of biogeochemistry.

Parent isotope	Decay mechanism	Half-life	E (MeV)	Daughter isotope
^{238}U	alpha	4.468×10^9 yr	4.27	^{234}Th
^{234}Th	beta	24.10 d	0.273	^{234}Pa
^{234}Pa	alpha	6.70 h	2.197	^{234}U
^{234}U	beta	245 500 yr	4.859	^{230}Th
^{230}Th	alpha	75 380 yr	4.77	^{226}Ra
^{226}Ra	alpha	1602 yr	4.871	^{222}Rn
^{222}Rn	alpha	3.8235 d	5.59	^{218}Po
^{218}Po	alpha 99.98%	3.10 min	6.115	^{214}Pb
	beta 0.02 %		0.265	^{218}At
^{218}At	alpha 99.90%	1.5 s	6.874	^{214}Bi
	beta 0.10 %		2.883	^{218}Rn
^{218}Rn	alpha	35 ms	7.263	^{214}Po
^{214}Pb	beta	26.8 min	1.024	^{214}Bi
^{214}Bi	beta 99.98%	19.9 min	3.272	^{214}Po
	alpha 0.02 %		5.617	^{210}Tl
^{214}Po	alpha	0.1643 ms	7.883	^{210}Pb
^{210}Tl	beta	1.30 min	5.484	^{210}Pb
^{210}Pb	beta	22.3 yr	0.064	^{210}Bi
^{210}Bi	beta 99.99987%	5.013 d	1.426	^{210}Po
	alpha 0.00013%		5.982	^{206}Tl
^{210}Po	alpha	138.376 d	5.407	^{206}Pb
^{206}Tl	beta	4.199 min	1.533	^{206}Pb
^{206}Pb	–	stable	–	–
^{232}Th	alpha	1.405×10^{10} yr	4.081	^{228}Ra
^{228}Ra	beta	5.75 yr	0.046	^{228}Ac
^{228}Ac	beta	6.25 h	2.124	^{228}Th
^{228}Th	alpha	1.9116 yr	5.52	^{224}Ra
^{224}Ra	alpha	3.6319 d	5.789	^{220}Rn
^{220}Rn	alpha	55.6 s	6.404	^{216}Po
^{216}Po	alpha	0.145 s	6.906	^{212}Pb
^{212}Pb	beta	10.64 h	0.57	^{212}Bi
^{212}Bi	beta 64.6%	60.55 min	2.252	^{212}Po
	alpha 35.4%		6.208	^{208}Tl
^{212}Po	alpha	299 ns	8.955	^{208}Pb
^{208}Tl	beta	3.053 min	4.999	^{208}Pb
^{208}Pb	–	stable	–	–

Some Plutonium and Uranium Fission Daughter Products

^{135}I		6.57 h		
^{135}Xe		9.14 h		
^{137}Cs		30.17 yr		
^{135}Cs		2.3×10^6 yr		
^{134}Cs		2.1 yr		
^{99}Tc		2.11×10^5 yr		
^{90}Sr		28.8 yr		

Rearranging Eq. (11.5):

$$-1/P \times dP = \lambda_P \times dt \qquad (11.6)$$

and integrating both sides gives:

$$-\ln(P) = \lambda_P \times t + C \qquad (11.7)$$

C is the constant of integration. If we consider this equation at time = 0, then the term $\lambda_P \times t = 0$, and $C = -\ln(P_0)$, where P_0 represents the amount of parent isotope at time $t = 0$. Substituting $-\ln(P_0)$ for C in equation 11.7:

$$-\lambda_P \times t = \ln(P) - \ln(P_0) \qquad (11.8)$$

Using the exponential function provides:

$$e^{-\lambda_P t} = P/P_0 \qquad (11.9)$$

and the amount of parent isotope remaining at time t is:

$$P = P_0 \times e^{-\lambda_P t} \qquad (11.10)$$

Focus Box 11.2

Exponential Decay Conceptually and Quantitatively

When minerals crystallize or organisms grow, they incorporate radioactive atoms into their structures. From this point forward, radioactive decay will deplete the amount of a radioactive isotope. After one half-life, 50% of the original parent isotope remains, and the amount of daughter isotope that has been produced will equal the amount of parent material (Figure 11.1) unless the daughter isotope is radioactive and also being depleted by radioactive decay. After two half-lives, 25% of the parent isotope remains, and after three half-lives, 12.5% remains, and so on. This type of exponential decay is described by the equation $P = P_0 \times e^{-\lambda_P \times t}$, where the decay constant λ is related to the half-life. Equation (11.10) is a classic exponential decay equation that applies to many fields of environmental science.

After one half-life, $P = 0.5 \times P_0$ (the amount of parent material is 50% of the original amount of parent isotope), and the relationship between half-life and decay constant is given by the following equations:

$$t_{1/2} = \ln(P_0/0.5 \times P_0) \times 1/\lambda = \ln 2/\lambda = 0.693/\lambda \quad (11.11)$$

The decay constant (λ) for ^{14}C to ^{14}N (by β^- decay) is 1.209×10^{-4} yr^{-1}, and the half-life is 5730 years. For ^{238}U, λ for the α decay to ^{234}Th is 1.551×10^{-10} yr^{-1}, and the half-life is 4.47×10^9 years (Faure 1986).

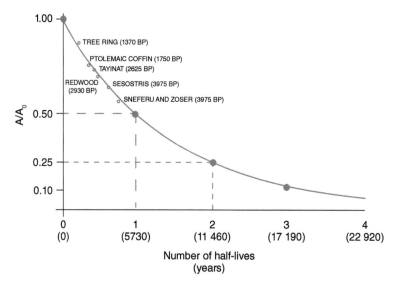

Fig. 11.1 Graphical representation of exponential decay, using ^{14}C as the example, showing half-lives and corresponding time elapsed on x-axis, and the proportion of the initial ^{14}C that remains plotted on the y-axis (A/A_0). Also shown are known ages that Arnold and Libby (1949) used when developing the radiocarbon dating method in the 1940s. All of their samples were wood of some sort, whether from an ancient palace in Syria (Tayinat), remains of a funerary boat (Sesostris), ancient tombs (Sneferu and Zoser), or trees.

Quantifying the radioactivity of a sample is typically expressed as the *activity* of the isotope in question, where activity (A) is defined as:

$$A = c \times \lambda \times N \qquad (11.12)$$

Activity measured by a Geiger counter or similar detector, and the factors that influence it are the decay constant λ, the number of radioactive atoms remaining in the sample (N), and a coefficient that accounts for the type of detector used in analysis (c).

The ratio A/A_0 expresses the activity of a given radioactive isotope in a given reservoir to the initial activity A_0 (e.g. $^{14}C/^{12}C$ of modern air fixed into plants). If $A/A_0 = 0.5$, the sample (of ancient wood, e.g.) has experienced one half-life, which in the case of ^{14}C is 5730 years (Figure 11.1).

11.1.3 Uranium and thorium decay series

The *radioactive decay series* that begins with ^{238}U (Figure 11.2) includes varied mechanisms and half-lives of radioactive decay. Recall that alpha decay results in the nucleus losing two protons and two neutrons, and that beta decay is gain of a proton and loss of a neutron. The first reaction in the long decay chain to ^{206}Pb is:

$$^{238}_{92}U \rightarrow\ ^{206}_{82}Pb + 8\,\alpha + 6\,\beta^- + E \qquad (11.13)$$

Observe the varied paths to arrive at ^{206}Pb from ^{238}U, and note that regardless of path, the gross quantity of radiation emitted is always eight alpha particles and six beta particles. The total energy emitted is 47.4 MeV per ^{238}U atom (Faure 1986). Half-lives and decay mechanisms of

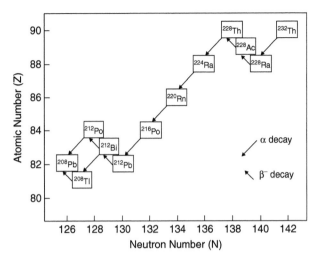

Fig. 11.3 Schematic representation of the decay series of ^{232}Th.

radionuclides in the ^{238}U, ^{235}U, and ^{232}Th series are presented in Table 11.1.

Two important radionuclides in environmental geochemistry in the ^{238}U decay series are ^{226}Ra (radium), a groundwater contaminant in some aquifers (both from natural mineral decay and from anthropogenic sources such as U mining) and ^{222}Rn, the isotope of radon that is a common indoor air contaminant in areas with ^{238}U-enriched bedrock. Granites and related rocks and soils are the most common source of ^{222}Rn.

The ^{232}Th decay series (Figure 11.3) is similar to that of ^{238}U with numerous alpha decay reactions, and the sum of the series from ^{232}Th to ^{208}Pb is:

$$^{232}_{90}Th \rightarrow\ ^{208}_{82}Pb + 6\,\alpha + 4\,\beta^- + E \qquad (11.14)$$

The total energy released is 39.8 MeV (Faure 1986). Like ^{226}Ra, ^{228}Ra is also a groundwater contaminant.

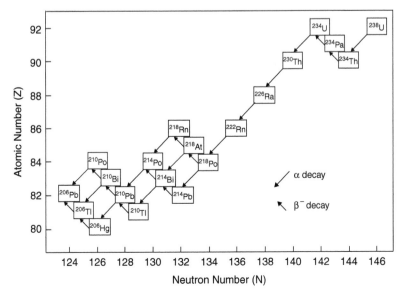

Fig. 11.2 Schematic representation of the decay series of ^{238}U.

The ^{238}U, ^{235}U, and ^{232}Th decay series each results in the formation of *stable isotopes of lead*, ^{206}Pb in the case of ^{238}U decay, ^{207}Pb in the case of ^{235}U decay, and ^{208}Pb in the case of ^{232}Th decay. Terminologically, these are stable, radiogenic isotopes of lead. Radioactive decay of uranium and thorium to lead over geological time helps to explain the relative abundance of lead in the crust of the earth relative to other heavy metals of similar atomic number (e.g. Bi, Hg, Pt). An example of Pb isotopes applied to deciphering the source of Pb in air pollution is presented immediately below in Section 11.2.

For an interesting and well-written history of the discovery of radioactive and stable isotopes, as well as a very thorough presentation of detailed decay mechanisms and the application of isotopes (both stable and radioactive) to geochemistry, the reader is referred to *Principles of Isotope Geology* by Gunter Faure (Faure 1986).

11.2 RADIONUCLIDE TRACERS IN ENVIRONMENTAL GEOCHEMISTRY

Many naturally occurring radioisotopes are applied in studies of earth surface systems and shallow crustal systems to measure age (e.g. dating of groundwater, rocks, or sediments) or as tracers of hydrological (e.g. groundwater flow paths), mineralogical, and biochemical processes. So too are certain anthropogenic isotopes produced in nuclear bomb blasts or accidents e.g. ^3H, ^{90}Sr, ^{131}I, ^{137}Cs. This section presents some radioactive isotopes (and isotope pairs) that serve as examples of the great breadth of isotopic analysis of geological-geochemical systems. This is followed by a section that examines selected radionuclides as contaminants in soils and groundwater.

11.2.1 $^{206}Pb/^{207}Pb$

Radiogenic lead isotopes, particularly the ratio $^{206}Pb/^{207}Pb$, are applied to determining sources of atmospheric lead contamination. Lead has four naturally occurring stable isotopes, three of which are produced by branched decay of uranium or thorium (Table 1.1; Figures 11.2 and 11.3):

$$^{232}Th \rightarrow {}^{208}Pb$$

$$^{235}U \rightarrow {}^{207}Pb$$

$$^{238}U \rightarrow {}^{206}Pb$$

The fourth, ^{204}Pb, is primordial and formed during nucleosynthesis.

In the late 1960s, Murozomi et al. (1969) determined that the concentration of lead in snow and ice in Greenland

had increased by a factor of 200 since preindustrial times, and Rosman et al. (1993) sought to determine the origin of the lead contamination through an analysis of lead isotope compositions. Previous work by geochemists such as Chow et al. (1975), Shirahata et al. (1980), and Sturges and Barrie (1987) had determined that $^{206}Pb/^{207}Pb$ ratios differ by region or country depending on the source of industrial lead, some of which was an additive in gasoline (as alkyl lead) that was a major source of atmospheric lead prior to the 1980s, and some of which is added to the atmosphere by ore refining and other industrial processes. $^{206}Pb/^{207}Pb$ ratios from sites suspected by Rosman et al. (1993) to be potential sources of lead in the Greenland environment are as follows: for Canada and Europe–Asia, $^{206}Pb/^{207}Pb =$ 1.15 ± 0.01; for the USA, $^{206}Pb/^{207}Pb$ ranged from 1.15 (pre-1970) to 1.21 (post-1970). The reason for the shift in US $^{206}Pb/^{207}Pb$ ratio is due to a switch to an industrial lead source (Mississippi Valley-type lead) that is more enriched in ^{206}Pb relative to other lead sources.

Results show that pre-1970 $^{206}Pb/^{207}Pb$ ratio values in Greenland snow and ice are 1.16 ± 0.1, but that the lead signature underwent a revealing increase in the $^{206}Pb/^{207}Pb$ ratio, culminating in an average value of 1.20 in 1976. Following the peak in $^{206}Pb/^{207}Pb$ in 1976, values progressively declined throughout the late 1970s and 1980s, reaching a value of 1.15 ± 0.01 in 1989 (Rosman et al. 1993). The increase in $^{206}Pb/^{207}Pb$ values in the late 1960s and early 1970s is attributed to the change in the isotopic composition of lead added to USA gasoline; the decrease in $^{206}Pb/^{207}Pb$ from 1976 through the late 1980s is attributed to the impact of the Clean Air Act in the USA which banned leaded gasoline. The 1989 value of 1.15 likely reflects the lower but relatively constant output from North America, Europe, and Asia.

11.2.2 $^{87}Sr/^{86}Sr$

Radioactive decay of ^{87}Rb produces ^{87}Sr, and the ratio $^{87}Sr/^{86}Sr$ (i.e. of the radiogenic isotope to a stable isotope of Sr) is a useful tracer of geochemical cycling, especially where Ca^{2+} is a focus. Sr^{2+} and Ca^{2+} have similar radii – Sr^{2+} (1.13 Å) is only 13% larger than Ca^{2+} (0.99 Å) – and both are divalent. Sr^{2+} substitutes for Ca^{2+} in calcium carbonate (more so in aragonite than in calcite), apatite, and plagioclase feldspar. It is a trace component of rocks and sediments, typically comprising $10–1000\ mg\ kg^{-1}$ (ppm), and has four naturally occurring isotopes, ^{88}Sr, ^{87}Sr, ^{86}Sr, and ^{84}Sr, which comprise 82.53%, 7.04%, 9.87%, and 0.56%, respectively, of all strontium. All four Sr isotopes are stable, but the amount

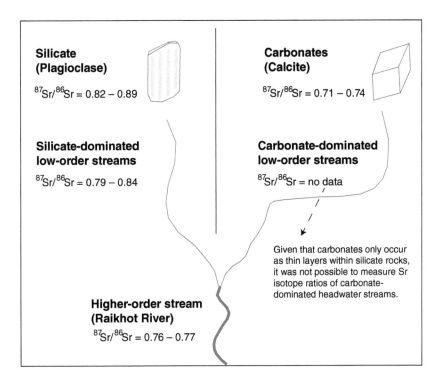

**Silicate
(Plagioclase)**

$^{87}Sr/^{86}Sr = 0.82 - 0.89$

**Carbonates
(Calcite)**

$^{87}Sr/^{86}Sr = 0.71 - 0.74$

**Silicate-dominated
low-order streams**

$^{87}Sr/^{86}Sr = 0.79 - 0.84$

**Carbonate-dominated
low-order streams**

$^{87}Sr/^{86}Sr = $ no data

Given that carbonates only occur
as thin layers within silicate rocks,
it was not possible to measure Sr
isotope ratios of carbonate-
dominated headwater streams.

**Higher-order stream
(Raikhot River)**

$^{87}Sr/^{86}Sr = 0.76 - 0.77$

Fig. 11.4   Schematic diagram of $^{87}Sr/^{86}Sr$ as tracer of chemical weathering showing (a) isotopic signature of source rocks, (b) streams at different positions in the watershed (after Blum et al. 1998).

of ^{87}Sr has increased slightly over geological time because it is radiogenic (derived from decay of ^{87}Rb).

Analysis of chemical weathering and soil water/groundwater flow paths is a good example of how $^{87}Sr/^{86}Sr$ measurements apply to environmental or geological analysis. Imagine a watershed where understanding nutrient flux is important, especially of Ca. Sr isotope analysis of plagioclase feldspar and calcite in soils and bedrock of the watershed reveals a difference in $^{87}Sr/^{86}Sr$ of plagioclase ($\sim$0.85) relative to calcite ($\sim$0.73), and little overlap (Figure 11.4). Assuming that the isotopic ratio of $^{87}Sr/^{86}Sr$ of the watershed is controlled by mineral weathering, the relative contribution of calcite weathering relative to plagioclase should be revealed by the $^{87}Sr/^{86}Sr$ ratio of soil water, shallow groundwater, or surface water.

Chemical weathering in the Himalayas was examined by strontium isotope analysis; differences between strontium isotope compositions of calcite in marble layers ($^{87}Sr/^{86}Sr = 0.71$–0.74) compared to silicate minerals ($^{87}Sr/^{86}Sr = 0.82$–0.89) were used to assess what type of weathering reactions were controlling surface water compositions (Figure 11.4). Results indicate that even in areas where silicate minerals in crystalline metamorphic and igneous rocks comprise $\geq$ 99% of the total mineral assemblage, the $^{87}Sr/^{86}Sr$ ratios of rivers draining watershed soils ($^{87}Sr/^{86}Sr$ $\sim$0.77) are closer to calcite than they are to plagioclase. This indicates that even where calcite is $\leq$1% of a hydrologic basin, weathering of calcite to Ca^{2+} and HCO_3^- should be considered as the dominant source of ecosystem-available calcium relative to plagioclase.

11.3 RADIONUCLIDES AS ENVIRONMENTAL CONTAMINANTS

The main controls on the fate and transport of ions and compounds in aqueous systems – e.g. pH, adsorption, coprecipitation, water : rock ratio, formation of polyatomic anions, ionic strength, residence time, complexation to organic matter, plant uptake, etc. – also apply to radionuclides. In addition to these controls, the mobility and behavior of radioactive and radiogenic species in aqueous systems is also controlled by properties and processes related to radioactive decay.

11.3.1 Controls on U, Th, and their decay products

Possible reservoirs of radionuclides associated with U and Th parent isotopes in environmental systems include:

1   in primary minerals (often as trace elements in calcite, zircon, apatite, and others);

2   fixed into secondary minerals, especially for inherently insoluble elements (e.g. any isotope of thorium);

3   fixed into organic matter – radionuclides may be taken up by organisms or soil organic matter, whether living or not;

4 sorbed to solids (minerals, humus), especially cations such as $_2^4\alpha^{2+}$, $^{226}Ra^{2+}$, and any isotope of Pb^{2+};

5 in aqueous solutions, especially the relatively soluble cations likely to exist in a dissolved state (e.g. $^{226}Ra^{2+}$, $^{90}Sr^{2+}$);

6 an inert gas (radon) with the potential to leave the aqueous system through exsolution (e.g. ^{222}Rn); and

7 a series of short-lived radionuclides with varied properties (e.g. ^{218}Po, ^{214}Pb, ^{214}Bi, ^{210}Pb) that are likely to sorb to particles and relatively rapidly decay to $\sim$ insoluble radiogenic lead.

Different isotopes of the same element may undergo decay at different rates, so some isotopes will persist longer, and some will decay at higher rates. Radium, for example, has three main isotopes in natural systems, ^{226}Ra (a member of the ^{238}U decay series), ^{228}Ra and ^{224}Ra (members of the ^{232}Th decay series), and ^{223}Ra (a member of the ^{235}U decay series). The varied half-lives of the radium isotopes indicate that the different isotopes will have widely varied residence times and radioactive toxicities. Atoms with short half-lives produce radioactive decay products (α, β, γ, and daughter isotopes) at greater rates than isotopes with longer half-lives, so species such as ^{224}Ra ($t_{1/2} = 3.6$ d) and ^{228}Ra ($t_{1/2} = 5.75$ y), both derived from ^{232}Th decay, will decay more rapidly and have greater radioactive toxicities than ^{238}U-derived ^{226}Ra ($t_{1/2} = 1602$ y), but they will also persist for shorter amounts of time. Given its divalent charge, Ra^{2+} behavior may be controlled by Ca^{2+} or Ba^{2+} (e.g. uptake into Ca or Ba minerals).

Uranium and thorium occur naturally in earth surface systems, and the radiogenic isotopes produced in the decay series of ^{238}U, ^{235}U, and ^{232}Th all can occur as natural constituents of soils, surface waters, and groundwater, in some cases at concentrations deemed unsafe for use as drinking water or in the production of agricultural crops. *Average concentrations of U and Th in continental crust are 2.7 and 9.6 mg kg^{-1} (ppm), respectively* (Wanty and Nordstrom 1993). Granites and shales tend to be elevated in U and Th relative to continental crustal averages, with uranium exhibiting average concentrations of 4.4 ppm (granite) and 3.8 ppm (shale), and thorium

exhibiting average concentrations of 16 ppm (granite) and 12 ppm (shale) (Wanty and Nordstrom 1993). Mafic and ultramafic rocks are depleted in U and Th (Table 1.1), with average uranium values of 0.75 ppm (basalt) and 0.002 ppm (ultramafic rock) and average thorium values of 3.5 ppm (basalt) and 0.005 ppm (ultramafics) (Turekian and Wedepohl 1961; Vinogradov 1962). So clearly, natural variability in rock, soil, or aquifer lithology will exert a strong control on the abundance of uranium, thorium, and their decay products in natural systems.

Beyond natural variability in the amounts of uranium and thorium in bedrock, soil, or sediment, the numerous controls on aqueous composition covered in Chapters 4 and 5 apply to any radionuclide in solution with the exception of radon gas (where the potential for the gas to escape from water or rock is an additional variable). Uranium in oxidizing environments is soluble as U^{6+} (as the UO_2^{2+} uranyl ion in various hydroxyl-based and carbonate-based polyatomic anions, e.g. $[UO_2]_2CO_3[OH]_3^-$, $UO_2[CO_3]_3^{4-}$), but the tetravalent U^{4+} (in reducing environments) and Th^{4+} (anywhere) are almost completely insoluble.

The very slow rate of decay of ^{238}U means that its toxicity is not so much derived from alpha decay (to ^{234}Th) as from the inherent toxicity of U as a trace metal, e.g. carcinogenic associations and potential to cause kidney disease. Spontaneous decay of uranium and thorium and their daughter products can result in elevated concentrations of radium, a relatively soluble component of many groundwater systems, as well as radon gas and alpha particles. These all are known to occur at elevated concentrations in many soils and groundwaters and as such are regulated by the WHO and US EPA.

Radium occurs as the Ra^{2+} cation and the two dominant isotopes in natural systems are ^{226}Ra and ^{228}Ra. The ^{224}Ra isotope is not abundant because its half-life so short (3.6 days) and the ^{223}Ra isotope is rare because the parent isotope at the head of the decay chain that produces ^{223}Ra, ^{235}U, is far less abundant than ^{238}U and ^{232}Th, the parents of ^{226}Ra and ^{228}Ra, respectively (Figures 11.2 and 11.3). Soluble Ra^{2+} behaves much like other alkali and alkaline earth metals in solution, particularly Ba^{2+}. Its relatively

---

Focus Box 11.3

Alpha Recoil and Radioactivity in Aqueous Solutions

Alpha recoil occurs when the high energy and spontaneity of alpha decay cause an alpha particle to be accelerated out from the nucleus. The high velocity of the alpha particle may enable it to be ejected from a mineral structure into aqueous solution, making alpha recoil (Figure 11.5) a consideration when studying radionuclide behavior in environmental systems.

Uranium in Surface Water and Groundwater

The relatively high solubility of uranium in natural waters produces concentrations of 2–4 µg l⁻¹ (ppb) in ocean water and, locally, > 1000 ppb in streams and groundwaters affected either by (i) uranium-rich soil, sediment, or bedrock or (ii) evaporative concentration in arid regions. In some areas, relatively low concentrations of uranium in aquifer sediments or bedrock (e.g. < 10 mg kg⁻¹ of uranium in bulk sediment or rock) still can produce elevated uranium in groundwater due to the effect of carbonate-enhanced uranium solubility (Kim et al. 2014). In areas with uranium-poor bedrock or soils, uranium might occur in trace amounts (e.g. ≤ 1 ppb) in streams and groundwaters, especially where the lack of carbonate or bicarbonate inhibits formation of soluble anions with uranyl. For reference, the US EPA maximum contaminant level (MCL) for uranium in public water supplies is 30 ppb, and the third edition of the World Health Organization (WHO) Guidelines for Drinking-water Quality (2004) indicated a provisional guideline value of 15 ppb.

Radionuclide Drinking Water Guidelines

WHO guidelines for radium in drinking water recommend maximum activities of 0.1 Bq l⁻¹ for ^{228}Ra and 1 Bq l⁻¹ for ^{226}Ra. The order of magnitude difference reflects the shorter half-life (and thus greater decay rate) of ^{228}Ra as compared to ^{226}Ra. The US EPA MCL of 5 pCi l⁻¹ (0.185 Bq l⁻¹) is for combined ^{228}Ra and ^{226}Ra and does not differentiate among the two isotopes. Alpha decay of ^{226}Ra produces ^{222}Rn, the isotope of radon that often occurs as an indoor air contaminant where buildings are constructed on soil or bedrock with elevated uranium. The WHO guideline for ^{222}Rn is 100 Bq m⁻³ (http://www.euro.who.int/document/aiq/8_3radon.pdf), while the US EPA recommended limit for ^{222}Rn in indoor air is 4 pCi l⁻¹ (where the radiation for WHO and EPA guidelines is measured per unit volume of air). Converting pCi to Bq and liters to m³, the US EPA value corresponds to 148 Bq m⁻³. Remediation for radon contamination is relatively simple and typically involves improved ventilation systems that replace Rn-enriched air with fresh air. The WHO guideline for gross alpha radiation in water is 0.5 Bq l⁻¹ while the EPA value is 15 pCi l⁻¹ (0.56 Bq l⁻¹). Guidelines for gross beta activity are 1 Bq l⁻¹ (WHO) and 4 millirems per year (US EPA).

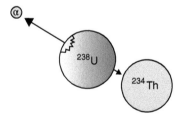

Fig. 11.5 Schematic illustration of alpha recoil occurring in response to alpha decay of the ^{238}U nucleus shown. The much greater mass of the radiogenic ^{234}Th nucleus means that it moves little when the decay takes place, compared to the low-mass alpha particle.

high solubility indicates that in some groundwater systems, Ra²⁺ will exhibit conservative behavior; however, in aquifers with dissolved barium (Ba²⁺) and sulfate (SO₄²⁻) at concentrations that approach saturation for barite (BaSO₄), chemical precipitation of barite will cause coprecipitation of Ra²⁺ into the barite structure, thus limiting

its mobility in the aquifer, as follows schematically:

$$999\,Ba^{2+}_{(aq)} + 1\,Ra^{2+}_{(aq)} + 1000\,SO_4^{2-}_{(aq)}$$
$$= 1000\,Ba_{0.999}Ra_{0.001}SO_{4(s)} \tag{11.15}$$

Also, aquifers with clay minerals or organic matter that possess high cation exchange capacities also can limit Ra²⁺ mobility by sorption.

^{226}Ra²⁺ decays to the inert gas ^{222}Rn⁰, by emission of an alpha particle:

$$^{226}Ra^{2+} \rightarrow\ ^{222}Rn^0 +\ _2^4\alpha^{2+} \tag{11.16}$$

When this decay reaction occurs, the behavior of the radionuclide shifts dramatically from a soluble divalent cation (Ra²⁺) to an inert gas that has the potential to exsolve from water into air; however, given the relatively short half-life of ^{222}Ra (3.8 days), this is only likely to happen in soils or in the shallow vadose zone. A given ^{222}Rn atom typically diffuses a few cm in water before

---

Focus Box 11.6

Radium in Groundwater of Southwest Jordan

High concentrations of naturally occurring radium occur in groundwater from Paleozoic Nubian sandstones that form the Disi aquifer in southwest Jordan (Vengosh et al. 2009). Dissolved Ra^{2+} is approximately an order of magnitude greater than typical standards for drinking water, and data suggest that release of Ra^{2+} into groundwater is primarily caused by alpha recoil from U and Th undergoing decay near crystal edges. In the shallow unconfined zone, average ^{226}Ra and ^{228}Ra activities are 0.53 and 0.91 $Bq\,l^{-1}$, respectively, whereas in the deep, confined aquifer, average ^{226}Ra and ^{228}Ra activities are 0.68 and 2.0 $Bq\,l^{-1}$, respectively. In the unconfined zone, $^{228}Ra/^{226}Ra$ activity ratios have an average value of 1.6 and are similar to the $^{228}Ra/^{226}Ra$ ratio measured in the host aquifer rocks

($\sim$1.6); however, in the deeper confined zone of this groundwater system, the $^{228}Ra/^{226}Ra$ ratio of groundwater (2.9) is higher than shallow groundwater and is also higher than host rock (1.6). Recall that ^{228}Ra is derived from ^{232}Th decay whereas ^{226}Ra is derived from ^{238}U. So the high $^{228}Ra/^{226}Ra$ ratio might indicate a higher Th/U ratio at depth, but another possible interpretation is that leaching loss of a soluble ^{238}U daughter product (e.g. ^{234}U) results in excess ^{232}Th over ^{230}Th (the direct parents of ^{228}Ra and ^{226}Ra, respectively), thus producing a higher $^{228}Ra/^{226}Ra$ ratio in groundwater. The absence of clay minerals in this coarse-grained, sandy aquifer system leads to very little sorption of Ra^{2+} to mineral surfaces, contributing to high radium concentrations in the groundwater.

Focus Box 11.7

Notes on Units of Radioactivity

1 curie (Ci) is equal to 3.7×10^{10} radioactive decays per second, and was originally defined based on the radioactivity produced by 1 g of ^{226}Ra. The unit applied to measuring radioactivity of water supplies in the United States is the picocurie per liter (pCi l^{-1}), where a pCi is equal to 3.7×10^{-2} radioactive decays per second (or the amount of radioactivity produced by 10^{-12} g (one picogram) of ^{226}Ra). The currently accepted SI unit of radioactivity is the becquerel, where one Bq is equal to 2.703×10^{-11} Ci, meaning

that 1 Bq = 27.03 pCi. A becquerel is also equal to 1 count per second. A millirem, the unit used by the US EPA to quantify beta radiation in drinking water, measures the radiation dose to tissue, attempting to factor in the different biological effects of different types of ionizing radiation. For this reason, it is difficult to apply a unit conversion for Bq (or pCi l^{-1}), which are direct measurements of radioactivity in water, to millirems. (http://www.lbl.gov/abc/wallchart/chapters/15/2.html)

---

decaying to ^{218}Po, so radon gas rarely will diffuse to the atmosphere from the saturated zone. In reality, the two main pathways by which radon can contaminate indoor air are (i) seepage from soil into buildings through cracks in foundations or old dirt cellars, or (ii) by pumping of radon-enriched groundwater from depth into a building, where ^{222}Rn then may enter air, especially when the water is used in showers (the misting action of showers increases the probability that Rn will exsolve into air). Following alpha decay of ^{222}Rn to ^{218}Po, relatively rapid decay of ^{218}Po and daughter products produces ^{206}Pb via branched decay series (Figure 11.2 and Table 11.1).

Alpha particles emitted by spontaneous decay of ^{238}U, ^{232}Th, ^{226}Ra, ^{222}Rn, and other radionuclides in groundwater are monitored (i) because they can be used to indicate the presence of radioactive elements in water supplies, and (ii) because they are a form of ionizing radiation

that is harmful to humans. Because of the high charge to small radius of alpha particles, they tend to strongly adsorb to solids.

11.3.2 Uranium ores, refining, and nuclear wastes

Uranium is mined from ore deposits that typically contain 0.1–1% uranium and are most commonly hosted in granitic rocks or diagenetically altered sandstones. In U-rich sandstone deposits, elevated concentrations of uranium occur at the boundary of oxidized and reduced sediment where oxidized fluids flow into reduced zones (Figure 11.6). These deposits (known as roll-front deposits) form when oxidized groundwater containing dissolved uranium carbonate complexes flow into reduced sediments (often caused by presence of organic matter, perhaps

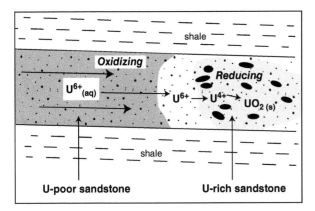

Fig. 11.6 Schematic sketch of how flow from oxidizing to reducing conditions in sandstone can lead to formation of a uranium roll-front deposit. The U^{6+} is likely part of a polyatomic anion with carbonate and or hydroxide. Arrows indicate direction of paleo-groundwater flow; black ellipses represent organic matter (e.g. decaying plant remains) within a reducing zone. Similar processes cause U and many trace metals to precipitate in wetlands.

a swamp buried in a sedimentary sequence), where soluble U^{6+} is reduced to insoluble U^{4+}. Reduction to U^{4+} results in precipitation of uranium into oxide minerals such as uraninite and pitchblende (both approximately UO_2) and the phosphate mineral autunite ($Ca[UO_2]_2[PO_4]_2 \cdot xH_2O$).

Only approximately 0.72% of naturally occurring uranium is the ^{235}U isotope (> 99% of natural uranium is ^{238}U), so it is necessary to enrich uranium ore so that ^{235}U concentrations reach 2–4% for nuclear power and to > 80% ^{235}U for weapons (although "dirty" nuclear bombs can be made with as little as 20% ^{235}U). Centrifuges are used to do this.

When ^{235}U undergoes fission in a nuclear reactor or explosion it forms two approximately equal-sized daughter products with atomic masses in the ranges of 90–100 and 130–145, examples of which include ^{90}Sr, ^{99}Mo, ^{99}Tc, ^{131}I, ^{135}Cs, ^{137}Cs, and ^{141}Ba (Table 11.1). Fission of ^{235}U also produces lighter isotopes such as ^{14}C and ^{3}H (tritium) plus neutrons and very large amounts of energy. Tritium atoms often combine with oxygen to form tritiated water that can escape into local air and water bodies surrounding nuclear power plants. Tritium has a half-life of 12.3 years and decays to ^{3}He by low-energy β decay – the low-energy β radiation lacks the energy needed to penetrate skin (as does high-energy β radiation) but tritium can be problematic when ingested or inhaled.

The fission of ^{235}U is initiated by neutron bombardment that first transforms ^{235}U into the highly unstable ^{236}U isotope, which undergoes spontaneous decay to many combinations of daughter products. The example

presented below shows the formation of the daughter products ^{90}Sr and ^{144}Xe plus two neutrons and energy:

$$^{235}_{92}U + {}^{1}_{0}n \rightarrow {}^{236}_{92}U \rightarrow {}^{90}_{38}Sr + {}^{144}_{54}Xe + 2{}^{1}_{0}n + E \quad (11.17)$$

Another example of a ^{235}U fission reaction, depicted to show the varied possibilities of fission products, is:

$$^{235}_{92}U + {}^{1}_{0}n \rightarrow {}^{236}_{92}U \rightarrow {}^{93}_{37}Rb + {}^{140}_{55}Cs + 3{}^{1}_{0}n + E \quad (11.18)$$

The liberation of neutrons by ^{236}U fission leads to a chain reaction in which fission-produced neutrons bombard other nearby ^{235}U atoms, which in turn transform to ^{236}U atoms, which then undergo fission to produce daughter elements and more neutrons (and much energy), and so on. In nuclear power plants, the fission reactions are controlled. Nuclear bomb explosions are uncontrolled fission reactions that proceed at a very high rate, releasing a tremendous amount of heat and radioactive daughter products nearly instantaneously.

The energy produced by nuclear fission is liberated in the form of heat that – as with coal combustion – converts liquid water to steam, which in turn runs turbines that produce electricity. In this way, the main difference between nuclear power and more conventional carbon-based energy sources is the origin of the heat – in the case of carbon-based fuel, the heat is produced by combustion.

After one to three years, ^{235}U-enriched fuel becomes unsuitable for energy production. This is because the accumulation of daughter products in the fuel rods adsorbs and deflects neutrons, thus decreasing the amount of energy that can be produced from a given rod. At this point, the rods are removed from the nuclear reactor and either stored or disposed of in a waste repository. However, the spent fuel is highly radioactive and produces high amounts of heat energy, so any disposal (or storage) options must prevent leakage of radioactive isotopes and their decay products until the harmful isotopes have been depleted by radioactive decay. Storage or disposal must also be able to account for heat produced by decay reactions within the spent fuel. In some cases, spent fuel can be reprocessed and used again as fuel in a nuclear reactor – this is the case in many countries but not the United States.

With half-lives in the range of 1–30 years, the relatively short-lived daughter products of fission reactions (e.g. ^{90}Sr, ^{99}Tc, ^{106}Ru, ^{131}I, ^{134}Cs, and ^{137}Cs) cause the greatest amount of damage to humans and ecosystems when released into the environment. These short-lived isotopes are the main elements of concern in *high-level radioactive waste*. The two main reasons for their high toxicity and potential to cause damage are:

1 their decay rates are orders of magnitude greater than longer-lived isotopes such as ^{99}Tc, ^{233}U, ^{237}Np, ^{238}U, and ^{239}Pu, which contain half-lives in the range of 210 000 years (^{99}Tc) and 4.5 billion years (^{238}U); and,

2 they persist for far longer than some of the very short-lived daughter products such as ^{89}Sr, ^{133}Xe, and ^{103}Ru (while highly radioactive, their half-lives are $\leq$ 50 days). These highly unstable, very short-lived daughter isotopes decay so rapidly that they are very quickly (e.g. within one year) transformed into less hazardous isotopes, and really only pose a problem in bomb explosions or during uncontrolled fission and melting of radioactive fuel at a power plant (often referred to as a meltdown).

The fate and transport of the varied isotopes in high-level radioactive wastes are controlled by decay rates and mechanisms as well as by typical controls on aqueous species (pH, inherent solubility of the species, sorption, alkalinity, dissolved oxygen, etc.). Given their abundance in spent fuel and half-lives of 28 and 30 years, respectively, ^{90}Sr and ^{137}Cs are the main source of radioactivity in wastes that are less than 100 years old. Wastes older than 100 years will be depleted in ^{90}Sr and ^{137}Cs (by decay to less-radioactive daughter isotopes) relative to some of the more long-lived isotopes, particularly ^{237}Np, ^{239}Pu, ^{240}Pu, ^{241}Am, and ^{243}Am (Langmuir 1997). These isotopes of Np, Pu, and Am will then dominate radioactive decay reactions in the spent fuel for approximately 100 000 years, after which isotopes of lead and radium (e.g. ^{210}Pb, ^{226}Ra) become the main sources of radioactivity. By this time, emitted radiation would be close to background levels.

11.3.3 High-level radioactive wastes: geological disposal and considerations

According to the International Atomic Energy Agency (IAEA 2019), the 449 currently operating nuclear power reactors on earth provide 11% of global electricity demand and generate 10 000–20 000 metric tons of spent fuel per year, a quantity that would fill a football field to a depth of 1–2 m. While a relatively small volume ($\sim$10 000 m^3), environmental and security risks dictate that appropriate disposal is crucial. Currently, most wastes are being stored temporarily on site at nuclear plants in pools of water, yet there is a growing consensus that deep underground storage is the best long-term solution.

Finland, France, and Sweden are in the process of creating geological repositories for the long-term disposal of spent nuclear fuel and high-level waste. Sweden appears to be the most advanced in this respect, featuring a site 150 km north of Stockholm near the small city of

Östhammar. It is in operation, capable of accepting 600 metric tons per year for decades to come. The radioactive waste deposited in the repository known as SFR is low-and medium-level waste, meaning that unlike spent nuclear fuel, it need not be cooled and isotopes are relatively short-lived. The repository is situated 50 m below the bottom of the Baltic Sea in crystalline metamorphic rock with km-long tunnels that provide access from land. Other rock types being considered in Europe for deep repositories are thick clay deposits (Belgium, France, Germany, Spain, and Switzerland) and salt deposits (Germany).

While many actinides occur in only one oxidation state in natural systems (e.g. Am^{3+}, Cm^{3+}, Th^{4+}), Np and Pu can exist in several oxidation states (and recall that uranium can occur as U^{4+} and U^{6+}). In oxidizing groundwaters, neptunium mainly occurs as Np^{5+} in the soluble and mobile species NpO$_2$$^+$ below a pH of 10 (Langmuir 1997). The low ratio of ionic charge to radius of NpO$_2$$^+$ indicates that it will not be strongly adsorbed to mineral surfaces. In very strongly oxidizing water, Np can occur as Np^{6+} in the less soluble NpO$_2$$^{2+}$, but this species is far less common than NpO$_2$$^+$.

Plutonium occurs in three common oxidation states in soils and groundwaters, Pu^{4+}, Pu^{5+}, and Pu^{6+} – there is a trivalent ion (Pu^{3+}), but it only occurs in extremely reducing or highly acidic environments. PuO$_2$$^+$ is the dominant soluble plutonium ion in most natural waters. The pentavalent cation XO$_2$$^+$ (where X is Np or Pu) comprises approximately 95% of soluble neptunium and plutonium species dissolved in nonalkaline waters where carbonate concentrations are $\leq 10^{-4}$ M (< 10 mg l^{-1}). However, they differ in behavior because the solubility of plutonium is commonly limited to 10^{-8} to 10^{-6} M (0.2–20 μg l^{-1} or ppb) by reduction of Pu^{5+} to Pu^{4+} (a reaction which can occur in relatively oxidizing, slightly alkaline groundwater), and this facilitates formation of the insoluble species Pu(OH)$_4$ (Figure 11.7). Neptunium, however, cannot be reduced to a tetravalent ion and thus does not form an insoluble hydroxide, so its concentration in solution (as NpO$_2$$^+$) can be as high as 10^{-4} M, or $\sim$20 mg l^{-1} (Choppin 2007).

Accordingly, plutonium is relatively insoluble compared to most actinides (with the exception of the notoriously insoluble Th^{4+}). Nonetheless, formation of colloidal plutonium Pu(OH)$_4$ can enable transport of plutonium tens to thousands of meters in groundwater (McCarthy and Zachara 1989), illustrating the complex relationships between aqueous phases, solid phases, and the potential to transport extremely fine-grained solids (i.e. colloids with diameters of 10^{-8} to 10^{-6} m) in suspension. In this case, although insoluble, the particles of Pu(OH)$_4$ are examples

---

Focus Box 11.8

Fission Products Released from 2011 Fukushima Disaster

In the wake of the 2011 earthquake/tsunami-caused disaster at Fukushima Daiichi in Japan, ^{131}I and ^{137}Cs were the main short-lived radionuclides released to the air and sea. They are the most highly volatile of the fission products, and because Fukushima was not affected by a major fire (as was the case at Chernobyl), most other less-volatile radionuclides were contained on site. The Fukushima disaster released much less radioactivity compared to Chernobyl, yet emissions of ^{131}I and ^{134}Cs and

^{137}Cs are still of serious concern because they are easily taken up by humans and other organisms (e.g. ^{131}I tends to concentrate in the thyroid where it undergoes decay and can cause cancer). Research continues into land and ocean cycling, including transport of short half-life ^{134}Cs and ^{137}Cs ($t_{1/2} = 2.06$ and 30.17 years, respectively) across the Pacific Ocean. These long-transport isotopes do not appear to be present at levels that would cause adverse health effects.

Focus Box 11.9

The Example of the Yucca Mountain Repository Site

In 2008, the US Department of Energy (DOE) applied to the US Nuclear Regulatory Commission (NRC) for authorization to construct a deep geologic repository in Miocene-age volcanic tuffs at Yucca Mountain, Nevada, to host high-level radioactive waste from sites across the United States (NRC 2019, available at http://www.nrc.gov/waste/hlw-disposal.html). The site is geologically and hydrologically favorable. The arid location contributes to a very deep water table (500–750 m below land surface, Montazer and Wilson 1984) which should help to avoid water-based corrosion of storage containers. The abundance of smectite clays and zeolites with high cation exchange capacities (Bish et al. 2003) in the tuffs will likely retard migration of cations such as Cs^+, $Pu^{4+,5+,6+}$, and Sr^{2+}. However, critics point out that fractures in the tuff may permit water to penetrate downward toward the repository, potentially leading to interaction with the waste, although this point is countered by the expectation that infiltrating water is likely to be driven away from the wastes by vaporization caused by the heat of the wastes themselves (Langmuir 1997).

Corrosion of waste-bearing containers within the first 100 years of storage of spent fuel could potentially release

soluble $^{90}Sr^{2+}$ and $^{137}Cs^+$. The more likely scenario is that corrosion would occur after ^{90}Sr and ^{137}Cs have decayed to insignificant amounts, the point at which the actinides (mainly Np, Pu, and Am) plus ^{99}Tc and ^{129}I will be the dominant radioactive species in the spent fuel. Many of these isotopes exist as soluble anions that are less likely to sorb to smectites and zeolites in the volcanic tuffs than would be cationic species (e.g. Cs^+, Sr^{2+}). For this reason ^{99}Tc, ^{129}I, ^{233}U, ^{234}U, and ^{233}Np are assumed to be soluble and the most likely to contaminate groundwater or reach the surface before the approximately 100 000 years required to decay to background radioactivity levels. Examples of the soluble species of these isotopes include TcO_4^-, I^-, the various polyatomic anions formed by uranium with hydroxide and carbonate anions, and $Am(CO_3)_2^-$. Given the alkaline character of the groundwater in the arid Yucca Mountain region, these radionuclides are likely to form OH^- and CO_3^{2-} compounds. As of 2019, regional opposition to the project and lack of federal funding has put a halt to any progress at the Yucca site.

---

of solids that are so fine-grained that geochemists must consider their potential for transport as a colloidal phase, particularly in neutral to alkaline conditions that favor formation of $Pu(OH)_4$.

11.4 GEOCHRONOLOGY

Shortly following the discovery of radioactivity by Henri Becquerel in 1896, and amid the rapid advances in understanding of radioactivity by Pierre and Marie Curie, Ernest

Rutherford and others at the turn of the century, the decay of ^{238}U to ^{206}Pb was first applied to geological dating. The first published paper, by British geologist Arthur Holmes, provided the first geochemically quantitative determination of the age (370 ma) of a rock, in this case a granite in Norway (Holmes 1911). Since then, the U–Pb method has been refined and many others have been developed, including other decay pairs or chains, radiocarbon dating, and dating of surficial earth features by cosmogenic radionuclides. The following section

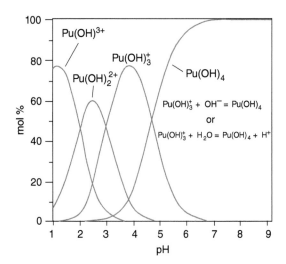

Fig. 11.7 Speciation of Pu^{4+} as a function of pH in mildly oxidized waters, a redox state that is typical of shallow ground water. Note that Pu solubility is pH-dependent, where decreasing pH results in transformation of the insoluble $Pu(OH)_4$ to relatively soluble $Pu(OH)_3^+$ and other cationic plutonium hydroxides. Modified after Choppin (2007).

presents some of the main principles, methods, and some selected applications of radionuclides applied to dating of geological and biological materials.

11.4.1 ^{14}C, cosmogenic radionuclides, and earth-surface dating techniques

Cosmogenic radionuclides are produced in the atmosphere by nuclear reactions initiated by bombardment of stable atoms by cosmic rays, and also at the surface of the Earth in rocks, soils, and sediments that are exposed to cosmic rays. These two modes of formation mean that cosmogenic isotopes can be used to date (i) deposition of sediments in lakes, ponds, or other surficial environments; and (ii) exposure ages of boulders, bedrock surfaces, sediments, and other materials that can determine when rocks or sediments where first exposed to cosmic ray bombardment.

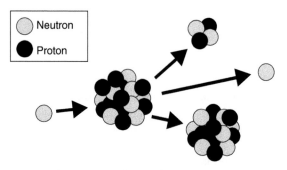

Fig. 11.8 Schematic sketch of spallation reaction where a neutron crashes into a nucleus of an atom (^{40}Ca, e.g.), yielding a smaller atom (e.g. the cosmogenic radionuclide ^{36}Cl), an alpha particle (2P, 2N), and a neutron.

When cosmogenic radionuclides which form in the atmosphere fall to the surface in precipitation, they may be incorporated into aquifers, snow, glacial ice, or sediments, enabling dating of materials as varied as groundwater and deep-ocean water to layers in glacial ice and sediments. Cosmic ray bombardment of rock or sediment exposed at the Earth's surface forms cosmogenic isotopes that can date the amount of time elapsed since exposure of the rock or sediment at the surface, allowing dating of surface processes such as landslides, river erosion, fault activity, glacial ice retreat, and cooling/crystallization of lava flows.

Cosmogenic isotopes mainly form by nuclear *spallation*, the ejection of protons or neutrons from the nucleus (Figure 11.8). It is a reaction at the level of the nucleus, one triggered by bombardment of atoms in the atmosphere by high-energy, fast-moving cosmic rays, neutrons, or protons. The cosmogenic radionuclides form when the nucleus of some parent atom (Table 11.2) collides with neutrons or protons in the atmosphere or at the earth surface. The reactions in the "Origin" column of Table 11.2 are examples of spallation reactions, as are Eqs. (11.19) and (11.22). Some researchers abbreviate a reaction like that which forms ^{26}Al as $^{28}Si(n,n2p)^{26}Al$. Although not shown, some cosmogenic radionuclides form by spallation of a parent argon atom, e.g. $^{40}Ar(p,n\alpha)^{36}Cl$.

Table 11.2 Common cosmogenic radionuclides used in earth science and their half-lives, spallation origins, and potential host minerals for the nuclide.

Nuclide	Half-life	Origin	Host Mineral
^3H (tritium)	12.3 yr	$^{14}N + _0^1n = {}^{12}C + _1^3H$	–
^3He	– (stable)	O, Si spallation	olivine, pyroxene
^{10}Be	1.38×10^6 yr	O, Si spallation	quartz
^{14}C	5730 yr	$^{16}O + _0^1n = {}^{14}C + _0^1n + 2\,_1^1p$	quartz
^{21}Ne	– (stable)	Mg, Ca, Fe spallation	olivine, plagioclase
^{26}Al	0.717×10^6 yr	$^{28}Si + _0^1n = {}^{26}Al + + _0^1n + 2\,_1^1p$	quartz
^{36}Cl	0.301×10^6 yr	^{40}Ca spallation (Eq. (11.22))	plagioclase, calcite

---

Focus Box 11.10

^{14}C Decay: Parent, Daughter, and Measuring ^{14}C Activity

Radiocarbon dating, unlike most radioactive dating methods, does not involve measuring the ratio of parent isotope (^{14}C) to daughter isotope (^{14}N). The abundance of ^{14}N in the atmosphere and at the earth surface and shallow subsurface means that it is effectively impossible to distinguish between ^{14}N derived from ^{14}C and the large reservoir of ^{14}N in the environment (i.e. in atmosphere as N_2, in soils as component of organic matter, in water as NO_3^-, etc). So, for this reason, age measurements are made by counting the decay of individual C atoms by e.g. gas proportional counting (de Vries and Barendsen 1953) or liquid scintillation counting (Noakes et al. 1967), or, more recently, by the advent of accelerator mass spectrometry (AMS). The older methods are still appropriate for dating samples with either high C content or young ages, but the advantage of AMS is that the ratio of ^{14}C : ^{12}C atoms can be directly measured. AMS facilitates dating of materials previously not possible, including small samples that contain only a few mg of C (e.g. lake sediments with small amounts of organic matter or shells with small amounts of incorporated organic carbon), or older samples with ages > 30 000 years (i.e. where small amounts of remaining ^{14}C make it difficult to measure precisely).

11.4.1.1 ^{14}C (radiocarbon)

Radiocarbon dating was developed by Willard Libby and University of Chicago colleagues in the late 1940s, in part theoretically and in part empirically (Figure 11.1). Radiocarbon dating was the first cosmogenic method to be developed and it is now widely applied in geology and allied fields like archeology and biology. For his pioneering work, Libby was awarded the Nobel Prize in chemistry in 1960. Since its inception, radiocarbon dating has become one of the most commonly applied methods for dating carbon-bearing materials with ages < 50 000 years.

^{14}C is a cosmogenic radionuclide because it forms as the result of the interaction of cosmic rays and their products with nitrogen in the atmosphere. When cosmic rays (mainly originating from the sun) collide with atoms of atmospheric gases, one of the consequences is that neutrons or protons can be ejected from the nucleus. The reaction of atmospheric nitrogen (nitrogen atoms in N_2 gas) to ^{14}C occurs according to this reaction:

$$^{14}N + {_0^1}n \rightarrow {^{14}}C + {^1}H + E \qquad (11.19)$$

The ^{14}C generated in the atmosphere very rapidly reacts with O_2 to produce $^{14}CO_2$, and atmospheric circulation causes rapid dispersion and mixing of $^{14}CO_2$, so there is little or no variability in concentration of ^{14}C as a function of latitude. The $^{14}CO_2$ can be incorporated into plants by photosynthesis or it can combine with H_2O to form $H_2{}^{14}CO_3$ delivered to the earth surface in precipitation (just like "normal" CO_2 with ^{12}C), from where it can be incorporated into living organisms, minerals, or water bodies. The ratio of $^{14}C/{^{12}}C$ in CO_2 or H_2CO_3 is governed mainly by the atmospheric ratio of these two isotopes.

When an organism dies, photosynthesis stops and the ratio of ^{14}C : ^{12}C progressively decreases as ^{14}C undergoes radioactive decay. The most common material dated by radiocarbon dating is organic matter (especially wood, seeds, pine cones and other readily identifiable materials). Considering that atmospheric CO_2 can also dissolve in water, radiocarbon dating is sometimes useful in groundwater studies where the ^{14}C : ^{12}C ratio, initially in contact with air, begins to decrease once waters have flowed deep enough to be isolated from exchange with the atmosphere or soil organic matter. Groundwater age analysis contributes to understanding recharge rates and sustainable use of groundwater resources.

^{14}C decays to ^{14}N according to the following spontaneous beta decay reaction:

$$^{14}C \rightarrow {^{14}}N + \beta^- + \underline{\nu} + E \qquad (11.20)$$

where, as above, $\underline{\nu}$ is an antineutrino and E represents energy released during the decay reaction.

The decrease in the amount of ^{14}C with time is given as:

$$[{^{14}}C]_t = [{^{14}}C]_0 \times e^{-\lambda_P \times t} \qquad (11.21)$$

This is a more specific form of the exponential decay equation where $\lambda = 1.209 \times 10^{-4}$ y^{-1}, corresponding to $t_{1/2} = 5730$ y. This relatively short half-life is the reason that materials older than ~50 000 y cannot be dated by the radiocarbon method – after 10 half-lives (57 300 y), only about 0.1% of the original ^{14}C remains in the material to be dated. However, the positive attribute of the short $t_{1/2}$ is that it enables relatively precise dating of geologically young materials.

A few important premises and recognition of potential sources of error in radiocarbon age analysis are:

1 Any variability in the rate of neutron production by cosmic radiation will alter the rate of ^{14}C production, and these changes must be accounted for or incorporated into analytical uncertainty. Examples of factors that tend to increase the rate of neutron flux, and hence ^{14}C production, include increases in sun spot activity and solar wind, and decreases in the magnetic moment of Earth's magnetic field. The opposites tend to decrease neutron flux. Greater rates of ^{14}C production from some time in the past compared to the modern rate would yield relatively high ^{14}C concentrations compared to the sample's actual age, thus likely producing anomalously young ages, and vice versa.

2 Radiocarbon dating assumes an atmosphere that is spatially well-mixed (homogeneous) with respect to CO_2 concentration. This is a good assumption (and one that does not hold true for non-gaseous cosmogenic radionuclides).

3 Changes in atmospheric CO_2 that have occurred during the late Pleistocene and Holocene are quantifiable and can be applied to calibration of ^{14}C ages. A cooler Earth means cooler oceans and greater oceanic uptake of CO_2; this decreases atmospheric CO_2, which in turn could increase the concentration of ^{14}C in atmospheric CO_2 (because the N parent isotope is not changing). Variations in atmospheric ^{14}C concentration, including increases of approximately 2% during stages of the Little Ice Age (~1300–1850 C.E.) are known as the *de Vries effect*, after deVries et al. (1958).

4 Lastly, once the material to be dated has been removed from the carbon source (e.g. an organism dies and no longer incorporates CO_2), it should be a closed system with respect to ^{14}C. There should be no exchange of ^{14}C with its surroundings, because any gains or losses of ^{14}C by a pine needle or a fragment of charcoal in a buried soil will bias the age determination.

Specimens that contain inorganic carbon that has formed by chemical or biological precipitation can also be dated by the ^{14}C method. In general, the carbon in these materials is contained in the minerals calcite or aragonite, polymorphs of $CaCO_3$. The inorganic carbon incorporated in these instances consists of dissolved CO_2 or carbonic acid (CO_2 dissolved in water), bicarbonate (HCO_3^-) and carbonate (CO_3^{2-}) in seawater, soil water, groundwater, or fresh surface water (e.g. lakes or ponds), and occurs in shells of organisms (e.g. mollusks) as calcium carbonate or as chemically precipitated carbonate minerals in soil water, groundwater, or sea water. Although inorganic, the carbonate in these materials is datable if it

was in equilibrium with ^{14}C from the atmosphere. This assumption is often where analytical errors arise. For example, contributions of inorganic carbon from weathered limestone could contain bicarbonate with no ^{14}C, and this would dilute the initial concentration of ^{14}C in the shell when it formed, producing a raw radiocarbon date that is anomalously old. This is known as the *reservoir effect*, and the inorganic carbon source with no ^{14}C is often known as "dead carbon."

The organic carbon in petroleum also contains no ^{14}C (the petroleum is far too old to contain ^{14}C dating back to the initial burial of organic matter), so the vast quantities of CO_2 from fossil-fuel combustion have introduced large quantities of dead carbon to the atmosphere since the 1800s, lowering atmospheric ^{14}C by about 2%. This phenomenon is known as the "Suess effect" (Suess 1955). Age bias can also be introduced by slow exchange of CO_2 between atmosphere and the waters from which shells or carbonate minerals form, particularly those from deep marine or deep groundwater regimes, and by isotope fractionation, a process whereby an organism or mineral may exhibit preference for one isotope of an element over another. For example, plants tend to preferentially incorporate ^{12}C over ^{13}C and ^{14}C, and this type of fractionation must be accounted for in age determinations.

Figure 11.9 illustrates the relationship between raw ages and radiocarbon years (both adjusted to pre-1950 scale) corrected for factors such as atmospheric variability and the de Vries effect. Calibration of radiocarbon ages includes analysis of historical materials of known age (e.g. 10 000-year old trees, wood from Egyptian tombs) as well as comparison to coral records (coral produces annual growth rings) and coral ages determined by $^{230}Th/^{234}U$ and ^{14}C. For detailed information on calibration of radiocarbon dates the reader is referred to Stuiver et al. (1998), Hughen et al. (2004), and Reimer et al. (2004).

11.4.1.2 Cosmogenic ^{10}Be, ^{26}Al, and ^{36}Cl

A search for information on cosmogenic radionuclides will likely lead you to ^{10}Be, ^{26}Al, ^{36}Cl, and other isotopes (Table 11.2) that provide approaches to dating samples without dateable organic matter or other ^{14}C possibilities. Each of the three isotopes highlighted in this section are produced in a manner similar to ^{14}C in the atmosphere, but in many ways they are fundamentally different than ^{14}C.

1 Their half-lives are two to three orders of magnitude greater than ^{14}C (^{10}Be is 1.5×10^6 yr; ^{26}Al is 7.16×10^5 yr; ^{36}Cl is 3.1×10^5 yr), and thus they are particularly well

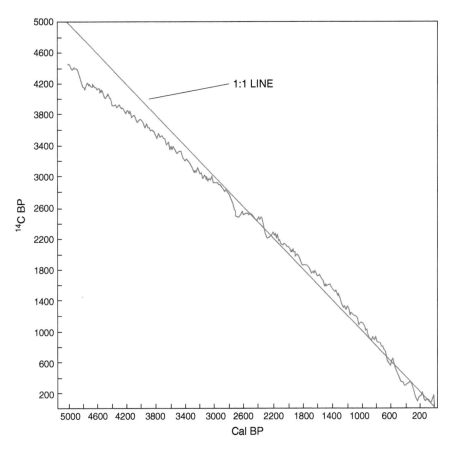

Fig. 11.9 Calibrated radiocarbon ages (x-axis) plotted against raw radiocarbon ages (ages here were calibrated against tree ring and marine coral records). The 1:1 line is presented for reference to illustrate deviations, and the axes are relative to 1950, by convention (see text for explanation). Compiled from data in Stuiver et al. 1998.

suited for dating materials that are one to two orders of magnitude older than those dateable by ^{14}C. This provides mechanisms for dating Earth surface and shallow crustal materials with ages on the order of 10^5 to 10^6 years.

2 Unlike ^{14}C, they are not partitioned into a gas phase and thus they are not well mixed in the atmosphere. This complicates matters, requiring assumptions and corrections for production and deposition rates.

3 ^{10}Be, ^{26}Al, and ^{36}Cl occur in inorganic materials such as quartz and feldspar, and this highlights their difference relative to ^{14}C, which requires (usually) dateable organic

matter. I wonder how many geomorphologists, ecologists, and geochemists have said, "let's hope these sediments contain pine cones or sticks so we can get a radiocarbon date."

4 ^{10}Be, ^{26}Al, and ^{36}Cl can date inorganic materials that lack carbon needed for ^{14}C, and given the paucity or absence of organic matter or datable inorganic carbon in many sediments, waters, and rocks, this is an important advantage over ^{14}C.

A representative spallation reaction is the formation of ^{36}Cl by collision of a proton with a ^{40}Ca atom (in

---

Focus Box 11.11

Reporting Radiocarbon Ages

Raw radiocarbon ages are commonly reported relative to years before present (B.P.), where the present is defined by convention as the year 1950 CE. If a pine cone retrieved from lake sediments in the year 2010 contained a ^{14}C content that yielded a date of 4070 years old (before 2010), its age would be adjusted by 60 years (2010–1950) to 4010 B.P. N-cal (to make the date relative to 1950 and indicate that the reported age is a calibrated one). Variability in atmospheric ^{14}C through the Holocene and Pleistocene caused

by the mechanisms described in Section 11.4.1.1 – plus potential sources of dead carbon – makes it necessary to calibrate raw radiocarbon ages. Thus, when interpreting radiocarbon dates, be aware that raw dates differ from (and typically are older than) calibrated dates, in some cases by up to 10% or more (Figure 11.9). Some researchers report results relative to CE/BCE reference, typically as calibrated ages.

particular, with the nucleus of the atom).

$$^{40}_{20}\text{Ca} + ^{1}_{0}\text{n} \rightarrow ^{36}_{17}\text{Cl} + 2^{1}_{0}\text{n} + 3^{1}_{1}\text{p} \qquad (11.22)$$

In addition to the cosmogenic ^{36}Cl atom, products also include two neutrons and three protons, although the nuclear detritus could also be one neutron and three alpha particles, i.e. $^{40}\text{Ca}(n,n\alpha)^{36}\text{Cl}$, depending on how the nucleus of the Ca atom breaks apart when struck by the neutron (Figure 11.8).

Cosmogenic radionuclides formed in the atmosphere are rapidly removed from air with precipitation (e.g. within one week of formation) because, unlike ^{14}C, which forms CO_2 and becomes well mixed in the atmosphere, ^{10}Be and ^{26}Al adsorb onto aerosols that act as condensation nuclei for rain droplets. ^{36}Cl dissolves in rainwater and is also removed from air relatively fast, preventing global mixing. One consequence of their rapid deposition out of the atmosphere is that cosmogenic nuclides are not deposited equally over the Earth's surface – their low residence times in the atmosphere prevent thorough mixing, so that is a variable in these methods.

Cosmogenic radionuclides are produced in unequal amounts spatially and temporally. Latitudinal variation in the magnetic field and influx of solar radiation means that production rate varies with latitude (and elevation), and because they fall out of the atmosphere rapidly, cosmogenic isotopes are unevenly delivered to the Earth's surface. Unlike ^{14}C, a constant deposition rate across the surface of the earth for a given period in time cannot be assumed. Sites at high latitude and low elevation receive lower concentrations of cosmogenic radionuclides than high elevation sites closer to the equator. The same applies to production of cosmogenic radionuclides in rocks exposed at the Earth's surface. Like ^{14}C, production rates of ^{10}Be, ^{26}Al, and ^{36}Cl are also prone to temporal variations due to changes in cosmic ray flux and variations in Earth's geomagnetic field, and production rates are site specific.

11.4.1.3 Dating groundwater with ^{14}C and ^{36}Cl

A study that sought to assess age of groundwater in different compartments of a sandstone aquifer system in the Amadeus Basin in central Australia used a combination of cosmogenic ^{14}C and ^{36}Cl (Cresswell et al. 1999). Relatively high $^{36}\text{Cl}:\text{Cl}_{total}$ ratios occur in shallow groundwater ($\sim 175 \times 10^{-15}$) but appreciably lower $^{36}\text{Cl}:\text{Cl}_{total}$ ratios (as low as 8×10^{-15}) are present in groundwater from hundreds of meters of depth. Qualitatively, the decrease in ^{36}Cl with depth reflects radioactive decay of ^{36}Cl to ^{36}Ar (or less frequently, to ^{36}S) that occurs as water travels

to deeper portions of the aquifer system. Quantitatively, the lower ratios at depth indicate that the deep waters infiltrated the aquifer as long ago as 400 000 years, following dipping sedimentary bedding planes along a 40–50 km flow path. This ^{36}Cl-determined groundwater age is consistent with what is known about the aquifer hydraulic conductivity (0.1 m per year), radiocarbon ages, and groundwater flow paths.

The absolute age of groundwater is determined by the equation

$$^{36}\text{Cl} = \; ^{36}\text{Cl}_o \times e^{-\lambda_p t} \qquad (11.23)$$

where ^{36}Cl is the amount of ^{36}Cl remaining at time t, $^{36}\text{Cl}_o$ is the amount of ^{36}Cl at time t = 0 (i.e. when the water infiltrated the aquifer), and λ_p is the decay constant for Cl (2.25 $\times 10^{-6}$ yr.$^{-1}$).

The determination of $^{36}\text{Cl}_o$ requires certain assumptions:

1 the rate of production and deposition of $^{36}\text{Cl}_o$ from the atmosphere (latitude and elevation dependence) is known;

2 rate of transport of ^{36}Cl through soil to the aquifer is known;

3 Cl is not retained in soil (e.g. plant uptake or anion exchange) nor in the vadose zone, for significant periods of time;

4 potential additional sources of Cl that might contaminate the atmospheric deposition signal can be quantified or are negligible;

5 potential contamination from ^{36}Cl derived from rocks exposed at or near the surface can be estimated (i.e. cosmic ray bombardment of rocks at the surface may be an additional source of ^{36}Cl infiltrating the aquifer and thus must be accounted for); and

6 the abundance of Cl in coastal air derived from sea spray can be quantified because if not, it may dilute ^{36}Cl and produce low $^{36}\text{Cl}:\text{Cl}_{total}$ ratios. Understanding (or assumptions) of past weather systems and transport of coastal air masses must be factored into the age analysis.

For reasons like these, many studies rely on two or more isotope systems in their analyses to verify of the accuracy of results.

One factor that makes ^{36}Cl useful in dating groundwater is that chlorine in its common chloride oxidation state (Cl^-) exhibits *conservative* behavior in water, a term that describes its high solubility and tendency to remain dissolved. Conservative components like Cl^- are unlikely to be removed from the aqueous phase by mineral crystallization or ion exchange. Thus, ^{36}Cl that infiltrates the saturated zone of an aquifer is likely to remain dissolved, and analysis of the ratio of ^{36}Cl to total dissolved chloride will facilitate groundwater age analysis.

Unlike Cl^-, the cosmogenic isotopes Al^{3+} and Be^{2+} exhibit *nonconservative* behavior. They are effectively insoluble at typical soil and groundwater pH values and thus will adsorb onto surfaces of minerals or organic matter or – especially in the case of Al – become incorporated into low-temperature minerals such as the Al-hydroxide gibbsite or clays such as kaolinite or smectite.

The dating of buried layers in glacial ice cores is similar to dating of groundwater given that progressive deposition of new material removes sediments from active cosmic ray bombardment once buried to a depth of ~1 m. The nonconservative behavior of ^{10}Be and ^{26}Al is of benefit in sediments or glacial ice, where meltwater or porewater could redistribute the mobile ^{36}Cl – and blur the signal – but not the relatively immobile ^{10}Be and ^{26}Al.

11.4.1.4 Exposure age analysis

Determining exposure ages is useful because it can provide information on topics ranging from timing of landslides or glacial retreat to submergence of terraces from below sea level or deposition of an apron of volcanic debris. The amount of time elapsed since the surface became exposed to cosmic ray bombardment should be recorded, either in fallout isotopes or in situ isotopes. Cosmogenic isotopes rain out of the sky and progressively accumulate on the surface, allowing exposure age to be determined provided that rate of fallout is known, and that any possible loss (leaching, radioactive decay) can be quantified. In addition to reactions in the atmosphere, ^{10}Be, ^{26}Al, ^{36}Cl, and others also form by direct bombardment of soils and rocks at the earth's surface (to a depth of ~ 1 m). Spallation of Si and O in quartz produces ^{10}Be, spallation of Si in quartz or feldspar produces ^{26}Al, and spallation of K and Ca in feldspars produces ^{36}Cl, and more (Table 11.2). Quartz is especially valuable because it is abundant, chemically stable (and thus not prone to chemical weathering), and it lacks ^{27}Al, which comprises 10–18% of feldspar and thus

may serve as a contaminant in feldspar with respect to detection of ^{26}Al.

Minerals at depths greater than 1 m do not receive the cosmic radiation needed to form cosmogenic isotopes, and furthermore, any cosmogenic nuclide buried > 1 m will undergo radioactive decay which must be accounted for (except for stable ^3He and ^{21}Ne, e.g.). A typical depth-concentration curve for cosmogenic radionuclides in rocks exposed at the surface is shown in Figure 11.10.

Once exposed to cosmic rays, quartz grains begin to accumulate ^{10}Be and ^{26}Al, and initial production rates are high. However, net production rates decrease approaching saturation, the point where the rate of production of the radionuclide is equal to its rate of decay. Given that annual production rates of ^{10}Be and ^{26}Al in quartz are on the order of tens to hundreds of *atoms* per gram and that

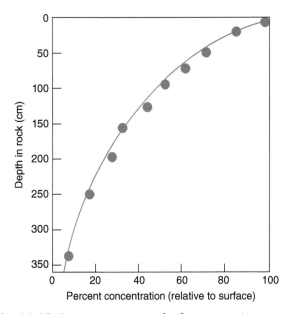

Fig. 11.10 Representative example of a cosmogenic radionuclide depth profile. Units of concentration on the x-axis are typically on the order of 10^2 to 10^5 atoms (of ^{10}Be, ^{26}Al, ^{36}Cl) per gram of mineral (e.g. quartz).

Focus Box 11.12

Atomic Bomb Tests and Groundwater Dating

In addition to eons of production by natural atmospheric reactions, ^{36}Cl was also produced in relatively large quantities during the 1950s by atmospheric nuclear bomb explosions that irradiated sea water, and this pulse is a useful means of identifying groundwater younger than 50 years old (groundwater greater than 50 years old lacks the signature of this pulse). A similar spike in ^3H (*tritium*) was produced in the atmosphere in the 1950s and 1960s in association with hydrogen bomb testing, with a peak at 1963, making tritium a similarly useful method for determining groundwater age. Groundwater with no ^3H (i.e. "tritium-dead water") likely recharged from surface through soils and into the aquifer in the 1950s or earlier.

Focus Box 11.13

^{10}Be Analysis of River Incision into a Bedrock Channel

Reusser et al. (2007) sought to determine the timing and rate of incision of the Susquehanna River into its bedrock channel in the foothills of the central Appalachian Mountains of eastern North America. Fluvial erosion has produced a series of strath (eroded bedrock) terraces with a total relief of 10 m that (i) have not been buried by sediment or soil, and (ii) apparently have been only minimally affected by post-fluvial weathering. In other words, the main erosive force appears to have been high-energy fluvial action that formed the terraces, and since then chemical and physical weathering of the terraces is negligible; the terraces are presently exposed at the surface and are assumed to have been exposed to cosmic rays (i.e. not buried) since the time of their original formation. ^{10}Be ages from 80 samples indicate exposure ages ranging from 36 000 years on the uppermost terraces to 14 000 years on the lowest terraces (those closest to the modern stream level). These data are interpreted to indicate rapid erosion into the bedrock channel during the last glacial cycle – 10 m of incision over 22 Ka – a process that the authors attribute to greater frequency and magnitude of floods during the late Pleistocene compared to the hydrologically calmer interglacial periods before and after the last glacial cycle.

Focus Box 11.14

^{10}Be Analysis of Tibetan Glaciers and Paleoclimate Link

The determination of changes in glacial mass balance responding to rapid shifts in the climate of western Tibet was studied using a combination of ^{10}Be surface exposure dating and surface mapping (Seong et al. 2009). Results from dating of boulders on moraines and sediment in depth profiles indicate that glaciers advanced and retreated at least 12 times over the past 17 ka with a periodicity of 1.5 ± 1 ka that seems to correlate with northern hemisphere climate oscillations and, to a lesser extent, the south Asian monsoon. The rapid response of climate in the Tibetan Plateau to global shifts may have implications for predicting future climate in the region.

---

saturation of ^{10}Be and ^{26}Al in quartz is typically reached at concentrations of 10^7 to 10^8 atoms per gram (at mid to high latitudes at sea level), rocks exposed at the surface will become saturated after approximately 10^6–10^7 years (Nishiizumi et al. 1986). Saturation is typically reached after a rock has been irradiated for approximately four half-lives (Faure 1998), so in the case of ^{10}Be, saturation is reached after approximately 6 million years, whereas saturation with respect to ^{26}Al is reached after approximately 3 million years, and the saturation time for ^{36}Cl is about 1 million years. Dunai and Stuart (2009) present guidelines for reporting of exposure age and erosion rate determinations in situ cosmogenic nuclide data.

For specimens below saturation, exposure age can be calculated by balancing production rate with decay rate:

$$P = R/\lambda \times (1 - e^{-\lambda_P t}) \qquad (11.24)$$

P is the net quantity of cosmogenic isotope produced (in atoms per gram), R = rate of production of cosmogenic isotope (in atoms per gram of sample per year), λ is the decay constant (in yr^{-1}), and t is the exposure time (in yr).

In addition to rate of production and decay of a cosmogenic isotope, analysis of surface exposure ages must also consider potential effects of physical or chemical erosion. Any removal of material from the surface of a boulder by wind abrasion, rolling, or chemical weathering will remove some portion of the cosmogenic atoms that have accumulated – the resistance to chemical erosion of quartz makes it a desirable mineral for exposure dating, but physical erosion and slow chemical erosion may still act upon even this resistant mineral. In cases where erosion has affected target minerals, cosmogenic isotope-derived exposure ages will be anomalously young. Because erosion removes the outer layer, with the greatest concentration of cosmogenic atoms, cosmogenic exposure ages will underestimate the actual exposure age. Lal (1991) developed an equation for correcting exposure ages based on estimates of erosion rate (refer to source for details):

$$P = R/(\lambda + \rho\varepsilon/\Lambda) \times (1 - e^{(-\lambda_P + \rho\varepsilon/\Lambda)t}) \qquad (11.25)$$

Error may also be introduced if boulders or other surface-exposed materials had been previously exposed and still contain cosmogenic atoms from prior exposure.

These latent atoms will result in anomalously old exposure ages.

^{10}Be, ^{26}Al, and ^{36}Cl (and ^{14}C) are the most common cosmogenic isotopes applied to dating Earth surface processes, but there are many others, examples of which include:

1 ^{3}He has certain advantages over the common cosmogenic isotopes ^{10}Be, ^{14}C, ^{26}Al, and ^{36}Cl. First, it is stable, i.e. does not undergo radioactive decay, meaning that it does not reach saturation in an exposed rock; second, it has the highest production rate of the cosmogenic radionuclides ($\geq 10^{2}$ atoms g^{-1} yr^{-1}), making it useful for dating surfaces as young as 600 years old (much younger than is datable using ^{10}Be, ^{26}Al, or ^{36}Cl methods; Kurz et al. 1990); and third, the production of ^{3}He from spallation of magnesium, calcium, and other elements in calcite, olivine, and pyroxene makes ^{3}He dating useful for carbonate and mafic rocks where other cosmogenic approaches are precluded by the absence of quartz (Cerling 1990; Licciardi et al. 1999; Amidon and Farley 2011).

2 ^{21}Ne is formed by bombardment of igneous olivine and plagioclase feldspar (e.g. ^{24}Mg(n,α)^{21}Ne,) the rate of which has proven useful in dating exposure ages of young volcanic rocks (Poreda and Cerling 1992).

3 ^{32}Si forms by bombardment of Ar in the atmosphere and then rapidly falls to the Earth's surface in precipitation. With a half-life is 140 ± 15 years, it has the potential to become an important isotope because its half-life is intermediate to short-lived ^{3}H (12.3 yr) and ^{210}Pb (22.3 yr) on the one hand, and the longer-lived ^{14}C (5730 yr) on the other hand. The ^{32}Si decay rate is well suited for dating materials with age ranges of ~50–2000 years, including glacial layers, groundwaters, ocean waters, and late Holocene sediments.

4 ^{129}I forms by spallation of ^{129}Xe in the atmosphere. ^{129}I, like ^{36}Cl, exhibits conservative behavior in aqueous systems. The long ^{129}I half-life (15.7 million years) has enabled it to date groundwaters with ages ranging from ~1 to 100 million years (Fabryka-Martin et al. 1985). ^{129}I is also produced by ^{238}U fission, so it exhibits a late 1950s – early 1960s nuclear bomb spike (much like ^{3}H and ^{36}Cl), and it was used in the mid-1980s to detect fallout from the Chernobyl nuclear power plant accident (Paul et al. 1987).

11.5 RADIOACTIVE DECAY METHODS OF DATING SEDIMENTS AND MINERALS

11.5.1 ^{210}Pb

The radioactive decay of ^{210}Pb, with a half-life of 22 years, is applicable to dating very young sediments or glacial ice generally with ages ≤ 150 years. ^{210}Pb forms in the atmosphere from decay of the noble gas ^{222}Rn via short-lived intermediate products that include ^{218}Po, ^{218}At, ^{218}Rn, and ^{214}Po, all with half-lives of milliseconds to minutes (Table 11.1). ^{210}Pb is a member of the ^{238}U radioactive decay series shown in Figure 11.2, and it is not a cosmogenic isotope because it forms via spontaneous decay in the ^{238}U series rather than from interactions with cosmic rays or their products. For this reason, it is referred to as a *fallout radionuclide*, a group that also includes ^{7}Be and ^{137}Cs. The rapid transition from gaseous ^{222}Rn to ^{210}Pb, combined with the tendency of ^{210}Pb to strongly sorb to aerosols, means that ^{210}Pb is rapidly rained out of the atmosphere (within 10 days of formation, much like cosmogenic isotopes). Because it is adsorbed to particle surfaces, it is assumed to rapidly settle out of low-energy water (lakes, swamps, marine bodies) and accumulate in sediments soon after it rains out of the atmosphere.

The activity of atmospherically derived ^{210}Pb (unsupported ^{210}Pb) in sediments or layers of glacial ice will decrease with time at a rate in accordance with its half-life, and typically the age of a given layer of sediment or ice is determined using the following formula:

$$t(\text{years}) = 1/\lambda \times \ln(A^{210}Pb_0/A^{210}Pb_h) \qquad (11.26)$$

where λ is the decay constant 0.0311 yr^{-1}, A^{210}Pb$_0$ is the unsupported lead activity in disintegrations per minute at time zero (the present), and A^{210}Pb$_h$ is the activity in disintegrations per minute at depth h.

Ideally, a plot of ^{210}Pb activity will exhibit a logarithmic decrease with depth. However, error can be introduced from ^{210}Pb that is derived from decay reactions of other radionuclides in the sediment, commonly the decay of ^{226}Ra to (eventually) ^{210}Pb. This lead is known as "Ra-supported ^{210}Pb" and because it is being added to the sediment after deposition, it will produce higher ^{210}Pb concentrations in the sediment than would be yielded by atmospherically derived ^{210}Pb alone. The presence of ^{210}Pb supported by ^{226}Ra decay in sediments or glacial ice is often evident in deeper layers where a log-normal plot of ^{210}Pb activity vs. depth strays from linearity (Figure 11.11).

Here, the supported ^{210}Pb is evident in the deeper, older sediments where atmospherically derived (unsupported) ^{210}Pb has already decayed to its daughter products, leaving ^{210}Pb derived from decay of ^{226}Ra in the sediment or ice. In this example, the supported ^{210}Pb level is 8–10 decays per minute per gram (dpm g^{-1}). In addition to correcting for supported ^{210}Pb, ^{210}Pb dates can be confirmed or adjusted by ^{137}Cs dating, which takes advantage of a

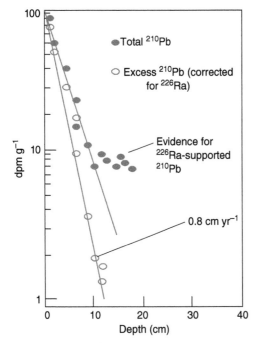

Fig. 11.11 Example of the type of curve showing deviation from linearity in ^{210}Pb activity with depth produced by $^{226}Ra \rightarrow {}^{210}Pb$ decay. This non-atmospherically derived ("supported") ^{210}Pb must be accounted for in age determinations. In this hypothetical example, the corrected ^{210}Pb activities give a sedimentation rate of 0.8 cm yr^{-1}.

notable spike in ^{137}Cs deposition in late 1950s and early 1960s related to nuclear bomb testing.

An example of how ^{210}Pb dating is used in environmental analysis (the impact of acid deposition on a watershed) is presented in Chapter 9 (Figure 9.14). In that case, ^{210}Pb was used to determine ages of layers in a sediment core back to the year 1850, and the dates obtained helped to determine timing of lake acidification and changes to the calcium–aluminum chemistry of the watershed.

11.5.2 K–Ar

The decay of radioactive ^{40}K to ^{40}Ar is a well-established method for dating K-rich minerals such as illite, muscovite, biotite, and K–feldspar in rocks ranging from shales and sandstones to igneous and metamorphic rocks. The half-life of ^{40}K is 1.25×10^{10} years, and the method has been applied to dating sites as young as Pleistocene (e.g. basalts at Olduvai Gorge) to as old as Precambrian.

When a K-bearing mineral (e.g. a mica) crystallizes from solution, it does not incorporate Ar into the structure. A noble gas, Ar does not form ions, and thus is not present in K-minerals when they form, so all ^{40}Ar present in a K-bearing mineral is derived from ^{40}K decay.

The decay of ^{40}K to ^{40}Ar occurs by electron capture:

$$^{40}_{19}K + {}^{0}_{-1}\beta^- \rightarrow {}^{40}_{18}Ar + \gamma\ (1.46\ MeV) \qquad (11.27)$$

This process accounts for 11% of the decay of radioactive ^{40}K.

The remaining 89% of ^{40}K atoms undergo transformation to stable ^{40}Ca by beta decay, sometimes referred to as negatron emission (Faure 1986):

$$^{40}_{19}K \rightarrow {}^{40}_{20}Ca + {}^{0}_{-1}\beta^- + 1.32\ MeV \qquad (11.28)$$

However, because ^{40}Ca is the most abundant isotope of calcium, the amount produced by ^{40}K decay relative to the nonradiogenic ^{40}Ca in the mineral makes ^{40}K-derived ^{40}Ca difficult to determine. Thus, K–Ar dating is far more common than the K–Ca method.

The decay constants for the two spontaneous decay reactions (Steiger and Jäger 1977) are:

$$
\begin{aligned}
&^{40}K \rightarrow {}^{40}Ar\ \ \lambda(^{40}Ar) = 0.581 \times 10^{-10}\,yr^{-1} \\
&\underline{^{40}K \rightarrow {}^{40}Ca\ \ \lambda(^{40}Ca) = 4.962 \times 10^{-10}\,yr^{-1}} \\
&\quad\quad\quad Total\ \lambda_{tot} = 5.543 \times 10^{-10}\,yr^{-1}
\end{aligned}
\qquad (11.29)
$$

The total decay constant for the branched decay of ^{40}K to ^{40}Ar and ^{40}Ca yields a half-life of:

$$
\begin{aligned}
t_{1/2} &= \ln(2)/(5.543 \times 10^{-10}\ yr^{-1}) \\
&= 0.693/(5.543 \times 10^{-10}\ yr^{-1}) \\
&= 1.250 \times 10^{10}\ yr
\end{aligned}
$$

The K–Ar date of the mineral provides a valid crystallization age if the following criteria are met (or addressed) – they mainly require that the mineral remains a closed system after the mineral crystallizes.

1 The mineral must contain no inherited ^{40}Ar introduced into the crystal structure after crystallization.

2 The mineral must not lose ^{40}Ar following mineral crystallization. This criterion is harder to achieve because uncharged ^{40}Ar will not be electrostatically attracted to crystallographic sites that are suited to bonding with K$^+$ and may leak out above the *closure temperature* (T$_c$) or *blocking temperature* (for K–Ar, the temperature below which a mineral will retain Ar). Closure temperatures vary and depend on mineral structure.

3 Argon does not diffuse out of the mineral lattice at temperatures below the T$_c$.

4 Mineral crystallization should occur rapidly, because slow crystallization tends to foster argon loss. However, cooling rates of igneous plutons or metamorphic rocks can be measured if K–Ar ages are obtained for a suite of minerals with different closure temperatures – as the

region cools through the closure temperatures of various minerals, they should record an age progression, with high T_c minerals (e.g. hornblende) recording the oldest dates and low T_c minerals (e.g. fine-grained illite) recording youngest dates (Figure 11.12).

5 No new potassium can be incorporated into the mineral once it has crystallized.

11.5.3 Ar–Ar analysis

One of the means of dealing with partial loss of Ar is to make use of the *Ar–Ar method* of dating. In this method, the total amount of stable ^{39}K in the mineral of interest is measured and then the ^{39}K is partially converted to unstable ^{39}Ar ($t_{1/2} = 269$ yr) by neutron irradiation in a nuclear reactor. Although unstable, ^{39}Ar is sufficiently long-lived for lab analyses that take place on the order of days or weeks that its decay is negligible, and ultimately the age is determined by the ratio of ^{40}Ar produced by decay of ^{40}K (like in the K–Ar method) to ^{39}Ar produced by irradiation of ^{39}K in the nuclear reactor. The specimen is heated stepwise and the $^{40}Ar/^{39}Ar$ ratio of the unknown specimen is compared to a standard with known $^{40}Ar/^{39}Ar$ ratio, K concentration and age that has been simultaneously analyzed with the unknown specimen. In this way, the Ar–Ar age is a relative age compared to a standard, rather than a result of a direct measurement of parent and daughter isotope, but the stepwise nature of the Ar release yields information about the distribution of K and Ar in the mineral structure that can inform researchers about Ar loss.

11.5.4 Rb–Sr

Rubidium is an alkali metal (Rb^+) and its behavior in geochemical systems is much like the common alkali metals sodium and potassium, so it substitutes into minerals in place of K^+. The identical charge and the similarity in ionic radii (12-fold coordination) of Rb^+ (1.72 Å) and K^+ (1.64 Å) (Table 2.1) facilitate Rb for K substitution. The method is very similar to K–Ar but the decay product ^{87}Sr is divalent and Sr^{2+} is held more strongly in the mineral lattice than Ar^0. For this reason, closure temperatures for Sr are higher than for Ar, and Rb–Sr ages tend to be closer to crystallization ages than K–Ar ages. Also, Sr^{2+} tends not to diffuse out of crystal matrices as does Ar, so Rb–Sr dates are less prone to resetting by subsequent thermal events; for this reason, K–Ar and Rb–Sr have been applied in conjunction to date paleohydrology/fluid flow in sedimentary basins (Clauer et al. 2003).

11.5.5 U–Th–Pb

Uranium in nature consists primarily of ^{238}U (99.27%) with the remainder as ^{235}U (0.72%) and trace amounts of ^{234}U (0.0057%). ^{238}U and ^{235}U have long half-lives of 4.47×10^9 yr and 7.04×10^8 yr, but ^{234}U is less stable and decays to ^{230}Th by alpha decay with a half-life of 2.47×10^5 yr (Table 11.1).

The main naturally occurring isotope of thorium is ^{232}Th with a very long half-life of 1.41×10^{10} yr. The other thorium isotopes (^{227}Th, ^{228}Th, ^{230}Th, ^{231}Th, ^{234}Th) are all radiogenic, produced in the decay chains of ^{232}Th, ^{235}U, and ^{238}U. Half-lives of the naturally occurring radiogenic thorium isotopes are 18.7 days (^{227}Th), 1.92 years (^{228}Th), 7.54×10^4 years (^{230}Th), 25.5 hours (^{231}Th), and 24.1 days (^{234}Th).

Tetravalent uranium and thorium have similar radii in 12-fold coordination ($U^{4+} = 1.17$ Å, $Th^{4+} = 1.21$ Å) and readily cosubstitute in mineral structures as trace elements in minerals such as zircon, sphene, monazite, and apatite.

Focus Box 11.15

Closure Temperatures for K–Ar Analysis

The temperature above which Ar escapes from mineral structures is an important facet of K–Ar analysis. Values of closure temperature (T_c) range from as low as 130 °C for microcline (Harrison and McDougall 1982) to 200–250 °C for low crystallinity illites, 300–350 °C for biotites, 350–400 °C for muscovites (Dahl 1996), and 500–600 °C for hornblende (Faure 1986). Closure temperatures increase with increasing crystal size, reflecting longer diffusion pathways and slower Ar loss. While Ar loss during heating by geological fluids can produce K–Ar ages younger than mineral crystallization ages, the tendency of minerals to lose Ar above their T_c values can also be an advantage to geochemists as it can reveal timing of geological events that produce heat flow such as orogenies (Elliott and Aronson 1987), hydrothermal events (Bish and Aronson 1993), and fluid flow along faults (Ylagan et al. 2002).

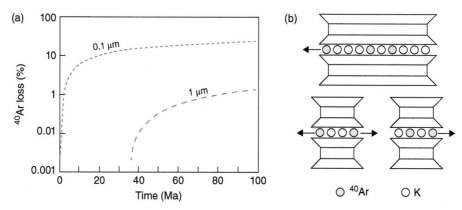

Fig. 11.12 Diffusion of radiogenic ^{40}Ar out of illite crystals. (a) Grain size exerts a strong control on Ar loss, where finer grained crystals (e.g. 0.1 µm effective radius) are more prone to Ar loss than are coarser grained crystals (e.g. 1 µm effective radius). This is because fine grained crystals have shorter diffusion domains than do coarser crystals (Modified after Clauer et al. 2003); (b) the smaller the crystal, the more likely it is to have a greater proportion of Ar atoms at crystal edges where diffusion out of the crystal lattice is more likely to occur.

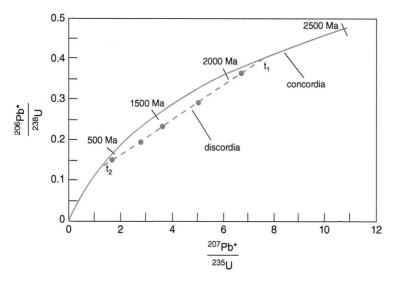

Fig. 11.13 Representative concordia–discordia diagram. The older date (t_1) is typically interpreted to be the date when the system became closed (e.g. timing of crystallization of zircon) and the younger date (t_2) corresponds to the age of an event (commonly metamorphism) that caused Pb loss. In this case, one possible interpretation would be that the zircon first crystallized 2200 Ma (e.g. during cooling of granitic magma) and then was affected by metamorphism at ~400 Ma.

Radioactive decay of U and Th eventually results in stable isotopes of Pb, and these decay series (Figures 11.2 and 11.3) are the basis of U–Th–Pb dating methods. The only nonradiogenic lead isotope, ^{204}Pb, is used for reference. The decay rates of the parent ^{238}U, ^{235}U, and ^{232}Th are many orders of magnitude greater than their intermediate daughter products, meaning that the abundance of the parent isotope remains effectively unchanged for many half-lives of the daughter isotopes.

Dating by U–Th–Pb methods relies on two important assumptions:

1 Secular equilibrium exists (this occurs when the ratio of parent isotope to daughter isotope reaches a constant value).

2 The decay chain occurs within a closed system – this is because loss of daughter products would (i) alter the secular equilibrium and (ii) produce erroneously low

measurements of the daughter product needed to quantify age (e.g. ^{206}Pb), which would result in an anomalously young age. However, minerals used for U and Th series dating tend to be refractory (stable) silicates such as zircon ($ZrSiO_4$) and the phosphate monazite ([CeLaNdTh][PO_4]). Other than zircon, where the crystal structure allows substitution of U^{4+} and Th^{4+} for Zr^{4+} but inhibits incorporation of Pb^{2+}, an initial ratio of inherited radiogenic lead (i.e. present at the time of crystallization) must also be assumed. This is one of the reasons why zircon is the most well-suited mineral for U–Th–Pb dating.

The three equations used in dating igneous or metamorphic minerals using U or Th decay are:

$$^{206}Pb/^{204}Pb = (^{206}Pb/^{204}Pb)_i + (^{238}U/^{204}Pb) \times e^{-\lambda_a t}$$

$$(11.30)$$

$$^{207}\text{Pb}/^{204}\text{Pb} = (^{207}\text{Pb}/^{204}\text{Pb})_i + (^{235}\text{U}/^{204}\text{Pb}) \times e^{-\lambda_b t}$$
(11.31)

$$^{208}\text{Pb}/^{204}\text{Pb} = (^{208}\text{Pb}/^{204}\text{Pb})_i + (^{232}\text{Th}/^{204}\text{Pb}) \times e^{-\lambda_c t}$$
(11.32)

λ_a, λ_b, and λ_c correspond to decay constants for ^{238}U, ^{235}U, and ^{232}Th, respectively, and the first term to the right of the equal sign represents the initial ratio of inherited radiogenic lead (^{206}Pb, ^{207}Pb, or ^{208}Pb, depending on parent isotope) to stable ^{204}Pb present in the mineral when it crystallized.

Ideally, the decay chains of $^{238}\text{U} \rightarrow {}^{206}\text{Pb}$, $^{235}\text{U} \rightarrow {}^{207}\text{Pb}$, and $^{232}\text{Th} \rightarrow {}^{208}\text{Pb}$, respectively, should result in identical mineral crystallization ages, but this is sometimes not the case, and these ages are *discordant*. The main cause of discordance is loss of daughter isotopes in the long decay chains from U or Th parent isotope to the formation of stable isotopes of lead at the end of the decay chains. Some of the daughter products are reasonably mobile, including the various isotopes of radon, which can be lost from the system as a gas, and radium, which is moderately soluble in many fluids. The loss of daughter products from the mineral results in anomalously low lead concentrations and dates that are younger than the crystallization age. Radioactive decay also causes damage to crystal lattices and this too can lead to loss of daughter products. Alpha decay, for example, produces a high-energy particle with an atomic mass of 4 Da, and the migration of an α particle will disrupt the crystal lattice and lead to greater solubility that enhances loss of certain daughter products.

The benefit of multiple approaches (e.g. U–Th or ^{238}U–^{235}U) is that it provides internal checks on age determinations. Figure 11.13 shows data from two isotope pairs that yield individual ratios (ages), and loss of lead is indicated by points that fall below the concordia curve. If discordant ages from zircons or similar U-bearing minerals in a rock plot as a straight line, the upper intercept of the discordia line with the concordia curve corresponds to the crystallization age of the mineral (e.g. initial igneous crystallization age). Intersection of the discordia line with the concordia curve at the low-age end may represent a younger event that caused Pb loss, e.g. metamorphism.

11.5.6 $^{234}\text{U}/^{238}\text{U}$ and ^{234}U disequilibrium

While the U–Th–Pb methods are applicable for determining crystallization ages of igneous, sedimentary, and metamorphic minerals millions to billions of years old, U–Th decay can also be applied to low-temperature

systems such as soils and groundwater where the ages in question are thousands to hundreds of thousands of years. These methods are based on the disequilibrium that can develop between parent and daughter isotopes in the long decay chains of ^{238}U and ^{232}Th, where the varied geochemical behaviors of the many decay products of ^{238}U and ^{232}Th mean that some of the more soluble species can become separated from their parent isotopes, particularly in aqueous systems.

One of the more common low-temperature geochronometers and tracers of geochemical processes in the U and Th decay series is the $^{234}\text{U}/^{238}\text{U}$ method, an approach based on the disequilibrium that develops between ^{238}U and its mobile daughter product ^{234}U.

^{234}U is the first product of ^{238}U decay in a series that includes intermediate isotopes of thorium and protactinium whose half-lives are 24.1 days and 6.7 hours, respectively (Table 11.1). Therefore, the decay of ^{238}U to ^{234}U via ^{234}Th and ^{234}Pa occurs with a combined half-life of a mere 24.4 days.

$$^{238}\text{U} \rightarrow {}^{234}\text{Th} + \alpha, \text{then } ^{234}\text{Th} \rightarrow {}^{234}\text{Pa} + \beta^-,$$

$$\text{then } ^{234}\text{Pa} \rightarrow {}^{234}\text{U} + \beta^- \quad (11.33)$$

Unlike the conditions of secular equilibrium that tend to occur in minerals older than a few million years old (where the ratio of $^{234}\text{U}/^{238}\text{U}$ becomes constant), disequilibrium between ^{234}U and ^{238}U tends to develop when uranium-bearing minerals undergo spontaneous decay in soil or groundwater. An important term applied to describe systems with respect to secular equilibrium or disequilibrium is the *activity ratio* of parent to daughter, which in the case of a ^{234}U–^{238}U system in secular equilibrium is defined as:

$$\text{AR} = \lambda(^{234}\text{U}) \times \text{N}(^{234}\text{U})/\lambda(^{238}\text{U}) \times \text{N}(^{238}\text{U}) = 1$$
(11.34)

Given the decay constants for ^{234}U and ^{238}U of 2.81×10^{-6} yr^{-1} and 1.55×10^{-10} yr^{-1}, respectively, the ratio of atoms of ^{234}U and ^{238}U is approximately 5.5×10^{-5} under conditions of secular equilibrium, i.e. when AR = 1.

Upon decay of ^{238}U to ^{234}U, secular equilibrium tends to be lost due to preferential release of ^{234}U into solution (where U^{6+} species are soluble), and this increases the $^{234}\text{U}/^{238}\text{U}$ activity ratio of the newly liberated uranium (in solution) to values that are > 1, and in some cases as high as AR = 10. On the other hand, preferential loss of ^{234}U from minerals can result in minerals with AR values < 1, which is one of the potential sources of error in the U–Th–Pb methods described in Section 11.5.5.

^{234}U tends to be partitioned into the aqueous phase at a greater rate than ^{238}U for the following reasons:

1 Damage to the crystal lattice caused by alpha decay facilitates leaching of newly formed ^{234}U out of crystals into solution;

2 Radiation produced by decay of ^{238}U to ^{234}U via ^{234}Th and ^{234}Pa can oxidize ^{234}U to the soluble U^{6+} ion, enhancing its migration into solution; and

3 The kinetic energy of alpha particles emitted upon spontaneous decay of ^{238}U can cause newly formed ^{234}Th daughter product to be blasted out of the crystal and into solution – this particularly applies to cases where the parent ^{238}U atom is located along the margins of a crystal. Rapid decay of ^{234}Th to ^{234}U through ^{234}Pa (Table 11.1) then can result in dissolved ^{234}U in solution and ^{238}U that remains locked in the crystal lattice. A schematic sketch of alpha recoil is provided in Figure 11.5.

Flow of groundwater into the oceans either directly or via streams delivers uranium to the oceans with a homogenized $^{234}U/^{238}U$ activity ratio of 1.15 (Miyake et al. 1966). When calcium carbonate (aragonite or calcite) precipitates from ocean water it incorporates uranium (typically in the range of 1 ppm), and provided we assume a $^{234}U/^{238}U$ AR (typically 1.15), the decay of ^{234}U to ^{230}Th by alpha decay can be used to date crystallization of marine carbonates. With a half-life of 2.48×10^5 years, ^{234}U decay is well suited for dating coral reefs and carbonates with ages of tens of thousands to hundreds of thousands of years. The important assumptions here are (i) that the system has remained closed to uranium, and (ii) that the initial ratio of $^{234}U/^{238}U$ is known.

The rapid decay of ^{234}U to ^{230}Th as compared to the slow decay of ^{238}U to ^{234}U (via ^{234}Th and ^{234}Pa isotopes) means that with time, the activity ratio of $^{234}U/^{238}U$ exponentially decreases from 1.15 to 1.0. When the $^{234}U/^{238}U$ AR reaches a value of 1.0, the isotope pair will have returned to a state of secular equilibrium (Figure 11.14). For an initial $^{234}U/^{238}U$ AR of 1.15, reaching AR = 1.0

requires approximately 1 million years, thus bracketing the applicability of this method to specimens with maximum age of 1 million years.

Unlike uranium, which typically comprises 1 ppm of marine carbonates, thorium is excluded from carbonates when they initially crystallize, meaning that any ^{230}Th in a carbonate mineral got there by decay of ^{234}U. Assuming an initial $^{234}U/^{238}U$ activity ratio of 1.15, the decay of both inherited (excess) ^{234}U and ^{238}U to ^{230}Th aids in determining the age of carbonate crystallization. These methods and their assumptions and corrections were developed by Broecker (1963) and are presented in detail by Faure (1986).

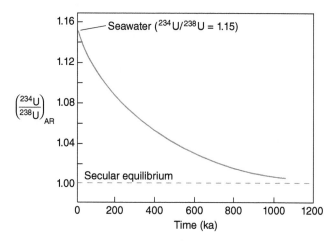

Fig. 11.14 Decrease in ratio of $^{234}U/^{238}U$ with time. The decrease is related to relatively rapid decay of ^{234}U compared to ^{238}U, and the decreasing ratio with time facilitates dating of marine carbonates (e.g. corals).

The radioactive isotopes presented here include examples of radionuclides as environmental contaminants as well as common methods for dating sediments, waters,

---

Focus Box 11.16

Using $^{234}U/^{238}U$ to Trace Groundwater Recharge and Flow, Yucca Mountain

In a study designed to examine groundwater flow patterns in and around the proposed Yucca Mountain high-level radioactive waste repository in Nevada, USA, Paces et al. (2002) noted relatively low uranium concentrations (0.8–2.5 ppb) and high $^{234}U/^{238}U$ ratios (7–8) in groundwater from the aquifer beneath Yucca Mountain. The values are in contrast to groundwater from surrounding aquifers, which contain relatively high and variable uranium concentrations (0.6–10 ppb) and lower $^{234}U/^{238}U$ ratios (1.5–6). These differences in [U] and $^{234}U/^{238}U$ appear to indicate that the

groundwater beneath Yucca Mountain is isolated from surrounding aquifers. Furthermore, $^{234}U/^{238}U$ ratios were very similar in saturated zone and unsaturated zone groundwater beneath Yucca Mountain, implying that recharge of the aquifer beneath Yucca Mountain is controlled by localized downgradient flow. The apparent isolation of groundwater beneath Yucca Mountain from other regional aquifers would comprise an important factor when considering site suitability for waste storage or disposal.

organic matter, and rocks. Some of the dating methods not included in this chapter include the Sm–Nd, Lu–Hf, and Re–Os methods, fission track age analysis, and various methods pertaining to specific isotope pairs (e.g. ^{230}Th–^{226}Ra, ^{230}Th–^{232}Th) contained within the lengthy ^{238}U, ^{235}U, and ^{232}Th decay chains. Information on these methods is presented in more detailed accounts by Faure (1986) and Dickin (2005).

QUESTIONS

11.1 Complete and balance (protons and neutrons) the following radioactive decay reactions by inserting alpha or beta decay products in correct molar proportions.

$$^{234}Pa \rightarrow {}^{234}U$$

$$^{226}Ra \rightarrow {}^{222}Rn$$

$$^{234}Th \rightarrow {}^{234}Pa$$

$$^{26}Al \rightarrow {}^{26}Mg$$

$$^{235}U \rightarrow {}^{231}Th$$

$$^{226}Ra \rightarrow {}^{210}Pb$$

11.2 List, describe, and indicate the differences among the three main forms of radiation released by radioactive decay.

11.3 Would you predict uranium to be more mobile in shallow, oxidizing groundwater or deeper, reducing groundwater? How does U redox chemistry lead to formation of roll-front U ore deposits?

11.4 Explain why the insolubility of thorium makes $^{234}U/^{238}U$ dating possible.

11.5 Compare the expected mobility of two classes of high-level radioactive wastes that could leak out of storage areas, cationic forms (e.g. $^{90}Sr^{2+}$ and $^{137}Cs^+$) and anionic forms (e.g. $^{99}TcO_4^-$ and $^{131}I^-$). If their inherent solubilities are approximately equal, consider two other factors that will influence their mobilities.

11.6 Describe the main principles behind K–Ar dating, including the K–Ar decay reaction, important assumptions, and limitations to the method.

11.7 What is a spallation reaction? What instigates spallation? Provide an example and diagram it, being sure to account for all protons and neutrons.

11.8 How does ^{14}C form and how is it incorporated into living material? What is the daughter product of ^{14}C, and how are radiocarbon ages determined if the decay product escapes as a gas?

11.9 Describe the benefits of using ^{36}Cl to date groundwater. In particular, indicate how it forms, how it decays, its sorption potential, and behavior in solution. Of ^{10}Be, ^{26}Al, and ^{36}Cl, which is the best option for dating groundwater. Explain.

11.10 What are three advantages to ^{10}Be, ^{26}Al, and ^{36}Cl dating methods compared to ^{14}C dating?

11.11 What correction is necessary when determining sediment depositional ages by the ^{210}Pb method? In your reply, include an equation and brief description.

REFERENCES

Amidon, W.H. and Farley, K.A. (2011). Cosmogenic ^{3}He dating of apatite, zircon and pyroxene from Bonneville flood erosional surfaces. *Quaternary Geochronology* 6: 10–21.

Arnold, J.R. and Libby, W.F. (1949). Age determinations by radiocarbon content: checks with samples of known age. *Science* 110 (2869): 678–680.

Bish, D.L. and Aronson, J.L. (1993). Paleogeothermal and paleohydrologic conditions in silicic tuff from Yucca Mountain, Nevada. *Clays and Clay Minerals* 41: 148–161.

Bish, D.L., Vaniman, D.T., Chipera, S.J., and Carfey, J.W. (2003). The distribution and importance of zeolites at Yucca Mountain, Nevada, USA. *American Mineralogist* 88: 1889–1902.

Blum, J.D., Gazis, C.A., Jacobson, A.D., and Chamberlain, C.P. (1998). Carbonate versus silicate weathering in the Raikhot watershed within the high Himalayan crystalline series. *Geology* 26: 411–414.

Broecker, W.S. (1963). A preliminary evaluation of uranium series inequilibrium as a tool for absolute age measurements on marine carbonates. *Journal of Geophysical Research* 68: 2817–2834.

Cerling, T.E. (1990). Dating geomorphic surfaces using cosmogenic ^{3}He. *Quaternary Research* 33: 148–156.

Choppin, G.R. (2007). Actinide speciation in the environment. *Journal of Radioanalytical and Nuclear Chemistry* 273: 695–703. https://doi.org/10.1007/s10967-007-0933-3.

Chow, T.J., Snyder, C.B., and Earl, J.L. (1975). Isotope ratios of lead as pollutant source indicators. *Proceedings of the International Atomic Energy Agency (Vienna)* 191: 95–108.

Clauer, N., Zwingmann, H., and Gorokhov, I. (2003). Postdepositional evolution of platform claystones based on a simulation of thermally induced diffusion of radiogenic ^{40}Ar from diagenetic illite. *Journal of Sedimentary Research* 73: 58–63.

Cresswell, R.G., Jacobson, G., Wischusen, J., and Fifield, L.K. (1999). Ancient groundwaters in the Amadeus Basin, Central Australia: evidence from the radio-isotope ^{36}Cl. *Journal of Hydrology* 223: 212–220.

Dahl, P.S. (1996). The crystal-chemical basis for Ar retention in micas: inferences from interlayer partitioning and implications for geochronology. *Contributions to Mineralogy and Petrology* 123: 22–39. https://doi.org/10.1007/s004100050141.

Dickin, A.P. (2005). *Radiogenic Isotope Geology*, 2e. Cambridge University Press.

Dunai, T.J. and Stuart, F.M. (2009). Reporting of cosmogenic nuclide data for exposure age and erosion rate determinations. *Quaternary Geochronology* 4: 437–440.

Elliott, W.C. and Aronson, J.L. (1987). Alleghanian episode of K-bentonite illitization in the southern Appalachian Basin. *Geology* 15: 735–739.

Fabryka-Martin, J., Bentley, H., Elmore, D., and Airey, P.L. (1985). Natural iodine-129 as an environmental tracer. *Geochimica et Cosmochimica Acta* 49: 337.

Faure, G. (1986). *Principles of Isotope Geology*, 2e. New York: Wiley.

Faure, G. (1998). *Principles and Applications of Geochemistry*, 2e. New Jersey, USA, Prentice-Hall: Upper Saddle River.

Gauthier-Lafaye, F., Holliger, P., and Blanc, P.-L. (1996). Natural fission reactors in the Franceville basin, Gabon: a review of the conditions and results of a "critical event" in a geologic system. *Geochimica et Cosmochimica Acta* 60: 4831–4852.

Harrison, T.M. and McDougall, I. (1982). The thermal significance of potassium feldspar K–Ar ages inferred from ^{40}Ar-^{39}Ar age spectrum results. *Geochimica et Cosmochimica Acta* 46: 1811–1820.

Holmes, A. (1911). The association of lead with uranium in rock-minerals, and its application to the measurement of geological time. *Proceedings of the Royal Society of London A* 85: 248–256.

Hughen, K., Lehman, S.J., Southon, J.R. et al. (2004). ^{14}C activity and global carbon cycle changes over the past 50,000 years. *Science* 303: 202–207.

International Atomic Energy Agency (2019). *Nuclear Power Reactors*. https://www.iaea.org/topics/nuclear-power-reactors. (Accessed 5 April 2019).

Kim, J., Ryan, P.C., Klepeis, K. et al. (2014). Tectonic evolution of a Paleozoic thrust fault influences the hydrogeology of a fractured rock aquifer, northeastern Appalachian foreland. *Geofluids* 14: 266–290. https://doi.org/10.1111/gfl.12076.

Kurz, M.D., Colodner, D., Trull, T.W. et al. (1990). Cosmic-ray exposure dating with in situ produced cosmogenic ^{3}He: results from young Hawaiian lava flows. *Earth and Planetary Science Letters* 97: 177–189.

Lal, D. (1991). Cosmic ray labeling of erosion surfaces: in situ production rates and erosion models. *Earth and Planetary Science Letters* 104: 424–439.

Langmuir, D. (1997). *Aqueous Environmental Geochemistry*. Upper Saddle River, New Jersey, USA: Prentice Hall.

Licciardi, J.M., Kurz, M.D., Clark, P.U., and Brook, E.J. (1999). Calibration of cosmogenic ^{3}He production rates from Holocene lava flows in Oregon, USA, and effects of the Earth's magnetic field. *Earth and Planetary Science Letters* 172: 261–271.

McCarthy, J.F. and Zachara, J.M. (1989). Subsurface transport of contaminants: Mobile colloids in the subsurface environment may alter the transport of contaminants. *Environmental Science and Technology* 23: 496–502.

Miyake, Y., Sugimura, Y., and Uchida, T. (1966). Ratio U^{234}/U^{238} and the uranium concentration in seawater in the western North Pacific. *Journal of Geophysical Research* 71: 3083–3087.

Montazer, P. and Wilson, W.E. (1984). Conceptual Hydrologic Model of Flow in the Unsaturated Zone, Yucca Mountain, Nevada. *USGS Water-Resources Investigations Report* 84-4345.

Murozomi, M.T., Chow, J., and Patterson, C.C. (1969). Chemical concentration of pollutant lead aerosols, terrestrial dusts and sea salts in Greenland and Antarctic snow strata. *Geochimica et Cosmochimica Acta* 33: 1247–1294.

Nishiizumi, K., Lai, D., Klein, J. et al. (1986). Production of ^{10}Be and ^{26}Al by cosmic rays in terrestrial quartz in situ and implications for erosion rates. *Nature* 319: 134–136.

Noakes, J.E., Kim, S.M., and Akers, L.K. (1967). Recent improvements in benzene chemistry for radiocarbon dating. *Geochimica et Cosmochimica Acta* 31: 1094–1096.

Nuclear Regulatory Commission (2019). *High-Level Waste Disposal*. https://www.nrc.gov/waste/hlw-disposal.html. (Accessed 5 April 2019).

Paces, J.B., Ludwig, K., Peterman, Z., and Neymark, L. (2002). ^{234}U/^{238}U evidence for local recharge and patterns of ground-water flow in the vicinity of Yucca Mountain, Nevada, USA. *Applied Geochemistry* 17: 751–779.

Paul, M., Fink, D., Hollos, G. et al. (1987). Measurement of ^{129}I in the environment after the Chernobyl reactor accident. *Nuclear Instruments and Methods in Physics Research* B29: 341–345.

Poreda, R.J. and Cerling, T.E. (1992). Cosmogenic neon in recent lavas from the western United States. *Geophysical Research Letters* 19: 1863–1866.

Reimer, P.J., Baillie, M.G.L., Bard, E. et al. (2004). IntCal04 terrestrial radiocarbon age calibration, 0-26 cal kyr BP. *Radiocarbon* 46: 1029–1058.

Reusser, L., Bierman, P., Pavich, M. et al. (2007). An episode of rapid bedrock channel incision during the last glacial cycle, measured with ^{10}Be. *American Journal of Science* 306: 69–102.

Rosman, K.J.R., Chisholm, W., Boutron, C.F. et al. (1993). Isotopic evidence for the source of lead in Greenland snows since the late 1960s. *Science* 362: 333–335.

Seong, Y.B., Owen, L.A., Yi, C., and Finkel, R.C. (2009). Quaternary glaciation of Muztag Ata and Kongur Shan: evidence for glacier response to rapid climate changes throughout the Late Glacial and Holocene in westernmost Tibet. *Geological Society of America Bulletin* 121: 348–365. https://doi.org/10.1130/B26339.1.

Shirahata, H., Elias, R.W., and Patterson, C.C. (1980). Chronological variations in concentrations and isotopic compositions of anthropogenic atmospheric lead in sediments of a remote subalpine pond. *Geochimica et Cosmochimica Acta* 49: 149–162.

Steiger, R.H. and Jäger, E. (1977). Subcommission on geochronology: convention on the use of decay constants in geo-

and cosmochronology. *Earth and Planetary Science Letters* 36: 359–362.

Stuiver, M., Reimer, P.J., Bard, E. et al. (1998). IntCal98 radiocarbon age calibration, 24,000–0 cal BP. *Radiocarbon* 40: 1041–1083.

Sturges, W.T. and Barrie, L.A. (1987). Lead 206/207 isotope ratios in the atmosphere of North America as tracers of US and Canadian emissions. *Nature* 329: 144–146. https://doi.org/10.1038/329144a0.

Suess, H.E. (1955). Radiocarbon concentration in modern wood. *Science* 122: 415–417.

Turekian, K.K. and Wedepohl, K.H. (1961). Distribution of the elements in some major units of the Earth's crust. *Geological Society of America Bulletin* 72: 175–192.

Vengosh, A., Hirschfeld, D., Vinson, D.S. et al. (2009). High naturally occurring radioactivity in fossil groundwater in the Middle East. *Environmental Science and Technology* 43: 1769–1775. https://doi.org/10.1021/es802969r.

Vinogradov, A.P. (1962). Average contents of chemical elements in the principal types of igneous rocks of the Earth's crust. *Geochemistry* 7: 641–664.

de Vries, H. and Barendsen, G.W. (1953). Radiocarbon dating by a proportional counter filled with carbon dioxide. *Physica* 19: 987–1003.

de Vries, H., Barendsen, G.W., and Waterbolk, H.T. (1958). Groningen radiocarbon dates II. *Science* 127: 129–137.

Wanty, R.C. and Nordstrom, D.K. (1993). Natural radionuclides. In: *Regional Ground-Water Studies* (ed. W.M. Alley), 423–441. New York: Van Nostrand Reinhold.

Ylagan, R.F., Kim, C.S., Pevear, D.R., and Vrolijk, P.J. (2002). Illite polytype quantification for accurate K-Ar age determination. *American Mineralogist* 87: 1536–1545.

Appendix I

Case Study on the Relationships Among Volatile Organic Compounds (VOCs), Microbial Activity, Redox Reactions, Remediation, and Arsenic Mobility in Groundwater

This case study presents an interesting example of interactions and reactions among inorganic species (e.g. iron hydroxides, arsenic), organic compounds (e.g. organochlorine solvents), and biotic factors (e.g. microbially mediated mineral dissolution). In order to provide the reader with the opportunity to analyze and interpret, results are presented prior to – and separately from – interpretations. This case study was selected because it integrates concepts presented throughout the text, including kinetics and thermodynamics, the roles of minerals and microbes in elemental cycling, naturally derived inorganic and anthropogenic organic contaminants in aqueous systems, oxidation–reduction and sorption–desorption reactions, stable isotopes (ongoing investigation) and numerous analytical approaches (X- ray diffraction [XRD], X-ray absorption spectroscopy [XAS], inductively coupled plasma mass spectrometry [ICP-MS], and ion and liquid chromatography).

The information is mainly from US EPA reports (e.g. Ford et al. 2008) and a pair of peer-reviewed journal articles (He et al. 2010; Ford et al. 2011), with additional updates from conference abstracts (Hildum et al. 2012; Ahmed et al. 2013). It is worth examining the original sources for the data and interpretations they contain, and also because they illustrate ways in which scientific results are published (e.g. focused 12-page journal article vs. comprehensive 193-page EPA report vs. 300-word conference abstract).

I.1 SITE INFORMATION, CONTAMINANT DELINEATION

The site is located at Fort Devens, a former military base in central Massachusetts, USA (Figure I.1). This appendix focuses on two areas: (i) Shepley's Hill Landfill (Figure I.2a), a disposal site for household waste and construction debris from 1917 to 1992; and (ii) an airfield two km north of the landfill where volatile organic compounds (VOCs) (chlorinated solvents and fuels) were stored in underground storage tanks (USTs) (Figure I.2b). The airfield is situated in a Pleistocene terrace.

In 1991 the EPA began investigations to delineate contamination:

1 Monitoring wells (MWs) drilled through the *landfill* into the sandy aquifer below, and also in groundwater downgradient from the landfill, revealed elevated concentrations of numerous inorganic trace elements, including As (as high as 15 000 $\mu g\,l^{-1}$; Hildum et al. 2012), Cd, Fe, Mn, and Pb. For reference, the US EPA maximum contaminant level (MCL) for As in drinking water is 10 $\mu g\,l^{-1}$ (10 ppb). VOCs were also detected at relatively low levels in

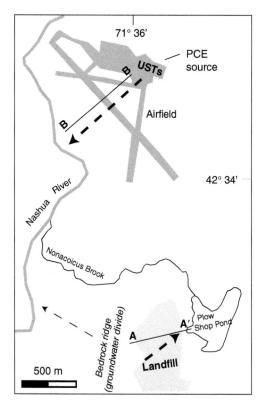

Fig. I.1 Location map of Shepley's Hill landfill (bottom center) and the airfield at Fort Devens US EPA Superfund site. Note cross-sections A-A' and B-B', which are shown in Figure I.2.

groundwater below the landfill in the 1990s: dichlorobenzenes comprised $11\,mg\,l^{-1}$ in monitoring well SHL-20, and 1,2-dichloroethane was reported at $9.9\,\mu g\,l^{-1}$ in monitoring well SHM-93-10C. The landfill occupies a low point in the pre-landfill topography, and some of the landfill is below the water table (Figure I.2a), suggesting reducing conditions in the deep part of the landfill. Interestingly, sediment cores into the aquifer below the landfill do not indicate elevated As in the solids of the aquifer, only in the groundwater itself (i.e. As is not elevated in the solid phase but is in the aqueous phase). Elevated As, Cd, Mn, Fe, Mn, and Pb also occur in sediments and deep water of Plow Shop Pond, located downgradient of the landfill.

2 Monitoring wells installed in the shallow sandy aquifer downgradient of the *VOC* source at the *airfield* documented the presence of perchloroethylene (PCE) as well as PCE-derived chlorinated metabolites (trichloroethylene [TCE], dichloroethene, vinyl chloride, dichloropropane, methylene chloride, and dichloroethane). Close to the leaking UST, PCE concentration in groundwater was as high as $30\,000\,\mu g\,l^{-1}$ (the EPA MCL is $5\,\mu g\,l^{-1}$). Concentrations of PCE in soil were as high as 5.5 micrograms per kilogram ($\mu g\,kg^{-1}$ or ppb) at 3 m below the surface

and $3.2\,\mu g\,kg^{-1}$ in groundwater sediments 15 m below surface. Given that PCE is synthetic, all PCE detected is anthropogenic (in contrast to As, which could be either naturally occurring or introduced by human activities).

I.2 REMEDIATION EFFORTS

Remediation at Shepley's Hill Landfill includes improving the cap on the landfill (to limit infiltration) and a pump-and-treat system downgradient from the *landfill* to remove As from groundwater. In order to address VOC contamination at the *airfield* a soil vapor extraction (SVE) system was installed adjacent to the former UST area in 1993 and later, after documenting the PCE plume migrating downgradient (Figure I.2b), an enhanced reductive dechlorination (ERD) remediation system was installed to address VOC contamination (refer to Figure 3.26 and adjacent section in Chapter 3 for background on reductive breakdown of VOCs).

After many years of ERD treatment for VOCs at the airfield, monitoring showed the following (He et al. 2010):

1 *Upgradient of ERD injection (and within the PCE plume).* Concentrations of redox-sensitive elements dissolved in groundwater are: As $< 1\,\mu g\,l^{-1}$, Fe $< 0.1\,mg\,l^{-1}$, Mn $= 0.1\,mg\,l^{-1}$, and SO$_4$ $= 16.4\,mg\,l^{-1}$ (MW-u, Figure I.2b);

2 *Downgradient of ERD injection (MW-1d).* As, Fe, and Mn were significantly elevated (As $= 680\,\mu g\,l^{-1}$, Fe $= 420\,mg\,l^{-1}$, and Mn $= 5.0\,mg\,l^{-1}$), whereas SO$_4$ had decreased to $5.0\,mg\,l^{-1}$.

3 *Far downgradient of ERD injection (MW-2d).* Conditions are similar to the upgradient MW: As, Fe, and Mn are undetectable and SO$_4$ $= 17.7\,mg\,l^{-1}$.

I.3 SOURCES OF PCE AND ARSENIC

The source of the *VOCs* (PCE and its metabolites) is clear – leaking USTs and disposal of used PCE into a dry well allowed these compounds to seep down into soil and then into groundwater, where the relatively high solubility of PCE (Table 3.4) and potential to behave as a DNAPL (dense non-aqueous phase liquid) (Figure 3.25) facilitated its dispersion in the aquifer.

The source of elevated *As and other inorganic elements* below and downgradient of the landfill, and in the vicinity of the ERD treatment, is open to interpretation (Ahmed 2014). If As, Cd, Fe, Mn, and Pb were among the materials thrown into the landfill, then leachate from the landfill itself could be an important source. Unconsolidated sand and gravel aquifers with 1–$20\,mg\,kg^{-1}$ As are known to

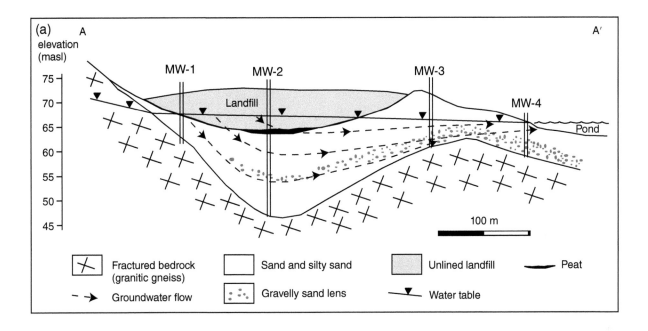

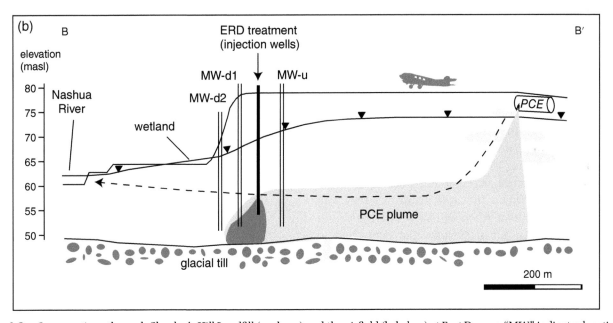

Fig. I.2 Cross-sections through Shepley's Hill Landfill (a, above) and the airfield (b, below) at Fort Devens. "MW" indicates locations of some of the dozens of monitoring wells. Source: From Ford et al. (2008).

produce groundwater with concentrations of dissolved As like those seen in the Shepley's Hill landfill plume, so the possibility that the As and other inorganics may be derived from the natural sediments or bedrock of the aquifer system must also be considered.

The chemical composition of landfill leachate would be a useful dataset (Strudwick 1999; Christensen et al. 2001) to help with interpretation.

Landfill leachate characteristics:

1 As tends to occur within the range of 10–1000 µg l^{-1};
2 Cl$^-$ is often indicative of leachate – it tends to occur in the range of 150–4500 mg l^{-1}; and
3 organic compounds tend to be elevated – aromatic hydrocarbons (e.g. benzene) usually comprise 10–10 000 µg l^{-1} and organochlorines (e.g. PCE, vinyl chloride) tend to range from 3 to 4000 µg l^{-1}.

Comparing leachate data to groundwater below the landfill:

1 As concentrations as high as $15\,000\ \mu g\,l^{-1}$ in the groundwater below the landfill, suggesting that landfill leachate is not the sole source of As;

2 the presence of VOCs confirms infiltration of leachate through the peat layer and into groundwater; and

3 Cl^- concentrations $< 40\ mg\,l^{-1}$ suggest that landfill leachate is a minor component of the groundwater below the landfill.

The evidence from MWs installed in the aquifer below the landfill suggest that leachate from the unlined landfill has infiltrated the aquifer, but also that the leachate is not responsible for transporting the high levels of As and other inorganics from landfill into aquifer. Rather, it appears that processes within the aquifer may be causing As and other inorganics to be mobilized.

I.4 MOBILIZATION OF ARSENIC

High amounts of dissolved organic carbon (DOC; from decomposing food, paper, septic/sewage waste, etc.) in landfill leachate support diverse microbial populations and foster reducing/anaerobic conditions in aquifers below landfills (Christensen et al. 2001). In reducing environments like this, iron hydroxides are thermodynamically unstable and undergo microbially mediated reductive dissolution. The iron hydroxide is used by the microbes as an electron acceptor and if As, Pb, or other elements are sorbed to the iron hydroxide, they will be released to solution when $Fe(III) \rightarrow Fe(II)$ and iron hydroxide dissolves. Kinetic factors that contribute to dissolution rate are abundance and species of microbes, abundance of DOC, and other aqueous factors such as pH and detachment rate/solubility of resulting Fe species.

Sequential chemical extraction and XAS indicate that As is sorbed to Fe-hydroxides in Fort Devens aquifer sediments (He et al. 2010), so a likely mechanism for release of As into solution (in the presence of microbes) is:

$$4Fe_{0.99}As_{0.01}(OH)_3 + CH_2O + 7H^+ \rightarrow 3.96Fe^{2+}$$

$$+ 0.04As(III) + 10H_2O + CO_2 + OH^- \qquad (I.1)$$

The heavy loading of organic matter (here shown as CH_2O) from the landfill, perhaps augmented by organic matter seeping out of the peat sediments below the landfill, is a probable cause of the high concentrations of As observed. The effect of reduction on iron hydroxide is shown in the Eh–pH diagram in Figure I.3, and given

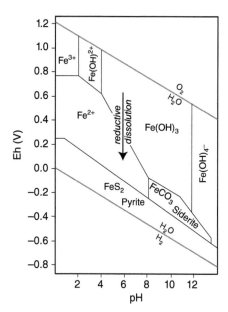

Fig. I.3 Eh–pH diagram illustrating destabilization of Fe-hydroxides associated with shift from oxidizing to reducing conditions. Refer to Figure 4.15 for details.

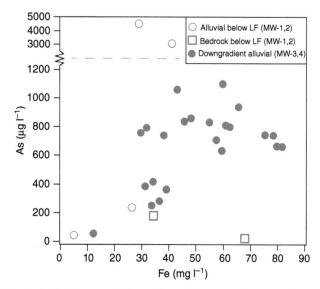

Fig. I.4 Positive correlation of Fe and As is consistent with dissolution of iron hydroxides as source of As in alluvial (sand and gravel) aquifer. "MW" = monitoring wells, locations of which are shown in Figure I.2. "LF" = landfill.

that Cd and Pb commonly sorb to Fe-hydroxides, this too could explain their high values. High Mn in the aquifer is probably due to reductive dissolution of Mn oxides by a process similar to that shown in Eq. (I.1).

Reductive dissolution of iron hydroxides as an important As source is also supported by the positive correlation of Fe and As in aquifer sediments (Figure I.4).

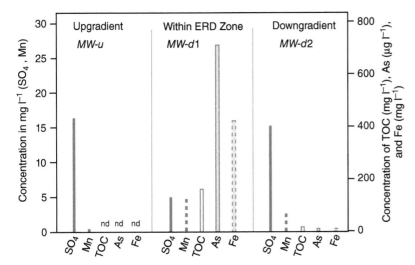

Fig. I.5 Concentrations in solution of redox-sensitive species SO_4, Mn, TOC, As, and Fe, from upgradient of ERD injection (MW-u), through the main ERD injection area (MW-d1), and downgradient of ERD injection (MW-d2). Decreasing SO_4 in conjunction with increasing Mn, TOC, As, and Fe are indication of strongly reducing conditions that cause $SO_4 \rightarrow S^{2-}$ and reduction of Fe-hydroxides and Mn-oxides. Far downgradient (MW-d2), Fe-hydroxides are stable and As is sorbed out of solution. Source: From data in He et al. (2010).

The fate of As mobilized below the landfill is also related to redox processes. Given that groundwater below the landfill flows into Plow Shop Pond (Figure I.1, right side of Figure I.2a) or into Nonacoicus Brook, both of which are relatively oxidizing environments compared to the groundwater, iron hydroxide precipitates in sediments at the interface of groundwater and pond water (Ford et al. 2011). Reaction (I.1) proceeds in reverse, incorporating As, Pb, and other trace elements into solid iron hydroxides at the redox boundary (of reducing groundwater and oxidizing surface water).

The effect of ERD on concentrations of SO_4, Mn, TOC, As, and Fe in groundwater is shown in Figure I.5, which depicts a transect through the ERD injection area, from upgradient of the injection to within the ERD zone to downgradient. The ingredient used to stimulate ERD was molasses injected into wells installed into the PCE plume (Figure I.2b) – molasses is the source of organic carbon that fosters reducing conditions and stimulates microbial activity.

The role of microbes in enhancing reduction of Fe-hydroxides is supported by the experiments of He et al. (2010) which reproduced bacteria-facilitated reduction of Fe-hydroxide in the presence of molasses. The enhanced mobility of As following ERD is not surprising – the EPA recognizes this as a potential consequence of ERD – but when faced with a complicated problem like a deep plume of PCE migrating in the direction of a river, sometimes there are no foolproof solutions, as is often the case with complex environmental problems.

REFERENCES

Ahmed, S. (2014). Stable Isotopic Study of Groundwater Arsenic Contaminated Plume at Shepley's Hill Landfill. M.Sc. Thesis, Boston College, Massachusetts, USA. https://dlib.bc.edu/islandora/object/bc-ir:101228/datastream/PDF/view

Ahmed, S., Hon, R., Hildum, B., and Xie, Y. (2013). Stable isotopes (δD and $\delta^{18}O$) fractionation in groundwater in the vicinity of an arsenic contaminant landfill plume in central Massachusetts. *Geological Society of America* 45 (1): 67.

Christensen, T.H., Kjeldsen, P., Bjerg, P.L. et al. (2001). Biogeochemistry of landfill leachate plumes. *Applied Geochemistry* 16 (7–8): 659–718.

Ford RG, Scheckel KG, Acree SD et al. 2008. Final Report Arsenic Fate, Transport and Stability Study Groundwater, Surface Water, Soil and Sediment Investigation, Fort Devens Superfund Site, Devens, Massachusetts.

Ford, R.G., Acree, S.D., Lien, B.K. et al. (2011). Delineating landfill leachate discharge to an arsenic contaminated waterway. *Chemosphere* 85 (9): 1525–1537.

He, Y.T., Fitzmaurice, A.G., Bilgin, A. et al. (2010). Geochemical processes controlling arsenic mobility in groundwater: a case study of arsenic mobilization and natural attenuation. *Applied Geochemistry* 25: 69–80.

Hildum, B., Hon, R., Ahmed, S., and Simeone, R. (2012). Arsenic speciation and groundwater chemistry at Shepley's Hill landfill, Devens, Massachusetts. *Geological Society of America* 44 (2): 46.

Strudwick, D.G., 1999. Leachate chemistry of two modern municipal waste landfills in Melbourne, Victoria. Masters thesis, University of Melbourne.

Case Study of PFOA Migration in a Fractured Rock Aquifer: Using Geochemistry to Decipher Causes of Heterogeneity

Unlike the volatile organic compounds (VOCs) in the Appendix A case study, PFOA (perfluorooctanoic acid) is a very stable and very soluble organic compound. The strong covalent C–F bonds in PFOA make it resistant to decay (Section 3.1.9), largely because the C atoms are virtually inaccessible to microorganisms who seek an electron donor, and this prevents microbial degradation. In early 2016, PFOA was detected in private water wells at concentrations that exceed the state health-based action level of $20\,ng\,l^{-1}$ (parts per *trillion*) near a manufacturing plant in southwestern Vermont, USA. (Figure II.1) Within a month, 29 out of 34 wells tested within a 1.5 km radius of the now-disused plant had PFOA levels above $20\,ng\,l^{-1}$. More testing of wells in a progressively wider area ($40\,km^2$) had determined that 289 out of 537 (54%) wells contain PFOA levels greater than $20\,ng\,l^{-1}$, many with values $> 1000\,ng\,l^{-1}$. The problem is not static either – some wells that initially contained < 20 ppt PFOA now contain levels above this threshold.

As an emerging contaminant, relatively little is known about PFOA fate and transport, especially in the complex fractured rock aquifer (FRA) systems that are a common drinking water source in northeastern North America (as well as large areas on all continents). Several factors contribute to the significance of this case study in SW Vermont. First, the means by which PFOA was introduced to the land surface and aquifer – airborne dispersion of vapor emissions, then nearby deposition onto the land surface, then leaching through soils – is a common pathway of

PFOA release to the environment, and is one that is the cause of PFOA-contaminated groundwater at hundreds of sites globally. Therefore, results from this case study should be applicable to predicting how other aquifers contaminated by PFOA will behave. Second, a 2018 Google Scholar search for "PFOA Fractured Rock Aquifer" revealed a very limited body of work, emphasizing the need for predictive models to understand PFOA fate and transport in FRAs. This latter point becomes especially important when considering the high toxicity of PFOA and related perfluorinated chemicals (PFCs), expanding recognition of PFOA-contaminated watersheds, and the reality that the high stability and solubility of PFOA mean that it will persist and migrate over time. Predicting behavior can enable informed management of water resources and drinking water supplies.

The sources of information for this case study include government documents (e.g. Vermont Geological Survey reports and maps), conference abstracts, as-yet unpublished data, and peer-reviewed articles (e.g. Xiao et al. 2015; Yingling 2015; Hu et al. 2016) that provide information on PFOA sources and behavior in soils and unconsolidated aquifers. This SW Vermont case study is the result of a multidisciplinary collaboration headed up by Jonathan Kim of the Vermont Geological Survey with colleagues from local universities (Bennington College, U. Massachusetts, Middlebury College, SUNY-Plattsburgh) and government agencies (EPA, USGS). It integrates multiple approaches, including determination of PFOA

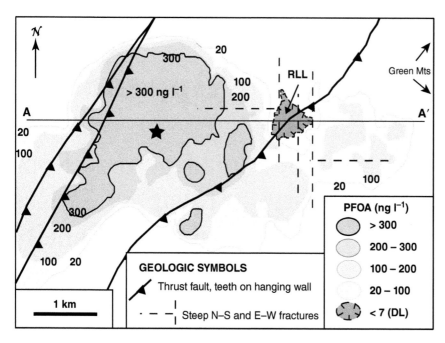

Fig. II.1 Contour plot of PFOA concentration in the fractured rock aquifer of the Bennington area, southwestern Vermont, USA. The star is the point source of PFOA vapor vented to air from an industrial site, and "RLL" is the Rice Lane Low, an anomalous cluster of wells with no detectable PFOA. The A-A′ line corresponds to the cross-section sketch in Figure II.4. DL refers to the 7 ng l^{-1} detection limit for PFOA measurements. Geologic features are from Kim (2017a) and PFOA values are contoured from Vermont DEC data, both available at: https://dec.vermont.gov/geological-survey/groundwater/town-gw/bennington.

concentration in water and soil, inorganic chemistry of groundwater and surface water, hydrogen and oxygen isotope analysis of groundwater and surface water, age-dating of groundwater by chlorofluorocarbons (CFCs) and tritium (^{3}H), and geologic mapping and geophysics to map fractures and flow paths.

II.1 GEOLOGIC FRAMEWORK

The FRA is situated in a broad valley at the base of the Green Mountains. The aquifer is composed of faulted, folded, and fractured Paleozoic limestones, dolostones, shales, and quartzites (Kim 2017a, b). Thrust faults dip shallowly to the east; fractures are nearly vertical, mainly oriented either N–S or E–W (Figure II.1). The aquifer is overlain by surficial deposits of glacial till, outwash, and floodplain sediments, much deposited during and after retreat of the continental glaciers at the end of the Pleistocene epoch. Variability in permeability and sorption potential likely has influenced the rate of PFOA migration from the surface to the aquifer below (Schroeder et al. 2018). The aforementioned geologic structures also likely exert a strong control on groundwater flow.

II.2 INORGANIC CHEMISTRY OF GROUNDWATER

Table II.1 presents data from groundwater constituents that help to differentiate groundwater types in the aquifer.

The groundwater in high-PFOA wells is characterized by a Ca–Mg–HCO$_3$ signature that is typical of aquifers in carbonate rocks (N = 12). In the western part of the study area, low-PFOA groundwater in shaly rocks is distinguished by relatively high SiO$_2$, SO$_4$, and Sr derived respectively from clay minerals, oxidation of pyrite, and calcite veins (Sr^{2+} for Ca^{2+} substitution) (N = 7). A significant anomalous groundwater signature occurs in a series of wells with no PFOA (the "Rice Lane Low," RLL) that have high K$^+$ and SiO$_2$ (H$_4$SiO$_4$) and low Na$^+$, Mg^{2+}, Ca^{2+}, Cl$^-$, SO$_4$$^{2-}$, and PFOA (relative to the dominant carbonate rock signature) (N = 8). Figure II.1 highlights the location of the RLL, which is notable because it is a PFOA-free zone surrounded by high-PFOA wells, all of which occur within the boundaries of PFOA air deposition. Also notable is that the RLL wells produce water from carbonates, yet the chemical signature imparted by bedrock is different from the carbonates in the high-PFOA areas (Ryan et al. 2018).

One graphical way to assess compositional differences (or similarities) is to create spider plots, an approach often used in igneous geochemistry, and one used here to visually depict differences in groundwater types. Spider plots show data normalized to a reference sample; in some cases, this might be a certified standard, or merely a representative sample that works for the analysis. The spider diagram in Figure II.2 was created by dividing values in Table II.1 corresponding to Carbonates and RLL by the corresponding value from the average groundwater for the region (bottom row in Table II.1). For example, if the average total dissolved solid (TDS) value for the carbonates (247 mg l^{-1}) were divided by the reference value of

Table II.1 Mean values for certain groundwater types in the study area (with standard deviations below in italics).

	PFOA	TDS	HCO_3	pH	Ca	Mg	Sr	Na	Cl	NO_3	SO_4	K	SiO_2
Carbonates	768	247	249	7.3	68.5	30.2	92.9	3.8	14.2	1.8	14.5	0.8	8.7
	939	*25*	*41*	*0.1*	*4*	*1.7*	*3.8*	*1.5*	*9*	*0.4*	*2.5*	*0.2*	*1.6*
Rice Ln low	< 7	119	153	7.4	34.0	17.7	72.4	1.8	< 1.0	0.2	6.0	1.3	13.2
	NA	*5.7*	*7.3*	*0.2*	*1.6*	*1.3*	*7.3*	*0.8*	*NA*	*0.1*	*2.9*	*0.1*	*3.3*
Shales	12.0	112	182	7.5	63.2	14.4	334	6.9	6.7	0.0	30.2	0.5	15.3
	8.7	*14.6*	*28.9*	*0.1*	*5.1*	*4.6*	*57.4*	*2.7*	*6.0*	*0.0*	*6.3*	*0.1*	*1.7*
Regional avg	269	203	221	7.4	63.3	25.4	119	24.4	46.3	0.8	14.9	1.1	8.8

Units are mg l^{-1} except for PFOA (ng l^{-1}) and Sr (μg l^{-1}). Nitrate is NO$_3$-N.

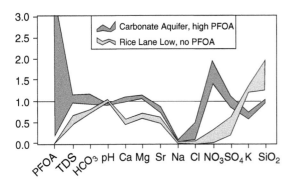

Fig. II.2 Spider plot as visual means of discriminating water types; here, the RLL is distinct from regional PFOA-rich carbonate groundwater. The gray band encompasses values for a cluster of six wells with high PFOA, and the RLL band encompasses data for five samples. (Values for all samples are plotted, so where the range is high, the band is thick, e.g. PFOA in carbonate signature wells.) PFOA in the carbonate aquifer actually reaches eightfold greater than the reference (269 ng^{-1}), so the highest PFOA values are off the chart.

203 mg l^{-1}, it would plot as 1.2 on the spider diagram. If the sample were to have the same value as the reference standard, it would plot as 1.0 on the spider plot. Nitrate in the carbonate aquifer is more than double nitrate in the regional average reference standard, so it plots accordingly. The spider plot helps to reinforce the interpretation that the groundwater in the RLL (i) has no PFOA, and (ii) is compositionally distinct from surrounding carbonate groundwater with high PFOA.

What is the cause of the RLL geochemical anomaly? RLL groundwater is depleted in most constituents, including anthropogenic ones (PFOA, likely Na, Cl, and NO$_3$) and rock-derived ones (Na, Mg, Ca, and SO$_4$). K and SiO$_2$ are the only constituents with higher concentrations in RLL water relative to the regional carbonate signature. Upgradient watersheds in the Green Mountains that feed the PFOA-contaminated area are underlain by crystalline rocks with likely enough potassium feldspar and micas to

supply the K–Si signature of the RLL, but they are > 10 km away from the RLL. The low Ca–Mg–HCO$_3$ of the RLL water indicates that it has not interacted to any great extent with the carbonate rocks that it is pumped out for drinking water, yet high SiO$_2$ suggests high residence time (assuming silicate weathering is the Si source).

II.3 STABLE ISOTOPE COMPOSITIONS OF GROUNDWATER

Stable isotopes of hydrogen and oxygen are excellent tracers of groundwater flow because they are more conservative than even classically conservative tracers like soluble ions (e.g. Na$^+$, K$^+$, Ca^{2+}, Sr^{2+}, Cl$^-$, Br$^-$, NO$_3^-$, SO$_4^{2-}$, UO$_2$[CO$_3$]$_2^{2-}$) or soluble and relatively stable organic compounds (e.g. tracer dyes, PFOA). The δD and δ^{18}O signatures come from the water molecules themselves, making isotopic analysis a very good tracer of water source and cycling, from atmosphere to surface water and groundwater.

The δD and δ^{18}O values of precipitation are controlled by the temperature of condensation, where in a practical sense, isotopically lighter (more negative) δD and δ^{18}O values correspond to precipitation that falls at higher elevations compared to heavier water that falls at lower elevations (Hoefs 2009). Applying this reasoning to tracing flow paths in the broad valley aquifer of this case study, water that infiltrates at higher elevations – and would arrive via intermediate to regional flow paths – carries isotopically lighter δD and δ^{18}O signatures than isotopically heavier precipitation that falls in valleys and recharges the aquifer via local flow.

The results of isotopic analysis (N = 24) are compelling. Groundwater coming from wells in the RLL is isotopically the lightest water in the region, with δD from −68‰ to −74‰ and δ^{18}O from −10.5‰ to −11‰. These values are comparable to the isotopic composition of springs in the higher elevations of the Green Mountains to the east

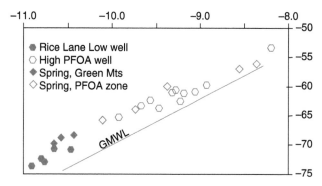

Fig. II.3 Isotopic compositions of hydrogen and oxygen in water from PFOA-free wells (RLL), high-PFOA wells, springs in the higher elevations of the Green Mountains to the east, and springs from the lower elevation valley where PFOA deposition was greatest. (x-axis is $\delta^{18}O$ and y-axis is δD, both VSMOW ‰).

(Figure II.3). The RLL values are distinct from heavier groundwater coming from high-PFOA wells in carbonates and surficial deposits (Figure II.3), with δD from $-53‰$ to $-65‰$ and $\delta^{18}O$ values from $-8.2‰$ to $-10‰$. Springs in the valley where the high-PFOA wells occur have isotopic compositions that overlap with the high-PFOA wells (Belaval et al. 2018).

The similarity in δD and $\delta^{18}O$ of RLL water and springs in the Green Mountains is consistent with the *hypothesis* that RLL water – geochemically distinct and isotopically similar to high elevation springs – is derived from regional flow, from a recharge area in the Green Mountains along pathways that impart a K–Si signature. In contrast, high-PFOA Ca–Mg–HCO$_3$ water appears to recharge locally, in the valley, where PFOA percolates downward in water that carries the isotopically light signature of valley precipitation and recharge, and the compositional signatures of the carbonates that dominate the valley where the main PFOA deposition occurred. This is a *hypothesis that is testable* by determining groundwater ages. The RLL water – if this hypothesis is true – should be much older than locally recharged water. RLL water presumably would have followed much longer flow paths than the high-PFOA locally recharged groundwater.

II.4 GROUNDWATER AGE-DATING

Concentrations of CFCs, sulfur hexafluoride (SF$_6$), and tritium (^3H) in groundwater from the RLL and the high-PFOA carbonate signature wells provide valuable information on recharge ages (Shanley et al. 2018). In a concise summary, CFCs and SF$_6$ are synthetic compounds that have varied in atmosphere concentration over time,

so analysis of multiple CFCs (e.g. CFC-11, CFC-12, and CFC-113) in combination with SF$_6$ is a useful tool for determining recharge age, i.e. when water seeped into soil and began to infiltrate the groundwater system. (Ages are often an average, depending on how much mixing in the subsurface occurs between waters of different recharge ages.) As one of the radioactive decay products of nuclear bomb tests, ^3H first appeared above background levels in 1952 and peaked in 1963–1964. With a half-life of 12.3 years, concentrations of ^3H in precipitation have decreased since then.

Groundwater in the RLL has very low CFC and SF$_6$ concentrations that correspond to recharge ages of 56 to 70 years ago ($\sim$1950s), as well as very low ^3H concentrations (≤ 1 tritium unit, TU) that are consistent with recharge prior to the early 1960s. High-PFOA carbonate signature wells have CFC recharge ages of 26 to 32 years ago (mid-1980s) that are corroborated by TU values of 4.6–6.5. Thus, groundwater age analysis indicates that RLL water infiltrated two or three decades before industrial use of PFOA began, whereas the high-PFOA carbonate signature wells have average recharge ages that correspond to the period when PFOA was emitted to air, i.e. late 1970s to 2002.

II.5 CONCEPTUAL MODEL FOR THE GROUNDWATER SYSTEM

Understanding high PFOA levels in the Ca–Mg–HCO$_3$ wells is relatively straightforward: PFOA was deposited on the land surface and then infiltrated downward through carbonate-bearing sediments and into the carbonate aquifer (Figure II.4). Deciphering the cause of the RLL anomaly is more complicated, but important given that it is a PFOA-free water source in an area otherwise contaminated by PFOA. RLL water differs from surrounding Ca–Mg–HCO$_3$ water by its higher amounts of dissolved K and SiO$_2$ and lower δD and $\delta^{18}O$; the former is consistent with weathering in phyllitic watersheds in the Green Mountains to the east, and the latter an isotopic signature of high elevation recharge. Upland crystalline bedrock and overlying tills are a significant source of recharge to valley aquifers in northeastern North America (Boutt 2017), and older recharge dates of RLL water ($\sim$1955) compared to the Ca–Mg–HCO$_3$ water ($\sim$1985) are consistent with relatively long flow paths of RLL water.

Although the RLL waters are old (for this area) – i.e. they have high residence time in the subsurface, – they appear to be far below saturation with respect to Ca, Mg, and HCO$_3$. This means that the groundwater in the RLL

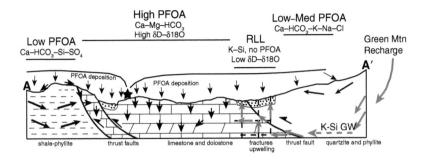

Fig. II.4 Conceptual model of groundwater flow paths for cross section A-A′ in Figure II.1 resulting in high PFOA in Ca–Mg–HCO$_3$.

has spent little time in the carbonate rocks where the RLL is located. Upgradient rocks to the east are phyllites and quartzites (Figure II.4), a likely source of old, K-rich, HCO$_3$-poor, upland-derived water. Geologic structures clearly play an important role: the fractures associated with the RLL (and mapped at the surface) appear to be major conduits to flow and have been identified as producing zones in wells (Romanowicz and Kim 2018).

Even if RLL water is derived from an area where PFOA was deposited at the surface, it may be decades before RLL wells become contaminated, based on recharge ages. RLL water appears to have infiltrated the subsurface 20 to 30 years before PFOA emissions began in the late 1970s. These results are still preliminary, and data collection and analysis is ongoing, but information gleaned thus far serves as an interesting example of environmental geochemistry applied to contaminant fate and transport and water resource assessment.

REFERENCES

Belaval, M., Boutt, D.F., Schroeder, T. et al. (2018). Characterizing the groundwater-suface water system in a PFOA-contaminated fractured rock aquifer using radon and stable isotopes. *Geological Society of America Abstracts with Programs* 50: https://doi.org/10.1130/abs/2018NE-310939.

Boutt, D.F. (2017). Assessing hydrogeologic controls on dynamic groundwater storage using long-term instrumental records of water table levels. *Hydrogeological Processes* 2017: 1–20. https://doi.org/10.1002/hyp.11119.

Hoefs, J. (2009). *Stable Isotope Geochemistry*, 6e. Berlin: Springer-Verlag.

Hu, X.C., Andrews, D.Q., Lindstrom, A.B. et al. (2016). Detection of poly- and perfluoroalkyl substances (PFASs) in U.S. drinking water linked to industrial sites, military fire training areas, and wastewater treatment plants. *Environmental Science and Technology Letters* 3: 344–350. https://doi.org/10.1021/acs.estlett.6b00260.

Kim, J.J. (2017a). (Draft) Preliminary Bedrock Geologic Map of the Bennington Area, Vermont. *Vermont Geological Survey Open File Report* VG2017-4A, scale 1:12,000.

Kim, J.J. (2017b). (Draft) Preliminary Fracture Map of the Bennington Area, Vermont. *Vermont Geological Survey Open File Report* VG2017-4B, scale 1:12,000.

Romanowicz, E. and Kim, J.J. (2018). Relationship between fractures, vertical water flow and geologic structures on the distribution of PFOA in domestic wells, Bennington, Vermont. *Geological Society of America Abstracts with Programs* 50: https://doi.org/10.1130/abs/2018NE-311104.

Ryan, P.C., Kim, J.J., Norris, E.D., and Allen, D. (2018). Tracing groundwater flow by inorganic hydrogeochemistry: a tool to understanding PFOA migration in a fractured rock aquifer. *Geological Society of America Abstracts with Programs* 50: https://doi.org/10.1130/abs/2018NE-311303.

Schroeder, T., Kim, J.J., and Ryan, P.C. (2018). Widespread PFC contamination by aerosol deposition in Bennington, Vermont: a long-term problem due to retention in vadose zone soils. *Geological Society of America Abstracts with Programs* 50: https://doi.org/10.1130/abs/2018NE-311080.

Shanley, J.B., Mack, T.J., and Levitt, J.P. (2018). Groundwater age-tracers shed light on the nature of PFOA transport in the N. Bennington, Vermont bedrock aquifer. *Geological Society of America Abstracts with Programs* 50: https://doi.org/10.1130/abs/2018NE-311118.

Xiao, F., Simcik, M.F., Halbach, T.R., and Gulliver, J.S. (2015). Perfluorooctane sulfonate (PFOS) and perfluorooctanoate (PFOA) in soils and groundwater of a U.S. metropolitan area: Migration and implications for human exposure. *Water Research* 72: 64–74.

Yingling, V. (2015). Karst influence in the creation of a PFC megaplume. *14th Sinkhole Conference, NCKRI Symposium* 5.

Appendix III

Instrumental Analysis

Environmental geochemistry is characterized by dozens of methods of instrumental analysis applied to determining concentrations of ions or molecules in solution, of mineral types or compositions, of speciation of trace elements in minerals, and of isotopic compositions of minerals, waters, or gases, and more. The following section provides some fundamental concepts and applications of instrumental analysis in geochemistry. It is not all-inclusive but does provide insight into some common methods.

III.1 ANALYSIS OF MINERALS AND CRYSTAL CHEMISTRY

Minerals in soils and sediments are commonly too fine-grained to be identified in hand specimen or by standard petrographic microscopes. Instead, their identification and analysis requires high-powered electron microscopy for imaging of crystal forms and diffraction and vibrational spectroscopy for determination of crystallographic structures. Regardless of crystal size, elemental analysis is achieved using approaches that typically measure elemental absorbance, emission, fluorescence, or mass. Some methods require destruction of the sample for analysis, particularly those that employ strong acids to dissolve minerals into solution; while these methods require dissolution into an aqueous solution for chemical analysis of solids, their benefit is that they are also capable of determining compositions of natural waters (e.g. in

soils, streams, aquifers, seas). Other methods, like electron microscopy, X-ray fluorescence (XRF) and X-ray diffraction (XRD), are capable of analyzing mineral powders and hence are not sample-destructive, but they are not capable of measuring dissolved ions in water. The following sections present some basic information on analysis of minerals and elements by instrumental methods.

III.1.1 Electron microscopy (SEM, TEM, and many other acronyms)

Samples analyzed by electron microscopy commonly consist of single mineral grains, polished sections of rock, or dispersed grains of sediment or soil that are coated with carbon, copper, or gold. There are two main approaches to electron microscopy, both of which involve directing a high-energy electron beam at the sample. The electron beam is scattered by the sample in *scanning electron microscopy* (SEM) and it is transmitted through the sample in *transmission electron microscopy* (TEM).

One of the main advantages of SEM and TEM is the imaging of morphological properties such as crystal shape, pore size, and spatial relationships of crystals and other solids (including microbes and amorphous material). Electron microscopy, particularly high-resolution transmission electron microscopy (HRTEM) also facilitates microanalysis of single mineral grains, even when they are exceedingly fine-grained, an attribute lacking

Environmental and Low-Temperature Geochemistry, Second Edition. Peter Ryan.
© 2020 John Wiley & Sons Ltd. Published 2020 by John Wiley & Sons Ltd.

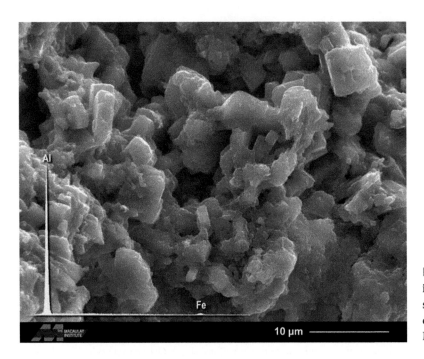

Fig. III.1   Scanning electron microscope (SEM) image and energy dispersive X-ray (EDX) spectrum (in yellow) of gibbsite. Source: Image obtained with assistance of Evylene Delbos, Macaulay Institute (now Hutton Institute).

Fig. III.2   TEM image of tropical soil minerals, including kaolinite (note hexagonal crystal form) and smectite (lower center). Compare the scales of these SEM and TEM images – 2000 Å = 0.2 μm, so the field of view in Figure III.1 (SEM) is ~50 times that of the TEM image. Source: Image obtained with assistance of María del Mar, Centro de Instrumentación Científica (Granada, Spain).

in other methods. Secondary X-rays produce an elemental spectrum that yields compositional information that enables determination of major element compositions of minerals – this type of microanalysis is known as energy dispersive X-ray analysis (variably abbreviated as EDX, EDS, and EDAX). An SEM image of gibbsite [Al(OH)$_3$] from a tropical soil and its elemental spectrum are presented in Figure III.1. Note the form of the 1–5 μm gibbsite crystals as well as the spectrum indicating trace amounts of Fe associated with the gibbsite.

Resolution of images produced by TEM is much greater than SEM. In Figure III.2, individual hexagonal crystals of 0.1 μm kaolinite are visible.

Electron microprobe analysis (EMP or EMPA) is similar to EDX analysis in an SEM or TEM. A solid sample (powder or rock chip) is bombarded with an electron beam, and the emitted X-rays occur at wavelengths characteristic to given elements.

III.1.2 XRD

XRD provides a method for identification of crystalline phases, not by direct imaging as in electron microscopy, but rather by diffraction peaks produced by constructive interference of coherently scattered X-rays. It was the method used by Linus Pauling in the 1920s and 1930s to determine the structures of micas and other phyllosilicates, and by Watson and Crick in the 1950s to decipher the structure of DNA. In modern environmental geochemistry its main application is the determination of mineral assemblages of soils, sediments, and powdered rocks, hence it is sometimes referred to as X-ray powder diffraction or variations thereof, as opposed to single-crystal XRD methods used in crystallography. Sample preparation is generally minimal except where mineral quantification is desired.

III.1.2.1 Bragg's Law

The X-rays used in XRD are typically generated from a copper tube (sometimes cobalt or other metals), collimated and aimed at the sample. Bragg's Law, formulated in 1912 by the father–son team of W.H and W.L Bragg, describes how diffraction of an X-ray beam can be applied to analysis of crystal structure. The geometry of XRD as it pertains to Bragg's Law is presented in Figure III.3. Note that an incoming X-ray beam is depicted very schematically as only three in-phase rays that are diffracted by planes of atoms in a crystalline lattice.

Bragg's Law permits determination of d-spacing of a unit cell. Most minerals have a relatively unique set of d-spacings that, when combined with a relatively unique set of XRD peak intensities, provides data that enable mineral identification. When analyzing an unknown mineral, d is the variable (Figure III.3). We know θ (theta), the angle between the incident beam and the crystallographic planes in question, because the diffractometer has

a readout that displays the angle of incidence. In terms of this geometric approach, the diffracted beam is effectively reflected, so the angle θ^* is equal to θ, and the angles EAB and BAF are also equal to θ. We also know the wavelength λ, which for CuKα radiation is 1.5418 Å (Cu tubes are the most common X-ray source in powder XRD).

In order for the incoming, in-phase X-ray beam to be diffracted coherently (as opposed to destructively, where out-of-phase X-rays cancel and there is no peak), the reflected rays must be in-phase – in this example, they are depicted as in-phase when they cross line O–O′, so a diffraction peak will be produced. In order for this to occur, rays 2–2* and 3–3* (and others) must travel one, two, three (or some other whole integer) wavelength(s) more than is traveled by ray 1–1* – this is the case in Figure III.3, where ray 2–2* travels two complete wavelengths between points E and F, and ray 3–3* travels four complete wavelengths between points D and G.

All that remains to derive Bragg's Law is to determine the relationship between λ, θ, and d. Working with triangles AEB and AFB (keeping in mind that AB = d-spacing), $\sin\theta = $ EB/AB and $\sin\theta^* = $ BF/AB. Rearranged slightly, AB·$\sin\theta$ = EB and AB·$\sin\theta^*$ = BF. Given that AB = d and that $\sin\theta = \sin\theta^*$, then $2d\sin\theta = $ EB + BF. Finally, for coherent diffraction, EB + BF must be equal to some integer multiple of λ, so EB + BF = nλ, yielding Bragg's Law:

$$n\lambda = 2d\sin\theta \qquad \text{(III.1)}$$

With modern diffractometers and software there is no need to sit down with a calculator to determine d, but

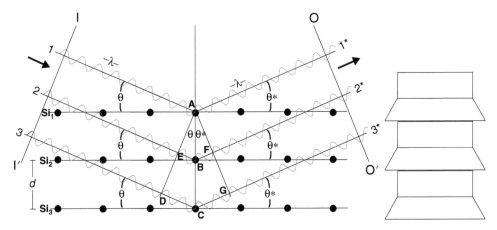

Fig. III.3 Schematic representation of diffraction from three rows of atoms at equivalent positions in three unit cells (where the equivalent positions are conceptually represented here as planes of Si atoms in kaolinite unit cells). The numerals 1, 2, and 3 represent rays of the incident beam, and 1*, 2*, and 3* are rays from the diffracted beam. θ is the angle of incidence, λ is wavelength, and d is d-spacing along the c-axis. I and I′ represent a plane perpendicular to the incident beam and O and O′ represent a plane perpendicular to the outgoing (diffracted or reflected) beam.

understanding Bragg's Law makes XRD analysis more than a mere "black-box" approach.

III.1.2.2 Mineral Identification by XRD

Figure III.4 contains an XRD pattern of the minerals kaolinite and illite that includes information on d-spacings (in Å) and peaks (in Miller indices). At most angles from 2° to 40° 2θ, X-rays are diffracted by the specimen out of phase, producing destructive interference and hence no peaks – only at very specific angles are peaks produced. In this case, platy kaolinite and illite crystals are oriented perpendicular to the c-axis, so the only peaks are from d-spacings in the c-crystallographic direction. The convention in powder diffraction uses the angle of diffraction as 2θ rather than θ. Also note the 7.16 Å d-spacing of kaolinite (along the c-axis) and the 10 Å spacing of mica, each indicated by the 001 peak – these are the same spacings as indicated schematically in Figure III.3.

Given the geometry of diffraction and conditions which produce coherent diffraction, peaks also occur at angles which are whole integer multiples of the angle where the 001 peak occurs, so for kaolinite we observe the 001 peak at 12.3° 2θ (7.16 Å), the 002 peak at 24.8° 2θ (3.58 Å), and the 003 peak at 37.7° 2θ (2.38 Å), and would expect to see higher-order peaks at (1.79, 1.43 Å, etc.). Similarly, illite produces peaks at 10, 5.0, 3.33, 2.5, 2.0 Å, and so on. The intensity of the peaks is controlled by mineral composition as well as the structure factor and other more detailed aspects of diffraction not covered here.

When smectites are heated to $\geq 100\,°C$, interlayer water is driven out, producing a dehydrated smectite with d(001) of ~10 Å. The ability of water (and hydrated cations) to enter and leave the smectite interlayer is the source of much of this mineral group's high cation exchange capacity. It also is a characteristic property that aids in the identification of this mineral group. While smectites, chlorites, and vermiculites all can have d(001) values of ~14 Å, only smectite will fully collapse to 9–10 Å when heated to 250 °C, and in the presence of the polar organic compound ethylene glycol, smectites expand to 17 Å.

For further information on XRD and mineral analysis the reader is referred to an excellent and thorough treatment of the topic presented in Moore and Reynolds (1997).

III.1.3 FTIR

Infrared radiation (IR) occurs in the electromagnetic spectrum between the long-wavelength end of the visible spectrum (0.75 μm) and the short-wavelength part of the microwave spectrum (~1000 μm). Units of IR wavelength are typically presented either as units of length, e.g. nm or μm, or as wavenumber (cm^{-1}), which is the inverse of wavelength (i.e. $1/\lambda$). So, a wavelength of 500 μm (0.5 cm) corresponds to a wave number of 2 cm^{-1}, and a wavelength of 1 μm (10^{-3} cm) equals 1000 cm^{-1}. The common unit in mineralogical analyses is the wavenumber, so that will be used from here on, with wavelengths in parentheses.

The infrared spectrum is often subdivided as follows:

1 The near-infrared, from ~13 000 to 4000 cm^{-1} (0.075–0.25 μm);

2 The mid-infrared, from ~4000 to 400 cm^{-1} (0.25–2.5 μm); and

3 The far-infrared, from ~400 to 10 cm^{-1} (2.5–100 μm).

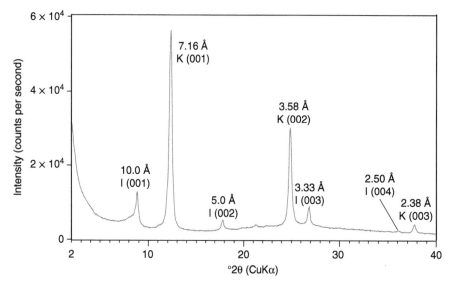

Fig. III.4 X-ray diffraction pattern (spectrum) of the < 2 μm (clay size) fraction of a hydrothermally altered schist. The presence of illite is indicated by the series of 10, 5, 3.33, and 2.50 Å peaks. Kaolinite is indicated by the 7.16, 3.58, and 2.38 Å peaks. Note that, as dictated by Bragg's Law, d-spacing decreases as the angle 2θ increases (to the right). The data in this figure were generated from an oriented clay preparation, so other than a very weak dioctahedral 02,11 peak at ~21.3° 2θ, the only peaks visible are those of the 00/ series for kaolinite and illite.

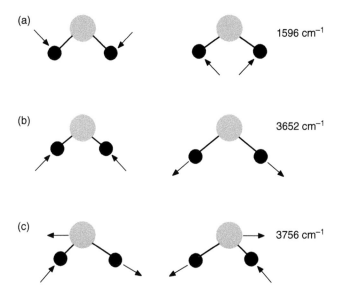

Fig. III.5 Schematic sketches of the ways in which water absorbs IR radiation: (a) flexing, (b) symmetrical stretching, and (c) asymmetrical stretching/rotation. Molecular distortions are exaggerated to make them more apparent.

Mid-infrared absorption can be used to study fundamental vibrations and associated rotational–vibrational structures, and is the region mainly used to study minerals. Figure III.5 summarizes the different ways that components of minerals, soils, and other materials (e.g. H_2O) stretch, vibrate, and rotate in the presence of IR. Water molecules absorb at many points in the IR spectrum; the examples shown in Figure III.5 include absorption at $1595\,cm^{-1}$ ($0.627\,\mu m$) due to flexing of the O–H bond, $3652\,cm^{-1}$ ($0.274\,\mu m$) due to symmetrical stretching of the O–H bond, and $3756\,cm^{-1}$ ($0.266\,\mu m$) due to asymmetrical stretching of the O–H bond (Figure III.5). In Figure III.6, note that water adsorbed to mineral surfaces produces a broad peak between ~3800 and $3100\,cm^{-1}$.

Absorption of IR at relatively precise wavelengths provides data on the types of components in minerals (Figure III.6). The region from 3800 to $3400\,cm^{-1}$ is known as the OH stretching region, a part of the spectrum which provides information on bonds between hydroxyls and cations (and which also includes the broad H_2O band mentioned in the previous paragraph). The hydroxides goethite and gibbsite contain peaks in this region, as do all of the clay minerals shown in Figure III.6. One distinctive peak at $3620\,cm^{-1}$ is produced by inner hydroxyl groups located between the tetrahedral and octahedral sheets. Calcite and quartz lack hydroxyls, so spectra for these minerals do not contain OH stretching peaks. The region from 1300 to $400\,cm^{-1}$ is known as the OH bending region. The composition of the octahedral sheet in clay minerals strongly influences the position of OH bending bands; for example, in dioctahedral minerals such as kaolinite and illite, absorption is centered around the $950–800\,cm^{-1}$ region, whereas OH absorption in trioctahedral minerals (e.g. chlorite) occurs in the $700–600\,cm^{-1}$ range. Al–OH–Al bending bands of kaolinite occur at ~$915\,cm^{-1}$, while Mg–OH–Mg bending bands of trioctahedral clays like chlorite and serpentine occur at ~$650–600\,cm^{-1}$. Mineral composition can also be assessed; for example, substitutions in the octahedral sheet of clay minerals or hydroxides (e.g. Fe or Mg for Al) can be determined based on positions and intensities of IR peaks.

Many modern IR spectrometers are Fourier Transform Infrared radiation (FTIR) spectrometers that provide greater resolution than older IR spectrometers. For further reading on FTIR in mineral analysis, the reader is referred to the classic book by pioneering researcher Farmer (1974) as well as the more recent, undergraduate-oriented chapter by Schroeder (2002).

III.1.4 X-ray absorption spectroscopy (XAS) techniques (EXAFS, XANES)

X-ray absorption spectroscopy (XAS) uses a synchrotron to produce an intense and adjustable source of X-rays, and the main application is determination of speciation of elements, particularly oxidation state and bond lengths, bond angles, and coordination numbers. This can be very useful when trying to determine the speciation of trace elements.

X-ray absorption fine-structure spectroscopy (XAFS) refers to the ways in which X-rays are absorbed by an atom at energies just below, at, and immediately above the binding energies of the atom. The two main regions of the XAS spectrum from which useful information can be obtained are known as X-ray absorption near-edge spectroscopy (XANES) and extended X-ray absorption fine-structure spectroscopy (EXAFS). Figure III.7a illustrates the difference between the two using a XAS spectrum of iron oxide (FeO). The region between XANES and EXAFS is the near extended X-ray absorption fine-structure spectroscopy region (NEXAFS). When the incident X-ray energy is low (left side of Figure III.7a), little to no absorption of X-rays occurs. As the energy is increased to a point, a stepwise increase in absorption occurs; this is known as the absorption edge, and it corresponds to increased absorption of X-rays by an excited atom whose core electron structure (core electrons are those that are not valence

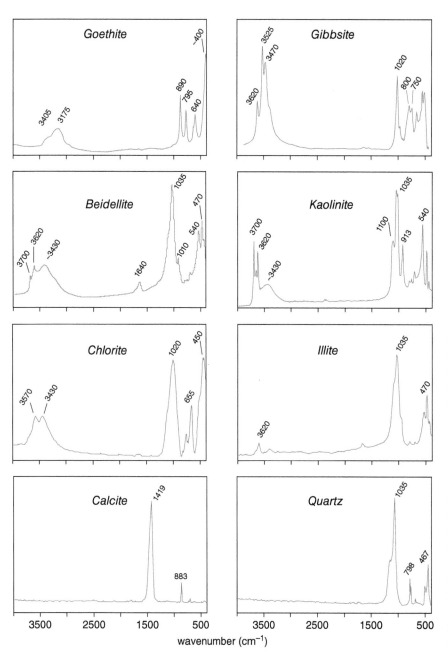

Fig. III.6 FTIR spectra for selected minerals. Peaks in the OH stretching region (3800–3400 cm^{-1}) are from cation–OH bonds. All silicate minerals contain an Si–O stretching band, which in the examples provided here occurs at 1035–1020 cm^{-1} – note that this peak is absent in goethite and calcite. A peak in this range does occur in gibbsite because Al–OH bending vibrations absorb IR in this range. In quartz and other silicates, the 798 cm^{-1} peak is produced by Si–Si stretching and the ~470 cm^{-1} peak is from O–Si–O bending. 3430 and 1640 cm^{-1} peaks in beidellite and, to a lesser extent, in kaolinite and illite, are H–O–H stretching bands produced by adsorbed water. The 1419 and 883 peaks in calcite are from vibrations in the CO_3^{2-} molecule.

electrons) has been altered by the incident X-rays. Beer's Law describes absorption of X-rays as follows:

$$I = I_0 e^{-\mu t} \tag{III.2}$$

I_0 is the intensity (energy) of the incident X-ray beam, t is the sample thickness, μ is the absorption coefficient, and I is the intensity of the X-ray beam once it has passed through the sample. Important controls on μ are atomic number (Z) and I_0, which are related approximately as:

$$\mu \sim Z^4/I_0^{\;3} \tag{III.3}$$

At low energies, I is roughly equal to I_0, but as I_0 increases, μ decreases, causing I (intensity of transmitted X-rays) to decrease. This is the absorption edge. Given that Z is raised to the fourth power, small differences in atomic mass produce values of μ that are very different for different elements, thus enhancing resolution.

The location of the edge (i.e. x-axis position) and its intensity are related to the binding energies of the central atom and its neighbors. Every atom has core-level electrons with well-defined binding energies, and by adjusting the energy of the X-ray beam, analysts can select the element of interest. The XANES part of the

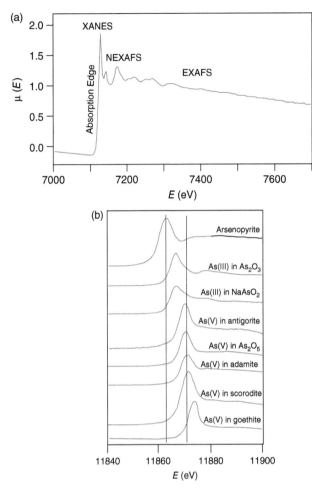

increasing energy – this is the EXAFS region. The oscillations are controlled by the coordination number (i.e. number of neighboring atoms), the type of neighboring elements, and bond distance.

One way in which XANES and EXAFS are applied to problems in environmental geochemistry is illustrated in Figure III.7b, which shows how the oxidation state of arsenic can be determined.

III.2 CHEMICAL ANALYSIS OF ROCKS AND SEDIMENTS: XRF

XRF spectrometry exposes powdered solids to short-wavelength X-radiation which causes ionization of atoms in the sample. When ionization occurs, one or more electrons are ejected from the atom, often from inner orbitals. When this occurs, the electron configuration of the atom is unstable, and electrons in higher orbitals move into the lower orbital to stabilize the atom. The released energy occurs in the form of a photon whose energy is equal to the difference in energy of the two orbitals involved. In this way, individual atoms in the sample emit radiation characteristic of the atoms present, and the intensity of radiation at a given energy level is proportional to the abundance of that particular element. Fluorescence describes the two-step process in which an atom absorbs energy (in this case, from X-rays), and then re-emits radiation of a different energy that is typically of a lower level than the initially absorbed energy (see Figure III.8).

Fig. III.7 XAFS spectra: (a) FeO spectrum that shows the absorption edge, XANES, NEXAFS, and EXAFS regions, as well as the oscillations in the EXAFS region useful in determining coordination number, neighboring elements, and bond distance. The y-axis is the absorption coefficient – note marked increase across absorption edge; (b) XANES region showing spectra of arsenic-bearing compounds. Scorodite is $FeAsO_4 \cdot 2H_2O$ and adamite is $Zn_2(AsO_4)(OH)$, both minerals which contain As(V), i.e. As^{5+}. Arsenopyrite contains $\sim As^0$. Source: Image is adapted from Newville (2004) and from data presented in Foster et al. (1997) and Hattori et al. (2005).

spectrum occurs near the absorption edge, and provides information on the oxidation state of the absorbing atom. Figure III.7b depicts the location of the absorption edge in the XANES region for various arsenic compounds. For a given element, higher absorption-edge energies generally indicate higher oxidation states because the greater (more positive) charge makes it more difficult to photoionize core electrons. This is visible in Figure III.7b, where As(V) is indicated by higher absorption edge than As(III). On the higher-energy (right-hand) side of the absorption edge, absorbance decreases in an oscillatory manner with

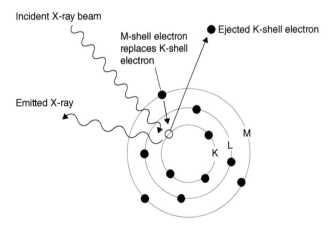

Fig. III.8 Schematic representation of fluorescence caused by ejection of a K-shell electron by an incident X-ray beam, followed by an M-shell electron falling into the K-shell to replace the ejected electron. The energy of the X-ray emitted by the falling electron is characteristic of a given element and is used in determining chemical composition by XRF.

III.3 ELEMENTS OR COMPOUNDS IN SOLUTION

III.3.1 Elements in solution by ICP-AES, ICP-MS, AAS

The compositions of rocks, soils, sediments, and minerals can be determined by spectrometry provided that the solid material can be dissolved into solution (typically some form of acid). Three common techniques are inductively coupled plasma-atomic emission spectrometry (ICP-AES), which is also known as inductively coupled plasma-optical emission spectrometry (ICP-OES) and inductively coupled argon plasma spectrometry (ICAPS); inductively coupled plasma-mass spectrometry (ICP-MS); and atomic absorption (AA) and graphite furnace atomic absorption (GFAA). These methods are also useful for analysis of aqueous solutions such as surface waters, groundwaters, soil water, and waste water.

ICP spectrometers contain a very hot argon plasma ($\sim$7000 °C) into which a solution is injected at a constant rate by a peristaltic pump. At the extremely high temperature of the argon plasma, all compounds are atomized and the resulting electronically excited atoms emit electromagnetic radiation at wavelengths characteristic of individual elements. For example, chromium is commonly detected by radiation emitted with a wavelength of 267.7 nm and nickel by radiation at 231.6 nm. The intensity of the emitted radiation is proportional to the concentration of the element in the sample, and the instrument is calibrated by running standards of known concentration and creating a calibration curve like the one shown for calcium in Figure III.9. Given that many elements produce spectra with overlapping peaks, it is necessary to correct for interferences, a task made possible by the fact that the spectrum for each element is unique. One of the great advantages of ICP-AES is the ability to analyze dozens of elements simultaneously, a feat not possible with XAS or AA. One of the disadvantages is that certain elements, notably arsenic, lead, and uranium, are difficult to detect at low concentrations because they emit relatively weak signals in the ICP compared to other metals and metalloids. For this reason, AA or ICP-MS are commonly used to detect As, Pb, and U.

ICP-MS uses an argon plasma in the same manner as ICP-AES, but rather than a bank of detectors calibrated to measure emitted radiation from excited atoms, as is the case with ICP-AES, atoms in an ICP-MS are accelerated toward detectors calibrated to receive atoms of discrete atomic mass. Many ICP-MS units contain a quadrupole magnet that separates elements based on mass-to-charge ratio, and each detector is bombarded by ions of known mass whose signal is proportional to concentration. Given that ICP-MS quantifies based on mass, it can be used in isotopic analysis, a distinct advantage over ICP-AES. Detection limits are also typically lower in ICP-MS than they are in ICP-AES. AA spectroscopy is, in a way, the inverse of ICP-AES because elemental abundance is determined by absorption, rather than emission, of radiation, and for this reason, certain elements with weak emission signals (e.g. As, Pb, U) can be detected at lower concentrations with AA than with ICP-AES. Like ICP spectrometers, AA commonly is applied to measurement of concentration of ions in solution, where solutions are aspirated into a chamber. Lamps which emit radiation of known wavelengths are used for individual elements known to absorb radiation at that given wavelength, and the greater the absorption, the greater the amount of element in question.

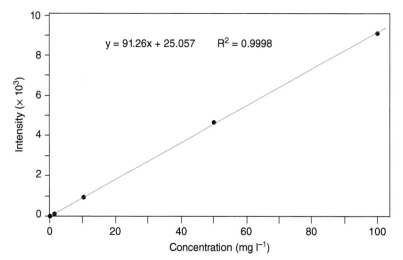

Fig. III.9 Calibration curve for calcium analysis (peak at 317.9 nm) by ICP-AES. Note that points become slightly nonlinear at higher concentrations, i.e. while the linear fit produces an excellent R^2 value, the intensity at 50 mg l^{-1} is slightly above the line, the intensity at 100 mg l^{-1} is slightly below the line, and the y-intercept is not zero.

III.3.2 Chromatography

Anions and volatile organic compounds (VOCs) are commonly analyzed by chromatography. Ion chromatography (IC) is commonly used in analysis of nitrate, chloride, sulfate, phosphate, and other anions (e.g. using US EPA method 300.0) (see Figure III.10), whereas gas chromatography (GC) is used for VOCs. In both cases samples are injected into columns packed with beads (e.g. resin or silica), and the rate at which compounds pass through the column depends on charge and molecular mass. With certain columns and beads, small, low-charge ions (e.g. F^-, Cl^-) pass through the column more rapidly than larger molecules with the same charge (e.g. NO_3^-), and those that move most slowly through the column are even higher-charged molecules such as HPO_4^{2-} and SO_4^{2-}. Controls on the rate at which organic compounds flow through a column include volatility (more volatile = more rapid flow through column), polarity and presence/absence of functional groups (OH groups are attracted to resin beads and thus retard flow), and molecular mass (larger molecules tend to migrate more slowly). Common detectors for GC are gas chromatography-electron capture detector (GC-ECD) and gas chromatography-mass spectrometer (GC-MS), gas chromatography-flame ionization detector (GC-FID), and gas chromatography-photoionization detector (GC-PID).

III.4 ISOTOPIC ANALYSIS: MASS SPECTROMETRY

The main analytical method for isotope ratios and concentrations is mass spectrometry (Figure III.11). The left-hand

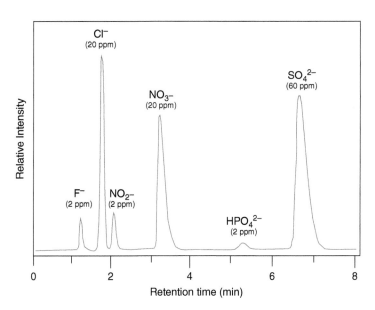

Fig. III.10 Ion chromatogram showing controls on retention time and relative intensity of peaks for different ions (note concentrations), using a Dionex AS4a column (P/N 37041). Source: Adopted from Pfaff (1993), US EPA Method 300.0.

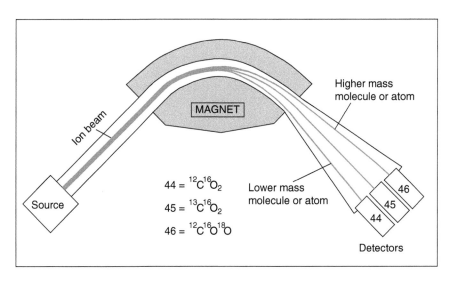

Fig. III.11 Schematic diagram of a mass spectrometer illustrating the effect of the magnet on the trajectories of CO_2 molecules (in ion beams) of different masses. Note that the rare oxygen isotope ^{17}O is not included in the example of isotopes in CO_2.

side of the diagram shows the source of the material to be analyzed – this source typically consists of either ionized gas (e.g. CO_2 in the case of carbon isotope analysis) produced by strong electromagnetic field or a hot filament, or in the case of multicollector-inductively coupled plasma-mass spectrometry (MC-ICP-MS), as ionized atoms produced in a $10\,000\,°C$ plasma.

The gaseous ions are accelerated by a vacuum toward the magnet depicted in the central part of the diagram, and the magnet bends or pulls the components of the ion beam to an extent that is determined by their respective masses. Lighter isotopes are more greatly affected by the magnet and thus are bent to a greater extent. Shown in the lower right-hand side of the diagram are the collectors which measure the quantity of atoms or molecules in the beams (corresponding to the different isotopes) as a current that is amplified, allowing determination with high precision. The positions of the collectors are very precisely calibrated to the isotopes in question, and the two main types of detectors are Faraday cups (or Faraday cages), which measure ion current incident upon a flat surface, and ion counters that measure individual ions impacting the surface of the counter.

Very low concentrations of isotopes such as ^{14}C, ^{10}Be, and ^{26}Al are detected by tandem accelerator mass spectrometers. As is the case with traditional mass spectrometry, the ion source produces a beam of ions, but for users the really important application is that tandem accelerator mass spectrometers can work with very small samples (e.g. a few milligrams of a fragment of a pine needle or other solids). They make use of multiple magnets to hone in on the target isotope; the tandem accelerator part of the mass spectrometer contains two accelerating gaps with high voltage in the center, and individual ions are counted by a gas ionization detector.

Two recent advances in stable isotope mass spectrometry applicable to stable isotope analysis are thermal ionization mass spectrometry (TIMS) and MC-ICP-MS, the development of which provide greater precision analysis for traditional isotope systems (e.g. hydrogen, carbon, nitrogen, oxygen, sulfur) while also fostering new advances in systems (e.g. metal stable isotopes) previously limited by detection limits or precision of analysis.

REFERENCES

Farmer, V.C. (1974). *The Infrared Spectra of Minerals*. Monograph 4. London: Mineralogical Society 539 pp.

Foster, A.L., Brown, G.E. Jr., Parks, G.A. et al. (1997). XAFS determination of As(V) associated with Fe(III) oxyhydroxides in weathered mine tailings and contaminated soil from California, U.S.A. *Supplément au Journal de Physique IV de France* 7: 815–816.

Hattori, K., Takahashi, Y., Guillot, S., and Johanson, B. (2005). Occurrence of arsenic (V) in forearc mantle serpentinites based on X-ray absorption spectroscopy study. *Geochimica et Cosmochimica Acta* 69: 5585–5596.

Moore, D. and Reynolds, R.C. Jr. (1997). *X-Ray Diffraction and the Identification and Analysis of Clay Minerals*, 2e. New York: Oxford University Press.

Newville, M. (2004). *Fundamentals of XAFS*. Chicago, Illinois, USA: University of Chicago.

Pfaff, J.D. (1993). *US EPA Method 300.0: Methods for the Determination of Inorganic Substances in Environmental Samples, EPA-600/R-93-100, NTIS PB94-121811*. Cincinnati, Ohio, USA: Environmental Monitoring Systems Laboratory, Office of Research and Development, U.S. Environmental Protection Agency.

Schroeder, P.A. (2002) Infrared Spectroscopy in clay science: In *CMS Workshop Lectures, Volume 11, Teaching Clay Science* (eds. A. Rule and S. Guggenheim). Aurora, Colorado, USA: The Clay Mineral Society. 181–206.

Appendix IV

Table of Thermodynamic Data of Selected Species at 1 atm and 25 °C

Species	State	Name	H (kJ mol⁻¹)	S (J mol K⁻¹)	G (kJ mol⁻¹)	Source
Al^{3+}	aq	Aluminum ion	−531.4	−321.7	−485.3	Dean (1979)
Al_2O_3	s	γ-corundum	−1656.9	59.8	−1562.7	Dean (1979)
Al_2O_3	s	α-corundum	−1675.3	50.9	−1582.0	Dean (1979)
$Al(OH)_3$	s	Gibbsite	−1281.4	70.1	−1143.7	Dean (1979)
$AlOOH$	s	Boehmite	−987.4	48.4	−912.7	Dean (1979)
$AlOOH$	s	Diaspore	−1000.0	35.3	−920.5	Dean (1979)
$Al_2Si_2O_5(OH)_4$	s	Halloysite	−4079.8	203.3	−3759.3	Dean (1979)
$Al_2Si_2O_5(OH)_4$	s	Kaolinite	−4098.6	202.9	−3778.2	Dean (1979)
$Al_2(SO_4)_3$	s	Aluminum sulfate	−3435.1	239.3	−3506.6	Dean (1979)
$H_2AsO_4^-$	aq	Dihydrogen arsenate ion	−909.6	117.2	−753.3	Dean (1979)
$HAsO_4^{2-}$	aq	Hydrogen arsenate ion	−906.3	−1.7	−714.7	Dean (1979)
As_2O_3	s	Arsenolite	−657.0	107.4	−576.0	Dean (1979)
AsS	s	Realgar	−71.3	63.5	−70.3	Dean (1979)
As_2S_3	s	Orpiment	−169.0	163.6	−168.4	Dean (1979)
$FeAsS$	s	Arsenopyrite	−41.8	121.3	−50.2	Dean (1979)
Ba^{2+}	aq	Barium ion	−537.6	9.6	−560.7	Dean (1979)
$BaCO_3$	s	Witherite	−1216.3	112.1	−1137.6	Dean (1979)
$BaSO_4$	s	Barite	−1473.2	132.2	−1362.3	Dean (1979)
C	s	Diamond	1.9	2.4	2.9	Dean (1979)
C	s	Graphite	0	5.7	0	Dean (1979)
$C_6H_5CH_3$	l	Toluene	12.0	221.0	113.8	Dean (1979)
C_6H_6	l	Benzene	49.0	173.3	124.3	Dean (1979)
$CH_3(CH_2)_6CH_3$	l	Octane	−250.0	357.7	7.4	Dean (1979)
$CH_3(CH_2)_6CH_3$	g	Octane	−208.4	466.7	16.4	Dean (1979)
CO	g	Carbon monoxide	−110.5	197.9	−137.3	Dean (1979)

Environmental and Low-Temperature Geochemistry, Second Edition. Peter Ryan.
© 2020 John Wiley & Sons Ltd. Published 2020 by John Wiley & Sons Ltd.

Species	State	Name	H (kJ mol^{-1})	S (J mol K^{-1})	G (kJ mol^{-1})	Source
CO_2	g	Carbon dioxide	−393.5	213.7	−394.4	Dean (1979)
CO_2	aq	(Undissociated)	−413.8	117.6	−386.0	Dean (1979)
H_2CO_3	aq	Carbonic acid	−699.1	189.3	−623.1	Drever (1997)
HCO_3^-	aq	Bicarbonate	−692.0	91.2	−586.8	Dean (1979)
CO_3^{2-}	aq	Carbonate ion	−677.1	−56.9	−527.9	Dean (1979)
Ca^{2+}	aq	Calcium ion	−542.8	−53.1	−553.5	Dean (1979)
$Ca_5(PO_4)_3(OH)$	s	Hydroxyapatite	−6738.5	390.4	−6337.1	Wolery and Jove-Colon (2007)
$Ca_5(PO_4)_3F$	s	Fluorapatite	−6872.2	387.9	−6491.5	Dean (1979)
$Ca(OH)_2$	s	Portlandite	−986.1	83.4	−898.5	Wolery and Jove-Colon (2007)
$CaCO_3$	s	Aragonite	−1207.4	88.0	−1127.8	Robie et al. (1979)
$CaCO_3$	s	Calcite	−1207.4	91.7	−1128.8	Robie et al. (1979)
$CaMg(CO_3)_2$	s	Dolomite	−2326.3	155.2	−2163.5	Dean (1979)
$CaSiO_3$	s	Wollastonite	−1630.0	82.0	−1544.8	Wolery and Jove-Colon (2007)
$CaSO_4$	s	Anhydrite	−1434.1	106.7	−1321.8	Wolery and Jove-Colon (2007)
$CaSO_4 \cdot 2H_2O$	s	Gypsum	−2022.6	194.1	−1797.2	Wolery and Jove-Colon (2007)
$CaMgSi_2O_6$	s	Diopside	−3206.2	142.9	−3032.1	Dean (1979)
$CaAl_2Si_2O_8$	s	Anorthite			−4012.8	Dean (1979)
$Ca_{0.165}Al_{2.33}Si_{3.67}O_{10}(OH)_2$	s	Ca–beidellite		244.1	−5357.3	Wolery and Jove-Colon (2007)
$Ca_{0.165}Mg_{0.33}Al_{1.67}Si_4O_{10}(OH)_2$	s	Ca–montmorillonite		250.4	−5323.3	Wolery and Jove-Colon (2007)
$Ca_{0.165}Fe_2Al_{0.33}Si_{3.67}O_{10}(OH)_2$	s	Ca–nontronite		282.6	−4513.2	Wolery and Jove-Colon (2007)
$Ca_{0.165}Mg_3Al_{0.33}Si_{3.67}O_{10}(OH)_2$	s	Ca–saponite		260.8	−5591.6	Wolery and Jove-Colon (2007)
Cl^-	aq	Chloride ion	−167.2	56.5	−131.3	Dean (1979)
F^-	aq	Fluoride ion	−255.6	145.5	−262.3	Dean (1979)
Cu^+	aq	Copper(I) ion	71.7	40.6	50.0	Dean (1979)
Cu^{2+}	aq	Copper(II) ion	64.8	−99.6	65.5	Dean (1979)
$CuCO_3 \cdot Cu(OH)_2$	s	Malachite	−1051.4	186.2	−893.7	Dean (1979)
$CuSO_4 \cdot 5H_2O$	s	Chalcanthite	−2279.7	300.4	−1880.1	Dean (1979)
Cu_2S	s	Chalcocite	−80.1	120.8	−86.9	Dean (1979)
Fe^{2+}	aq	Ferrous ion	−89.1	−137.7	−78.9	Dean (1979)
Fe^{3+}	aq	Ferric ion	−48.5	−315.9	−4.6	Dean (1979)
$Fe(OH)^{2+}$	aq		−291.2		−233.2	Drever (1997)
$Fe(OH)_2^+$	aq		−548.9		−450.5	Drever (1997)
$Fe(OH)_3$	s	Iron hydroxide	−832.6	104.6	−696.5	Wolery and Jove-Colon (2007)
$FeOOH$	s	Goethite	−558.1	60.4	−488.6	Wolery and Jove-Colon (2007)
Fe_2O_3	s	Hematite	−824.2	87.4	−742.2	Dean (1979)
Fe_3O_4	s	Magnetite	−1118.4	146.4	−1015.5	Dean (1979)

Species	State	Name	H (kJ mol^{-1})	S (J mol K^{-1})	G (kJ mol^{-1})	Source
$FeCr_2O_4$	s	Chromite	−1458.6	146.9	−1355.9	Wolery and Jove-Colon (2007)
Fe_7S_8	s	Pyrrhotite	−736.4	485.8	−748.5	Dean (1979)
FeS	s	Pyrrhotite	−100.0	60.3	−100.4	Dean (1979)
FeS	s	Mackinawite			−93.0	Drever (1997)
FeS_2	s	Pyrite	−178.2	52.9	−166.9	Dean (1979)
Fe_2SiO_4	s	Fayalite	−1479.9	145.2	−1379.0	Dean (1979)
$FeSO_4 \cdot 7H2O$	s	Melanterite	−3014.6	409.2	−2510.3	Dean (1979)
$Fe_{4.8}(SO4)_6(OH)_2 \cdot 20H_2O$	s	Ferricopiapite	−12045.1	1449.2	−10089.8	Majzlan et al. (2006)
H^+	aq	Hydrogen ion	0	0	0	Dean (1979)
H_2O	g	Water vapor	−241.8	188.7	−228.6	Dean (1979)
H_2O	l	Liquid water	−285.8	69.9	−237.2	Dean (1979)
K^+	aq	Potassium ion			−282.6	Dean (1979)
$KAlSi_3O_8$	s	Microcline			−3732.7	Dean (1979)
$K_{0.6}Mg_{0.25}Al_{2.3}Si_{3.5}O_{10}(OH)_2$	s	Illite		266.4	−5455.7	Wolery and Jove-Colon (2007)
$K_{0.33}Al_{2.33}Si_{3.67}O_{10}(OH)_2$	s	K–beidellite			−5360.3	Wolery and Jove-Colon (2007)
$KFe(SO_4)_2(OH)_6$	s	Jarosite	257??		−3299.5	Arslan and Arslan (2003)
$MgCO_3$	s	Magnesite	−1113.3	65.1	−1029.5	Robie et al. (1979)
MnO_2	s	Pyrolusite	−520.0	53.1	−465.1	Dean (1979)
$MnCO_3$	s	Rhodochrosite	−889.3	100.0	−816.0	Dean (1979)
Na^+	aq	Sodium ion			−261.7	Dean (1979)
$NaAlSi_3O_8$	s	Albite (low albite)	−3935.1	207.4	−3711.7	Robie et al. (1979)
$Na_{0.33}Al_{2.33}Si_{3.67}O_{10}(OH)_2$	s	Na–beidellite		249.6	−5354.2	Wolery and Jove-Colon (2007)
$Na_{0.96}Al_{0.96}Si_{2.04}O_6 \cdot H_2O$	s	Analcime	−3296.9	226.7	−3077.2	Wolery and Jove-Colon (2007)
$NaHCO_3$	s	Nahcolite	−936.2	101.2	−852.9	Dean (1979)
$H_2PO_4^-$	aq	Dihydrogen phosphate ion	−1296.3	90.4	−1130.3	Dean (1979)
HPO_4^{2-}	aq	Hydrogen phosphate ion	−1292.0	−33.5	−1089.1	Dean (1979)
PO_4^{3-}	aq	Phosphate ion	−1284.4	−221.0	−1025.5	Dean (1979)
$PbCO_3$	s	Cerussite	−699.2	131.0	−625.3	Robie et al. (1979)
PbS	s	Galena	−977.1	91.4	−960.8	Robie et al. (1979)
H_4SiO_4	aq	Dissolved silica	−1460	180	−1316	Drever (1997)
SiO_2	s	Quartz	−910.7	41.5	−856.3	Robie et al. (1979)
SiO_2	s	Cristobalite	−908.4	43.4	−854.5	Robie et al. (1979)
SO_4^{2-}	aq	Sulfate anion	−907.5	17	−742.6	Ebbing (1990), Nordstrom (2013)
$ZnCO_3$	s	Smithsonite	−812.8	82.4	−731.5	USGS
ZnS	s	Sphalerite	−206.9	58.7	−202.5	USGS

REFERENCES

Arslan, C. and Arslan, F. (2003). Thermochemical review of jarosite and goethite stability regions at 25 and 95 °C. *Turkish Journal of Engineering and Environmental Science* 27: 45–52.

Dean, J.A. (1979). *Lange's Handbook of Chemistry*, 12e. New York: McGraw-Hill.

Drever, J.I. (1997). *The Geochemistry of Natural Waters: Surface and Groundwater Environments*, 3e. Upper Saddle River, NJ, USA: Simon and Schuster.

Ebbing, D.D. (1990). *General Chemistry*, 3e. Houghton Mifflin.

Majzlan, J., Navrotsky, A., McCleskey, R.B., and Alpers, C.N. (2006). Thermodynamic properties and crystal structure refinement of ferricopiapite, coquimbite, rhomboclase, and $Fe_2(SO_4)_3(H_2O)_5$. *European Journal of Mineralogy* 18: 175–186.

Nordstrom, D.K. (2013). Improving internal consistency of standard state thermodynamic data for sulfate ion, portlandite, gypsum, barite, celestine, and associated ions. *Procedia Earth and Planetary Science* 7: 624–627.

Robie, R.A., Hemingway, B.S., Fisher, J.R. (1979). Thermodynamic properties of minerals and related substances at 298.15 K and 1 bar 105 pascals pressure and at higher temperatures. *United States Geological Survey Bulletin* 1452. Accessed online 5 December 2013: http://pubs.usgs.gov/bul/1452/report.pdf.

Wolery, T. and Jove-Colon, C. (2007). *Qualification of Thermodynamic Data for Geochemical Modeling of Mineral–Water Interactions in Dilute Systems*. Report Prepared for US Department of Energy, ANL-WIS-GS-000003 REV 01. Accessed online 5 December 2013: http://pbadupws.nrc.gov/docs/ML0907/ML090770163.pdf

Index

Environmental and Low-Temperature Geochemistry, Second Edition. Peter Ryan.
© 2020 John Wiley & Sons Ltd. Published 2020 by John Wiley & Sons Ltd.

Plate 1 Oxidized mine tailings showing colors typical of hematite (right) and goethite and more-disordered Fe-hydroxides (left). These oxides and hydroxides formed from oxidation of pyrrhotite and chalcopyrite.

Plate 3 Photograph of chalcopyrite ($CuFeS_2$)-rich rock that is weathering into iron hydroxide (orange coating) and chalcanthite ($CuSO_2 \cdot 5\ H_2O$). Both minerals are secondary – the iron hydroxide forms when iron liberated from chalcopyrite is oxidized; the chalcanthite (tan or crystals) forms on the rock surface when water containing dissolved Cu and sulfate (derived from chalcopyrite oxidation) is drawn upward through fractures and then evaporates, producing a solution saturated with respect to Cu and SO_4.

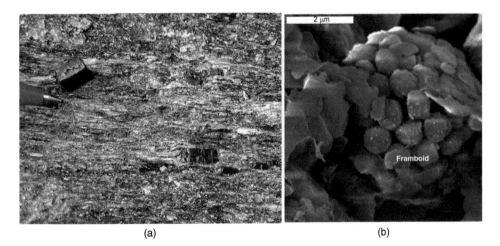

(a)   (b)

Plate 2 Macroscopic cubic pyrite crystals in Cambrian schist (below, left) and microscopic framboidal pyrite (below, right) occurring in the saturated zone of a landfill. (a) Source: Photo by J. Kim, Vermont Geological Survey. (b) Source: Photo by G. Breit, US Geological Survey.

Environmental and Low-Temperature Geochemistry, Second Edition. Peter Ryan.
© 2020 John Wiley & Sons Ltd. Published 2020 by John Wiley & Sons Ltd.

Plate 4 Humic and/or fulvic acids imparting a tea-like light-brown color to river water tumbling over a waterfall. Source: Photograph by E.C. Ryan.

FeS rich layers

Plate 6 Black layers of poorly crystalline iron sulfide interbedded with tan sand. The colors of these layers are indicative of their respective redox states – black reduced sediment surrounded by relatively oxidized tan sediment. Source: Photo by Michele Tuttle, US Geological Survey.

Plate 5 Orange iron hydroxide formed when dissolved Fe(II) in reduced groundwater seeps out at the surface and is oxidized to Fe(III).

2 cm

Plate 7 Effect of mineral dissolution rate displayed in calcite and dolomite layers in Ordovician carbonate rock. The more-resistant dolomite layers have positive relief relative to calcite layers, which decompose more rapidly and hence are recessed.

Plate 8 Algal bloom in a eutrophic lake. Photograph by Nara Souza, Florida Fish and Wildlife Commission.

Plate 9 Effect of naturally derived acid rain on the flanks of Volcan Poás in Costa Rica. Prevailing winds transport nitric and sulfuric acid emission westward (to the left), creating a barren landscape.

Plate 10 Public display of air quality in Granada, Spain, showing acceptable levels of CO, NO_2, O_3, and SO_2 but unhealthy levels of particulates.

Plate 11 Air inversion in southern Spain. Cold air has sunk into the valley, causing condensation (fog) and trapping of contaminants. The warm air above the inversion layer prevents air circulation. Air temperature where the photograph was taken was about 5 °C warmer than air below. Source: Photo by F. Javier Huertas, Instituto Andaluz de Ciencias de la Tierra.

Plate 12 Graphite-rich phyllitic rocks in the Sierra Nevada Mountains, Spain. The graphite was likely once organic matter on the seafloor that was buried then metamorphosed under reducing conditions in the crust. Mosses and lichens are beginning the weathering process.

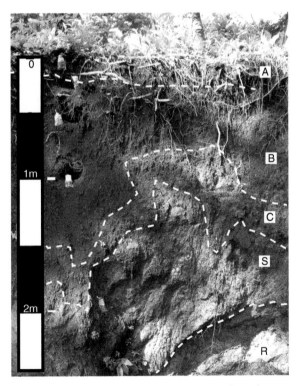

Plate 13 Soil profile formed via weathering of basalt in northwestern Costa Rica. The soil is a tropical Alfisol and horizons are indicated by A-B-C-S-R on the right, where S = saprolite and R = rock. Note the wavy nature of horizon boundaries and the progressive increase up-profile in red colors indicating formation of iron hydroxides (e.g. goethite) via liberation of Fe from silicate minerals.

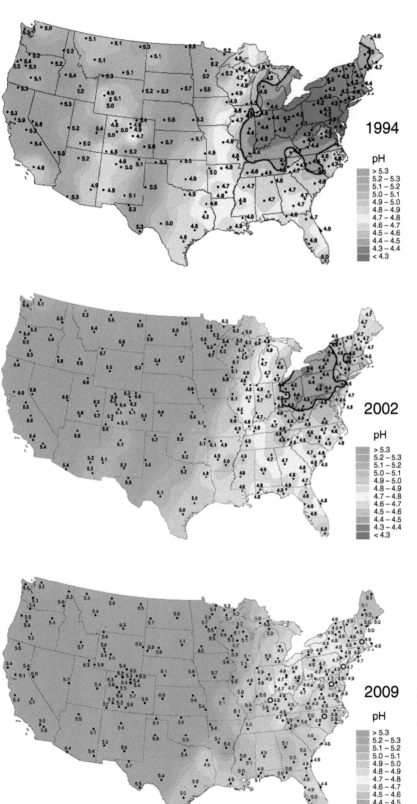

Plate 14 Average annual pH of precipitation in the conterminous United States of America for the years 1994, 2002, and 2009. The black contour line highlights areas with mean annual pH < 4.5. Source: Adopted from US EPA NADP program: National Atmospheric Deposition Program (NRSP-3). 2019. NADP Program Office, Wisconsin State Laboratory of Hygiene, 465 Henry Mall, Madison, WI 53706.

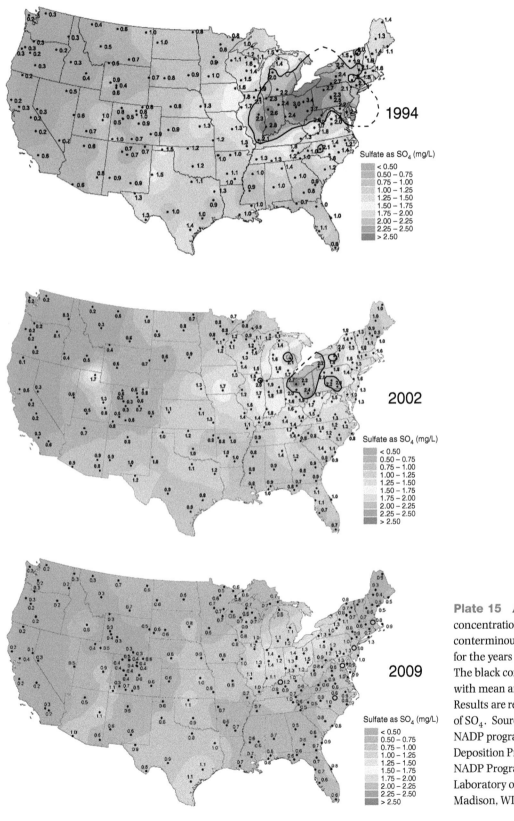

Plate 15 Average annual sulfate concentration of precipitation in the conterminous United States of America for the years 1994, 2002, and 2009. The black contour line highlights areas with mean annual $SO_4 > 2.0\,\mathrm{mg\,l^{-1}}$. Results are reported relative to the mass of SO_4. Source: Adopted from US EPA NADP program: National Atmospheric Deposition Program (NRSP-3). 2019. NADP Program Office, Wisconsin State Laboratory of Hygiene, 465 Henry Mall, Madison, WI 53706.

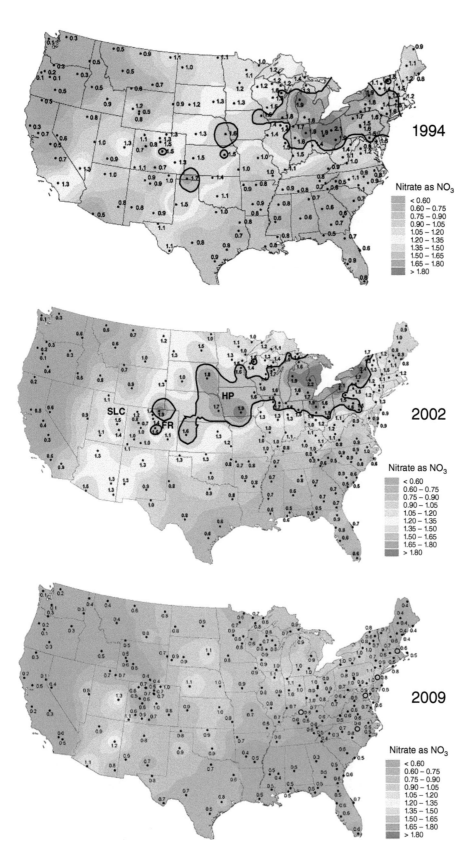

Plate 16 Average annual nitrate concentration of precipitation in the conterminous United States of America for the years 1994, 2002, and 2009. The black contour line highlights areas with mean annual NO_3 concentration $>1.5\,\mathrm{mg\,l^{-1}}$. Results are reported relative to the mass of NO_3 and not NO_3-N. Source: Adopted from US EPA NADP program: National Atmospheric Deposition Program (NRSP-3). 2019. NADP Program Office, Wisconsin State Laboratory of Hygiene, 465 Henry Mall, Madison, WI 53706.